Microbial Mitigation of Stress Response of Food Legumes

Microbial Mitigation of Stress Response of Food Legumes

Edited by
N. Amaresan
Senthilkumar Murugesan
Krishna Kumar
A. Sankaranarayanan

CRC Press
Taylor & Francis Group
Boca Raton London New York

CRC Press is an imprint of the
Taylor & Francis Group, an **informa** business

CRC Press
Taylor & Francis Group
6000 Broken Sound Parkway NW, Suite 300
Boca Raton, FL 33487-2742

Printed on acid-free paper

International Standard Book Number-13: 978-0-367-46024-2 (Hardback)

Visit the Taylor & Francis Web site at
http://www.taylorandfrancis.com

and the CRC Press Web site at
http://www.crcpress.com

Contents

Section II Crop Responses to Stress-Soybean

Section III Crop Responses and Microbe-Mediated Stress Management: Groundnut

Section IV Crop Responses to Microbe-Mediated Abiotic Stress Management

Section V Crop Responses to Microbe-Mediated Biotic Stress Management

Preface

Grain legumes are the key component in improving nutritional security, as a rich source in protein, minerals, and vitamins. They play an important role in sustaining soil health through symbiotic nitrogen fixation, a signature feature of legumes. Grain legumes are cultivated in marginal lands with limited input supply, and risk of abiotic and biotic stresses leads to yield reduction up to 30%–100% based on severity. The United Nations declared 2016 as the International Year of Pulses (IYP) to position pulses as a primary protein source and to accelerate research cooperation at the global level to improve pulse production. The demand for agriculture yield has been increased rapidly due to an increase in the world's population. Increased use of agrochemicals leads to degradation of soil quality and fertility in the agricultural fields. Hence, the attention should be made for sustainable agricultural productivity with safe practices.

Stress-tolerant varieties along with better management strategies can improve the productivity and climate resilience of grain legumes. Microbes have devised a sophisticated signaling system for eliciting an adaptive response to stresses. Such microbes can be exploited as a successful strategy to protect against deleterious effects caused by soilborne and seedborne deleterious plant pathogens. Decades of research have shown that development and utilization of microbial bioinoculants are a low-input, sustainable, and environmentally friendly technology to mitigate abiotic and biotic stresses of plants. Little information is available concerning the management of abiotic and biotic stresses in grain legumes including pulse crops. The present book will guide researchers by providing reviewed research findings from renowned laboratories related to microbe-mediated plant stress management under one roof.

This book is composed of five sections with 20 chapters. Sections I through III represent in-depth insights into crop responses and microbial-mediated mitigation in legume, soybean, and groundnut plants. Section IV deals with crop responses to microbial-mediated abiotic stress management, and Section V provides deep insight into crop responses to microbial-mediated abiotic stress management including the role of *cry* genes in pest management. It also discusses the current and future status of microbial formulations, commercialization, and product regulation for making the strategy a successful technology. This book will be useful to graduate students, research scholars, scientists, and teachers of different disciplines (plant microbiology, plant pathology, agronomy, botany, and biotechnology) and other professionals for understanding and updating their knowledge on microbe-mediated management of abiotic and biotic stresses in grain legumes.

N. Amaresan
Senthilkumar Murugesan
Krishna Kumar
A. Sankaranarayanan

Editors

Dr. N. Amaresan is an assistant professor at C.G. Bhakta Institute of Biotechnology, Uka Tarsadia University, Gujarat. He is a microbiologist and obtained his PhD in endophytic PGP bacteria from Bharathidasan University, Tamil Nadu. Dr. Amaresan has more than 13 years of experience in teaching and research and made several original and novel discoveries, especially in various allied fields of microbiology—mainly plant–microbe interactions, bioremediation, plant pathology, and others. For his original discoveries on agriculturally important microorganisms, he has been awarded young scientist awards by the Association of Microbiologists of India and the National Academy of Biological Sciences. He has also been awarded a visiting scientist fellowship from the National Academy of India to learn advanced techniques. Dr. Amaresan has handled two institute-funded and four externally funded projects from DST, GEMI, DBT, etc. He has published more than 50 research articles and books of national and international reputes. He also deposited over 350 bacterial 16S rDNA and fungal ITS rDNA sequences in the Genbank (NCBI, EMBL, and DDBJ) and also preserved over 150 microbial germplasm in various culture collection centers of India.

Dr. Senthilkumar Murugesan is currently working as principal scientist (Microbiology) at ICAR-Indian Institute of Pulses Research, Kanpur. He obtained a PhD (Microbiology) from the Indian Agricultural Research Institute, New Delhi, and received a post-doctoral fellowship from the Korean Atomic Energy Research Institute, South Korea. Dr. Kumar has 20 years of research experience in different aspects of plant-microbe interactions including symbiotic nitrogen fixation, biological control, and microbial-mediated abiotic stress management in pulses. He has explored the diversity of endophytic bacteria from wild and cultivated soybean varieties for biological control of soybean charcoal rot disease. He purified antimicrobial peptides from soybean endophytic bacteria. Dr. Kumar has developed bioinoculants and microbial consortia for improving moisture-deficit stress tolerance of rainfed chickpea. He has published more than 20 research papers in peer-reviewed journals. He is a fellow of the Indian Society of Pulses Research and Development (ISPRD).

Dr. Krishna Kumar presently works as a dean of the Pandit Deendayal Upadhyay College of Horticulture & Forestry, Dr. Rajendra Prasad Central Agricultural University, Muzaffarpur, Bihar, India. He obtained his PhD in plant pathology from Narendra Deva University of Agriculture & Technology, Faizabad, Uttar Pradesh. He worked as a research associate at IARI, New Delhi; assistant professor at SKUAST, Jammu; and then served as a senior scientist and principal scientist at the Central Agricultural Research Institute, Port Blair. Dr. Kumar served as a principal scientist at the Indian Agricultural Research Institute, New Delhi, India. He has also worked as a head of the Division of Crop Protection at the Indian Institute of Pulses Research, Kanpur, India. He has more than 20 years of research, teaching, and extension experience in mushroom production, biocontrol, microbial diversity, integrated disease management, diagnosis of unknown plant diseases, and suitable biocontrol agents. He is a recipient of the Fakhruddin Ali Ahmed Award for Outstanding Research in Tribal Farming Systems (2010) by the Indian Council of Agricultural Research, New Delhi, India; the SPPS Meritorious Scientist Award-2011 by the Society of Plant Protection Sciences, New Delhi, India; and Distinguished Service Award-2010 for his contribution in plant pathology by the BIOVED Research Society, Allahabad, Uttar Pradesh, India. He is a fellow of three prestigious professional societies and a lifetime member of 12 scientific societies. Dr. Kumar also deposited 42 microorganisms at the National Bureau of Agriculturally Important Microorganisms (NCBI) and submitted 400 sequences of them to NCBI. He delivered several radio talks and TV shows on different topics for the benefit of the farming community.

Dr. A. Sankaranarayanan has been associated with the C.G. Bhakta Institute of Biotechnology, Uka Tarsadia University, Surat of Gujarat state of India, since 2015. He obtained his PhD in aquatic microbiology from Manonmaniam Sundaranar University, Tirunelveli, Tamil Nadu. He has experience in the fields of antimicrobial activity of herbal and nanoparticles against MDR pathogens and fermented food products. His current research focus is on microbes in fermented food products and removal of bacteria from food by Dielectrophoresis. He has published 15 book chapters, 50 research articles in international and national journals of repute, and he authored 5 books which published by International publishers, guided 5 PhDs and 16 MPhil scholars, and operated five externally funded minor projects and two Institute-funded projects. From 2002 to 2015, he worked as an assistant professor and head in the Department of Microbiology, K.S.R. College of Arts & Science, Tiruchengode, India. He has been awarded with an Indian Academy of Sciences (IASc), National Academy of Sciences (NAS), and The National Academy of Sciences (TNAS)-sponsored summer research fellowship for young teachers consecutively for three years. His name is included as a mentor in the DST-Mentors/Resource persons for summer/winter camps and other INSPIRE initiatives, Department of Science & Technology, Government of India, New Delhi. He is a grant reviewer in the British Society for Antimicrobial Chemotherapy (BSAC), UK. He has involved himself in the organization of various national/international seminars/symposia. He is actively involved as an editor/editorial board member in journals, as a reviewer in various international/national reputed journals, and acted as an external examiner to adjudicate the PhD thesis of various universities in India.

Contributors

A. O. Adekiya
Crop and Soil Department
Landmark University
Omu-Aran, Nigeria

K. A. Adegbite
Crop and Soil Department
Landmark University
Omu-Aran, Nigeria

Ali Akbar
Faculty of Life Sciences
Department of Botany
Government College University Faisalabad
Faisalabad, Pakistan

E. T. Alori
Crop and Soil Department
Landmark University
Omu-Aran, Nigeria

N. Amaresan
C.G. Bhakta Institute of Biotechnology
Uka Tarsadia University
Surat, India

Dhruti Amin
C.G. Bhakta Institute of Biotechnology
Uka Tarsadia University
Surat, India

Wahiba Amri-Tiliouine
Division Biotechnologies et Amélioration des plantes, Équipe Amélioration des légumineuses
Institut National de la Recherche Agronomique d'Algérie
Algiers, Algeria

C. O. Aremu
Crop and Soil Department
Landmark University
Omu-Aran, Nigeria

Rajesh Arora
Department of Phyto Analytical Chemistry and Toxicology
and
Nutrition, Biochemistry, Exercise Physiology and Yoga
Defence Institute of Physiology & Allied Sciences
Defence Research and Development Organization
Lucknow, Delhi

Muhammad Arslan Ashraf
Faculty of Life Sciences
Department of Botany
Government College University Faisalabad
Faisalabad, Pakistan

Maurizio Badiani
Dipartimento di Agraria
Università Mediterranea di Reggio Calabria
Reggio Calabria, Italy

D. J. Bagyaraj
Centre for Natural Biological Resources and Community Development
Bangalore, India

Abhishek Bharti
Department of Microbiology
Barkatullah University
Bhopal, India

and

ICAR-Indian Institute of Soybean Research
Indore, India

Utkarsh M. Bitla
School of Edaphic Stress Management
ICAR-National Institute of Abiotic Stress Management
Pune, India

Hassan Etesami
Department of Soil Science
University College of Agriculture and Natural Resources
University of Tehran
Tehran, Iran

Chaitanya Ghalagi
Department of Crop Physiology
University of Agricultural Sciences
Bangalore, India

Difo Voukang Harouna
Department of Food Biotechnology and Nutritional Sciences
Nelson Mandela African Institution of Science and Technology (NM-AIST)
and
Centre for Research, Agricultural Advancement, Teaching Excellence and Sustainability in Food and Nutrition Security (CREATES-FNS)
Nelson Mandela African Institution of Science and Technology
Arusha, Tanzania

Iqbal Hussain
Faculty of Life Sciences
Department of Botany
Government College University Faisalabad
Faisalabad, Pakistan

Elmugheira M. Ibrahim Mohammed
Department of Sustainable Agriculture, Biodiversity and Ecosystems Management
Nelson Mandela African Institution of Science and Technology
and
Faculty of Forest Sciences & Technology
Department of Forest Management Sciences
University of Gezira
Wad Madani, Sudan

Muhammad Iqbal
Faculty of Life Sciences
Department of Botany
Government College University Faisalabad
Faisalabad, Pakistan

Mohammad Israil Ansari
Department of Botany
University of Lucknow
Lucknow, India

A. Joseph
Crop and Soil Department
Landmark University
Omu-Aran, Nigeria

Paltu Kumar Dhal
Department of Life Sciences and Biotechnology
Jadavpur University
Kolkata, India

Sandeep Kumar
Division of Crop Protection
Indian Institute of Pulses Research
Kanpur, India

Krishna Kumar
Division of Crop Protection
Indian Institute of Pulses Research
Kanpur, India

Pranamita Kunda
Department of Life Sciences and Biotechnology
Jadavpur University
Kolkata, India

and

Agricultural and Ecological Research Unit
Biological Sciences Division
Indian Statistical Institute
Giridih, India

Mourad Latati
Laboratoire d'Amélioration Intégrative des Productions Végétales (C2711100)
Département de Productions Végétales
Ecole Nationale Supérieure Agronomique (ENSA)
El Harrach, Algeria

Hemant S. Maheshwari
ICAR-Indian Institute of Soybean Research
Indore, India

and

Ecophysiology Laboratory
Faculty of Science and Engineering
GELIFES-Groningen Institute for Evolutionary Life Sciences
Groningen, the Netherlands

Ajay Kumar Mathur
Department of Plant Biotechnology
Central Institute of Medicinal and Aromatic Plants (CIMAP)
Council of Scientific and Industrial Research
Lucknow, India

Kamlesh K. Meena
School of Edaphic Stress Management
ICAR-National Institute of Abiotic Stress Management
Pune, India

Siva Satya Mohan Jampala
Department of Biosciences
Sardar Patel University
Anand, India

Abhishek Mukherjee
Agricultural and Ecological Research Unit
Biological Sciences Division
Indian Statistical Institute
Giridih, India

Senthilkumar Murugesan
Division of Basic Sciences
Indian Institute of Pulses Research
Kanpur, India

Fahima Nabi
Laboratoire Amélioration Intégrative Des Productions Végétales (AIPV)
Département de Productions végétales
Ecole Nationale Supérieure Agronomique (ENSA)
Algiers, Algeria

Sonika Pandey
Division of Crop Protection
Indian Institute of Pulses Research
Kanpur, India

Abida Parveen
Faculty of Life Sciences
Department of Botany
Government College University Faisalabad
Faisalabad, Pakistan

Prittesh Patel
C.G. Bhakta Institute of Biotechnology
Uka Tarsadia University
Surat, India

Nafisa Patel
Department of Microbiology
NaranLala College of Professional and Applied Sciences
Navsari, India

Anil Prakash
Department of Microbiology
Barkatullah University
Bhopal, India

B. Mohan Raju
Department of Crop Physiology
University of Agricultural Sciences
Bangalore, India

Rizwan Rasheed
Faculty of Life Sciences
Department of Botany
Government College University Faisalabad
Faisalabad, Pakistan

Utkarsh Singh Rathore
Division of Crop Protection
Indian Institute of Pulses Research
Kanpur, India

Sanket Ray
Department of Microbiology
NaranLala College of Professional and Applied Sciences
Navsari, India

A. Sankaranarayanan
Division of Crop Protection
Indian Institute of Pulses Research
Kanpur, India

Saeid Shafiei
Faculty of Agriculture
Department of Soil Science
University of Jiroft
Jiroft, Iran

Mohammed Shariq Iqbal
Amity Institute of Biotechnology
Amity University Uttar Pradesh
Lucknow, India

Abhishek Sharma
C.G. Bhakta Institute of Biotechnology
Uka Tarsadia University
Surat, India

Kusum Sharma
Division of Basic Sciences
Indian Institute of Pulses Research
Kanpur, India

Mahaveer P. Sharma
ICAR-Indian Institute of Soybean Research
Indore, India

Akash L. Shinde
School of Edaphic Stress Management
ICAR-National Institute of Abiotic Stress Management
Pune, India

Narendra P. Singh
School of Edaphic Stress Management
ICAR-National Institute of Abiotic Stress Management
Pune, India

Agostino Sorgona
Dipartimento di Agraria
Università Mediterranea di Reggio Calabria
Reggio Calabria, Italy

Ajay M. Sorty
School of Edaphic Stress Management
ICAR-National Institute of Abiotic Stress Management
Pune, India

G. K. Sujayanand
Division of Plant Biotechnology
ICAR-Indian Institute of Pulses Research
Kanpur, India

Sihem Tellah
Département de Productions végétales, Laboratoire Amélioration Intégrative Des Productions Végétales (AIPV), Ecole Nationale Supérieure Agronomique (ENSA), Algiers, Algeria

and

Dipartimento di Agraria
Università Mediterranea di Reggio Calabria
Reggio Calabria, Italy

Section I

Crop Responses and Microbe-Mediated Stress Management

Legumes

1

Simulation of the Daily Exchange of Carbon and Nitrogen between Microorganisms, Plant and Symbionts of Intercropped Legume with Cereal in Mediterranean Agro Ecosystem

Mourad Latati

CONTENTS

1.1 Introduction

The soil quality is one of the most important limiting factors for improving agroecosystem services and sustains plant productivities, which stimulate and increase nutrient use efficiency by cropping system (Cong et al., 2015; Latati et al., 2017). The superficial soil layers of most agroecosystems contain the highest reserves of organic carbon (C) and nitrogen (N), which promotes plant growth and increases bioavailability of essential nutrients (Ladd et al., 1992). C and N resources are considered as essential elements of organic matter (OM), which plays a key role in maintaining and restoring soil fertility by its strong effects on physical, chemical, and biological properties and more particularly on sequestration of soil C and N stocks (Pansu et al., 2018).

Microbial biomass (MB) is considered as one of the most biological indicators for monitoring agro-environmental changes. It is also defined as an important pool for the management of land and agricultural practices especially for C and N resources (Srivastava and Singh, 1989; Latati et al., 2019). The importance of soil organic matter (SOM) pools depends on the input of crop residues and losses of C and N. These losses are caused either by heterotrophic respiration during the decomposition of SOM by MB or by autotrophic respiration of symbiotic bacteria (Ibrahim et al., 2013).

Commonly used cultural practices such as fertilization, pesticide treatments, and monoculture contribute significantly to the degradation of environmental fertility by reducing biological diversity and by increasing the emissions of CO_2 sequestered in the atmosphere (Horrigan et al., 2002). Therefore, the ecosystem approach remains the surest way to avoid the harmful consequences of these practices in conventional agrosystems. This includes, in particular, the ecological services of atmospheric nitrogen fixation and the processes of complementarity and facilitation of legumes by promoting their interactions with soil microorganisms (Latati et al., 2016, 2017). Indeed, a positive effect on cereals when intercropped with legumes has been demonstrated with biomass accumulation and yield increase (Betencourt et al., 2012; Latati et al., 2013, 2014). This positive effect was confirmed for cowpea-maize (Latati et al., 2014), chickpea-durum wheat (Betencourt et al., 2012), and bean-maize intercropping (Dahmardeh et al., 2010; Latati et al., 2016). However, the increase in cereal yield in intercropping is not due only to the increase in N resources via symbiotic N_2 fixation but also to other mechanisms that remain unknown to this day (Betencourt et al., 2012).

The model focusing on microorganisms and plant stocks of C and N in complex cropping covers systems appears as a strong tool to assess the exchanges of the lithosphere organisms with plant, soil, and atmosphere pools. It proposes essential recommendations to better improve agricultural and environment systems as simulation of C and N cycles (Pansu et al., 2018).

Recent research studies have modeled and simulated the changes in C and N stocks under different practices of agroecosystems management and use over mid- to long-term periods. In these studies, C and N data are carried at long-term field experiences, such as 10 years (Vance et al., 1987) and 30 years (Kintché et al., 2010) under crop rotation system, 5 years (Romanyà et al., 2000) and 18 years (Palosuo et al., 2012) for forest ecosystem (Nieto et al., 2010). In the case of short-term experiences, Ibrahim et al., (2016), Pansu et al. (2018), and Latati et al. (2019) proposed a modeling method centered on the functioning of living organisms in order to calculate flow parameters using data on N and C stocks in decomposers, plant organs, symbiotic microorganisms, and the soil compartments. The finding reported in these last researchers' studies suggests that the mechanist models l will be a strong tool of simulation of C and N transfers between microorganisms and plant compartments, which will greatly improve simulations of the reserves linked to the earth life and acting on climate change.

In Algerian soils, a first quantification of N, following that of C in an agroecosystem cereal production of Setif, revealed a fragility of the reserves and the need to change agricultural practices with a soil preservation management (Latati et al., 2016, 2017, 2019). Under different environmental conditions, several modeling studies of stock variations of C and N in several production systems (Pansu et al., 2009, 2010) are available and allow a better understanding and adapting of the models.

In contrast with other published literature, which need long-term comparisons to quantify the C and N exchanges, the objective of this chapter is to present information on the validation at short term of the MOMOS model (Micro-Organismes et MatièreOrganique du Sol, i.e., soil microorganism and organic matter model) on the basis of the CN data collected in the cereal agroecosystem (Latati et al., 2019) of the Mediterranean region (e.g., Setif region in the northeast of Algeria). It presents the genesis of the MOMOS model, which is centered on microbial functioning and appears very sensitive to meteorological, edaphic, and biological conditions.

1.2 Increase in Symbiosis Efficiency for Intercropped Legumes with Cereals

Recent research studies on legume-cereal intercropping system have shown the efficient use of environmental resources by stimulating plant growth and yield in low-input soils (Cong et al., 2015; Latati et al., 2016), compared with fallow-cereal rotation practice. The legumes could increase the availability of nutrients such as N (Betencourt et al., 2012) and P (Latati et al., 2014) in the rhizosphere of the intercropped cereals, improving grain yield, nutrient uptake and efficiency in use of rhizobial symbiosis (EUR), especially in low P conditions (Houassine et al., 2019). Intercropping can also improve growth and nutrient use efficiency through the stimulation of biological N_2 fixation by nodules of the intercropped legumes (Bargaz et al., 2012) (Figure 1.1). Latati et al. (2017) confirm that N uptake in legume-cereal intercropping is greater as a result of a higher EURS under low soil input (e.g., P and N). Xu et al. (2013) reported that the changes in the C:N ratio of soil microorganisms (fungi and bacteria) have also been attributed to the relative demand of soil microbes for C and N. These changes may have a major effect on N and C cycling.

Previous literature on intercropping have shown that little information is available on the effects of cereal-legume intercropping on microbial mediated processes and the relation between intercropping and biological

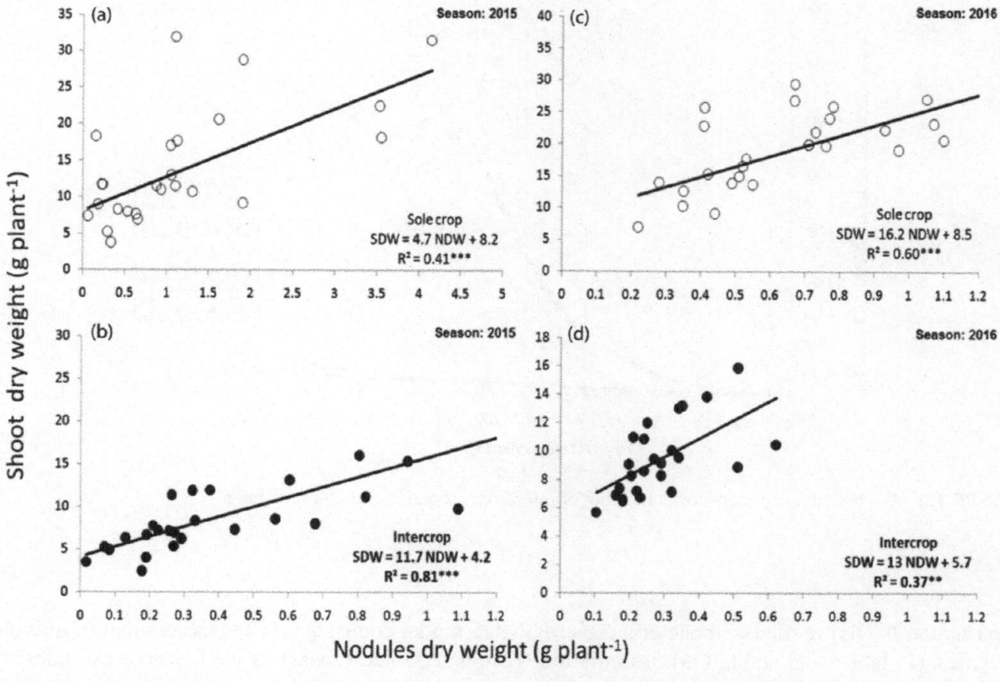

FIGURE 1.1 Efficiency in the use of rhizobial symbiosis of chickpea as sole crop (A and C) and intercrops (B and D) in 2015 and 2016. The equations inserted in the graphs are the regression functions. "** and ***" denote significance at $p < 0.01$ and $p < 0.001$, respectively. (Latati et al., 2020)

N_2 fixation by the intercropped legumes. Microbial biomass (MB) and EURS are considered as strong biological indicators for monitoring agroecosystem changes in rhizosphere of intercropped legume (cowpea, common bean, and chickpea) with cereals (maize and durum wheat). It is also considered as important for planning appropriate land use and for the management of agriculture practices (Latati et al., 2016, 2019).

1.3 Use and Application of MOMOS Model in Intercropping Agroecosystem

The MOMOS model (soil microorganism and organic matter model) is one of the most mechanistic models, which was developed and improved recently to simulate the daily exchanges of both C and N cycles within the soil-plant-microorganisms system. The MOMOS model was first developed and validated in such a way as to focus on crop system parameters that were studied under controlled conditions (e.g., incubation experiments) with the use of isotopic tracing (Pansu et al., 2004). No application on an open complex agroecosystem with regular flows of field conditions has yet been published except for Ibrahim et al. (2013, 2016), Pansu et al. (2018), and Latati et al. (2019). Indeed, MOMOS model parameters are strongly related to both soil moisture and temperature to better study the agroecosystem's sustainability and resilience in the face of climate change.

1.4 Role of the MOMOS Model in the Characterization of C and N Dynamics in Agroecological Systems

According to the MOMOS validation at agroecosystem scales, a recent research study demonstrated the possibility of predicting the daily interactions of CN stocks between the different organs of the plant (e.g., shoot, root, and nodules), soil, and atmosphere by simulating the functional role of microorganisms,

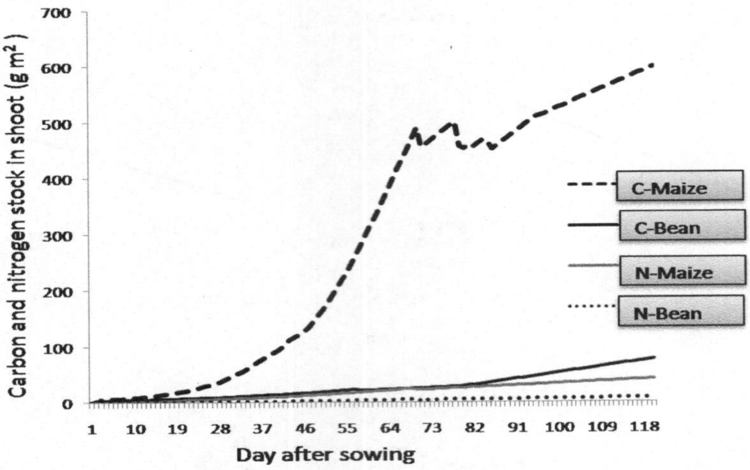

FIGURE 1.2 Carbon and nitrogen stocks in shoot of intercropped maize and common bean.

in particular N_2-fixing rhizobial bacteria. The MOMOS model contributes to the assessment of microbial activities via labile and stable OM assimilation (Humus and necromass) as well as root exudates with a simulation of C and N fluxes between all compartments of complex system soil-plant-microorganism (Thornton et al., 2004; Pansu et al., 2018; Latati et al., 2019).

The different simulations relative to daily exchange of C and N stocks in shoots of both intercropped common bean and maize (Figure 1.2) show significant adjustment between measured and simulated data. Recent research studies reported that the MOMOS model appears as a strong tool for assessment and analysis of various ecophysiological parameters, and to predict correctly an important data in the case of biogeochemical cycles studies (Ibrahim et al., 2013).

1.5 Microbial Biomass in the Rhizosphere of Intercropped Legumes

The MOMOS model was also used to simulate soil MB activities during crop cycle of either a maize-common bean (Latati et al., 2019) or wheat-faba bean (Ibrahim et al., 2016; Pansu et al., 2018) intercropping system. MB is characterized by rapid turnover compared to other OM compartments (e.g., necromass, residues, humus stable) and in particular as compared to the total C and N in the soil (Butler et al., 2004; Pansu et al., 2010).

MB is considered as a key indicator to assess agricultural practices' sustainability in an agroecosystem (Sparling et al., 1998; Barthès et al., 2008). The daily simulation of the MB-C and MB-N exchanges of microbial biomass in the rhizosphere of intercropped legumes-cereals (Figure 1.3) revealed the role of soil microorganisms in stimulating and controlling CN transfers in these intercropped species grown in a mixture crop. However, the C stocks in MB increase simultaneously with the increase in N-MB from the sowing to harvest time of the crop cycle (Molina and Smith, 1998; Cong et al., 2015), the high rate of this increase was systematically observed during full flowering stage of intercropped legumes (Figure 1.3).

Latati et al. (2017) reported a significant relation between biological N_2 fixation and the microbial biomass activity in the rhizosphere of common bean grown intercropped with maize. These authors confirm

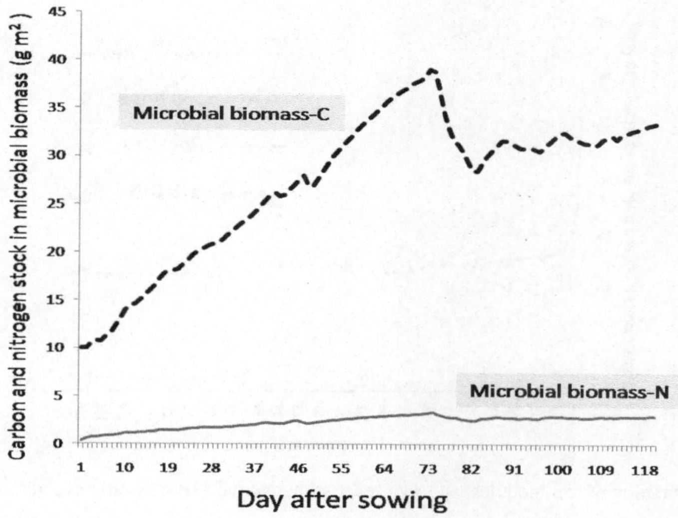

FIGURE 1.3 The simulation of carbon and nitrogen in rhizosphere microbial biomass of both intercropped maize and common bean.

that the C and N may also be transferred from the nodules of intercropped legume to the rhizosphere where they were assimilated by the microorganisms during their growth stage. This constitutes an important N and C stock in the rhizosphere MB for intercropped cereals.

1.6 Relationship between Microbial Biomass Activity and Legumes Symbiosis

The relationship between soil respiration and C-N stock in MB under a legumes-cereals intercropping system was evaluated in a recent research study reported by Latati et al. (2017). There was a strong dependence on cropping system and soil nutrient availability (e.g., N and P). Recent research studies reported that soil respiration can increase as a result of MB activity simulations via a shift in the nutritional (e.g., C, N, and P) balance of soil microorganisms (Xu et al., 2013). Cong et al. (2015) confirmed that the soil microbial community in return ensures an increase to the intercropped species' nutrients availability, thanks to microbial activities (mineralization, storage, and recycling).

However, the C:N ratio of the MB is considered as a key indicator to study nutritional balance of soil microbes. The high rate of C:N ratio explains that N contributes to limit microorganisms' growth (Jones and Darrah 1992; Bazot et al., 2005). Latati et al. (2017) confirm that the high root respiration in either intercropped legumes or cereals may also be due to the high rate of N_2 fixation of legumes, which is strongly linked to both rhizobial symbiotic activity as regulated by the permeability of the nodules to O_2 diffusion (Latati et al., 2014) and to other mechanisms controlling root and microorganisms' respiration such as enzyme activities (e.g., cellobiohydrolase, chitinase, and phosphatase activities).

The validation of the MOMOS model under field conditions in the Mediterranean basin (e.g., Setif region, Algeria) revealed a contrasting dynamic between C:N ratio of MB and root respiration (with nodules respiration). This relation was observed during the full flowering stage of intercropped legume (Figure 1.4). The possibility to simulate nodule respiration via the MOMOS model has significantly contributed to enrich the little information that is available in the literature on the effect of this intercropping on changes in MBC:N ratio, soil respiration, and nodule C.

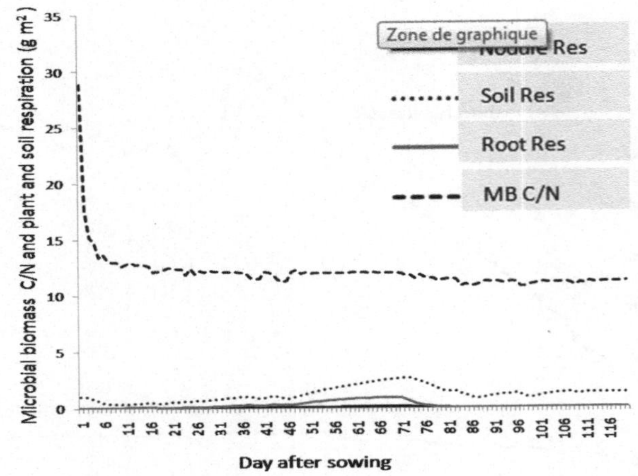

FIGURE 1.4 The simulation of soil, nodules, and root respiration and the MB C:N ratio of intercropped common bean with maize.

1.7 Relationship between the Rate of Nitrogen Fixation and Nodule Respiration

A previous research study has demonstrated the positive effect of N_2 fixation by intercropped legumes on aboveground parameters of their respective intercropped cereals (Zhang et al., 2004). It results in biomass accumulation and grain yield increase through N-fixed resource via facilitation processes (Robertson et al., 2002).

Several studies that were carried in a hydroaeroponic-controlled environment reported an increase of oxygen nodular permeability, which was correlated with the amount of biological nitrogen fixed by legume under P deficiency conditions (Saber et al., 2008; Alkama et al., 2009).

The last recent simulation (Ibrahim et al., 2016; Latati et al., 2019) of nodules respiration by the MOMOS model under field conditions contributed to define and compare this respiration activity with the rate of N fixed by intercropped legumes. MOMOS application on data collected in common bean-maize intercropping allows the simulation of the nodular respiration as well as the quantity of N fixed by intercropped common bean (Figure 1.5) of N_2. The simulation results show a strong and similar dynamic between these two parameters, in particular during the critical phase of symbiotic fixation.

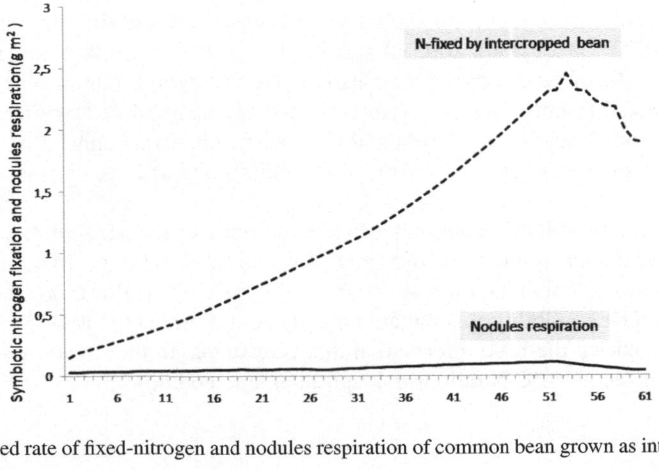

FIGURE 1.5 Simulated rate of fixed-nitrogen and nodules respiration of common bean grown as intercropping with maize.

ACKNOWLEDGMENTS

The authors thank «La Direction Génerale de la Recherche Scientifique et du Développement Technologique: DGRSDT-MESRES» for the financial support in the case of the FNRSDT-ARIMNET2-SEMIARID project. This work was also funded by the PRFU project: D04N01ES160320190001 the Ministry of Higher Education MESRES-Algeria.

REFERENCES

Alkama N, Bolou Bi Bolou E, Vailhe H, Roger L, Ounane SM, Drevon JJ (2009) Genotypic variability in P use efficiency for symbiotic nitrogen fixation is associated with variation of proton efflux in cowpea rhizosphere. *Soil Biol Biochem* 41: 1814–1823.

Bargaz A, Ghoulam C, Amenc L, Lazali M, Faghire M, Abadie J, Drevon JJ (2012) A phosphoenol pyruvate phosphatase transcript is induced in the root nodule cortex of *Phaseolus vulgaris* under conditions of phosphorus deficiency. *J Exp Bot* 63(13): 4723–4730.

Barthès BG, Brunet D, Hien E, Enjalric F, Conche S, Freschet G, d'Annunzio R, Toucet-Louri J (2008) Determining the distributions of soil carbon and nitrogen in particle size fractions using near-infrared reflectance spectrum of bulk soil samples. *Soil Biol Biochem* 40: 1533–1537.

Bazot S, Mikola J, Nguyen C, Robin C (2005) Do defoliation-induced changes in C allocation of field-grown *Lolium perenne* affect C availability, microbes and microbial feeders in soil? *Funct Ecol* 19(5): 886–896.

Betencourt E, Duputel M, Colomb B, Desclaux D, Hinsinger P (2012) Intercropping promotes the ability of durum wheat and chickpea to increase rhizosphere phosphorus availability in a low P soil. *Soil Biol Biochem* 46: 21–33.

Butler JL, Bottomley PJ, Griffith SM, Myrold DD (2004) Distribution and turnover of recently fixed photosynthate in ryegrass rhizospheres. *Soil Biol Biochem* 36: 371–381.

Cong WF, Hoffland F, Li L, Six J, Sun JH, Bao XG, Zhang FS, Van DerWerf W (2015) Intercropping enhances soil carbon and nitrogen. *Glob Change Biol* 21: 1715–1726.

Dahmardeh M, Ghanbari A, Syahsar BA, Ramrodi M (2010) The role of intercropping maize (*Zea mays* L.) and Cowpea (*Vigna unguiculata* L.) on yield and soil chemical properties. *African J Agric Res* 5(8): 631–636.

Horrigan L, Lawrence RS, Walker P (2002) How sustainable agriculture can address the environmental and human health harms of industrial agriculture. *Environ Health Perspect* 110: 445–456.

Houassine D, Latati M, Rebouh YN, Gerard F (2019) Phosphorus acquisition processes in the field: Study of faba bean cultivated on calcareous soils in Algeria. *Arch Agron Soil Sci*. doi: 10.1080/03650340.2019.1605166.

Ibrahim H, Hatira A, Pansu M (2013) Modelling the functional role of microorganisms in the daily exchanges of carbon between atmosphere, plants and soil. *Proc Env Sci* 19: 96–105.

Ibrahim H, Pansu M, Blavet D, Hatira A, McDonald P, Bernoux M, Drevon JJ (2016) Modelling the continuous exchange of carbon between living organisms, the soil and the atmosphere. *Plant Soil* 398: 381–397. https://doi.org/10.1007/s11104-015-2665-4.

Jones DL, Darrah PR (1992) Re-sorption of organic components by roots of *Zeamays* L. and its consequences in the rhizosphere. I. Re-sorption of 14C-labelled glucose, mannose and citric acid. *Plant Soil* 143: 259–266.

Kintché K, Guibert H, Sogbedji JM, Levêque J, Tittonell P (2010) Carbon losses and primary productivity decline in savannah soils under cotton-cereal rotations in semiarid Togo. *Plant Soil* 336: 469–484.

Ladd JN, Jocteur-Monrozier L, Amato M (1992) Carbon turnover and nitrogen transformations in an alfisol and vertisol amended with [U-14C] glucose and [15N] ammonium sulfate. *Soil Biol Biochem* 24: 359–371.

Latati M, Dokukin P, Aouiche A, Rebouh NY, Takouachet R, Hafnaoui E, Hamdani FZ, Bacha F, Ounane SM (2020) Species interactions improve above-ground biomass and land use effeciency in intercropped wheat and chickpea under low soil inputs. *Agronomy* 765. doi:10.3390/agronomy9110765. In press.

Latati M, Aouiche A, Rebou YN, Laouar M (2019) Modeling the functional role of the microorganisms in the daily exchanges of carbon and nitrogen in intercropping system under Mediterranean conditions. *Agron Res* 17: https://doi.org/10.15159/AR.19.050.

Latati M, Aouiche A, Tellah S, Laribi A, Benlahrech S, Kaci G, Ounan SM (2017) Intercropping maize and common bean enhances microbial carbon and nitrogen availability in low phosphorus soil under Mediterranean conditions. *Eur J Soil Biol* 80: 8–19.

Latati M, Bargaz A, Belarbi B, Lazali M, Benlahrech S, Tellaha S, Kaci G, Drevon, JJ, Ounane SM (2016) The intercropping common bean with maize improves the rhizobial efficiency, resource use and grain yield under low phosphorus availability. *Eur J Agron* 72: 80–90.

Latati M, Blavet D, Alkama N, Laoufi H, Pansu M, Drevon JJ, Gerard F, Ounane SM (2014) The intercropping cowpea-maize improves soil phosphorus availability and maize yields in an alkaline soil. *Plant Soil* 385: 181–191.

Latati M, Pansu M, Drevon JJ, Ounane SM (2013) Advantage of intercropping maize (*Zea mays* L.) and common bean (*Phaseolus vulgaris* L.) on yield and nitrogen uptake in Northeast Algeria. *Int J Res Appl Sci* 1: 1–7.

Molina JE, Smith P (1998) Modeling carbon and nitrogen processes in soils. *Adv Agron* 62: 253–298.

Nieto OM, Castro J, Fernández E, Smith P (2010) Simulation of soil organic carbon stocks in a Mediterranean olive grove under different soil management systems using the RothC model. *Soil Use Manage* 26(2): 118–125.

Palosuo T et al. (2012) A multi-model comparison of soil carbon assessment of a coniferous forest stand. *Environ Modell Softw* 35: 38–49.

Pansu M, Bottner P, Sarmiento L, Metselaar K (2004) Comparison of five soil organic matter decomposition models using data from a 14C and 15N labeling field experiment. *Glob Biogeochem Cycles* 18: GB4022.

Pansu M, Ibrahim H, Hatira A, Ibrahim N, Drevon JJ, Harmand JM, ChotteJL, Blavet D (2018) Modelling the continuous exchange of nitrogen between microbial decomposers, the organs and symbionts of plants, soil reserves and the atmosphere. *Soil Biol Biochem*. 125: 185–196.

Pansu M, Martineau Y, Saugier B (2009) A modelling method to quantify in situ the input of carbon from roots and the resulting C turnover in soil. *Plant Soil* 317: 103–120.

Pansu M, Sarmiento L, Rujano MA, Ablan M, Acevedo D, Bottner P (2010) Modelling organic transformations by micro-organisms of soils in six contrasting ecosystems: Validation of the MOMOS model. *Glob Biogeochem Cycles* 24: GB1008.

Robertson MJ et al. (2002) Simulation of growth and development of diverse legume species in APSIM. *Aust J Agric Res* 53: 429–446.

Romanyà J, Casals P, Cortina J, Bottner P , Coêteaux MM, Vallejo VR (2000) CO_2 efflux from a Mediterranean semi-arid forest soil. II. Effects of soil fauna and surface stoniness. *Biogeochemistry* 48: 283–306.

Saber S, Alkama N, Abdelly C, Drevon JJ (2008) Proton efflux by nodulated roots varies among common-bean genotypes (*Phaseolus vulgaris*) under phosphorus deficiency. *J Plant Nutr Soil Sci* 171: 242–248.

Sparling G, Vojvodić-Vuković M, Schipper LA (1998) Hot-water-soluble C as a simple measure of labile soil organic matter: The relationship with microbial biomass C. *Soil Biol Biochem* 30: 1469–1472.

Srivastava SC, Singh JS (1989) Effect of cultivation on microbial biomass C and N of dry tropical forest soil. *Soil Biol Fert Soils* 8: 343–348.

Thornton B, Paterson E, Midwood AJ, Sim A, Pratt SM (2004) Contribution of current carbon assimilation in supplying root exudates of Lolium perenne measured using steadystate 13C labelling. *Physiol Planta* 120(3): 434–441.

Vance ED, Brookes PC, Jenkinson DS (1987) An extraction method for measuring microbial biomass C. *Soil Biol Biochem* 19: 703–707.

Xu X, Thornton PE, Post WM (2013) A global analysis of soil microbial biomass carbon, nitrogen and phosphorus in terrestrial ecosystems. *Glob Ecol Biogeo* 22: 737–749.

Zhang F, Zhang S, Zhang J, Zhang R, Li F (2004) Nitrogen fertilization on uptake of soil inorganic phosphorus fractions in the wheat root zone. *Soil Sci Soc Am J* 68: 1890–1895.

2

Biotic and Abiotic Stress Responses of Domesticated Vigna Legumes: A Comprehensive Review

Difo Voukang Harouna

CONTENTS

2.1 Introduction

Legumes (*Leguminosae* or *Fabaceae*) represent the second most important plant family in agriculture after the *Poaceae* or grass family. They are highly coveted by both humans and animals, as they provide the largest single source of vegetable protein in human diets and livestock feed. The two widely consumed types of legumes belong to two different genera, namely the *Phaseolus* and the *Vigna*, commonly refer to as common beans and peas, respectively (Garcia et al., 1997; Gepts, 2001). Among these two genera of legumes, there exist many domesticated edible species of legumes as well as underexploited wild species.

Biotic and abiotic stresses have been recognized to be the major constraints in enhancing the productivity of legume food crops in general.

The genus *Vigna* is categorized into seven subgenera and 16 sections (Fatokun, Perrino, & Ng, 1997). The seven genera include *Ceratotropis*, *Haydonia*, *Lasiocarpa*, *Macrorhycha*, *Plectotropis*, *Sigmoidotropis*, and *Vigna* (Boukar, Bhattacharjee, Fatokun, Kumar, & Gueye, 2013). All domesticated and cultivated *Vigna* varieties belong to only three subgenera, namely *Ceratotropis*, *Plectotropis*, and *Vigna* (Pandiyan et al., 2012). The subgenus *Ceratotropis*, which is well known as Asian Vigna, and the subgenus *Vigna*, commonly called African Vigna, are the most known subgenera containing the most popular legumes like cowpeas, black gram, and green gram (Boukar et al., 2013; Pandiyan et al., 2012).

This review is aiming at examining the specific biotic and abiotic factors influencing the growth and development of the ten domesticated *Vigna* legumes.

2.2 Biotic Stress Responses of *Vigna* Legumes

It is reported that out of the more than 100 species of the genus *Vigna* (Leguminosae plant family) existing, and widely distributed in the tropical and subtropical areas of the world (Tomooka, Vaughan, Moss, & Maxted, 2002), only 10 species have been domesticated (Harouna, Venkataramana, Ndakidemi, & Matemu, 2018). Biotic stress is defined as a stress that is caused in plants due to damage instigated by other living organisms, including fungi, bacteria, viruses, parasites, weeds, insects, and other native or cultivated plants. All around the world in drylands, the major food crops cultivated and consumed are grain legumes. In this section, a brief review of the ten domesticated species with respect to their responses to biotic stress will be discussed.

2.2.1 Adzuki Bean

Adzuki bean (*Vigna angularis* var. *angularis*) is a warm-season legume, widely cultivated in China, and grown in more than 30 countries of the world (Wang, Zhu, Li, Wang, & Wu, 2018; Yang et al., 2015). Its seeds are an important source of starch, digestible protein, mineral elements, and vitamins, and the legume is widely used in foods for at least one billion people. Recently, it has been identified as the second most important legume in Japan after soybean (Kaga, Isemura, Tomooka, & Vaughan, 2008). Research from a wide range of literature databases shows that experiments involving investigations on the effects of biotic stress on the adzuki bean are very scant or limited to unpublished or grey literature. This could be due to less influence of this crop by biotic factors.

2.2.2 Bambara Groundnut

Bambara groundnut (*Vigna subterranea* (L.) Verdc) is a domesticated legume originating from West Africa and possessing considerable genetic diversity (Mohammed, Shimelis, & Laing, 2016). It is widely grown and used in semi-arid Africa. In many parts of Africa, it represents the third most important legume after groundnut (*Arachis hypogaea*) and cowpea (Basu et al., 2007). It consists of two taxonomic forms: var. *spontanea* comprising the wild forms, and var. *subterranea* comprising the cultivated forms, found in many parts of the tropics and particularly sub-Saharan Africa (Basu et al., 2007).

A number of reviews of Bambara groundnut have been published previously, with the first one published by Linnemann (1987) while the most recent one focused on aspects of commodity marketing (Hillocks, Bennett, & Mponda, 2012).

According to most of the reviews, biotic factors do not influence yield in Bambara groundnut seriously, and the causes of yield reduction could be attributed to other factors. In general, it is noticed that the common biotic factors that affect Bambara groundnut can be grouped into hemiparasites living on the roots of the plant, fungi (the most prevalent attack), viruses, nematodes, and some common pests of the crop in the field such as leafhoppers, as well as some common pests of stored seeds like *Collosobruchus maculates*.

2.2.3 Black Gram

Black gram (*Vigna mungo* var., *mungo* (L.) Hepper) is an important food legume crop originating from central Asia and now found in many tropical areas of Asia, Africa, and Madagascar, while in the United States and Australia it is mostly cultivated as a fodder crop (Jansen, 2006). Its global growing area is expected to be higher than 5 Mha, with India being the largest producer (about 3 Mha), followed by Myanmar (about 1 Mha) and Pakistan (0.5 Mha) (Kaewwongwal et al., 2015).

The crop is damaged by a number of insect pests during successive stages of the growth, right from root nodules to flowers and pods; however, the whitefly *Bemisia tabaci* appears as the major pest of black gram, besides mung bean and soybean (Taggar, Gill, Gupta, & Sandhu, 2012). The avoidable losses due to *B. tabaci* and other insect pests in black gram have been reported to range from 17.42 to 71 percent at different locations.

Currently, the information on biochemical mechanisms and induced resistance in black gram in response to whitefly is very limited and scarce (Taggar et al., 2012). In a recent study, investigations were carried out to compare the enzymatic responses of resistant versus susceptible genotypes of black gram to feeding by whitefly with a purpose to interpret mechanisms that will facilitate efforts to breed black gram varieties with durable whitefly resistance (Taggar et al., 2012).

Seeds of black gram are severely damaged by bruchids (*Callosobruchus* spp.), reducing seed quality and increasing yield losses. Most of the cultivated genotypes of black gram are susceptible to bruchids; however, a few tolerant genotypes have also been identified—but the mechanism of tolerance is poorly understood.

The defense response in black gram due to bruchid infestation has not yet been elucidated, although there are reports on availability of both mild resistant genotypes and wild relatives with complete resistance to bruchids. For the first time, a recent study investigated defense-related gene transcript dynamics during the bruchid-black gram interaction using Suppression Subtractive Hybridization library (Baruah et al., 2017).

2.2.4 Cowpea

Cowpea (*Vigna unguiculata* (L.) Walp.) is one of the most important grain legume crops in the world with a larger zone of occurrence and cultivation that covers the semiarid regions of Africa, Asia, southern Europe, the southern United States, and Central and South America (Diouf, 2011; Singh, 2006; Timko, Ehlers, & Roberts, 2007). Its cultivation covers 14.5 million hectares with an annual production of 5.5 million tons worldwide. Presently, Nigeria is the largest producer of cowpea followed by Niger, Burkina Faso, Myanmar, Cameroon, and Mali (Simon, Benko-Iseppon, Resende, Winter, & Kahl, 2007). It is cultivated not only for its seed but also for its leaves, which serve both as a human food as well as animal feed.

The production of cowpea is limited by several biotic stresses. The biotic stresses include insect pests, diseases, parasitic weeds, and nematodes. Every stage in the life cycle of the crop has at least one major insect pest that is responsible for yield losses (Boukar, Close, Roberts, Fatokun, & Huynh, 2016). Aphid (*Aphiscraccivora*) attacks provoke the highest amount of damage to the plants, mainly at the seedling stage. Flower bud thrips (*Megalurothrips sjostedti*) at the flowering stage cause the destruction of flower buds and failure of pod formation. Pods and young shoots are destroyed by pod borers (*Maruca vitrata*) while a complex of pod-sucking bugs (*Clavigralla tomentosicollis, Anoplocnemis curvipes, Nezara viridula*) penetrate and damage seeds in pods. In storage, bruchid weevils (*Callosobruchus maculatus*) cause serious damage to the grains.

In the case of diseases, cowpea is attacked by bacteria, viruses, fungi, and nematodes. Bacterial blight caused by *Xanthomonas vignicola* and bacterial pustule (*Xanthomonas* sp.) are the main bacterial diseases of cowpea. Several viruses infect cowpea including *Cowpea aphid-borne mosaic virus* (CABMV, genus *Potyvirus*, family *Potyviridae*); *Bean common mosaic virus*—blackeye cowpea mosaic strain (BCMV—BlCM, genus *Potyvirus*, family *Potyviridae*); *Cowpea mosaic virus* (CPMV, genus *Comovirus*, family *Secoviridae*); *Southern bean mosaic virus* (SBMV, genus *Sobemovirus*); *Cowpea mottle virus* (CPMoV, genus *Carmovirus*, family *Tombusviridae*); *Cucumber mosaic virus* (CMV, genus *Cucumovirus*, family

Bromoviridae); *Cowpea mild mottle virus* (CPMMV, genus *Carlavirus*, family *Betaflexiviridae*); and *Cowpea golden mosaic virus* (CGMV, genus *Begomovirus*, family *Geminiviridae*). The main fungi known to cause diseases in cowpea plants include *Colletotrichum* sp. causing anthracnose and brown blotch, *Macrophomina phaseolina* causing charcoal rot, ashy stem blight and stem canker, *Cercospora canescens* causing cercospora leaf spot, *Elsinoe phaseoli* causing scab, and *Rhizoctonia solani* causing web blight. Plant-parasitic nematodes, especially root-knot nematodes (*Meloidogyne* spp.), damage cowpea root systems and cause yield suppression in many cowpea production areas. These diseases may occur singly or in combinations of two or more pathogens, and in some cases are strongly influenced by the growing environment, for example, ashy stem blight caused by Macrophomina is much more prevalent and severe in drought-stressed cowpea plantings.

Parasitic weeds, *Striga gesnerioides* and *Alectra vogelii*, can cause significant damage to cowpea production. Striga is mainly present in the dry savannah areas of West and Central Africa, while Alectra is found predominantly in eastern and southern Africa.

2.2.5 Mung Bean

Mung bean (*Vigna radiata* (L.) Wilczek) is an important grain legume as well, especially in Asia (India, Southeast Asia, and East Asia), but also eaten in southern Europe and in the southern United States (Heuzé, Tran, Bastianelli, & Lebas, 2015). It is mainly produced in Asia (90%), with India being the largest producer (more than 50% of world production) and consumer of its entire production. China also produces large amounts, representing 19% of its legume production, but Thailand remains the main exporter of its production, which increased by 22% per year between 1980 and 2000 (Lambrides & Godwin, 2006). Although this crop is also produced in many African countries, it is known as a minor crop there (Mogotsi, 2006).

Some biotic stresses, specifically pest and diseases, are important constraints in realizing the full yield potential in mung bean production. The yield of mung bean is affected by several biotic factors. Among the biotic factors, mungbean yellow mosaic virus (MYMV), urdbean leaf crinkle virus (ULCV), mungbean leaf curl virus (MLCV), and Bean Common Mosaic Virus (BCMV) are the most predominant in crop yield reduction.

Some of the most common pests associated with mung bean are: *Alcidodes collaris* Pasc, *Amsacta albistriga* W., *Anarsia* spp., *Anoplocnemis phasiana* Fabr., *Anticarsia irrorata* (F.), *Aphis craccivora* Koch., *Apion ampulum* Fst., *Aspongopus janus* Fabr., *Attractomorpha crenulata* Fabr., *Callosobruchus* spp., *Catechrysops cnejus* Fabricius, *Chrotogonus* spp., *Colemania sphenarioides* B., *Coptosoma cribria* Fabr., *Cyrtozemia dispar* Pascoe, *Dolycoris indicus* Stal., *Empoasca kerri* Pruthi, *Eucosma melanaula* Meyr, *Helicoverpa armigera* (Hubner), *Herse convolvuli* L., and *Hyalospila leuconeurella* Rag. (Swaminathan, Singh, & Nepali, 2012).

2.2.6 Moth Bean

Moth bean (*Vigna aconitifolia* (Jacq.) Marechal) is a minor legume crop and considered to be the most drought- and heat-tolerant cultigen among Asian *Vigna* (Tomooka et al., 2011). It is widely grown in India and the Far East and has been qualified as a possibly more significant food source for the future (Adsule, 1996). It is considered to have been domesticated in India, Pakistan, Myanmar, or Ceylon (NARO Genebank, 2017). India's driest state, Rajasthan, is the major moth bean-growing state, contributing almost 86% area of the country's cultivation (Gupta, Shrivastava, Singh, & Bhagyawant, 2016).

Very limited or scarce published literature on the biotic stress aspects of moth bean exist. Some fungal diseases have been reported on *Vigna aconitifolia* in India, with descriptions of symptoms caused by *Alternaria alternata* (leaf spot), *Colletotrichum truncatum* (causing leaf spot and pod infection), *Curvularia lunata* [*Cochliobolus lunatus*] (leaf spot), *Fusarium moniliforme* [*Gibberella fujikuroi*] (seedling rot), and *Myrothecium roridum* (pod infection) (S. K. Singh & Srivastava, 1989). The susceptibility of moth bean to yellow mosaic virus (YMV) has been recently reported (Yaqoob et al., 2015).

2.2.7 Rice Bean

Rice bean (*Vigna umbellata* (Thunb.) Ohwi & Ohashi) is recognized as a multipurpose legume as well as a neglected and underutilized legume (Joshi, Bhanduri, Gautam, Bajracharya, & Hollington, 2008). Compared with cowpea (*Vigna unguiculata*), adzuki bean (*Vigna angularis*), and mung bean (*Vigna radiata*), it is a less important crop. However, it represents a locally important contributor to human nutrition in parts of India and Southeast Asia (Joshi et al., 2008; Tomooka, Kaga, Isemura, & Vaughan, 2011).

It grows well on a range of soils and is also recognized as a pest-resistant crop. It has been reported to be resistant to YMV (Sehrawat & Yadav, 2014). However, a report showed that 26 insect pests of different groups were observed infesting *Vigna umbellata* at different stages of crop growth with seven of them present throughout the crop growth after their first appearance, while ten species occurred sporadically (Satyanarayana, Singh, & Singh, 1995). The crop also has the potential to produce large amounts of nutritious animal fodder and high-quality grain, and there is great scope for genetic improvement in this neglected crop.

2.2.8 Creole Bean

Creole bean (*Vigna reflexo-pilosa* var. *glabra* = *Vigna glabrescens*), minni payaru (*Vigna stipulacea*), and tuber cowpea (*Vigna vexillata*) are very little-known legumes due to the fact that they are not yet fully domesticated and well commercialized (Harouna et al., 2018). Therefore, very little information concerning their biotic stress responses exists.

2.3 Abiotic Stress Responses of *Vigna* Legumes

Nowadays, agricultural food production is facing enormous challenges, leading to an insufficient food supply with low productivity and quality standards. In addition to an ever-increasing demographic demand, alterations in weather patterns due to changes in climate are negatively impacting crop productivity globally. Factors known as abiotic factors such as harsh climatic conditions like water scarcity (drought), extreme temperatures (heat, freezing), photon irradiance, and contamination of soils by high ion concentration (salt, metals) are the major growth stressors that significantly limit productivity and quality of crop species worldwide (Fraire-Velazquez & Balderas-Hernandez, 2013).

During harsh abiotic conditions, plants generally respond by activating tolerance mechanisms at multiple levels of organization (molecular, tissue, anatomical, and morphological); by adjusting the membrane system and the cell wall structure, altering the cell cycle and rate of cell division; and by metabolic tuning (Fraire-Velazquez & Balderas-Hernandez, 2013). At a molecular level, many genes are induced or repressed by abiotic stress, involving a precise regulation of extensive stress-gene networks. Products of those genes may function in stress response and tolerance at the cellular level. Proteins involved in biosynthesis of osmoprotectant compounds, detoxification enzyme systems, proteases, transporters, and chaperones are among the multiple protein functions triggered as a first line of direct protection from stress. Moreover, activation of regulatory proteins (e.g., transcription factors, protein phosphatases, and kinases) and signaling molecules are essential in the concomitant regulation of signal transduction and stress-responsive gene expression. Early plant response mechanisms prevent or alleviate cellular damage caused by the stress and reestablish homeostatic conditions and allow continuation of growth. Equilibrium recovery of the energetic, osmotic, and redox imbalances imposed by the stressor are the first targets of plant immediate responses.

Domesticated legume crop production is seriously challenged by a number of biotic (diseases and pests) and abiotic stresses (heat, frost, drought, and salinity), edaphic factors (associated with soil nutrient deficits), and policy issues (where less emphasis is put on legumes compared to priority starchy staples) (Ojiewo et al., 2017). This led to a constant search and quest to find new varieties with resistance to the various factors in the genius *Vigna* through screening and updating of wild germplasm by scientists recently (Gore et al., 2019). In this section, a brief review of the ten domesticated species with respect to their responses to abiotic stress will be discussed.

2.3.1 Adzuki Bean (*Vigna angularis* var. *angularis*)

Several experiments demonstrating the biosynthesis of some secondary metabolites or the genes responsible for their production in correlation with their exposure to some abiotic responses have been reported (Chauhan, Tripathi, Prasad, Singh, & Singh, 2020; Fraire-Velazquez & Balderas-Hernandez, 2013). The effect of low temperature on adzuki beans membrane lipids by tissue culture for a better uniformity and lower external factors other than the stress induced has been reported (Badea & Basu, 2009). The study showed that, among other modifications, the changes in lipid species of Phosphatidylcholine, Phosphatidylethanolamine, and Phosphatidylinositol of phospholipids; of Monogalactosyldiacylglycerol and Digalactosyldiacylglycerol of glycolipids; and of free sterols were provoked.

Screening of adzuki bean (*Vigna angularis*) together with its cross-compatible wild relatives (seven species) to investigate sources of salt tolerance has been reported (Yoshida et al., 2016). The study was performed in both soil culture and hydroponic culture, and identified JP205833 of *V. riukiuensis* (strain "Tojinbaka") and JP107879 of *V. nakashimae* (strain "Ukushima") as the valuable source of salt tolerance. It revealed that the two strains had different salt tolerance mechanisms, where "Ukushima" prevented Na^+ accumulation in leaves by filtering them in roots and stems, while "Tojinbaka" accumulated Na^+ throughout the whole plant body. It further demonstrated "Tojinbaka" and "Ukushima" could retain photosynthesis even under salt stress.

In a recent study, the diversity of drought tolerance in the genus *Vigna* (including domesticated and wild relatives of *Vigna angularis*) was examined (Iseki, Takahashi, Muto, Naito, & Tomooka, 2018). The study revealed that domesticated species were mostly classified as drought susceptible but could be improved using tolerant conspecific wild ancestors with cross compatibility. It further demonstrated that tolerance was related with higher plant water status presumably due to small water consumption. However, the variation of drought tolerance could not be explained by simple tolerance factor alone, and other tolerance mechanisms such as deep rooting and increasing in root biomass were found in the tolerant accessions (Iseki et al., 2018).

Studies experimenting the responses of adzuki bean plant to abiotic stress responses in terms of metabolites produced or genes responsible for their production are still very scanty in this area.

2.3.2 Bambara Groundnut

Bambara groundnut (*Vigna subterranea* (L.) Verdc) is well known as a drought-tolerant, underutilized indigenous African legume and reported to produce relatively greater seed yield than other legumes under the same conditions (Chai, Massawe, & Mayes, 2016).

Limited information on temperature stress for Bambara groundnut is published despite the wide fluctuations in daily mean temperature in semi-arid Africa. This necessitates a thorough understanding of the effect of temperature stress on growth and development of Bambara groundnut. The response of high temperature stress has been studied in detail for groundnut, wheat, cowpea, and rice (Chai et al., 2016). Environmental stresses represent limiting factors for growth and yield of Bambara groundnut.

The effect of drought on the growth and development of Bambara groundnut (*Vigna subterranea* (L.) Verdc.) in controlled-environment glasshouses in the UK has been reported (Mwale, Azam-Ali, & Massawe, 2007). The study revealed that Bambara groundnut responded to drought by reducing the rate of leaf area expansion, final canopy size, and total dry matter (TDM) during vegetative growth. Drought also caused significant reductions in pod dry matter (PDM), pod number, seed weight, and harvest index (HI), leading to a decrease in final pod yield that was different between landraces. Across landraces, drought reduced mean pod yield from 298 to 165 g m^{-2}, representing 45% yield loss.

A more recent study examined the response of Bambara groundnut under increasing drought stress and the effects of cumulative mild drought on final yields (Chai et al., 2016). The study showed that drought stress reduced *stomatal conductance* significantly ($F_{(1,130)} = 2259.59$, $p < 0.01$), with variation observed between lines of the segregating population ($F_{(64,130)} = 16.27$, $p < 0.01$). It further demonstrated that drought causes higher *stomatal density* and reduced *leaf area* ($p < 0.01$). Furthermore, mild drought stress negatively influenced *100-seed weight* ($F_{(1,258)} = 19.4$, $p < 0.01$) and *harvest index* ($F_{(1,258)} = 12.87$, $p < 0.01$) by 8% and 15.6%, compared to the control irrigated treatment, respectively. Therefore, it concludes that Bambara groundnut uses a combination of mechanisms to tolerate drought stress, including stomatal regulation of gas exchange, reduction of leaf area, and maintenance of a relatively high leaf water status and relatively high levels of photosynthesis.

More research aimed at investigating the effect of other abiotic stress responses of Bambara groundnut is highly needed in this area; also needed is a look at the global food demand and food insecurity as well as climate change challenges.

2.3.3 Black Gram

Black gram (*Vigna mungo* var., mungo (L.) Hepper) is another important grain legume grown in the tropical and subtropical regions of the Indian subcontinent, which is mainly affected by saline soils. Genetic engineering methods have been used to mitigate such abiotic stress by inducing the overexpression of the Glyoxalase 1 (Gly 1) gene and producing salt-tolerant transgenics of *Vigna mungo* (Bhalla-Sarin, Bhomkar, Debroy, Sharma, & Saxena, 2004).

The effect of water stress on the growth and reproduction of black gram was examined (Baroowa, Gogoi, Paul, & Sarma, 2013; Dahanayake, Ranawake, & Senadhipathy, 2015). The studies demonstrated that plant height, wet and dry weight of shoots and roots, root length, number of pods, wet and dry weight of pods, and wet and dry weight of seeds were significantly affected by the water stress condition. Water stress at the vegetative stage seriously limited the plant height, wet and dry weight of shoots and roots and root length, whereas at the reproductive stage, it showed clear impact on the number of pods, wet and dry weight of pods, and wet and dry weight of seeds.

The effect of three temperature regimes for identifying suitable stress-tolerant genotypes of black gram was also examined (Dash & Shree, 2012). It was found that high percent germination (87.2%), root length (3.68 cm), and carbohydrate content (3.72 mg g^{-1} fresh tissue) of the studied genotypes were higher at 10°C–20°C temperature. Furthermore, high shoot length (13.39 cm), free amino acid content (3.73 mg/g fresh tissue), and protein content (9.54 mg/g fresh tissue) were found to be present when the genotypes were exposed to 20°C–30°C temperature.

2.3.4 Cowpea (*Vigna unguiculata* (L.) Walp.)

Abiotic stresses affecting cowpea production include drought, heat, and low soil fertility. Although the crop is known to be drought tolerant, its yield can be reduced significantly when exposed at seedling stage, midseason, or terminal drought. Heat may cause serious damage during the off-season cropping. High night temperatures lead to flower abortion thereby preventing pod formation and a consequent reduction in grain yield. Soils that are deficient in phosphorous, an element required for nitrogen-fixation in legume root nodules, may lower the productivity of cowpea.

Drought represents the most important abiotic stress affecting cowpea production (Agbicodo, 2009). Several factors and mechanisms operate independently or jointly to enable plants to cope with drought stress. Therefore, drought tolerance is manifested as a complex trait in plants. Generally, drought tolerance is defined as the ability of plants to live, grow, and produce yields satisfactorily with limited soil water supply or under periodic water deficiencies.

The mechanisms that plants use to cope with drought stress generally grouped into three categories as drought escape, drought avoidance, and drought tolerance. However, crop plants use more than one mechanism at a time to cope with drought.

Drought escape is the ability of a plant to complete its life cycle before serious soil and plant water deficits occur. This mechanism involves rapid phonological development (early flowering and early maturity), developmental plasticity (variation in duration of growth period depending on the extent of water deficit), and remobilization of pre-anthesis assimilates.

Drought avoidance is the ability of plants to maintain relatively high tissue water potential despite a shortage of soil moisture. Plants develop strategies for maintaining turgor by increasing root depth or developing an efficient root system to maximize water uptake, and by reducing water loss through reduced epidermal (stomatal and lenticular) conductance, reduced absorption of radiation by leaf rolling or folding, and reduced evapotranspiration surface (leaf area).

Drought tolerance is the ability of plants to withstand water deficit with low tissue water potential. The mechanisms of drought tolerance are maintenance of turgor through osmotic adjustment (accumulation of solutes in the cell), increased cell elasticity, and decreased cell size and desiccation tolerance by protoplasmic resistance.

The increased incidence of drought in some cowpea growing areas has caused a shift to early maturing varieties. Early maturity of cowpea cultivars is desirable and has proven to be useful in some dry environments and years because of their ability to escape drought. Such early cultivars can reach maturity in as few as 60–70 days in many of the cowpea production zones of Africa.

In cowpea, two types of drought tolerance have been described at the seedling stage using the wooden box technique (Agbicodo, 2009). It has been established that the "Type 1" drought-tolerant lines like TVu 11986 and TVu 11979 stopped growth after the onset of drought stress and maintained uniformity, but displayed a declining turgidity in all tissues of the plants including the unifoliates and the emerging tiny trifoliates for over two weeks (Agbicodo, 2009). All plant parts such as the growing tip, unifoliates, and epicotyl gradually died almost at the same time. The "Type 2" drought tolerant lines like Danila and Kanannado remained green for a longer time and continued slow growth of the trifoliates under drought stress. With continued moisture stress, the trifoliates of these varieties started wilting as well and died about four weeks after drought stress started (Agbicodo, 2009).

The two types of tolerance responses by cowpea seedlings to drought stress indicate that cowpea genotypes evolved different mechanisms to cope with prolonged drought encountered in the semiarid regions of Africa where the crop is believed to have originated. Closure of stomata to reduce water loss through transpiration and cessation of growth (for Type 1 drought avoidance) and osmotic adjustment and continued slow growth (drought tolerance in Type 2) have been suggested as the possible mechanisms for drought response in cowpea.

Other methods to explain the mechanism of drought tolerance in cowpea focused on the genetic dissection of drought tolerance through identification of markers defining quantitative trait loci (QTL) with effects on specific traits related to drought tolerance. Others have studied the relationship of the drought response and yield components, morphological traits, and physiological parameters. Other abiotic stress responses of cowpea need scientific attention.

2.3.5 Mung Bean (*Vigna radiata* (L.) Wilczek)

Among the abiotic stresses, salinity stress is more damaging that limits functional plant growth and yield worldwide. Due to the complex nature of salinity stress and lack of suitable techniques for introgression of desirable agronomic traits, little progress has been made in developing salt-tolerant lines in legumes, in general, and mung bean in particular.

Investigations have been made to screen mung bean lines for salinity tolerance (Manasa, Bindumadhava, Nair, Prasad, & Shankar, 2017). The study showed that considerable reduction in growth and yield performances of both tolerant and susceptible lines was observed, with few lines displayed relatively a better biomass and pod yield as compared to nonstressed control plants. Based on seedling and whole plant level tolerance, a few tolerant (EC 693357, 58, 66, 71 and ML 1299) lines were identified for further investigation.

Response of mung bean crop to different levels of applied iron and zinc was also investigated (Jamal, Khan, Tariq, & Fawad, 2018). The study revealed that application of Fe at the rate 5 kg ha^{-1} and Zn at the rate 10 kg ha^{-1} significantly increased biological yield, grain yield, straw yield, nodule numbers, and weight by 5624 kg ha^{-1}, 968 kg ha^{-1}, 4655 kg ha^{-1}, 35 and 0.67 g, respectively, whereas the interaction was found nonsignificant. Furthermore, the nitrogen content in grains and straw was also significantly increased by 2.22% and 3.56%, respectively, with application of Fe at 5 kg ha^{-1} and Zn at 10 kg ha^{-1}; however, their interaction was also found nonsignificant. Similarly, the plant nitrogen uptake was also significantly increased by 323.33 kg ha^{-1} with application of Fe at 5 kg ha^{-1} and Zn at 10 kg ha^{-1}. It was then concluded that Fe and Zn enhanced mung bean productivity.

Examination of the physiological response of contrasting mung bean (*Vigna radiata*) genotypes viz., T 44 & MH–96–1 (tolerant) and Pusa Baisakhi & MH–1K–24 (sensitive) under waterlogging conditions was also reported (Kumar, Pal, Joshi, & Sairam, 2013).

It has also been demonstrated that mobile phone radiation inhibits *Vigna radiata* (mung bean) root growth by inducing oxidative stress (Parkash Sharma, Pal Singh, Kohli, & RaniBatish, 2009).

2.3.6 Moth Bean

Moth bean (*Vigna aconitifolia* (Jacq.) Marechal), is the most drought- and heat-tolerant cultigens among Asian *Vigna*. A comparative transcriptome analysis of moth bean cultivar "Marumoth" under control and stress condition was studied (Tiwari, Kalim, Tyagi, et al., 2018). The study revealed that some stress inducible genes like Catalase, Cyt P450 monooxygenase, heat shock proteins (HSP 90 and HSP 70), oxidoreductase, protein kinases, dehydration responsive protein (DRP), universal stress protein, and ferridoxin NADH oxidoreductase genes were upregulated in a stressed sample. Furthermore, genes that might be involved in moisture stress tolerance in moth bean were identified, and these might be useful for stress tolerance breeding in moth bean and other related crops.

Genotypes of moth bean were subjected to heat stress to reveal the pattern of expression of genes (Tiwari, Kalim, Bangar, Kumar, et al., 2018). It was found that heat shock proteins (HSPs) and Super oxide dismutase (MnSOD) are differentially regulated in plants subjected to heat stress when 20-days-old moth bean seedlings were exposed to a temperature of 55°C for 30 minutes.

The effects of heat stress on relative water content, membrane stability index, proline, and chlorophyll contents using ten genotypes of moth bean have been reported (Tiwari, Kalim, Bangar, & Kumari, 2018). In the same line, the effect of low water potential generated by polyethyleneglycol on growth, sugar content, and stress-related enzymes (catalase, GPOX) at seedling level followed by gene expression using SDS-PAGE have been studied (Soni, Rizwan, Bhatt, Mohapatra, & Singh, 2011). It was established that drought-tolerant and sensitive varieties differed in their protein profiling, with newer genes showing their expression in tolerance once.

The effect of short-term heat stress on osmo-protectant and antioxidants in 37 genotypes (32 mutants and five varieties) of moth bean (*Vigna aconitifolia*) has also been studied (Harsh et al., 2016). A significant overaccumulation of total sugar and proline along with an increased activity of CAT, GPOX, and SOD was observed in most of the genotypes under heat stress. However, correlation analysis among heat stress-induced biochemical parameters suggests that none of these traits were associated with the level of thermo-tolerance except GPOX activity.

2.3.7 Rice Bean (*Vigna umbellata* (Thunb.) Ohwi & Ohashi)

Very limited studies attempting to elucidate the responses of rice bean to abiotic stress exist. A more recent one focused on the characterization of an inducible C_2H_2-type zinc finger transcription factor VuSTOP1 in rice bean (*Vigna umbellata*) and revealed differential regulation between low pH and aluminum tolerance mechanisms (Fan et al., 2015).

A study demonstrated that magnesium (Mg) can alleviate aluminum (Al) toxicity in rice bean [*Vigna umbellata* (Thunb.) Ohwi & Ohashi] more effectively than is expected from a non-specific cation response (Yang, 2007). Furthermore, micromolar concentrations of Mg alleviated the inhibition of root growth by Al but not by lanthanum, and neither strontium nor barium at the micromolar level alleviates Al toxicity.

2.3.8 Creole Bean

Creole bean (*Vigna reflexo-pilosa* var. glabra = *Vigna glabrescens*), Minni payaru (*Vigna stipulacea*), and Tuber cowpea (*Vigna vexillata*), are very little-known legumes due to the fact that they are not yet fully domesticated and well commercialized (Harouna et al., 2018). Therefore, very little information concerning their abiotic stress responses exists.

2.4 Biotic and Abiotic Management Strategies in *Vigna* Legumes

The productivity of crops is influenced by abiotic and biotic factors. In order to improve and sustain the domesticated *Vigna* legumes production, the agricultural stakeholders (researchers, scientists, farmers) need adoption and refinement of low-cost agronomic practices. Many chemicals, though, have been proven effective in controlling biotic stresses, but these are not cost-effective and environmentally friendly. Various nonmonetary inputs and techniques such as selection of crops, selection of drought-resistant/tolerant varieties, early-maturing/short-duration varieties, healthy and disease-free seeds, and seed treatment need to be adopted for mitigating abiotic and biotic stress for maximizing productivity and profitability of the dryland farmers. Better crop management practices such as nutrient, water and weed management, insect-pest and disease management are essential to withstand these stresses. Climate-resilient crop varieties along with other suitable adaptation and mitigation strategies will help overcome the adverse impact of climate change by lowering the yield losses under stress conditions.

In addition to the abovementioned management techniques, other modern and more sophisticated and cost-effective methods such as microbial mitigation methods are being experimented and developed.

2.5 Conclusion

Biotic and abiotic stresses are the major constraints in enhancing the productivity of legume food crops, in general, and of *Vigna* legumes in particular. However, very few scientific efforts have been addressing the productivity challenges of this category of food legumes despite their huge potential and genetic diversity coupled to the world food demand of the developing world. Therefore, modern mitigation techniques of biotic and abiotic responses of legumes need more scientific attention especially with this category of legumes.

ACKNOWLEDGMENTS

This review was carried out as part of the big project partially supported by the International Foundation for Science (IFS) through the grant number I-3-B-6203-1, which is hereby acknowledged. The authors acknowledge the Centre for Research, Agricultural Advancement, Teaching Excellence and Sustainability in Food and Nutritional Security (CREATES-FNS) through the Nelson Mandela African Institute of Science and Technology (NM-AIST) for partially supporting the project.

REFERENCES

Adsule, R. N. (1996). Moth bean (*Vigna aconitifolia* (Jacq.) Maréchal). In *food and feed from legumes and oilseeds* (pp. 203–205). https://doi.org/10.1007/978-1-4613-0433-3_21.

Agbicodo, E. M. (2009). *Genetic analysis of abiotic and biotic resistance in cowpea* [Vigna unguiculata *(L.) Walp.*]. Wageningen University, Wageningen NL.

Badea, C., & Basu, S. K. (2009). Mini review the effect of low temperature on metabolism of membrane lipids in plants and associated gene expression. *Plant Omics*, *2*(2), 78.

Baroowa, B., Gogoi, N., Paul, S., & Sarma, B. (2013). Morphological responses of pulse (*Vigna* spp.) crops to soil water deficit. *Journal of Agricultural Sciences, Belgrade*, *57*(1), 31–40. https://doi.org/10.2298/jas1201031b.

Baruah, I. K., Panda, D., Jagadale, M. V., Das, D. J., Acharjee, S., Sen, P., & Sarmah, B. K. (2017). Bruchid egg induced transcript dynamics in developing seeds of black gram (*Vigna mungo*). *PLoS ONE*, *12*(4), 1–17. https://doi.org/10.1371/journal.pone.0176337.

Basu, S., Mayes, S., Davey, M., Roberts, J. A., Azam-Ali, S. N., Mithen, R., & Pasquet, R. S. (2007). Inheritance of "domestication" traits in bambara groundnut (*Vigna subterranea* (L.) Verdc.). *Euphytica*, *157*(1–2), 59–68. https://doi.org/10.1007/s10681-007-9396-4.

Bhalla-Sarin, N., Bhomkar, P., Debroy, S., Sharma, N., Saxena, M., Upadhyaya, C. P., ... & Hohn, T. (2004). Transformation of Vigna mungo (blackgram) for abiotic stress tolerance using marker free approach. In *New Directions for a Diverse Planet: Proceedings of the 4th International Crop Congress Brisbane, Australia*, Oct (p. 454).

Boukar, O., Bhattacharjee, R., Fatokun, C., Kumar, P. L., & Gueye, B. (2013). Cowpea. In *Genetic and genomic resources of grain legume improvement* (pp. 137–156). https://doi.org/10.1016/B978-0-12-397935-3.00006-2.

Boukar, O., Close, T. J., Roberts, P. A., Fatokun, C. A., & Huynh, B.-L. (2016). Genomic tools in cowpea breeding programs: Status and perspectives. *Frontiers in Plant Science*, *7*(June), 1–13. https://doi.org/10.3389/fpls.2016.00757.

Chai, H. H., Massawe, F., & Mayes, S. (2016). Effects of mild drought stress on the morpho-physiological characteristics of a bambara groundnut segregating population. *Euphytica*, *208*(2), 225–236.

Chauhan, D. K., Tripathi, D. K., Prasad, S. M., Singh, S., & Singh, V. P. (2020). Plant genes for abiotic stress. In V. P. Singh, S. Singh, S. M. Prasad, D. K. Tripathi, & D. K. Chauhan, (Eds.), *Abiotic stress and legumes: Tolerance and management* . San Diego, CA: Elsevier Science Publishing Co.

Dahanayake, N., Ranawake, A. L., & Senadhipathy, D. D. (2015). Effects of water stress on the growth and reproduction of black gram (*Vigna mungo* L.). *Tropical Agricultural Research and Extension*, *17*(1), 45. https://doi.org/10.4038/tare.v17i1.5297.

Dash, M., & Shree, D. (2012). Temperature effect on morphobiochemical characters in some black gram (*Vigna mungo*) genotypes. *ISRN Biotechnology*, *2013*, 1–6. https://doi.org/10.5402/2013/942868.

Diouf, D. (2011). Recent advances in cowpea "omics" research for genetic improvement. *African Journal of Biotechnology*, *10*(15), 2803–2810.

Fan, W., Lou, H. Q., Gong, Y. L., Liu, M. Y., Cao, M. J., Liu, Y., … Zheng, S. J. (2015). Characterization of an inducible C2H2-type zinc finger transcription factor VuSTOP1 in rice bean (*Vigna umbellata*) reveals differential regulation between low pH and aluminum tolerance mechanisms. *New Phytologist*, *208*(2), 456–468.

Fatokun, C. A., Perrino, P., & Ng, N. Q. (1997). Wide crossing in African vigna species. In B. B. Singh, D. R. Mohan Raj, K. E. Dashiel, & L. E. N. Jackai (Eds.), *Advances in cowpea research—Google books* (pp. 50–57). Retrieved from https://books.google.co.tz/books?id=s_5Y5BFRU1EC&pg=PA50&lpg=PA50&dq=uses+of+'wild+VIGNA+SPECIES"&source=bl&ots=YyTY1f6Ix_&sig=KtTECLpPEezZDrpbl6AhBWUhQtw&hl=en&sa=X&ved=0ahUKEwjorYul2NrYAhVCWSwKHaeiCxA4FBDoAQgkMAA#v=onepage&q=use.

Fraire-Velazquez, S., & Balderas-Hernandez, V. E. (2013). Abiotic stress in plants and metabolic responses. In K. Vahdati & C. Leslie (Eds.), *Abiotic stress—Plant responses and applications in agriculture* (Vol. 2, pp. 27–48). New York: InTechOpen.

Gepts, P. (2001). Origins of plant agriculture and major crop plants. In MK Tolba (Ed.), *Our fragile world, forerunner volumes to the encyclopedia of life-supporting systems* (pp. 629–637). Oxford, UK: Eolss Publishers.

Gore, P. G., Tripathi, K., Pratap, A., Bhat, K. V., Umdale, S. D., Gupta, V., & Pandey, A. (2019). Delineating taxonomic identity of two closely related Vigna species of section Aconitifoliae: *V. trilobata* (L.) Verdc. and *V. stipulacea* (Lam.) Kuntz in India. *Genetic Resources and Crop Evolution*, *9*. https://doi.org/10.1007/s10722-019-00767-9.

Garcia, H. E., Pena-Valdivia, B. C., Rogelio, J. R. A., & Muruaga, S. M. J. (1997). Morphological and agronomic traits of a wild population and an improved cultivar of common bean (*Phaseolus vulgaris* L.). *Annals of Botany*, *79*, 207–213.

Gupta, N., Shrivastava, N., Singh, P. K., & Bhagyawant, S. S. (2016). Phytochemical evaluation of moth bean (*Vigna aconitifolia* L.) seeds and their divergence. *Biochemistry Research International*, 2016. https://doi.org/10.1155/2016/3136043.

Harouna, D. V., Venkataramana, P. B., Ndakidemi, P. A., & Matemu, A. O. (2018). Under-exploited wild Vigna species potentials in human and animal nutrition: A review. *Global Food Security*, *18*(September 2018), 1–11. https://doi.org/10.1016/j.gfs.2018.06.002.

Harsh, A., Sharma, Y. K., Joshi, U., Rampuria, S., Singh, G., Kumar, S., & Sharma, R. (2016). Effect of short-term heat stress on total sugars, proline and some antioxidant enzymes in moth bean (*Vigna aconitifolia*). *Annals of Agricultural Sciences*, *61*(1), 57–64. https://doi.org/10.1016/j.aoas.2016.02.001.

Heuzé, V., Tran, G., Bastianelli, D., & Lebas, F. (2015). Mung bean (*Vigna radiata*). Feedipedia, a programme by INRA, CIRAD, AFZ and FAO. https://www.feedipedia.org/node/235 Last updated on July 3, 2015.

Hillocks, R., Bennett, C., & Mponda, O. (2012). Bambara nut: A review of utilisation, market potential and crop improvement. *African Crop Science Journal*, *20*(1), 1–16. Retrieved from https://www.ajol.info/index.php/acsj/article/view/78601.

Iseki, K., Takahashi, Y., Muto, C., Naito, K., & Tomooka, N. (2018). Diversity of drought tolerance in the genus vigna. *Frontiers in Plant Science*, *9*, 1–18. https://doi.org/10.3389/fpls.2018.00729.

Jamal, A., Khan, M. I., Tariq, M., & Fawad, M. (2018). Response of Mung bean crop to different levels of applied Iron and Zinc. *Journal of Horticulture and Plant Research*, 13–22. https://doi.org/https://doi.org/10.18052/www.scipress.com/JHPR.3.13.

Jansen, P. C. M. (2006). *Vigna mungo* (L.) Hepper. [Internet] Record from PROTA4U. In M. Brink & G. Belay (Ed)., *PROTA (Plant Resources of Tropical Africa/Ressources végétales de l'Afrique tropicale)*, Wageningen, the Netherlands.

Joshi, K. D., Bhanduri, B., Gautam, R., Bajracharya, J., & Hollington, P. B. (2008). Ricebean: a multipurpose underutilised legume. In J. Smartt & N. Haq (Eds.), *New crops and uses: Their role in a rapidly changing world* (pp. 243–248). Southampton, UK: Centre for Underutilised Crops University of Southampton.

Kaewwongwal, A., Kongjaimun, A., Somta, P., Chankaew, S., Yimram, T., & Srinives, P. (2015). Genetic diversity of the black gram [*Vigna mungo* (L.) Hepper] gene pool as revealed by SSR markers. *Breeding Science*, *65*(2), 127–137. https://doi.org/10.1270/jsbbs.65.127.

Kaga, A., Isemura, T., Tomooka, N., & Vaughan, D. A. (2008). The genetics of domestication of the azuki bean (*Vigna angularis*). *Genetics*, *178*(2), 1013–1036. https://doi.org/10.1534/genetics.107.078451.

Kumar, P., Pal, M., Joshi, R., & Sairam, R. K. (2013). Yield, growth and physiological responses of mung bean [*Vigna radiata* (L.) Wilczek] genotypes to waterlogging at vegetative stage. *Physiology and Molecular Biology of Plants*, *19*(2), 209–220.

Lambrides, C. J., & Godwin, I. D. (2006). Mung bean. In C. Kole (Ed.), *Genome mapping and molecular breeding in plants* (Vol. 1), Cereals and millets, Berlin, Germany: Springer-Verlag.

Linnemann, A. R. (1987). Bambara groundnut (*Vigna subter-ranea* (L.) Verdc)—A review. *Abstracts on Tropical Agriculture*, *12*(7), p9–25.

Manasa, R., Bindumadhava, H., Nair, R. M., Prasad, T. G., & Shankar, A. G. (2017). Screening mungbean (*Vigna radiata* L.) lines for salinity tolerance using salinity induction response technique at seedling and physiological growth assay at whole plant level. *The International Journal of Plant, Animal and Environmental Sciences*, *7*(4). Retrieved from https://www.gov.uk/dfid-research-outputs/screening-mungbean-vigna-radiata-l-lines-for-salinity-tolerance.

Mogotsi, K. K. (2006). *Vigna radiata* (L.) R. Wilczek. *PROTA* (*Plant Resour. Trop. Africa/Ressources végétales de l'Afrique Tropicale*), Wageningen. https://prota4u.org/database/protav8.asp?H=M4&t=Vigna,mungo&p=Vigna+radiata#Synonyms/ (Accessed 15 January 2018).

Mohammed, M. S., Shimelis, H. A., & Laing, M. D. (2016). Phenotypic characterization of diverse Bambara groundnut (*Vigna subterranea* [L.] Verdc.) germplasm collections through seed morphology. *Genetic Resources and Crop Evolution*, *63*(5), 889–899. https://doi.org/10.1007/s10722-016-0374-3.

Mwale, S. S., Azam-Ali, S. N., & Massawe, F. J. (2007). Growth and development of bambara groundnut (*Vigna subterranea*) in response to soil moisture: 1. Dry matter and yield. *European Journal of Agronomy*, *26*(4), 345–353. https://doi.org/https://doi.org/10.1016/j.eja.2006.09.007.

NARO Genebank. (2017). *Illustrated plant genetic resources database (Moth bean)*. Retrieved 15 January 2018, from Genebank Project, NARO-JAPAN website: https://www.gene.affrc.go.jp/databases-plant_images_detail_en.php?plno=5420610017.

Ojiewo, C., Saxena, R. K., Monyo, E., Boukar, O., Mukankusi-mugisha, C., Thudi, M., … Varshney, R. K. (2017). *Genomics, genetics and breeding of tropical legumes for better livelihoods of smallholder farmers* (September, pp. 1–13). https://doi.org/10.1111/pbr.12554.

Pandiyan, M., Senthil, N., Anitha, M., Raveendran, M., Sudha, M., Latha, M., … Balasubramanian, P. (2012). Diversity analysis of *Vigna* sp through morphological markers. *Wudpecker Journal of Agricultural Research*, 1, pp. 335–340.

Parkash Sharma, V., Pal Singh, H., Kohli, R. K., & RaniBatish, D. (2009). Mobile phone radiation inhibits *Vigna radiata* (mung bean) root growth by inducing oxidative stress. *Science of the Total Environment*, *407*(21), 5543–5547.

Satyanarayana, J., Singh, K. M., & Singh, R. N. (1995). Pest complex of rice bean, *Vigna umbellata*. *Bulletin of Entomology (New Delhi)*, *36*(2), 28–33. Retrieved from https://www.cabdirect.org/cabdirect/abstract/19981101828.

Sehrawat, N., & Yadav, M. (2014). Screening and cross-compatibility of various *Vigna* species for yellow mosaic virus resistance. *Journal of Innovative Biologynnovative Biology*, *1*(1), 2002–2005.

Simon, M. V., Benko-Iseppon, A.-M., Resende, L. V., Winter, P., & Kahl, G. (2007). Genetic diversity and phylogenetic relationships in *Vigna* Savi germplasm revealed by DNA amplification fingerprinting. *Genome*, *50*(6), 538–547. https://doi.org/10.1139/G07-029.

Singh, B. B. (2006). Cowpea [*Vigna unguculata* (L) Walp. In J. R. Singh (Ed.), *Genetic resources, chromosome engineering, and crop improvement: Vegetable crops* (pp. 117–162). Boca Raton, FL: CRC Press.

Singh, S. K., & Srivastava, H. P. (1989). Some new fungal diseases of moth bean. *Indian Phytopathology*, *42*(1), 164–167. Retrieved from https://www.cabdirect.org/cabdirect/abstract/19912310858.

Soni, P., Rizwan, M., Bhatt, K. V, Mohapatra, T., & Singh, G. (2011). *In-vitro* response of Vigna aconitifolia to drought stress induced by PEG—6000. *Journal of Stress Physiology & Biochemistry*, *7*(3), 108–121.

Swaminathan, R., Singh, K., & Nepali, V. (2012). Insect pests of green gram *Vigna radiata* (L.) Wilczek and their management. In G. Aflakpui (Ed.), *Agricultural Science*. https://doi.org/10.5772/35176.

Taggar, G. K., Gill, R. S., Gupta, A. K., & Sandhu, J. S. (2012). Fluctuations in peroxidase and catalase activities of resistant and susceptible black gram (*Vigna mungo* (L.) Hepper) genotypes elicited by *Bemisia tabaci* (Gennadius) feeding. *Plant Signaling and Behavior*, *7*(10), 1321–1329. https://doi.org/10.4161/psb.21435.

Timko, M. P., Ehlers, J. D., & Roberts, P. A. (2007). Cowpea. In *Pulses, sugar and tuber crops* (pp. 49–67). https://doi.org/10.1007/978-3-540-34516-9_3.

Tiwari, B., Kalim, S., Bangar, P., & Kumari, R. (2018). Physiological, biochemical, and molecular responses of thermotolerance in moth bean (*Vigna aconitifolia* (Jacq.) Marechal). *Turkish Journal of Agriculture and Forestry*, (January), 176–184. https://doi.org/10.3906/tar-1709-1.

Tiwari, B., Kalim, S., Bangar, P., Kumar, S., Kumari, R., Barman, P., & Bhat, K. (2018). Expression analysis in response to heat stress in moth bean (*Vigna aconitifolia* (Jacq.) Marechal). *Electronic Journal of Plant Breeding*, *8*(4), 1293. https://doi.org/10.5958/0975-928x.2017.00185.5.

Tiwari, B., Kalim, S., Tyagi, E., Kumari, R., Bangar, P., Barman, P., … Author, K. V. B. (2018). Identification of genes associated with stress tolerance in moth bean [*Vigna aconitifolia* (Jacq.) Marechal], a stress hardy crop. *Physiology and Molecular Biology of Plants*, *24*(4), 551–561.

Tomooka, N., Kaga, A., Isemura, T., Vaughan, D., Srinives, P., Somta, P., ... Inthapanya, P. (2009). Vigna Genetic Resources. In *Proceedings of the 14th NIAS International Workshop on Genetic Resources: Genetic Resources and Comparative Genomics of Legumes* (*Glycine* and *Vigna*), Tsukuba, Japan, 14 September; pp. 11–21.

Tomooka, N., Kaga, A., Isemura, T., & Vaughan, D. A. (2011). Vigna. In C. Kole (Ed.), *Wild crop relatives: Genomic and breeding resources: Cereals* (pp. 291–311). New York, NY: Springer.

Tomooka, N., Vaughan, D. A., Moss, H., & Maxted, N. (2002). Introduction. In *The Asian Vigna* (pp. 1–7). https://doi.org/10.1007/978-94-010-0314-8_1.

Wang, L., Zhu, J., Li, X., Wang, S., & Wu, J. (2018). Salt and drought stress and ABA responses related to bZIP genes from *V. radiata* and *V. angularis*. *Gene*, *651*, 152–160. https://doi.org/10.1016/j.gene.2018.02.005.

Yang, J. L. (2007). Magnesium enhances aluminum-induced citrate secretion in rice bean roots (*Vigna umbellata*) by restoring plasma membrane H þ-ATPase activity. *48*(1), 66–73. https://doi.org/10.1093/pcp/pcl038.

Yang, K., Tian, Z., Chen, C., Luo, L., Zhao, B., Wang, Z., … Wan, P. (2015). Genome sequencing of adzuki bean (*Vigna angularis*) provides insight into high starch and low fat accumulation and domestication. *Proceedings of the National Academy of Sciences*, *112*(43), 13213–13218. https://doi.org/10.1073/pnas.1420949112.

Yaqoob, M., Khalil, I. A., Khan, N. U., Mansoor, M., Afridi, K., & Gul, S. (2015). Moth bean germplasm screening against yellow mosaic virus, II: Developmentof moth bean high-yielding seed and fodder cultivars. *Turkish Journal of Agriculture and Forestry*, *39*, 212–226. https://doi.org/10.3906/tar-1406-3.

Yoshida, Y., Marubodee, R., Ogiso-Tanaka, E., Iseki, K., Isemura, T., Takahashi, Y., … Tomooka, N. (2016). Salt tolerance in wild relatives of adzuki bean, *Vigna angularis* (Willd.) Ohwi et Ohashi. *Genetic Resources and Crop Evolution*, *63*(4), 627–637. https://doi.org/10.1007/s10722-015-0272-0.

3

Ecosystem Services of the Crop Wild Relatives, in Particular, Food Legume Landraces- Rhizobia *Association in Northern Africa and the Adaptation of Their Genetic Material to Agronomic and Environmental Objectives*

Sihem Tellah, Wahiba Amri-Tiliouine, Fahima Nabi, Agostino Sorgona, and Maurizio Badiani

CONTENTS

3.1 Introduction

The environmental problems raised from the increasing use of chemicals and synthetic fertilizers have prompted the scientific community to find biological alternatives for supplying mineral nutrients to crops. Legumes provide such opportunity through the ecosystem benefits they produce during their cultivation (Kannaiyan and Kumar, 2006). The key environmental benefits deriving from legumes cultivation rely on their ability to fix atmospheric nitrogen symbiotically under a broad spectrum of environmental conditions, enabling such crops to exert a positive effect in diversified crop rotations.

Legumes can be considered an "agronomical asset" for improving agricultural production and contributing to the maintenance and improvement of soil fertility. Hence, the challenge would be to implement a sustainable approach to promote their ecological services by engineering their interactions with soil microorganisms to the benefit of the cereal agrosystems (Banerjee et al., 2006).

In Algeria, as well as in the wider context of the Mediterranean Africa, most cultivated lands are located in arid and semiarid areas where climatic constraints contribute to the fragility of the environment and, consequently, of the agricultural productions. The choice of species and/or varieties to be cultivated, therefore, has to take into account the use of appropriate agronomical practices able to maintain environmental integrity of agroecosystems and to promote the sustainability of crop production. The answer to this requirement can be provided by the development and operation of the nitrogen-fixing symbiotic associations in these arid areas, which are further subjected to climate change (Tellah et al., 2016).

Algeria is characterized by enormous contrasts, in terms of both geographical characters and ecological farming practices. Water deficit, salinity, and nutrient deficiency of the soils are among the main factors limiting agricultural production, which are recurring conditions in the semiarid subtropics on a global scale. Circumvention of these factors ensures ecological intensification of agricultural systems and offers more promise to preserve the environment and contribute to agrosystem sustainability. In this respect, the introduction of nitrogen-fixing legumes in cereal cropping systems can to some extent improve the crop sustainable productivity while reducing the use of nitrogen and phosphate fertilizers responsible for environmental pollution.

As a general strategy, it is first necessary to select legume species or varieties that are adapted to the local ecological conditions. This choice has to be based on the utilization of genetic resources obtained from a preexisting representative collection of local germplasm, which can assist in the selection of species and varieties that better adapt to the edaphic and climatic conditions of each region.

For this, an integrative approach has been pursued in the last couple of decades by Algerian institutions involved in agricultural research and extension services, among which the Ecole Nationale Supérieure Agronomique (ENSA), combining a range of disciplines (biology, physiology, ecophysiology, agronomy, economics, and social sciences), along with active participation of farmers in reference areas, to collect and characterize local accessions of legumes and their associated bacterial strains and to identify and test valuable symbiotic combinations. The aim is to provide cultural rotations or associations that maximize the ecological profits of legumes. This approach is based on four assumptions: (1) the nitrogen fixed by legumes helps to enrich the soil with nitrogen, (2) it increases the nitrogen nutrition and yield and (3) reduces the use of nitrogen fertilizer by the crops following legumes in the rotation; (4) legumes act as a long-term improvement of the physicochemical and biological properties of soils.

In the following, a selection of results is reported obtained during the last decade from combining Algerian food legume landraces and indigenous rhizobacterial strains and studying the responses of the resulting symbionts to water deprivation and excess salinity.

3.2 Importance of Food Legumes and Biological Nitrogen Fixation (BNF)

Legumes are dicotyledonous plants belonging to the botanical family of the *Fabaceae*, which represents the third plant family by the number of species (namely 18,000 referenced) after the *Asteraceae* and *Orchids*. Most members of this family possess the ability to use atmospheric nitrogen (N_2) for producing their own quaternary metabolites. Such ability is due to the symbiosis with N_2-fixing soil bacteria—both α- and β-proteo-bacteria, popularly referred to as *rhizobia*—within specialized organs, namely the nodules, which grow in the root system. In particular, nitrogen fixation is a biochemical process in which by using metabolic energy and reducing power ultimately deriving from the light reaction of photosynthesis, dinitrogen is reduced to ammonia, which is then used for the synthesis of many bioorganic compounds. Eventually, such biologically fixed nitrogen can be directly taken up by plants, maintaining the environment almost "untouched." Hence, by this morphophysiological feature, unlike other cultivated species, the leguminous crops do not generally require nitrogen fertilizers to express optimal growth, if not during the early vegetative stages, to acquire photosynthetic competence and efficiency. Therefore, they represent a symbiotic nitrogen gateway (this is, nitrogen derived from symbiotic fixation) in agricultural production systems (Schneider et al., 2005).

The literature reports values of symbiotically fixed N under field conditions ranging from 15 (Schwenke et al., 1998) to 648 kg N ha^{-1} (Sprent et al., 1977). Such high variability results from variability in specific growth conditions, genotypic features, and differences in the methods used for determining BNF.

The first root nodules are set about two weeks after crop emergence. The highest nitrogen fixation rates are found after flowering (Vinther and Dahlmann-Hansen, 2005), when the pods and grain become strong active sinks for assimilates and fixed nitrogen. During this phase, daily rates have been estimated at 4–5 kg N ha^{-1} (Zapata et al., 1987; Hauser, 1987). Consequently, the amount of nitrogen fixed can be increased by all those agronomic practices, which ensure vigorous crop growth (Dantuma and Klein Hulze, 1979; Maskey et al., 2001; Peoples et al., 2002, 2009; Jensen et al., 2005; Hauggaard-Nielsen et al., 2007), resulting in higher grain yields.

By maintaining soil fertility through N_2 fixation, a legume crop can be of benefit to the subsequent one, i.e., during crop rotation. Crop rotation with legumes has been recognized to increase soil fertility and agricultural productivity since the times of ancient China and Rome. Further, by reducing the need for nitrogen fertilizers, dinitrogen fixation brought about by nodulated legumes decreases nitrate leaching from soils and, consequently, eutrophication of water bodies, which has become a major environmental problem worldwide (Nasir Khan and Mohammad, 2014).

The agronomic implications of root symbiosis in legumes have promoted the research on the molecular and physiological mechanisms regulating plant-host recognition, successful infection process, and establishment of nitrogen fixation, as well as on the characterization of *rhizobia* (Fernandez-Pascual et al., 2007), for the selection/engineering of better inoculants. For an inoculation to be successful, exogenous bacterial strains need to be improved in order to outcompete indigenous strains for root nodulation of legumes.

Sustainable agriculture has to overcome simplified crop rotations that arose as a result of intensification and increased use of high-energy inputs, particularly synthetic mineral N fertilizers, biocides, and intensive tillage. Introducing (or reintroducing) grain legumes into highly simplified agroecosystems contributes to their diversification (Koepke and Nemecek, 2010). In a survey by Krupl and Funk (2000) concerning why farmers in Europe use grain legumes, the first reason given in the 60% of the answers was "for the rotation" and "soil improvement." "Labor organization" and "on-farm use" as well as the "competitiveness of the crop" were mentioned by only 15% of the respondents. Grain legumes enable temporal diversification of the agroecosystem (Hauggaard-Nielsen et al., 2007), i.e., planned biodiversity of crops as well as associated diversity of wild flora, wild fauna, and soil microbes.

3.3 Food Legume Landraces of Peanut and Cowpea in Front of Water Deficit and Salinity Under Controlled Conditions

Food legumes, such as cowpea [*Vigna unguiculata* subsp. *unguiculata* (L.) Walp.] and peanut (*Arachis hypogaea* L.) are important crops in developing countries of the tropics and subtropics, especially in sub-Saharan Africa, Asia, and Central and South America (Singh et al., 1997), and in some temperate areas, including the Mediterranean region and the southern states of the United States (Pasquet, 2000). Their relevance stems from several reasons: cowpea and peanut produce seeds with high nutritive value (Ehlers and Hall, 1997), are well adapted to grow under high temperature and drought (Hall and Patel, 1985), tolerate low soil fertility due to their high rate of nitrogen fixation (Eloward and Hall, 1987), and are able to sustain an efficient symbiotic interaction (Kwapata and Hall, 1985). Therefore, cowpea and peanut can play an important role in agricultural development, particularly in the Maghreb and sub-Saharan regions, where drought and salinity frequently limit crop production.

In Afro-*Mediterranean* countries, despite the economic and cultural importance of the above legumes, both plant material and cultivation practices are neither uniform nor standardized. Instead, their cultivation is often based on the landraces, each of which occupies relatively few hectares. For these reasons, the traditional farmers in many areas, whether by choice or by need, conserve and develop the local phytogenetic resources by preserving landraces and associated local knowledge. Although not desirable in terms of yield and yield stability over time, the cowpea and peanut landraces, adapted to constant adaptive pressure of a challenging environment such as the Maghreb and sub-Saharan regions, can be regarded as a formidable genetic reservoir of useful traits for the resistance, both on a local basis and on a wider scale, e.g., Mediterranean, to drought and salt stress especially in the present era of climate changes.

3.3.1 Evaluating the Agro-Physiological, Morphological, and Molecular Mechanisms Involved in the Tolerance to Water Stress in Algerian Groundnut Landraces

Water has a fundamental role in plant life, as it allows and regulates physiological and metabolic activities and serves as a vector for mineral nutrients (Riou, 1993). Water availability is, therefore, the main limiting factor of crop production in arid and semiarid regions.

During their vegetative cycle, plants undergo the conditions of the environment in which they live. They react by adapting to the variation of such conditions when these become unfavorable. Variation in growth conditions results from stresses that are biotic or abiotic in nature. In arid and semiarid environments, abiotic stress such as water deficit imposes limits on the development of the plant. Resistance to these stresses is dependent on the ability of a given genotype to develop morphological, physiological, and/or biochemical mechanisms to escape (dodge), avoid, or tolerate stress (Levitt, 1982).

In nodulated legumes, water constraints apply not only to the plants, but also to their symbionts. Each partner reacts differently in response to water stress. Nodules, which do not have effective mechanisms of tolerance, are the first to be affected, and often their ability to fix nitrogen is irreversibly inhibited. In fact, water stress and the resulting osmotic disturbances can affect the *rhizobium*-legume symbiosis by reducing the growth and survival of *rhizobia* in the soil, by reducing the colonization of the rhizosphere, or by totally or partially inhibiting the process of infection and development of nodules.

Peanut is one of the most important oilseed crops grown in the world, particularly in the arid and semiarid regions of Africa (Wani et al., 1995). Plastic and rustic, peanut plants have the agronomic advantage of requiring few inputs, to cover very quickly and thus to protect the soil with a thick herbaceous carpet (on a height of 20–70 cm). Its success also resides in the fact that it is an oil-protein crop with many food and feed uses (seed, oil, cakes, tops, and hulls). Symbiotic dinitrogen fixation can provide up to 30% of the nitrogen needed by the plant, but the actual amount of nitrogen fixed is quite variable because the symbiotic activity is influenced by several factors such as the bacterial strain, the host plant, and the environmental conditions (Wani et al., 1995). Researchers have been looking long since for tolerant plant -*rhizobium* pairs capable of functioning under these difficult conditions, aiming to identify and understand the parameters and physiological mechanisms related to this trait, which may serve as a basis for the establishment of selection criteria.

Most of the work done on peanuts as part of the genetic improvement of tolerance to water stress have had for a long time the primary objective of increasing productivity, an approach based on agronomic performance. Currently, peanut improvement programs are also interested in genetic improvement of tolerance to water stress. This improvement requires the study, identification, and verification of phenological, morphophysiological, and biochemical characteristics related to yield under water stress (Pfeiffer et al., 2000). Similarly, looking for molecular markers of the mode of transmission and the heritability of the characters identified, as reliable indicators of tolerance to water stress, is necessary to facilitate the use of these characters in the selection programs (Pfeiffer et al., 2000).

In a previous study of ours (Tellah et al., 2014) we aimed to characterize the agro-morphological, physiological, biochemical, and molecular mechanisms underlying the tolerance to water stress of peanut landraces originating from different sites of Algeria when grown under controlled environmental conditions. We intended to identify the best performing peanut landraces in terms of capacity to tolerate the water deficit, which can occur at a critical stage of development of the species. In particular, the following objectives were pursued:

- Studying agro-physiological, morphological, biochemical responses of different groundnuts landraces toward water constraint
- Identifying the most drought-tolerant peanut landraces through the study of the rooting pattern, as an important determinant of adaptation to drought
- Evaluating antioxidant mechanisms potentially involved in tolerance to water stress
- Conducting gene expression analysis induced by water stress

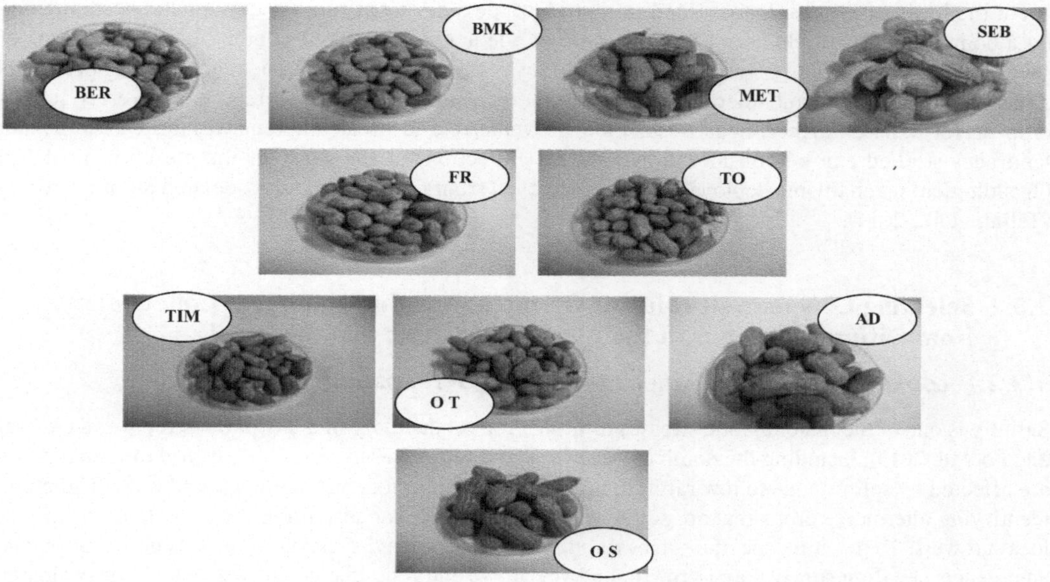

FIGURE 3.1 Morphological features of the seeds of the Algerian peanut (*Arachis hypogaea* L.) landraces studied here. For landraces acronyms, see the text. (Courtesy of Tellah, S., Etude des mécanismes agrophysiologiques, morphologiques et moléculaires impliqués dans la tolérance au stress hydrique chez quelques populations locales d'arachide (*Arachis hypogaea* L.), PhD Thesis., Ecole Nationale Supérieure Agronomique, 2016.)

The plant material under study included ten landraces of groundnut (Figure 3.1) obtained from a previous country-wide prospection operated by ENSA. The different Algerian sites from which the landraces were obtained can be geographically and climatically clustered into two main regions, according to the Algerian National Office of Meteorology, each of which hosts several sites of landraces collection:

1. The El Kala region, in Northeastern Algeria, from which the Berrihane (BER), Oum Tboul (OT), Tonga (TO), El Frin (FR), and Boumalek (BMK) peanut landraces were obtained;
2. The oases located in the Central and Eastern Algerian Sahara, from which the Adrar (AD), Timimoun (TIM), Ghardaïa from two sites, Metlili (MET) and Sebseb (SEB), and, finally, Oued Souf (OS) peanut landraces were obtained.

Peanut landraces were grown in pots from germinated seeds. Cultivation substrate was a mixture of sandy soil, garden soil, and washed quartz sand (50/20/30 percent in volume, respectively). Water stress was imposed by withholding irrigation for 10, 15, or 25 days, assumed to simulate, under the adopted experimental conditions, mild, moderate or severe water deficit, respectively.

Several morphophysiological traits were evaluated in the tested peanut landraces, and a number of them were found to be associated with the response to water deficit. By applying a multivariate statistical approach to the results obtained, such as the Principal Component Analysis (PCA), two biologically significant determinants were identified: (i) "root vigor," associated with a long and deep root system, with both a strong lateral root distribution containing several nodules, and pronounced growth of the stem; (ii) "physiological mechanisms," associated with the ability to maintain an adequate physiological state of the leaves and to lower the canopy temperature through permissive values of leaf water potential allowing a certain degree of stomatal permeability, even under severe water stress conditions (25 days of water deprivation). Further, PCA allowed to identify two subgroups of peanut landraces, namely BER, TO, AD, OT, and BMK, on one side, and MET, SEB, OS, FR, and TIM, on the other side, exhibiting the best and the worst biological performance, respectively, in the presence of water stress.

On the basis of the results obtained, a root ideotype able to optimize water uptake, accompanied by a comparatively ample aerial biomass and an adequate chlorophyll content under conditions of severe water stress, could be regarded as useful traits to be transferred in peanut breeding programs aimed at improving plant tolerance in areas where recurring water shortage can severely impair crop performance. Further, because of their provenance and great inherent variability, the peanut landraces studied can be considered as a useful collection for the study of the morphological and physiological mechanisms deployed by this species in front of drought and associated abiotic stresses (Tellah et al., 2014).

3.3.2 Selecting Cowpea Salt-Tolerant Germplasm within Landraces Collections from Different Regions of the Algerian Maghreb and the Sahel

3.3.2.1 Looking for Salt Tolerance at the Seed Germination Stage

Salinity is one of the major abiotic stresses limiting crop productivity in many parts of the world (Ashraf and Foolad, 2013), including the South of Algeria. These regions are predominantly arid to semiarid and are affected by salinity due to low rainfall, high evapotranspiration, and saline groundwater. Therefore, identifying alternative crops that are salt tolerant is imperative for sustaining the economic viability of local growers. To this aim, specific stages throughout the ontogeny of the plant, such as germination and emergence, seedling survival and growth, and vegetative and reproductive growth, should be evaluated separately from each other during the evaluation of germplasm for salt tolerance.

Germination of seeds, one of the most critical phases of plant life (Song et al., 2008), is greatly influenced by salinity. Salinity may cause significant reductions in the rate and final percentage of germination, which in turn may lead to uneven stand establishment and reduced crop yields (Genc et al., 2010). Ashraf and Foolad (2013) pointed out that genetic information regarding seed germination and related traits could help improving seedling emergence in saline soils through breeding programs. Rapid, uniform, and complete germination are a prerequisite for successful transplant production and stand establishment in vegetable crops (Genc et al., 2010).

Although the effects of salinity on the germination, vegetative growth, or yield of cowpea have been extensively studied (Murillo-Amador et al., 2000, 2002; Yildirim et al., 2006; Sehrawat et al., 2014; Zavariyan et al., 2015), there are no data about the salinity tolerance of current cowpea accessions from Algeria. Trying to fill such a knowledge gap, a study was undertaken to provide the first information on the tolerance of cowpea during the germination stage to salinity changes and to establish their possible range of salinity tolerance. Such information may help to carry out further research on breeding programs for genetic improvement of cowpea for saline environments (Nabi et al., 2017).

As the plant material, the seeds of five cowpea landraces collected from as many areas across Algeria were used (Ghalmi et al., 2009). The seeds were conserved ex situ (seeds dried and stored at −20°C and multiplied in a field collection) at the Institut Technique des Cultures Maraichères (ITCM Staouali, Algiers). Based on the locations in which they were collected, the cowpea landraces were assigned the following names: EK1, TZ7, A17, IS13, and B23, native of El Kala, Tizi Ouzou (Kabylia), Adrar (Sahara), In Salah (Sahara), and Bechar (Sahara), respectively. The selection of cultivated accessions was based on contrasting seed size, color, and origin. Kabylia is a region in the North of Algeria, part of the Tell Atlas located at the edge of the Mediterranean Sea. Its climate is subhumid with a temperate winter and rainfall >700 mm per year. Local soils, which are composed of clay-sand to marl, are often degraded on the slopes. El Kala is located in the extreme Northeast of Algeria in a triangle of coastal plains and hills bounded by the Mediterranean Sea and the Tunisian border. The climate of El Kala is humid to subhumid with mild winters with annual average precipitation of 910–1,300 mm. Several soil types are present: marshes, boggy soil, dunes, and wet meadows that become dry in the summer. The oases of the Sahara are characterized by a hyperarid climate with cold winters and very hot, dry summers. Seasonal variation may be huge, from over 50°C during summer days, to temperatures below freezing during winter nights. Daily variation is also very important. Not only does the Sahara receive very little rain, but precipitation events are also extremely irregular, ranging from 1 to 180 mm per year with a high degree of evapotranspiration. Local soils are mainly comprised of sand.

Previously sterilized and air-dried cowpea seeds were placed on "germitest" papers imbibed with distilled water or with aqueous sodium chloride in a proportion of 2.5 times the weight of the paper (Dantas et al., 2005). The NaCl levels tested were 0 (T0, control), 4 (T1), 8 (T2), 12 (T3), 16 (T4), and 20 (T5) dS m^{-1}, with the electrical conductivity of the imbibing solutions being measured by a conductivity meter. A germination test was run in a germinator at 25°C ± 0.5°C and 80% ± 1% relative humidity. Daily and final seeds germination percentage, as well as salt tolerance index and speed of germination, were determined for eight days.

The results obtained highlighted the variability of the salt tolerance within the tested group of Algerian cowpea landraces, which can be considered as a genetic resource for the improvement of salt tolerance. Germination percentage, final germination percentage, and speed of germination decreased significantly with increased NaCl concentration (from EC 12 dS m^{-1}). The genotype was found to be a significant factor in determining the output of all germination traits, which were all generally delayed by salt stress. Ranking salt tolerance indices led to grouping the cowpea landraces into tolerant—namely EK1, TZ7, and B23—and susceptible ones, namely A17 and IS13, so demonstrating that useful traits can be recovered from Algerian traditional cowpea ecotypes and possibly introduced into susceptible genotypes through breeding, for the genetic improvement of cowpea for saline soils. Further studies were recommended by Nabi et al. (2017) aimed at screening a higher number of salt-tolerant cowpea genotypes, suitable to grow in Algerian Sahara soils, and ascertain whether salt tolerance shown at the germination stage is maintained thereafter till crop maturity.

3.3.2.2 Salt Tolerant Germplasm Identified from Growth and Productivity Traits

Among the various abiotic stresses that plants are constantly exposed to, salt stress is the one that most affects plant growth and productivity worldwide (Veeranagamallaiah et al., 2007). Consequently, tolerance to the presence of salts such as sodium chloride is a highly appreciated trait in plants of agronomic interest (Alaoui et al., 2013).

Cowpea is one of the most largely cultivated food crops in the semiarid subtropics covering Asia, Africa, southern Europe, and Central America (Hadi et al., 2012). This source of concentrated proteins is cheaper than fish, meat, poultry, or dairy products and is very well integrated with cereals in human food, thus making an important contribution to subsistence agriculture (Hall, 2012). In Algeria, cowpea is grown on a small scale as a food crop in the most remote parts of the country (Kabylie mountains, Sahara oasis, El Kala coastline; Ghalmi, 2011).

Cowpea faces many cultivation constraints, including abiotic stress such as salinity, which severely affects its productivity (Lawlor, 2013). Several studies were carried out throughout the world in order to evaluate the response of cowpea to salinity conditions (Dantas et al., 2005; Murillo-Amador et al., 2006; Padilla et al., 2009; Nabi, 2009; Salih and Kia, 2013; Win and Oo, 2015).

In search of inherent tolerance traits within the vast array of biodiversity offered by the long-lasting cultivation of traditional ecotypes, the responses to increasing levels of salt stress were evaluated by a two-step procedure on a group of cowpea landraces collected in different regions of the Algerian Maghreb and the Sahel (Tellah et al., 2018). The following cowpea landraces were used: P1 and P2 from El Kala (see above); P3, P4, and P18 from Adrar (Sahara); P6 from Tizi Ouzou (Kabylia, see above); P19 from El Golea (Al-Manīa, Sahara); P31 and P32 from In Salah (Sahara); P33 from Bechar (Sahara); and NG1 from Niger (Figure 3.2).

The first selection step was a germination screening, aimed at identifying the most salt tolerant among the cowpea landraces according to their region of provenance. Four levels of NaCl, namely 0 (control, T0), 5 (T1), 10 (T2), or 15 (T3) g L^{-1}, assumed to represent mild, moderate, or severe salt stress, respectively, were compared for their effects on seed germination and kinetics during conventional seed germination experiments conducted in Petri dishes.

The results obtained indicated that among the 11 populations tested, the following four showed negligible levels of germination, probably due to inactivation following seeds conservation: P3 and P4 (0% germination at all treatments), P6 (10% at T0 and T1, 3.33% at T2 and 0% at T3), and P31 (10% at T0, 3.33% at T1 and T3, 0% at T2). Further, the experiments pointed out that the effect of NaCl was more pronounced in the case of severe salt stress (T3), which decreased the maximum germination rate and

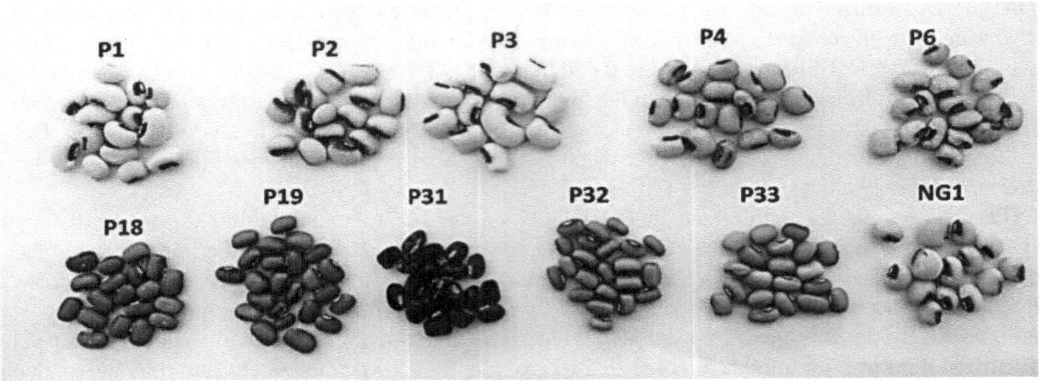

FIGURE 3.2 Seed morphological features of cowpea landraces from the Algerian Maghreb and the Sahel. (Courtesy of Tellah, S. et al., Influence of salt stress on germination, growth and productivity of some Algerian landraces of cowpea (*Vigna unguiculata* subsp. *unguiculata* (L.) Walp.), XXXVI convegno nazionale della società italiana di chimica agrariai, Agrariail ruolo della chimica agrariaper la gestione sostenibile delle risorse agrarie e forestali, Dipartimento di agraria – università mediterranea-24–26 Settembre 2018-Reggio Calabria-Italie, 2018.)

delayed germination kinetics (data not shown). The NG1, P32 and P33 cowpea seeds showed the greatest tolerance to salt, whose negative effects, in the first two landraces, became evident only under severe stress. The P33 landrace was slightly affected in the presence of moderate salt stress, i.e., T2, showing a maximum germination rate that was as high as 96.66% compared to a 100% value in its unstressed control. On the other side, the P2 landrace, a provenance of the El Kala region, showed the greatest sensitivity to salt stress, its maximum germination rate reaching 46.66% at T1 and T2, and only 16.66% at T3, as compared to a 60% germination rate in its unstressed control (T0; Figure 3.3).

The germination screening described above allowed for the selection of four cowpea landraces, namely P1, P32, P33, and NG1, on which whole-plant responses to salt stress were then evaluated. To such aim, cowpea plants grown in pots filled with soil, garden soil, and washed river sand (30/30/30 percent in volume, respectively) until the emission of their first trifoliate leaves were exposed to four levels of NaCl in the soil solution, namely 0 (control, T0), 3 (T1), 6 (T2), or 9 (T3) g L^{-1}, assumed to represent mild, moderate, or severe salt stress, respectively. Thirty days after salt stress imposition, its effects on plant growth, relative water content (RWC), total chlorophyll, accumulation of proline, foliar soluble sugars, and yield components were evaluated.

The stress-induced negative effects on plant growth—height of the stem, number of leaves, number of branches, and leaf area—on RWC and on total chlorophyll were observed to increase with the NaCl levels the plants were exposed to, and the same was apparent for the accumulation of proline and soluble sugars in the foliar tissues, which reached a maximum in the presence of severe salt stress (T3). Due to technical reasons, yield components such as number of pod per plant, number of seed per pod, number of seeds per plant, seed size, pod length per plant, and weight of 100 seeds were measured in only two of the four tested cowpea landraces, namely P1 and P33, in which similar effects of salt stress were observed.

However, the type and intensity of the observed effects varied in relation to the salt levels and the landrace. By applying multivariate statistical analysis to the biometric, physiological, and biochemical data sets obtained, it became possible to classify the four cowpea landraces according to their salt tolerance: P33 from Bechar was found to be the most tolerant to salinity, followed by P32 from In Salah. As mentioned above, these two cowpea landraces had already proven their salt tolerance during the seed germination tests. Conversely, the growth of the NG1 landrace from Niger and of the P1 landrace from El Kala revealed to be sensitive to salt stress. As far as the Algerian cowpea accessions are concerned, the southernmost provenances, namely P32 from In Salah and P33 from Bechar, both regions in which a desert climate predominates, and showed a greater degree of salt tolerance compared to the P1 landrace from El Kala, in the northern part of the country, where a Mediterranean-type climate prevails. This clearly suggests climate-driven adaptation to salt stress, and it further prompts the pursuit of a

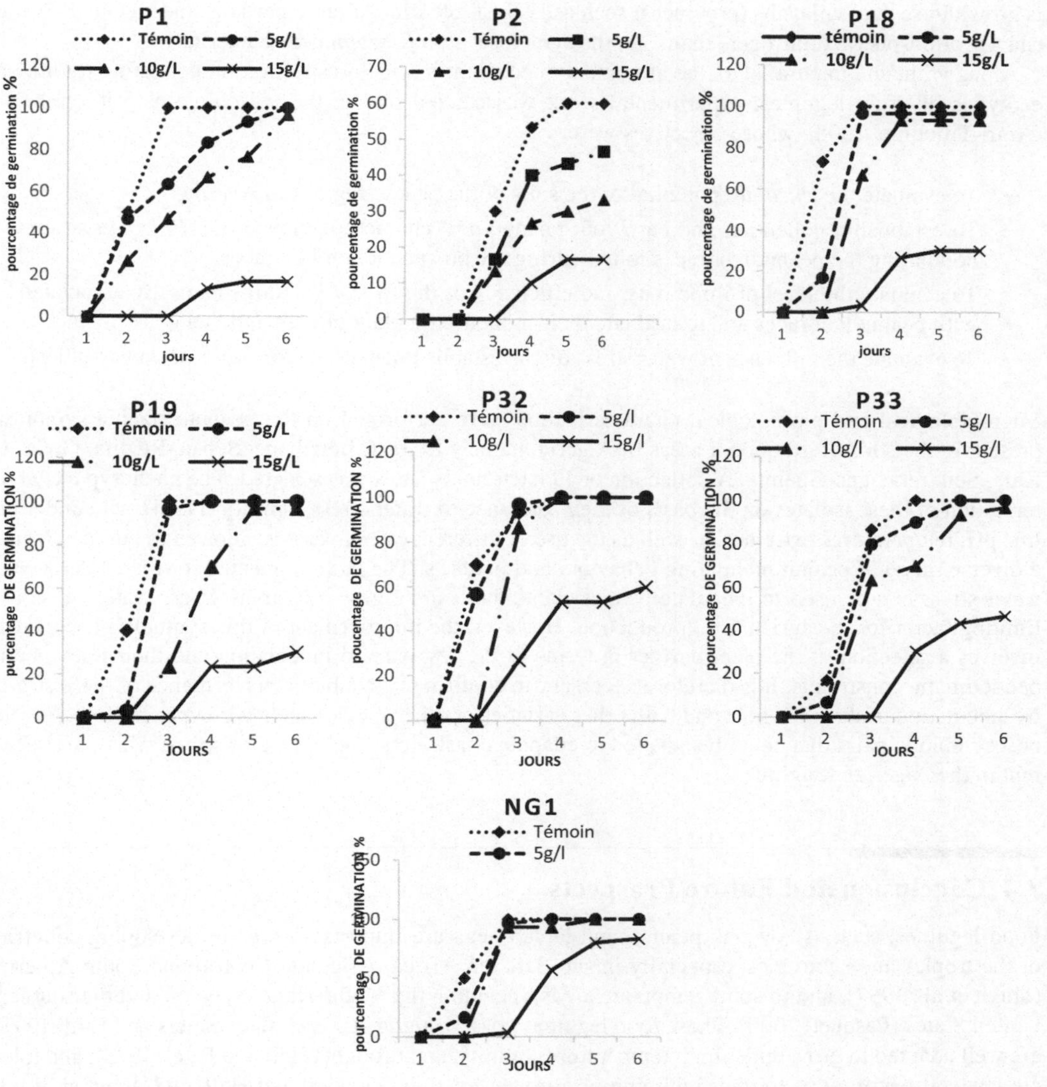

FIGURE 3.3 Effects of different concentrations of NaCl on seed germination kinetics of seven populations of cowpea landraces studied for six days. Each point represents the average of three replicates. (Courtesy of Tellah, S. et al., Influence of salt stress on germination, growth and productivity of some Algerian landraces of cowpea (*Vigna unguiculata* subsp. *unguiculata* (L.) Walp.). XXXVI convegno nazionale della società italiana di chimica agrariai. Agrariail ruolo della chimica agrariaper la gestione sostenibile delle risorse agrarie e forestali. Dipartimento di agraria—università mediterranea-24–26 Settembre 2018-Reggio Calabria-Italie, 2018.)

dynamic conservation strategy and a reasoned mining and exploitation of potentially useful tolerance traits embedded in the wide genetic diversity of traditional food legume landraces striving under challenging environments.

3.3.3 Rhizobium-Peanut Symbiosis: Microbiological, Physiological, and Agronomic Aspects

Peanut is one of the most important oilseeds cultivated around the world, particularly in the arid and semiarid regions of Africa. Its cultivation is widespread in Algeria (Benmohamed et al., 2004), where it is grown in several wilayahs (provinces) such as El Tarf, Skikda, Adrar, Ghardaia, and El Oued. Peanut can use atmospheric dinitrogen, thanks to the symbiosis with *Rhizobiaceae* bacteria.

Along with, and in parallel to, the aforementioned country-wide prospection for collecting traditional ecotypes of grain legumes, experimental work was carried out on the *rhizobium*-peanut symbiosis (Amri-Tiliouine, 2008), whose objectives were:

- To evaluate the *rhizobial* potential of the soils of the peanut regions in Algeria
- To establish a collection and carry out phenotypes' characterization of the *Rhizobia* strains nodulating the peanut landraces found during the prospection of landraces
- To evaluate the level of infectivity and efficiency of the *Rhizobia* strains naturally associated with peanut landraces and to evaluate their influence on plant growth and yield
- To evaluate the influence of water stress on the establishment of the *Rhizobia*-peanut symbiosis

Microbiological and physiological characterization were performed on the peanut-rhizobia symbiosis present in the soils of groundnut areas in Algeria, namely El Tarf, Berrihane, Sebaà, Réghia, Gréat, El Kala, Souarekh, and Guelma. A collection of 19 bacterial isolates was selected. The phenotypic characterization of these isolates on the basis of their tolerance to water stress, salinity (NaCl), acid and alkaline pH, temperatures extremes, as well as the use of different carbohydrates, allowed to put in evidence a diverse range of peanut nodulating behaviors and patterns. The strains selected for their tolerance to water stress can be used to inoculate peanut plants that can be grown in areas where water stress is a limiting factor for the agricultural production. However, the improvement of this symbiosis necessarily involves a selection of the plant partner in terms of the response to inoculation and their tolerance to pedoclimatic constraints. It is therefore necessary to confirm the symbiotic performance of these strains by inoculation trials in the field and with other varieties or other species such as cowpea for selecting the best symbiotic pairs that could be adapted to edaphic constraints, such as water stress, which are recurrent in the Algerian Maghreb.

3.4 Conclusion and Future Prospects

Food legumes, such as cowpea, peanut, and lablab bean are important crops in developing countries of the tropics and subtropics, especially in sub-Saharan Africa, Asia, and Central and South America (Singh et al., 1997), and in some temperate areas, including the Mediterranean region and the southern United States (Pasquet, 2000). These food legumes possess high nutritive value (Ehlers and Hall, 1997); are well adapted to grow under high temperature, salinity, and drought (Hall and Patel, 1985); and tolerate low soil fertility due to their high rate of nitrogen fixation (Eloward and Hall, 1987) and ability to form effective symbiotic mycorrhizae (Kwapata and Hall, 1985). Therefore, such species can play an important role in agricultural development, particularly in the Maghreb and sub-Saharan regions, where drought and salinity frequently limit crop production.

In Afro-Mediterranean countries, despite the economic and cultural importance of the above legumes, which are consumed not only as a source of food, but often also for their therapeutic properties, both plant material and cultivation practices are neither uniform nor standardized. Instead, their cultivation is often based on landraces, each of which occupies relatively few hectares. Traditional farmers in many areas, whether by choice or by need, conserve and develop local phytogenetic resources by preserving landraces and associated local knowledge. Although not desirable in terms of yield and yield stability over time, and hence of food availability and farmers' income, such a richness in biodiversity, arisen long since under the constant adaptive pressure of a challenging environment, can be regarded as a formidable genetic reservoir for the isolation and multidisciplinary characterization of useful traits to improve, both on a local basis and on a wider scale, e.g., *Mediterranean*, drought and salt stress tolerance in the present era of climate changes.

The studies presented here, and those still to come from the research institutions involved, implied and required the definition of a dynamic conservation strategy and a reasoned exploitation of the genetic diversity embedded in food legume landraces collected across the Algerian territory. The existing knowledge concerning the functioning role of the N-fixing symbiotic interaction among leguminous plant and *rhizobia*, together with the information, plant material, and experience obtained from previous studies, will hopefully contribute to the isolation and multidisciplinary characterization of useful traits to improve—both on a local basis and on a wider scale—i.e., Mediterranean, drought, and salinity tolerance in the era of climate changes. Major directions for future research are: (1) to evaluate the biological diversity in response to water and salt stress exhibited by collection of crop wild relatives and landraces traditionally grown in arid and semiarid environments, (2) to explore the relationships between multiple morphophysiological traits and drought/salinity tolerance, and (3) to identify useful traits that might be promising for breeding programs targeted to augment drought and salt stress tolerance in commercial lines. This will be achieved by an integrative approach, combining a range of disciplines (biology, physiology, ecophysiology, agronomy, economics, and social sciences) and active contributions from farmers in reference areas to identify all the factors that should be implicated.

ACKNOWLEDGMENTS

This chapter was supported by the Ecole Nationale Supérieure Agronomique (ENSA), d'Alger (Algeria), and the Università Mediterranea Di Reggio Calabria (Italy) and it contributes to the Framework Agreement of International Scientific Collaboration between these two institutions.

REFERENCES

Alaoui, M.M., El Jourmi, L., Ouarzane, A., Lazar, S., El Antri, S., Zahouily, M., and Hmyene, A. 2013. Effet du stress salin sur la germination et la croissance de six variétés marocaines de blé. *J. Mater. Environ. Sci.* 4:997–1004.

Amri-Tiliouine, W. 2008. Etude de la symbiose à rhizobium chez l'arachide (*Arachis hypogaea* L.): Aspects microbiologiques, physiologiques et biochimiques et agronomiques. MSc thesis, Algeria: Institut Nationale Agronomique-El-Harrach.

Ashraf, M., and Foolad, M.R. 2013. Crop breeding for salt tolerance in the era of molecular markers and marker-assisted selection. *Plant Breed.* 132:10–20.

Banerjee, M.R., Yesmin, L., and Vessey, J.K. 2006. Plant-growth-promoting rhizobacteria as biofertilizers and biopesticides. In *Handbook of Microbial Biofertilizers*, ed. M.K. Rai, 137–182. Boca Raton, FL: CRC Press/Taylor & Francis Group.

Benmohamed, A., Ounane, S.M., Ghalmi, N., Djebara, M., and Benamara, A. 2004. Etude de la diversité génétique de l'arachide local (*Arachis hypogaea* L.). Paper presented at the 2nd Mediterranean Conference of Rhizobiology, Oran 23–25 May.

Dantas, B.F., Ribeiro, D.S., and Aragão, C.A. 2005. Physiological response of cowpea seeds to salinity stress. *Revista Brasileira de Sementes* 27:144–148.

Dantuma, G., and Klein Hulze, J.A. 1979. Production and distribution of dry matter, and uptake, distribution and redistribution of nitrogen in *Vicia faba major* and *minor.* In *Some Current Research on Vicia faba in Western Europe*, ed. D.A. Bond, G.T. Scarascia-Mugnozza, M.H. Poulsen. Bruxelles, Belgium: Commission of the European Communities, EUR 6244 EN.

Ehlers, J.D., and Hall, A.E. 1997. Cowpea [*Vigna unguiculata* (L.) Walp.]. *Field Crops Res.* 53:187–204.

Eloward, H.O.A., and Hall, A.E. 1987. Influence of early and late nitrogen fertilization on yield and nitrogen fixation of cowpea under well-watered and dry field conditions. *Field Crops Res.* 15:229–244.

Fernandez-Pascual, M., Pueyo, J.J., Felipe, M.R., Golvano, M.P., and Lucas, M.M. 2007. Singular features of the *Bradyrhizobium-Lupinus* symbiosis. *Dyn. Soil Dyn. Plant* 1:1–16.

Genc, Y., Tester, M., and McDonald, G.K. 2010. Calcium requirement of wheat in saline and non-saline conditions. *Plant and Soil* 327:331–345.

Ghalmi, N. 2011. Etude de la diversité génétique de quelques écotypes locaux de *Vigna unguiculata* (L.) Walp. cultivé en Algérie. PhD thesis, Ecole National Supérieure Agronomique.

Ghalmi, N., Malice, M., Jacquemin, J.-M., Ounane. S.-M., Mekliche. L., and Baudoin. J.-P. 2009. Morphological and molecular diversity within Algerian cowpea (*Vigna unguiculata* (L.) Walp.) landraces. *Genet. Resour. Crop Evol.* 57:371–386.

Hadi, F., Hussain, F., and Arif, M. 2012. Growth performance and comparison of cowpea varieties under different NaCl salinity stresses. *Greener J. Phys. Sci.* 2:44–49.

Hall, A.E. 2012. Phenotyping cowpeas for adaptation to drought. *Front. Physiol.* 3:1–8.

Hall, A.E., and Patel, P.N. 1985. Breeding of resistance to drought and heat. In *Cowpea Research, Production and Utilization*, ed. S.R. Singh, and K.O. Rachie, 137–151. New York: Wiley.

Hauggaard-Nielsen, H., Ambus, P., Jensen, E.S., Bellostas, N., Boisen, S., Brisson, N., and Pristeri, A. (2007). Intercropping of pea and barley for increased production, weed control, improved product quality and prevention of nitrogen-losses in European organic farming systems. In *Book of Proceedings*, Part III, ed. M. Fotyma, and B. Kaminska, 53–60. Pulawy, Poland: Institute of Soil Science and Plant Cultivation, State Research Institute, Department of Plant Nutrition and Fertilization. Fragmenta Agronomica, No. 11.

Hauser, S. 1987. Schätzung der symbiotisch fixierten Stickstoffmenge von Ackerbohnen (*Vicia faba* L.) mit erweiterten Differenz methoden. PhD diss. University of Göttingen.

Jensen, E.S., Hauggaard-Nielsen, H., Kinane, J., Andersen, M.K., and Jørnsgård, B. 2005. Intercropping—the practical application of diversity, competition, and facilitation in arable and organic cropping systems. In *Researching sustainable systems*, ed. U. Köpke, U. Niggli, D. Neuhoff, P. Cornish, W. Lockeretz, H. Willer. Proceedings of the First Scientific Conference of the International Society of Organic Agriculture Research (ISOFAR), Adelaide, South Australia, 21–23 September 2005, 22–25.

Kannaiyan, S., and Kumar, K. 2006. Biological fertilizers for sustainable rice production. In *Handbook of Microbial Biofertilizers*, ed. M.K. Rai, 89–116. Boca Raton, FL: CRC Press/Taylor & Francis Group.

Koepke, U., and Nemecek, T. 2010. Ecological services of faba bean. *Field Crops Research* 115:217–233.

Krupl, C., and Funk, T. 2000. Agronomic and economic aspects of growing field peas in main production areas in the EU. Evaluation of farm surveys. (LINK Production Survey). *Eur. Assoc. Grain Leg. Res.* (AEP), Paris, France.

Kwapata, M.B., and Hall, A.E. 1985. Effects of moisture regime and phosphorus on mycorrhizal infection, nutrient uptake and growth of cowpeas (*Vigna unguiculata* (L.) Walp.). *Field Crops Res.* 12:241–250.

Lawlor, D.W. 2013. Genetic engineering to improve plant performance under drought: Physiological evaluation of achievements, limitations, and possibilities. *J. Exp. Bot.* 64:83–108.

Levitt, J. 1982. *Responses of Plants to Environmental Stresses*. New York: Academic Press.

Maskey, S.L., Bhattarai, S., Peoples, M.B., and Herridge, D.F. 2001. On-farm measurements of nitrogen fixation by winter and summer legumes in the Hill and Terai regions of Nepal. *Field Crops Res.* 70:209–221.

Murillo-Amador, B., and Troyo-Dieguez, E. 2000. Effect of salinity on the germination and seedlings characteristics of cowpea [*Vigna unguiculata* (L.) Walp.]. *Aust. J. Exp. Agric.* 40:433–438.

Murillo-Amador, B., Lopez-Aguilar, R., Kaya, C., Larrinaga Mayoral, J., and Flores-Hernandez, A. 2002. Comparative effects of NaCl and polyethylene glycol on germination, emergence and seedling growth of cowpea. *J. Agron. Crop Sci.* 188:235–247.

Murillo-Amador, B., Troyo-Dieguez, E., Hernandez, J.L.G., Lopez-Aguilar, R., Avila-Serrano, N.Y., Zamora-Salgado, S., Rueda-Puente, E.O., and Kaya, C. 2006. Effect of NaCl salinity in the genotypic variation of cowpea (*Vigna unguiculata*) during early vegetative growth. *Scientia Horticulturae* 108:423–431.

Nabi, F. 2009. Effet de la salinité sur la germination, la croissance et les composantes du rendement du *Vigna unguiculata* L. (Walp.). Mémoire en vue de l'obtention du diplôme de magister en biotechnologies végétales. MSc thesis, Ecole Nationale Supérieure Agronomique.

Nabi, F., Chaker-Haddadj, A., Tellah, S., Ghalem, A., Ounane, G., Ghalmi, N., Djebbar, R., and Ounane, S.M. 2017. Evaluation of Algerian Cowpea genotypes for salt tolerance at germination stage. *Adv. Environ. Biol.* 11:79–88.

Nasir Khan, M., and Mohammad, F. 2014. Eutrophication: Challenges and solutions. In *Eutrophication: Causes, Consequences and Control*, Vol. 2, ed. A.A. Ansari, and S.S. Gill, 1–16. Dordrecht, the Netherlands: Springer Science + Business Media.

Padilla, E.G., Lopez Sanchez, R.C., Eichler-Loebermann, B., Fernandez-Pascual, M., Barrero, K.T., and Martinez, L.A. 2009. Salt stress affects on cowpea (*Vigna unguiculata* L. Walp) varieties at different growing stages. Paper presented at the Conference of International Research on Food Security, National Resource Management and Rural Development, University of Hamburg, Hamburg, 1–4 October.

Pasquet, R.S. 2000. Allozyme diversity of cultivated cowpea *Vigna unguiculata* (L.) Walp. *Theor. Appl. Genet.* 101:211–219.

Peoples, M.B., Brockwell, J., Herridge, D.F., Rochester, I.J., Alves, B.J.R., Urquiaga, S., Boddey, R.M., et al. 2009. The contributions of nitrogen-fixing crop legumes to the productivity of agricultural systems. *Symbiosis* 48:1–17.

Peoples, M.B., Giller, K.E., Herridge, D.F., and Vessey, J.K. 2002. Limitations to biological nitrogen fixation as a renewable source of nitrogen for agriculture. In *Nitrogen fixation: Global Perspectives*, ed. T. Finan, M. O'Brian, D. Layzell, K. Vessay, and W. Newton, 356–360. Wallingford, UK: CAB International.

Pfeiffer, W.H., Sayre, K.D., and Reynolds, M.P. 2000. Enhancing genetic grain yield potential and yield stability in durum wheat. *Options Méditerranéennes* 40:83–93.

Riou, C. 1993. L'eau et la production végétale. *Sècheresse* 4:75–83.

Salih, H.O., and Kia, D.R. 2013. Effect of salinity level of irrigation water on cowpea (*Vigna unguiculata*) growth. *J. Agr. Veterinary Sci.* 6:37–41.

Schneider, A., Crépon, K., and Fénart, E. 2005. The perception of lupin in the European food industry. In *Optimised Processes for Preparing Healthy and Added Value Food Ingredients from Lupin Kernels, the European Protein Rich Grain Legume*, ed. A. Arnoldi, 9–20. Rome: Aracne Editrice.

Schwenke, G.D., Peoples, M.B., Turner, G.L., and Herridge, D.F. 1998. Does nitrogen fixation of commercial, dryland chickpea and faba bean crops in north-west New South Wales maintain or enhance soil nitrogen? *Aust. J. Exp. Agric.* 38:61–70.

Sehrawat, N., Yadav, M., Bhat, K.V., Sairam, R.K., and Jaiwal, P.K. 2014. Evaluation of mungbean genotypes for salt tolerance at early seedling growth stage. *Biocatal. Agric. Biotechnol.* 3:108–113.

Singh, B.B., Chambliss, O.L., and Sharma, B. 1997. Recent advances in cowpea breeding. In *Advances in Cowpea Research*, ed. B.B. Singh, D.R. Mohanraj, K.E. Dashiell, and L.E.N. Jackai, 30–49. Ibadan, Nigeria: International Institute of Tropical Agriculture—Japanese International Research Centre for Agricultural Sciences.

Song, J., Fan, H., Zhao, Y., Jia, Y., Du, X., and Wang, B. 2008. Effect of salinity on germination, seedling emergence, seedling growth and ion accumulation of a euhalophyte *Suaeda salsa* in an intertidal zone and on saline inland. *Aquat. Bot.* 88:331–337.

Sprent, J.I., Bradford, A.M., and Norton, C. 1977. Seasonal growth patterns in field beans (*Vicia faba*) as affected by population density, shading and its relationship with soil moisture. *J. Agric. Sci. Cambridge* 88:293–301.

Tellah, S. 2016. Etude des mécanismes agrophysiologiques, morphologiques et moléculaires impliqués dans la tolérance au stress hydrique chez quelques populations locales d'arachide (*Arachis hypogaea* L.). PhD thesis., Ecole Nationale Supérieure Agronomique.

Tellah, S., Badiani, M., Trifilò, P., Lo Gullo, M.A., Ounane, G., Ounane, S.M., and Sorgonà, A. 2014. Morpho-physiological traits contributing to water stress tolerance in a peanut (*Arachis hypogaea* L.) landraces collection from the Algerian Maghreb. *Agrochimica* 58:126–147.

Tellah, S., Nabi, F., Benselama, A., Alioua, H., Bouzenad, I., Sorgona, A., Badiani, M. 2018. Influence of salt stress on germination, growth and productivity of some Algerian landraces of cowpea (*Vigna unguiculata* subsp. *unguiculata* (L.) Walp.). XXXVI convegno nazionale della società italiana di chimica agrariai. Agrariail ruolo della chimica agrariaper la gestione sostenibile delle risorse agrarie e forestali. Dipartimento di agraria—università mediterranea- 24–26 Settembre 2018-Reggio Calabria-Italie.

Veeranagamallaiah, G., Chandraobulreddy, P., Jyothsnakumari, G., and Sudhakar C. 2007. Glutamine synthetase expression and pyrroline-5-carboxylate reductase activity influence proline accumulation in two cultivars of foxtail millet (*Setaria italica* L.) with differential salt sensitivity. *Environ. Exp. Bot.*, 60: 239–244.

Vinther, F.P., and Dahlmann-Hansen, L. 2005. Effects of ridging on crop performance and symbiotic N_2 fixation of faba bean (*Vicia faba* L.). *Soil Use Manage.* 21:205–211.
Wani, S.P., Rupela, O.P., and Lee, K.K. 1995. Sustainable agriculture in the semi-arid tropic through biological nitrogen fixation in grain legumes. *Plant and Soil* 174:29–49.
Win, K.T., and Oo, A.Z. 2015. Genotypic difference in salinity tolerance during early vegetative growth of cowpea (*Vigna unguiculata* L. Walp.) from Myanmar. *Biocatal. Agric. Biotechnol.* 4:449–455.
Yildirim, E., and Guvenc, I. 2006. Salt tolerance of pepper cultivars during germination and seedling growth. *Turk. J. Agric. For.* 30:347–353.
Zapata, F., Danso, S.K.A., Hardarson, G., and Fried, M. 1987. Nitrogen fixation and translocation in field-grown faba bean. *Agron. J.* 79:505–509.
Zavariyan, A., Rad, M., and Asghari, M. 2015. Effect of seed priming by potassium nitrate on germination and biochemical indices in *Silybum marianum* L. under salinity stress. *Int. J. Life Sci.* 9:23–29.

4

Role of Leguminous Microbiome in Plant Growth Promotion and Plant Protection

Prittesh Patel

CONTENTS

4.1 Introduction

Worldwide pulses crops are important for the nutritional security point of view of the cereal-based vegetarian diet of large scale of country. Pulses constitute a group of 12 crops that include mainly chickpea pigeon pea (*Cajanus cajan* L.), chickpea (*Cicer arietinum* L.), mung bean (*Vigna radiata* L. Wilczek), urdbean (*Vigna mungo* L. Hepper), lentil (*Lens culinaris* L.), and fieldpea (*Pisum sativum* L.). India is the largest producer of pulses, contributing 25.7% to the world production (Chauhan et al. 2016). Insect pests, diseases, and weeds are the major factors that are limiting productivity in pulses. The majority of the pulse crops are cultivated under rainfed conditions, hence less consideration is paid for the management of these crops and protection against diseases. The production losses have shown an increasing trend year after year. The present agriculture system is dealing with loss of soil fertility, unpredictable climatic factors, and increasing pathogen resistance. Sustainable farming and eco-friendly agricultural production depend more on natural approaches such as biofertilizers, biopesticides, and crop residue return. The beneficial effect of soil microbial inoculants emphasizes the need for further strengthening the research and their application in modern agriculture. Applications of microorganisms for improving agricultural yields, soil fertility, and crop health had been practiced for centuries. Such rhizobacteria can be grouped into the following types, depending on their proximity to the roots: (1) microorganisms associated roots and soil (rhizospheric microorganism), (2) microorganisms living in root tissue (endophytes), and (3) microorganisms living inside cells in specialized root structures called nodules (rhizobia) (Glick 1995).

Rhizobia comprise the majority of the root microbiome responsible for nodule formation and have been widely studied for decades. However, legume microbiomes from nodules are needed to evaluate in depth for better understanding of their functions and crop-specific diversity. Rhizobia comprise the genera *Azorhizobium, Allorhizobium, Bradyrhizobium, Rhizobium, Mesorhizobium*, and *Sinorhizobium.*

The Rhizobiaceae, Bradirhizobiaceae, and Phyllobacteriaceae root nodules harbor rhizobia, which induce plant growth and improve agricultural sustainability. Discovery of novel legume rhizobia can lead to the identification of many more species. Current advancements in the molecular tools for taxonomic categorization will further prove useful for sustainable agriculture. The rhizospheric microbial communities influence the plant growth in various ways by interacting with roots and in turn affect soil chemicals. All these microbial activities make the rhizospheric environment more dynamic for better plant health by making available nutrients from soil and antagonism against surrounding pathogens.

4.2 Plant Growth Promoting Traits of Leguminous Rhizobia

Application of PGP rhizobia for sustainable agriculture is gaining worldwide importance and recognition and will be a future trend. Rhizobia are beneficial bacteria that enhance materialization, colonize roots, stimulate growth, and enhance the crop production. Plant growth-promoting rhizobacteria (PGPR) with their properties are listed in Table 4.1. PGPR have been studied by metagenomic approaches, but the nodule microbiomes have been fundamentally studied by culture-dependent methods with the simple objective of selection of plant growth-promoting bacteria to be used in agronomic practices. Rhizosphere microbial communities, especially bacteria and fungi, play fundamental roles in nutrients cycling, plant growth, and plant health. The abundance of rhizosphere microorganisms was affected by various environmental and plant factors including nutrients, moisture, plant species, and plant age. Exudates release from roots includes various carbohydrates, amino acids, organic acids, and proteins. Structure of the bacterial diversity is determined by the plant species through composition and amounts of root exudates. Understanding of these is an important scientific aspect when applying bacteria as plant growth promoters (Drogue et al. 2012).

Rhizobia may promote the crop growth by direct or indirect mechanisms (Glick 2012). Direct mechanism of plant growth promotion occurs when rhizobacteria mobilize the essential nutrients from soil or modulate the level of hormones within a plant. Nutrient supplementation by rhizobacteria usually includes nitrogen, phosphorus, and iron. Rascovan et al. (2016) found different groups of bacteria including *Paraburkholderia* spp., *Pseudomonas* spp., and *Pantoea* spp. from soybean roots with important

TABLE 4.1
List of Plant Growth Promoting Rhizobia

PGP Rhizobia	Source Crop	PGP Activities	References
Rhizobium leguminosarum	*Phaseolus vulgaris* *Lens culinaris*	IAA, Cytokinin	Datta and Basu (2000), Noel et al. (1996)
Mesorhizobiums	*Acacia melanoxylon* *Sophora microphylla* *Cicer aritenium*	IAA, P solubilization siderophores, HCN, Ammonia, Exopolysaccharides	Wdowiak-Wróbel and Małek (2016), Zhu et al. (2015), De Meyer et al. (2015), Imen et al. (2015)
Allorhizobium	*Kummerowia striata* *Neptunia natans*	IAA	Ghosh et al. (2015)
Hizobium ciceri	*Cicer arietinum*	P solubilization	Rivas et al. (2006)
Rhizobium phaseoli	*Vigna radiate*	IAA	Zahir et al. (2010)
Bradyrhizobium japonicum	*Centrosema macrocarpum* *Erythrophleum fordii* *Phaseolus lunatus* *Vigna subterranean* *Vigna unguiculata* *Glycine max*	IAA Nitrogen fixation	Valdez et al. (2016), Subramanian et al. (2015)
Rhizobium meliloti	*Phaseolus vulgaris*	Siderophore	Arora et al. (2001)

plant growth promotion properties such as phytohormones IAA production, P solubilization, N fixation, and ACC deaminase. Gopalakrishnan et al. (2016) have demonstrated the PGP potentials of the *Pseudomonas plecoglossicida* SRI-156, *Brevibacterium antiquum* SRI-158, *Bacillus altitudinis* SRI-178, *Enterobacter ludwigii* SRI-211, *E. ludwigii* SRI-229, *Acinetobacter tandoii* SRI-305, and *Pseudomonas monteilii* SRI-360 in chickpea and pigeon pea under field conditions as a useful component of integrated plant health and disease management.

4.2.1 Legume—Non-legume Microbes Co-inoculation

Co-inoculation with rhizobia and non-legume PGPR is becoming an important agronomic practice in the development of sustainable agriculture. The main focus of such studies is the improvement of yield by increasing the nitrogen fixation by rhizobia. Most widely tested PGPR as co-inoculants with *Bradyrhizobium japonicum* include *Bacillus thuringiensis, B. subtilis, Serratia proteomaculans, Azospirillum brasiliense, Pseudomonas aureofaciens*, and *Seroratia liquefaciens*. Researchers support the idea that non-rhizobial PGPR can increase competitiveness of the rhizobial strains. Masciarelli et al. (2014) reported that co-inoculation of the *Bacillus amyloliquefaciens* LL2012 strain in soybean plants with the natural symbiont *Bradyrhizobium japonicum* altered plant growth parameters and significantly improved nodulation. It is reported that nodule endophytic strains of *Bacillus thuringiensis*, isolated from *Pueraria thunbergiana* and soybean, is responsible for an improvement in pea plant growth and nodulation and lentil in co-inoculation with *Rhizobium leguminosarum* (Mishra et al. 2009). *Rhizobium leguminosarum*-treated seed was reported to notably decrease root rot disease caused by *Fusarium solani* f.sp. *phaseoli* of bean and *Pythium* in pea plants (Das et al. 2017).

4.2.2 Nitrogen Fixation

The most widely exploited plant bacteria symbiotic relationship is the N-fixing rhizobia and legumes. Rhizobia can be used to enhance N fixation in soil, and research demonstrated their predominance in nodules even after 5–15 years of initial inoculation (Lindström et al. 1990), and this proved that they are effective colonizers persisting in soil even in the absence of their host (Sanginga et al. 1994). Their contribution in nitrogen fixation makes them an ideal biological nutrient source in sustainable agriculture. As soon as it gets to the plant, it finds a developing root hair and adheres to just behind the root tip, and it starts colonizing. Legumes make available rhizobia with reduced C and anaerobic surroundings required for nitrogenase activity, and on the other side, rhizobia supply the legumes with biologically available nitrogen. Because of this interaction, both rhizobia and legume undergo significant transformations. The legume forms a nodule, to shelter the rhizobia, and the rhizobia, in turn, changes from its free-living rod-shaped cell type to a branched, N-fixing bacteroid (Oke and Long 1999). Rhizobial N-fixation contributes significant amounts of N to global agricultural systems, with estimates ranging from 20 to 22 Tg N per year up to 40 Tg N per year (Herridge et al. 2008, Galloway et al. 2008). Symbiotic N fixation is a mechanism of transforming atmospheric N into plant assailable N, such as ammonia, through a cascade of reactions with the use of complex enzyme systems (Wilson and Burris 1947). The major part of legume fixed N is harvested as grains, while the soil and the subsequent crops also benefit from N in the form of root and shoot residues. Legume crops significantly decrease the N requirement from external sources such as chemical fertilizers (Bhattacharyya and Jha 2012). Pulse plants fix nitrogen at the expense of their own flowers and leaves. Although the process for nitrogen fixation is the same, different pulse crops need different strains of rhizobium to fix their own nitrogen. The bacterium multiplies at the growing point, heading toward the root system.

4.2.3 Phytohormone Production

Phytohormones such as indole acetic acid (IAA), cytokinins, and gibberellins producing capacity of several rhizobacteria were demonstrated by various researchers worldwide. The IAA is a phytohormone involved in plant growth promotion produced by diverse organisms in nature that is involved in modulation plant growth and development (Duca et al. 2014). GAs are involved in the formation of infection pocket and infection thread. The production of this auxin has been reported for classic rhizobia of genera *Rhizobium* (Datta and Basu 2000), *Mesorhizobium* (Wdowiak-Wróbel and Małek 2016), *Bradyrhizobium* (Valdez et al. 2016), and *Allorhizobium* (Ghosh et al. 2015). Environmental conditions also affect the phytohormone production. *Rhizobium* and *Bacillus* were reported to synthesize IAA at different cultural conditions such as pH, temperature, and in the presence of agro waste as substrate.

4.2.4 Phosphate Solubilization

Another important essential nutrient in plants is phosphorus (P). Large reserves of P lie in soils as non-soluble compounds, which cannot then be engrossed by plants and therefore limit the plant growth. Certain rhizobia are able to solubilize those P forms through acidification process (Richardson et al. 2009), chelation, or enzymatically (Hameeda et al. 2008). Bacteria such as *Azospirillum, Burkholderia, Bacillus, Erwinia, Rhizobium, Pseudomonas*, or *Serratia* are reported to have phosphate solubilizing capacity (Mehnaz and Lazarovits 2006, Patel et al. 2019). In the case of rhizobia, the *Mesorhizobium* species are the most active phosphate solubilizers (Verma et al. 2013, Imen et al. 2015). Apart from this, *Rhizobium leguminosarum* biovar *viceae* is also one of the most effective solubilizers of phosphate reported by Halder et al. (1990).

4.3 Leguminous Crop Diseases Management and Protection Mechanisms

Chickpea, pigeon pea, pea, and lentil are attacked by several diseases, some of which cause considerable crop damage. Chickpea is mainly affected by wilt (*Fusarium oxysporum* f.sp. *ciceris*), blight (*Mycosphaerella pinodes*), and rust (*Uromyces ciceris*). *Ascochyta blight*, a disease caused by a complex of *Ascochyta pinodes, Ascochyta pinodella, Ascochyta pisi*, and/or *Phoma koolunga*, is a critical problem in many field peas (*Pisum sativum* L.) growing regions, and it results in significant losses in grain yield. Ranjbar Sistani et al. (2017) reported that *Rhizobium* plays an effective role in resistance induction and yield enhancement strategy of seeds production. Many species of rhizobia promote plant growth and also inhibit the growth of certain pathogenic fungi and bacteria (Tables 4.2 and 4.3). The mechanisms of biocontrol by rhizobia include competition for nutrients (Arora et al. 2001), production of antibiotics (Bardin et al. 2004, Chandra and Pareek 2002), production of enzymes to degrade cell walls (Özkoc and Deliveli 2001), and production of siderophores (Carson et al. 2000, Deshwal et al. 2003) (Tables 4.2 and 4.3).

TABLE 4.2

List of Bacterial Diseases with Their Biocontrol Agents

Bacterial disease	Causal Pathogen	Crop	Biocontrol agent	References
Bacterial blight	*Pseudomonas syringae* *Xanthomonas fragariae* *Xanthomonas arboricola* *Xanthomonas campestris* *Xanthomonas axonopodis* pv. *vignicola*	*Cicer arietinum* *Vigna unguiculata*	*Pseudomonas aeruginosa* *P. fluorescens* *Lactobacillus plantarum*	Kanthaiah and Velu (2019), Corrêa et al. (2014) Daranas et al. (2019)
Bacterial wilt	*Curtobacterium flaccumfaciens*	*Phaseolus valgarius* L.	*Rhizobium leguminosarum* *Pantoea agglomerans*	Huang et al. (2007), Hsieh et al. (2005)
Halo leaf blight	*Pseudomonas phaseolicola*	*Phaseolus valgarius* L.	*P. fluorescence* *P. putida*	Eman and Afafa (2014)
Angular leaf spot	*Pseudomonas syringae* pv. *lachrymans* *Pseudocercospora griseola*	*Phaseolus vulgaris* L. *Cucumis sativus* L.	*Ochrobactrum pseudintermedium* *Pantoea agglomerans*	Akbaba and Ozaktan (2018)

TABLE 4.3
List of Fungal Diseases with Their Biocontrol Agents

Fungal Disease	Causal Pathogen	Crops	Biocontrol Agents	References
Anthracnose	*Colletotrichum truncatum* *Ceramica pisi* *Colletotrichum lindemuthianum*	*Cicer arietinum* *Lens culinaris* *Pisum sativum* *Glycine max*	*Trichoderma harzianum* *Trichoderma viride* *Pseudomonas fluorescens* *Bacillus ceresus*	Amin et al. (2014), Pandey et al. (2019)
Alternaria	*Alternaria alternata*	*Cicer arietinum* *Lens culinaris* *Pisum sativum* *Vicia faba* L.	*Trichoderma viride* *Trichoderma harzianum*	Surekha et al. (2013), Kayim et al. (2018)
Rust	*Uromyces appendiculatus*	*Phaseolus vulgaris* L.	*Pantoea agglomerans* *Stenotrophomonas maltophilia*	Yuen et al. (2001)
Ascochyta	*Ascochyta rabiei* *Ascochyta lentis*	*Cicer arietinum* *Lens culinaris*	*Pantoea agglomerans* *Bacillus* sp.	Liu et al. (2016)
Botrytis	*Botrytis cinerea*	*Cicer arietinum* *Lens culinaris*	*Trichoderma viride* *Pseudomonas fluorescens* *Penicillium citrinum* *Gliocladium roseum*	Khan et al. (2011), Sreevidya et al. (2015), Burgess et al. (1997)
Fusarium wilt	*F. oxysporum* f. sp. *pisi* *F. oxysporum* f. sp. *lentis* *F. oxysporum* f. sp. *ciceris*	*Pisum sativum* *Lens culinaris* *Cicer arietinum*	*Bacillus cereus* *Streptomyces* spp.	Corrêa et al. (2014), Anusha et al. (2019)
Charcoal rot	*Macrophomina phaseolina*	*Phaseolus vulgaris*	*Rhodococcus fascians* *Rhizobium meliloti*	Corrêa et al. (2014), Arora et al. (2001)
Sclerotinia disease	*Sclerotinia sclerotiorum* *S. trifoliorum,* *S. minor*	*Phaseolus vulgaris* L.	*B. subtilis* *P. fluorescens*	Khater (2010), Sabaté et al. (2018), Zhang and Xue (2010)
Stemphylium blight	*Stemphylium botryosum*	*Lens culinaris*	*Trichoderma viride* *Trichoderma harzianum*	Subedi et al. (2015)

Liu et al. (2016) isolated three biocontrol strains of *Bacillus* sp. and one of *Pantoea agglomerans* from pea-related niches, which were found to reduce the severity of disease under greenhouse and field conditions. Sreevidya et al. (2015) reported *Penicillium citrinum* for their antagonistic potential against *Botrytis cinerea* causing Botrytis gray mold disease in chickpea. *B. cinerea* is seed-borne because of poor germination and soft rot of the lower stem in chickpea. Burgess et al. (1997) evaluated *G. roseum* for control of *B. cinerea* on chickpea seed to assess its compatibility with rhizobium inoculant and to study the mechanism of action of biocontrol. Chakraborty and Purkayastha (1984) reported that some rhizobitoxine-producing strains of *Bradyrhizobium japonicum* protected soybeans from the infection of *Macrophomina phaseolina*, the charcoal rot fungus of leguminous crops.

Fusarium wilt caused by *F. oxysporum* is one of the most important diseases in chickpea *(Cicer arietinum* L.) throughout the world. It is one of the main yield-limiting factors in all chickpea-growing areas of the world (Dubey et al. 2007). General symptoms of Fusarium wilt is a serious threat in chickpea comprise drooping, yellowing, drying of the leaves and discoloration of vascular system. Upon infection, the pathogen enters the roots of the susceptible host, establishes in the vascular system of plants, followed by toxin(s) production that kills plants by blocking xylem vessels and restricts water transport (Anjaiah et al. 2003, Gopalakrishnan et al. 2005). Mung bean or green gram, *Vigna radiata* (L.) Wilczek, an important pulse crop, is an excellent source of low-cost and high-quality protein (Taylor et al. 2005). Since mung bean roots through symbiosis with nitrogen-fixing rhizobia fix the atmospheric N, this crop is important both economically as well as nutritionally (Yaqub et al. 2010). The low productivity and poor quality of mung bean can be attributed to several biotic constraints of which diseases caused by fungi are of great importance (Khan et al. 2001). Of these, root rot caused by the soil and seed-borne fungus *M. phaseolina* (Tassi) Goid is a major limiting factor in the mung bean production (Raguchander et al. 1993). It is a serious disease of many crops, inflicting up to 100% yield losses in mung bean under dry and hot conditions. Lentil (*Lens culinaris*) is one of the most significant pulse crops among the grain legumes. There are various biotic and abiotic stresses in lentil crop production including limiting factors such as poor crop management and diseases. Lentil crops are affected by a wide range of pathogens, among them fungal diseases being the most significant. Blight caused by *Stemphylium botryosum* is becoming a serious threat to lentil production. Subedi et al. (2015) reported *T. viride* and *T. harzianum* as the best antagonist that suppressed the growth of *S. botryosum* (Table 4.3). According to Xue (2003), *Clonostachys rosea* strain ACM941, a mycoparasite, is an effective bioagent in controlling pea root rot complex caused by *A. pinodes* and *Rhizoctonia solani*.

Rhizobial strains in soil compete for nutrients by inhibiting the pathogens. Colonization behavior of *P. fluorescens* and *S. meliloti* in alfalfa rhizosphere were reported as effective biocontrol agents to suppress pathogens (Villacieros et al. 2003). The synthesis of metabolites such as HCN, phenazines, viscoinamide pyrrolnitrin, and tensin by rhizobia were also reported as other mechanisms (Bhattacharyya and Jha 2012). HCN production by rhizobacteria is one of the potential and eco-friendly mechanisms for biological control in weeds (Kremer and Souissi 2001).

4.3.1 Siderophore

Iron in soil occurs in divalent or trivalent forms, depending on pH and the availability of other minerals (Bodek 1988). Because of the presence as insoluble hydroxides and oxyhydroxides, iron in soil becomes unavailable to both plants and microbes. This is made available by the synthesis of low molecular weight compounds called siderophores, which have a high affinity toward iron. Iron chelation by microbial siderophores from soil affect by pH, concentration, redox potential, stability constant, and receptor availability to exchange with phytosiderophores (Crowley 2006). Rhizobia produce high-affinity siderophores that inhibit the pathogens by limiting available iron (Arora et al. 2001). Microbes synthesize a variety of siderophores including catechols and hydroxamate. Many strains of *actinobacteria* have been reported as siderophore producers (Wang et al. 2014).

4.3.2 Lytic Enzymes

The majority of soil bacteria produce extracellular enzymes including lipases, glucanases, xylanases, proteinases, amylases, peptidases, chitinases, peroxidases, cellulases, pectinase, ligninases, hemicellulase, and keratinase. These enzymes are involved in biocontrol mechanisms against phytopathogens because of the chemical composition of their cell wall (Gupta et al. 1995, Fodil et al. 2011). *R. leguminosarum* bv. *trifolii, R. leguminosarum* bv. *viciae, R. meliloti, R. trifolii, S. meliloti*, and *B. japonicum* have been reported to produce antibiotics and cell wall-degrading enzymes that can reduce the growth of phytopathogens (Özkoc and Deliveli 2001, Chandra and Pareek 2002, Bardin et al. 2004).

4.3.3 Antibiotics

Antibiotics are toxins produced by microbes at low concentrations that can kill other microorganisms. Antibiotic production on the plant surface or within plant tissues by microbes is critical for biocontrol mechanism to control disease. *Bacillus subtilis* and *Bacillus amyloliquefaciens* are well recognized for their biocontrol of fungal and bacterial diseases. The most important mechanism is the synthesis of a variety of antibiotics with diverse activities by *Bacillus* sp. (Stein 2005). Among these antimicrobial compounds, cyclic lipopeptides (LPs) of the fengycin, iturin, and bacillomycin families exhibit strong in vitro antifungal activities against a broad variety of fungal pathogens. The ability to produce a range of antibiotics by microbes that can differentially reduce pathogen growth and disease condition is important for sustainable agriculture (Pal and Gardener 2006).

4.3.4 Induced Systemic Resistance (ISR)

Induced systemic resistance and enhanced expression of plant defense-related genes by rhizobia is one of the important mechanisms against pathogens to minimize disease in pulse crops. ISR is a key mechanism by which some rhizobacteria prime the whole plant for improved defense against a wide range of pathogens. Bacteria such as *Bacillus, Serratia, Pseudomonas, Streptomyces*, and *Stenotrophomonas* and fungi such as *Ampelomyces, Coniothyrium*, and *Trichoderma* are well-known microbes with beneficial influence on plant health. This level of resistance in plants is characterized by the activation of latent defense mechanisms that are expressed upon pathogen attack not only at the site of infection but also systemically in all plant parts spatially separated from the inducer (Pieterse et al. 2014). Inoculation of appropriate rhizobial strain leads to elicitation of phenolic compounds, isoflavonoid phytoalexins, and activation of enzymes such as L-phenylalanine ammonia lyase (PAL), peroxidase (POX), chalcone synthase (CHS), and polyphenol oxidase (PPO) (Das et al. 2017). Various signaling pathways that activate during pathogen challenge and stimulate the plant defense response by rhizobia may be expected to affect plant-pathogens interactions. Azpilicueta et al. (2004) reported that the systemic stimulation of phytoalexins production in peanut crops inoculated with *Bradyrhizobium* sp. SEMIA6144 is the key process for protection against *Aspergillus niger* and *Cladosporium cucumerinum* (Azpilicueta et al. 2004). In nutshell, rhizobia promote the growth of plants either directly through N2 fixation, phosphate solubilization, and synthesis of phytohormones, or indirectly as a biocontrol agent by inhibiting the growth of pathogens by producing siderophores, lytic enzymes, and secondary metabolites. The potential effect of leguminous microbiomes due to their multifaceted beneficial traits is likely to play an important role in modern, highly intensive agricultural practices.

REFERENCES

Akbaba, M. and H. Ozaktan. "Biocontrol of Angular Leaf Spot Disease and Colonization of Cucumber (*Cucumis sativus* L.) by Endophytic Bacteria." *Egyptian Journal of Biological Pest Control* 28, no. 1 (2018): 14.

Amin, M., S. Fitsum, T. Selvaraj, and N. Mulugeta. "Field Management of Anthracnose (*Colletotrichum lindemuthianum*) in Common Bean through Fungicides and Bioagents." *Advances in Crop Science and Technology* (2014): 1–7.

Anjaiah, V., P. Cornelis, and N. Koedam. "Effect of Genotype and Root Colonization in Biological Control of Fusarium Wilts in Pigeonpea and Chickpea by *Pseudomonas aeruginosa* PNA1." *Canadian Journal of Microbiology* 49, no. 2 (2003): 85–91.

Anusha, B.G., S. Gopalakrishnan, M.K. Naik, and M. Sharma. "Evaluation of *Streptomyces* Spp. and *Bacillus* Spp. for Biocontrol of Fusarium Wilt in Chickpea (*Cicer arietinum* L.)." *Archives of Phytopathology and Plant Protection* 52, no. 5–6 (2019): 417–442.

Arora, N.K., S.C. Kang, and D.K. Maheshwari. "Isolation of Siderophore-Producing Strains of *Rhizobium meliloti* and Their Biocontrol Potential against *Macrophomina phaseolina* That Causes Charcoal Rot of Groundnut." *Current Science-Bangalore* 81, no. 6 (2001): 673–77.

Azpilicueta, C.E., M.S. Zawoznik, and M.L. Tomaro. "Phytoalexins Synthesis Is Enhanced in Groundnut Plants Inoculated with *Bradyrhizobium* Sp. (Arachis)." *Crop Protection* 23, no. 11 (2004): 1069–1074.

Bardin, S.D., H.-C. Huang, J. Pinto, E.J. Amundsen, and R. Scott Erickson. "Biological Control of Pythium Damping-Off of Pea and Sugar Beet by *Rhizobium leguminosarum* Bv. *viceae*." *Canadian Journal of Botany* 82, no. 3 (2004): 291–296.

Bhattacharyya, P.N. and D.K Jha. "Plant Growth-Promoting Rhizobacteria (PGPR): Emergence in Agriculture." *World Journal of Microbiology and Biotechnology* 28, no. 4 (2012): 1327–1350.

Bodek, I. *Environmental Inorganic Chemistry: Properties, Processes, and Estimation Methods*. New York: Pergamon, 1988.

Burgess, D.R., T. Bretag, and P.J. Keane. "Biocontrol of Seedborne *Botrytis cinerea* in Chickpea with Gliocladium Roseum." *Plant Pathology* 46, no. 3 (1997): 298–305.

Carson, K.C., J.-M. Meyer, and M.J. Dilworth. "Hydroxamate Siderophores of Root Nodule Bacteria." *Soil Biology and Biochemistry* 32, no. 1 (2000): 11–21.

Chakraborty, U., and R.P. Purkayastha. "Role of Rhizobitoxine in Protecting Soybean Roots from *Macrophomina phaseolina* Infection." *Canadian Journal of Microbiology* 30, no. 3 (1984): 285–89.

Chandra, R., and R.P. Pareek. "Effect of Rhizobacteria in Urdbean and Lentil." *Indian Journal of Pulses Research* 15, no. 2 (2002): 152–155.

Chauhan, J.S., B.B. Singh, and S. Gupta. "Enhancing Pulses Production in India through Improving Seed and Variety Replacement Rates." *Indian Journal of Genetics* 76 (2016): 1–10.

Corrêa, B.O., J.T. Schafer, and A. B. Moura. "Spectrum of Biocontrol Bacteria to Control Leaf, Root and Vascular Diseases of Dry Bean." *Biological Control* 72 (2014): 71–75.

Crowley, D.E. "Microbial Siderophores in the Plant Rhizosphere." In *Iron Nutrition in Plants and Rhizospheric Microorganisms*, Larry L. Barton and Javier Abadia, Dordrecht, the Netherlands: Springer, (2006): 169–198.

Daranas, N., G. Roselló, J. Cabrefiga, I. Donati, J. Francés, E. Badosa, F. Spinelli, E. Montesinos, and A. Bonaterra. "Biological Control of Bacterial Plant Diseases with *Lactobacillus plantarum* Strains Selected for Their Broad-Spectrum Activity." *Annals of Applied Biology* 174, no. 1 (2019): 92–105.

Das, K., R. Prasanna, and A.K. Saxena. "Rhizobia: A Potential Biocontrol Agent for Soilborne Fungal Pathogens." *Folia Microbiologica* 62, no. 5 (2017): 425–435.

Datta, C., and P.S. Basu. "Indole Acetic Acid Production by a *Rhizobium* Species from Root Nodules of a Leguminous Shrub, *Cajanus cajan*." *Microbiological Research* 155, no. 2 (2000): 123–127.

De Meyer, S.E., H.W. Tan, P.B. Heenan, M. Andrews, and A. Willems. "*Mesorhizobium waimense* Sp. Nov. Isolated from *Sophora longicarinata* Root Nodules and *Mesorhizobium cantuariense* Sp. Nov. Isolated from *Sophora microphylla* Root Nodules." *International Journal of Systematic and Evolutionary Microbiology* 65, no. 10 (2015): 3419–3426.

Deshwal, V.K., P. Pandey, S.C. Kang, and D.K. Maheshwari. "Rhizobia as a Biological Control Agent against Soil Borne Plant Pathogenic Fungi." 41 no. 10 (2003): 1160–1164.

Drogue, B., H. Doré, S. Borland, F. Wisniewski-Dyé, and C. Prigent-Combaret. "Which Specificity in Cooperation between Phytostimulating Rhizobacteria and Plants?" *Research in Microbiology* 163, no. 8 (2012): 500–510.

Dubey, S.C., M. Suresh, and B. Singh. "Evaluation of *Trichoderma* Species against *Fusarium oxysporum* F. Sp. *ciceris* for Integrated Management of Chickpea Wilt." *Biological Control* 40, no. 1 (2007): 118–127.

Duca, D., J. Lorv, C.L. Patten, D. Rose, and B.R. Glick. "Indole-3-Acetic Acid in Plant–Microbe Interactions." *Antonie Van Leeuwenhoek* 106, no. 1 (2014): 85–125.

Eman, O.H. and ZAE Afafa. "Biocontrol of Halo Blight of Bean Caused by *Pseudomonas phaseolicola*." *International Journal of Virology* 10, no. 3 (2014): 235–242.

Fodil, D., A. Badis, B. Jaouadi, N. Zaraî, F.Z. Ferradji, and H. Boutoumi. "Purification and Characterization of Two Extracellular Peroxidases from *Streptomyces* Sp. Strain Am2, a Decolorizing Actinomycetes Responsible for the Biodegradation of Natural Humic Acids." *International Biodeterioration & Biodegradation* 65, no. 3 (2011): 470–478.

Galloway, J.N., A.R. Townsend, J.W. Erisman, M. Bekunda, Z. Cai, J.R. Freney, L.A Martinelli, S.P. Seitzinger, and M.A. Sutton. "Transformation of the Nitrogen Cycle: Recent Trends, Questions, and Potential Solutions." *Science* 320, no. 5878 (2008): 889–892.

Ghosh, P.K., T.K. De, and T.K. Maiti. "Production and Metabolism of Indole Acetic Acid in Root Nodules and Symbiont (*Rhizobium undicola*) Isolated from Root Nodule of Aquatic Medicinal Legume *Neptunia oleracea* Lour." *Journal of Botany* 2015 (2015).

Glick, B.R. "The Enhancement of Plant Growth by Free-Living Bacteria." *Canadian Journal of Microbiology* 41, no. 2 (1995): 109–117. "Plant Growth-Promoting Bacteria: Mechanisms and Applications." *Scientifica* 2012 (2012).

Gopalakrishnan, S., S. Vadlamudi, S. Samineni, and C.V. Sameer Kumar. "Plant Growth-Promotion and Biofortification of Chickpea and Pigeonpea through Inoculation of Biocontrol Potential Bacteria, Isolated from Organic Soils." *SpringerPlus* 5, no. 1 (2016): 1882.

Gopalakrishnan, S., M.H. Beale, J.L. Ward, and R.N. Strange. "Chickpea Wilt: Identification and Toxicity of 8-*O*-Methyl-Fusarubin from *Fusarium acutatum*." *Phytochemistry* 66, no. 13 (2005): 1536–1539.

Gupta, R., R.K. Saxena, P. Chaturvedi, and J.S. Virdi. "Chitinase Production by *Streptomyces viridificans*: Its Potential in Fungal Cell Wall Lysis." *Journal of Applied Bacteriology* 78, no. 4 (1995): 378–383.

Halder, A.K., A.K. Mishra, P. Bhattacharyya, and P.K. Chakrabartty. "Solubilization of Rock Phosphate by *Rhizobium* and *Bradyrhizobium*." *The Journal of General and Applied Microbiology* 36, no. 2 (1990): 81–92.

Hameeda, B., G. Harini, O.P. Rupela, S.P. Wani, and G. Reddy. "Growth Promotion of Maize by Phosphate-Solubilizing Bacteria Isolated from Composts and Macrofauna." *Microbiological Research* 163, no. 2 (2008): 234–242.

Herridge, D.F., M.B. Peoples, and R.M. Boddey. "Global Inputs of Biological Nitrogen Fixation in Agricultural Systems." *Plant and Soil* 311, no. 1–2 (2008): 1–18.

Hsieh, T.F., H.C. Huang, and R.S. Erickson. "Biological Control of Bacterial Wilt of Bean Using a Bacterial Endophyte, *Pantoea agglomerans*." *Journal of Phytopathology* 153, no. 10 (2005): 608–614.

Huang, H.C., R.S. Erickson, and T.F. Hsieh. "Control of Bacterial Wilt of Bean (*Curtobacterium flaccumfaciens* Pv. *flaccumfaciens*) by Seed Treatment with *Rhizobium leguminosarum*." *Crop Protection* 26, no. 7 (2007): 1055–1061.

Imen, H., A. Neila, B. Adnane, B. Manel, Y. Mabrouk, M. Saidi, and S. Bouaziz. "Inoculation with Phosphate Solubilizing *Mesorhizobium* Strains Improves the Performance of Chickpea (*Cicer aritenium* L.) under Phosphorus Deficiency." *Journal of Plant Nutrition* 38, no. 11 (2015): 1656–1671.

Kanthaiah, K. and R.K.Velu. "Characterization of the Bioactive Metabolite from a Plant Growth Promoting Rhizobacteria *Pseudomonas aeruginosa* VRKK1 and Exploitation of Antibacterial Behaviour against *Xanthomonas campestris* a Causative Agent of Bacterial Blight Disease in Cowpea." *Archives of Phytopathology and Plant Protection* (2019): 1–18.

Kayim, M., A.M. Yones, and A. Endes. "Biocontrol of Alternaria Alternata Causing Leaf Spot Disease on Faba Bean (*Vicia faba* L.) Using Some *Trichoderma harzianum* Isolates under In Vitro Condition." *Harran Tarım ve Gıda Bilimleri Dergisi* 22, no. 2 (2018): 169–178.

Khan, M.R., N. Khan, and S.M. Khan. "Evaluation of Agricultural Materials as Substrate for Mass Culture of Fungal Biocontrol Agents of Fusarial Wilt and Root-Knot Nematode Diseases." *Tests of Agrochemicals and Cultivars* 7, no. 22 (2001): 50–51.

Khan, M.R., M.A. Anwer, and S. Shahid. "Management of Gray Mold of Chickpea, *Botrytis cinerea* with Bacterial and Fungal Biopesticides Using Different Modes of Inoculation and Application." *Biological Control* 57, no. 1 (2011): 13–23.

Khater, M.M.M.N. "Biological Control of *Sclerotinia sclerotiorum*, the Causal Agent of White Basal Rot Disease of Beans (*Phaseolus vulgaris* L.)." Alexandria University, 2010.

Kremer, R.J. and T. Souissi. "Cyanide Production by Rhizobacteria and Potential for Suppression of Weed Seedling Growth." *Current Microbiology* 43, no. 3 (2001): 182–186.

Lindström, K., P. Lipsanen, and S. Kaijalainen. "Stability of Markers Used for Identification of Two *Rhizobium galegae* Inoculant Strains after Five Years in the Field." *Applied and Environmental Microbiology* 56, no. 2 (1990): 444–450.

Liu, N., S. Xu, X. Yao, G. Zhang, W. Mao, Q. Hu, Z. Feng, and Y. Gong. "Studies on the Control of Ascochyta Blight in Field Peas (*Pisum sativum* L.) Caused by *Ascochyta pinodes* in Zhejiang Province, China." *Frontiers in Microbiology* 7 (2016): 481.

Masciarelli, O., A. Llanes, and V. Luna. "A New Pgpr Co-Inoculated with *Bradyrhizobium japonicum* Enhances Soybean Nodulation." *Microbiological Research* 169, no. 7–8 (2014): 609–615.

Mehnaz, S. and G. Lazarovits. "Inoculation Effects of *Pseudomonas putida*, *Gluconacetobacter azotocaptans*, and *Azospirillum lipoferum* on Corn Plant Growth under Greenhouse Conditions." *Microbial Ecology* 51, no. 3 (2006): 326–335.

Mishra, P.K., S. Mishra, G. Selvakumar, J.K. Bisht, S. Kundu, and H.S. Gupta. "Coinoculation of *Bacillus thuringeinsis*-Kr1 with *Rhizobium leguminosarum* Enhances Plant Growth and Nodulation of Pea (*Pisum sativum* L.) and Lentil (*Lens culinaris* L.)." *World Journal of Microbiology and Biotechnology* 25, no. 5 (2009): 753–761.

Noel, T.C., C. Sheng, C.K. Yost, R.P. Pharis, and M.F. Hynes. "*Rhizobium leguminosarum* as a Plant Growth-Promoting Rhizobacterium: Direct Growth Promotion of Canola and Lettuce." *Canadian Journal of Microbiology* 42, no. 3 (1996): 279–283.

Oke, V. and S.R. Long. "Bacteroid Formation in the Rhizobium–Legume Symbiosis." *Current Opinion in Microbiology* 2, no. 6 (1999): 641–646.

Özkoc, İ. and M.H. Deliveli. "In Vitro Inhibition of the Mycelial Growth of Some Root Rot Fungi by *Rhizobium leguminosarum* Biovar *phaseoli* Isolates." *Turkish Journal of Biology* 25, no. 4 (2001): 435–445.

Pal, K.K. and B.M. Gardener. "Biological Control of Plant Pathogens." *The Plant Health Instructor* (2006): 1–25. DOI: 10.1094/PHI-A-2006-1117-02.

Pandey, N., S. Ravi, and V. M. Gusain. "In Vitro Efficacy of Biocontrol Agents and Fungicides against Anthracnose of French Bean Caused by Colletotrichum *lindemuthianum* (Sacc. and Magn.)." *International Journal of Communication Systems* 7, no. 3 (2019): 3266–3269.

Patel, P., R. Shah, B. Joshi, K. Ramar, and A. Natarajan. "Molecular Identification and Biocontrol Activity of Sugarcane Rhizosphere Bacteria against Red Rot Pathogen *Colletotrichum falcatum*." *Biotechnology Reports* 21 (2019): e00317.

Pieterse, C.M.J., C. Zamioudis, R.L. Berendsen, D.M. Weller, S.C. Van Wees, and P.A. Bakker. "Induced Systemic Resistance by Beneficial Microbes." *Annual Review of Phytopathology* 52 (2014): 347–375.

Raguchander, T., R. Samiappan, and G. Arjunan. "Biocontrol of Macrophomina Root Rot of Mungbean." *Indian Phytopathology* 46, no. 4 (1993): 379–382.

Ranjbar Sistani, N., H.-P. Kaul, G. Desalegn, and S. Wienkoop. "Rhizobium Impacts on Seed Productivity, Quality, and Protection of *Pisum sativum* Upon Disease Stress Caused by *Didymella pinodes*: Phenotypic, Proteomic, and Metabolomic Traits." *Frontiers in Plant Science* 8 (2017): 1961.

Rascovan, N., B. Carbonetto, D. Perrig, M. Díaz, W. Canciani, M. Abalo, J. Alloati, G. González-Anta, and M.P. Vazquez. "Integrated Analysis of Root Microbiomes of Soybean and Wheat from Agricultural Fields." *Scientific Reports* 6 (2016): 28084.

Richardson, A.E., J.-M. Barea, A.M. McNeill, and C. Prigent-Combaret. "Acquisition of Phosphorus and Nitrogen in the Rhizosphere and Plant Growth Promotion by Microorganisms." *Plant and Soil* 321, no. 1–2 (2009): 305–339.

Rivas, R., A. Peix, P.F. Mateos, M.E. Trujillo, E. Martínez-Molina, and E. Velázquez. "Biodiversity of Populations of Phosphate Solubilizing Rhizobia That Nodulates Chickpea in Different Spanish Soils." *Plant and soil* 287, no. 1–2 (2006): 23–33.

Sabaté, D.C., C.P. Brandan, G. Petroselli, R. Erra-Balsells, and M.C. Audisio. "Biocontrol of *Sclerotinia sclerotiorum* (Lib.) De Bary on Common Bean by Native Lipopeptide-Producer *Bacillus* Strains." *Microbiological Research* 211 (2018): 21–30.

Sanginga, N., S.K.A. Danso, K. Mulongoy, and A.A. Ojeifo. "Persistence and Recovery of Introduced Rhizobium Ten Years after Inoculation on *Leucaena leucocephala* Grown on an Alfisol in Southwestern Nigeria." *Plant and Soil* 159, no. 2 (1994): 199–204.

Sreevidya, M., S. Gopalakrishnan, T.M. Melø, N. Simic, P. Bruheim, M. Sharma, V. Srinivas, and G. Alekhya. "Biological Control of *Botrytis cinerea* and Plant Growth Promotion Potential by *Penicillium citrinum* in Chickpea (*Cicer arietinum* L.)." *Biocontrol Science and Technology* 25, no. 7 (2015): 739–755.

Stein, T. "*Bacillus subtilis* Antibiotics: Structures, Syntheses and Specific Functions." *Molecular Microbiology* 56, no. 4 (2005): 845–857.

Subedi, S., S.M. Shrestha, K.C. G. Bahadur, R.B. Thapa, S.K. Ghimire, S. Neupane, and B. Nessa. "Botanical, Chemical and Biological Management of *Stemphylium botryosum* Blight Disease of Lentil in Nepal." *Indian Phytopath* 68, no. 4 (2015): 415–423.

Subramanian, P., K. Kim, R. Krishnamoorthy, S. Sundaram, and T. Sa. "Endophytic Bacteria Improve Nodule Function and Plant Nitrogen in Soybean on Co-Inoculation with *Bradyrhizobium japonicum* Mn110." *Plant Growth Regulation* 76, no. 3 (2015): 327–332.

Surekha, C.H., N.R.R. Neelapu, G. Kamala, B. Siva Prasad, and P. Sankar Ganesh. "Efficacy of *Trichoderma viride* to Induce Disease Resistance and Antioxidant Responses in Legume *Vigna mungo* Infested by *Fusarium oxysporum* and *Alternaria alternata*." *International Journal of Agricultural Science Research* 3, no. 2 (2013): 285–294.

Taylor, R.S., D.B. Weaver, C. Wood, and E. van Santen. "Nitrogen Application Increases Yield and Early Dry Matter Accumulation in Late-Planted Soybean." *Crop Science* 45, no. 3 (2005): 854–858.

Valdez, R.A., B. Soriano, G. Prado, D. Zavaleta, M. Matsubara, D. Zúñiga, P. Dion. et al. "Symbiotic and Agronomic Characterization of Bradyrhizobial Strains Nodulating Cowpea in Northern Perú." In *Biological Nitrogen Fixation and Beneficial Plant-Microbe Interaction*, ed. Fernando González-Andrés; Euan James, 195–212, Cham, Switzerland: Springer, 2016.

Verma, J.P., J. Yadav, K.N. Tiwari, and A. Kumar. "Effect of Indigenous *Mesorhizobium* Spp. and Plant Growth Promoting Rhizobacteria on Yields and Nutrients Uptake of Chickpea (*Cicer arietinum* L.) under Sustainable Agriculture." *Ecological Engineering* 51 (2013): 282–286.

Villacieros, M., B. Power, M. Sánchez-Contreras, J. Lloret, R.I. Oruezabal, M. Martín, F. Fernández-Piñas, et al. "Colonization Behaviour of *Pseudomonas fluorescens* and *Sinorhizobium meliloti* in the Alfalfa (*Medicago sativa*) Rhizosphere." *Plant and Soil* 251, no. 1 (2003): 47–54.

Wang, W., Z. Qiu, H. Tan, and L. Cao. "Siderophore Production by Actinobacteria." *Biometals* 27, no. 4 (2014): 623–631.

Wdowiak-Wróbel, S. and W. Małek. "Properties of *Astragalus* Sp. Microsymbionts and Their Putative Role in Plant Growth Promotion." *Archives of Microbiology* 198, no. 8 (2016): 793–801.

Wilson, P.W., and R.H. Burris. "The Mechanism of Biological Nitrogen Fixation." *Bacteriological Reviews* 11, no. 1 (1947): 41.

Xue, A.G. "Biological Control of Pathogens Causing Root Rot Complex in Field Pea Using *Clonostachys rosea* Strain ACM941." *Phytopathology* 93, no. 3 (2003): 329–335.

Yaqub, M., T. Mahmood, M. Akhtar, M.M. Iqbal, and S. Ali. "Induction of Mungbean [*Vigna radiata* (L.) Wilczek] as a Grain Legume in the Annual Rice-Wheat Double Cropping System." *Pakistan Journal of Botany* 42, no. 5 (2010): 3125–3135.

Yuen, G.Y., J.R. Steadman, D.T. Lindgren, D. Schaff, and C. Jochum. "Bean Rust Biological Control Using Bacterial Agents." *Crop Protection* 20, no. 5 (2001): 395–402.

Zahir, Z.A., M.K. Shah, M. Naveed, and M.J. Akhter. "Substrate-Dependent Auxin Production by *Rhizobium phaseoli* Improves the Growth and Yield of *Vigna radiata* L. Under Salt Stress Conditions." *Journal of Microbiology and Biotechnology* 20, no. 9 (2010): 1288–1294.

Zhang, J.X. and A.G. Xue. "Biocontrol of Sclerotinia Stem Rot (*Sclerotinia sclerotiorum*) of Soybean Using Novel *Bacillus subtilis* Strain Sb24 under Control Conditions." *Plant Pathology* 59, no. 2 (2010): 382–391.

Zhu, Y.J., J.K. Lu, Y.L. Chen, S.K. Wang, X.H. Sui, and L.H. Kang. "*Mesorhizobium acaciae* Sp. Nov., Isolated from Root Nodules of *Acacia melanoxylon* R. Br." *International Journal of Systematic and Evolutionary Microbiology* 65, no. Pt 10 (2015): 3558.

5

Plant Growth-Promoting Bacteria Modulate Biotic and Abiotic Stress Tolerance in Legumes

Muhammad Arslan Ashraf, Rizwan Rasheed, Muhammad Iqbal, Iqbal Hussain, Abida Parveen, and Ali Akbar

CONTENTS

5.1 Introduction

The role of microorganisms in nitrogen fixation in legumes was discovered in the nineteenth century. The symbiosis of microorganisms in legumes is displayed as formation of nodules on stems or roots. Nitrogen fixation is carried out in nodules where bacteria transformed in bacteroids reside. The first bacterial isolate, namely, *Bacillus radicicola* from nodule suspension was able to form nodules in *Vicia* and *Pisum*, which was later referred to as *Rhizobium leguminosarum* (Velázquez et al., 2010). Legumes served as the major host for rhizobial species. Plant growth and productivity are directly related to the availability of nitrogenous compounds. The process where microorganisms convert atmospheric nitrogen to plant usable forms of nitrogen is known as biological nitrogen fixation (BNF). The major portion of soil nitrogen is replenished by BNF that decreases the demand for chemical nitrogen fertilizers (Herridge et al., 2008). BNF approximately contributes 50–70 million tons in agricultural systems.

The major bacterial genera in BNF process include *Sinorhizobium*, *Rhizobium*, *Mesorhizobium*, *Bradyrhizobium*, and *Azorhizobium*, jointly known as rhizobia (Franche et al., 2009). Recently, a number of new bacterial species with significant input in BNF have been described, e.g., *Cupriavidus* (Ralstonia) (Chen et al., 2001), *Burkholderia* (Moulin et al., 2001), *Phyllobacterium* (Valverde et al., 2005), *Ochrobactrum* (Zurdo-Piñeiro et al., 2007), *Herbaspirillum* (Valverde et al., 2003), *Devosia* (Rivas et al., 2002), and *Methylobacterium* (Sy et al., 2001). Bacteria as bacteroids reside inside the nodules developed on roots and stems of legumes and thereby convert N_2 to ammonia (Gibson et al., 2008). Legumes get reduced nitrogenous compounds in the exchange of organic acids in the rhizobial symbiotic association. The symbiotic associations of rhizobial bacteria with legumes contribute to nearly 80% of N obtained from BNF (Graham and Vance, 2003). Rhizobia develop symbiosis with nearly 18,000 legume species, with 100 agricultural legume species spanning various geographic regions of the world and thereby contribute to 50% of N from BNF (Graham and Vance, 2003). The N enrichment, sustainability, and productivity of agricultural systems were significantly improved as a result of legume rotation with non-nitrogen-fixing plants. The improvement in readily mineralizable organic nitrogen reserves is mainly derived from legumes sources compared with chemical fertilizers nitrogen input. Thus, BNF serves as renewable N reserve to replace N fertilizer that enriches soil N fertility (Peoples et al., 1995).

Rhizobia may develop nodulation in a wide range of legume hosts. For example, *Rhizobium* sp. NGR234 nodulate more than 112 hosts (Pueppke and Broughton, 1999). By contrast, *R. leguminosarum* bv. trifolii nodulates only *Trifolium* spp., while its close neighbor *R. leguminosarum* bv. Viciae nodulates sweet pea (*Lathyrus* spp.), lentil (*Lens* spp.), vetch (*Vicia* spp.), and pea (*Pisum* spp.) (Perret et al., 2000). The symbiotic interaction occurs as a result of molecular signals. For instance, roots exude a variety of compounds such as non-flavonoid inducers, isoflavonoids, and flavonoids in the rhizosphere that attracts rhizobial strains for symbiotic association (Dharmatilake and Bauer, 1992). These chemical compounds mediate expression of nodulation genes and bacterial growth (Hungria and Stacey, 1997). This also enhances the biosynthesis of lipochitin oligosaccharides referred to as nodulation factor (Lerouge et al., 1990). A single rhizobial strain may produce a wide range of metabolites that function as Nod factors (Spaink et al., 1995). Nodulation genes are divided into three groups. First, nodABC, common nodulation genes exist in almost all bacteria except for bradyrhizobia (Giraud et al., 2007). Nod phenotypes arise as a result of mutations in these genes (Debellé et al., 1986). Biosynthesis of hexameric backbone from b-1,4-linked *N*-acetyl-D-glucosamin etrimeric is mediated by NodC, while NodA and B carry out the acetylation and deacetylation of glucosamine, respectively (D'Haeze and Holsters, 2002). A second group of genes involves host-specificity nodulation genes (nodPQ, nodG, nodH, nodFE besides several others), which mediate the rate of nodule formation (Schwedock and Long, 1989). The third category of genes belongs to nodD genes (Spaink, 2000). Nod factors expression is induced by flavonoid compounds exuded by plant roots. Plant roots exude a diverse range of sugars, acids, amines, amino acids, and several other low molecular weight compounds including growth regulators, vitamins, alkaloids, steroids, and flavonoids (Bertin et al., 2003), and thus induce biotic and abiotic stress tolerance in plants. Soil bacterial population is significantly affected by these compounds. Some of these exuded compounds—including amino acids, acids, and sugars—serve as energy and C sources for microorganisms (Jaeger et al., 1999). Other compounds involving flavonoids mediate the exchange of signal transduction between antipathogen plant defense system or two symbionts (Phillips and Kapulnik, 1995). Some of the plant root exudates significantly mediate quorum sensing and thereby affects bacterial communication systems (Teplitski et al., 2000). Plants are recognized as soil engineers through the selection of suitable microbes (Simms and Taylor, 2002). The energy sources and C in root exudation can be effectively enhanced by rhizosphere microorganisms that indicate the existence of feedback in plant-rhizobium interactions mediation bacterial nutrition (Phillips et al., 2004).

5.2 Flavonoids and Non-flavonoids Signaling Mediate Nodulation in Legume

The effective rhizobium-legume associations are determined by flavonoids. Chalcone synthase carries out condensation of 4-coumaryl-CoA to produce flavonoids (Dixon and Paiva, 1995). Various alterations in this metabolic pathway yield different groups of flavonoids entailing anthocyanidines, chalcones, coumestans, isoflavonoids, flavones, and flavanones (Harborne and Williams, 2000). Higher plants produce more than 4,000 flavonoids that are released in the rhizosphere (Perret et al., 2000). However, the age and physiological state of plants may affect the variety of flavonoids being released in the rhizosphere (Harborne and Williams, 2000). The flavonoids are released in greater quantity at the site of rhizobium infection (Zuanazzi et al., 1998). Flavonoids play a significant role in the various steps of nodulation. Primarily flavonoids induce the expression of nod genes through the stimulation of NodD protein (Brencic and Winans, 2005). In this context, 4,70-dihydroxyflavone and luteolin act as the inducers in *S. meliloti* (Peck et al., 2006). Likewise, genistein, isoflavone inhibit nod expression in *S. meliloti*, whereas nod gene expression is stimulated by these flavonoid compounds in *B. Japonicum* (Kosslak et al., 1987). Naringenin and hespertin act as protein inducers in *R. leguminosarum* (Begum et al., 2001). Flavonoids mediate the synthesis of Nod factors in infection threads, and regulate initiation of nodule primordial cell division and auxin transport (Zhang et al., 2009). Different functions of flavonoids mediating the development of nodules have been reported recently (Wasson et al., 2006; Zhang et al., 2009). In a study, silencing of flavonoid biosynthesis in *M. truncatula* resulted in complete loss of nodulation by *S. meliloti*, while the limited nodulation is seen in flavone-deficient conditions. Deficiency of isoflavone did not appear to affect nodulation in *M. truncatula* (Zhang et al., 2009). Complete restoration of nodulation was seen upon the addition of 7,40-dihydroxyflavone in flavone-deficient roots. The sustained induction of Nod factors results in the production of Dhf that inhibit auxin transport and thereby result in development, cell division, and nodule initiation (Zhang et al., 2009). By contrast, auxin transport increased alongside the reduced nodule development in soybean roots as a result of isoflavone biosynthesis silencing that suggests the essential functions of isoflavones in soybean nodule formation. Isoflavone appeared crucial for sustained induction of Nod-factor synthesis in the roots (Subramanian et al., 2006).

Legume growth improvement is achieved when flavonoids mediate the nod gene transcription. Malvidin glucoside and quercetin increased nodulation by *R. leguminosarum* bv. Phaseoli (Hungria and Phillips, 1993). Similarly, nodulation in alfalfa was significantly enhanced by luteolin (Kapulnik et al., 1987). Genistein pretreatment in *B. japonicum* resulted in an increase in total protein, grain yield, and nodulation in soybean under laboratory and field conditions (Zhang and Smith, 1995, 1996). Naringenin and flavanones hespertin induced *R. leguminosarum* for nodulation and induced increase in plant dry mass in lentil and pea compared with untreated *R. leguminosarum* under controlled growth chamber conditions (Begum et al., 2001). Seed exudates, being the mixture of flavonoids, act as inducers for nod genes, while flavonoid inhibitors of nod gene also exist in seed exudates. Literature showed the evidence for compounds analogous to genistein or diadzein as the effective inhibitors for nod gene in *B. japonicum*. Nod gene inhibitors compete with inducers at a common target site (Kosslak et al., 1990). Bean and clover root exudates are efficient nod gene inducers in *R. leguminosarum* compared with that of flavonoids alone, reflecting that nod gene inducers act synergistically (Maj et al., 2010). Better induction of nod genes is seen in the case of multiple inducers that exert early symbiotic association (Cooper, 2007). However, the specific interaction of NodD protein and flavonoids cannot be ruled out (Peck et al., 2006).

The expression of nod genes can also be induced by non-flavonoid compounds including simple phenolic compounds, xanthones, betains, tetronic acid, erythronic acid, and jasmonates. The only difference with these compounds compared with flavonoids is the high concentration required to induce nod genes and nod factors in legumes (Cooper, 2007). Jasmonates produced in plants are linolenic acid derivatives via the octadecanoic pathway. Jasmonates act as essential signaling molecules mediating disease resistance besides the activation of important physiological activities. Jasmonic acid (100 mM) decreased nodulation in *S. meliloti* inoculated *Medicago truncatula* when applied exogenously. Low levels of jasmonates decreased Ca effect on Nod factor, whereas further low levels of jasmonates (10–50 mM) modifies Ca signaling as a result of a decline in frequency of spiking. Changes in Ca signaling affects colonization where Ca spiking is essential to induce a Nod factor signaling pathway (Gutjahr and Paszkowski, 2009).

5.3 Microbes-Legume Interactions

Nitrogen (N) is the main limiting nutrients for growth and productivity of non-crop and crop species. BNF is an important mechanism like that of photosynthesis for plant growth and development. A variety of biological associations significantly contribute to BNF in aquatic and soil systems. BNF is the sole characteristic of certain bacteria, namely, diazotrophs containing nitrogenase that mediated the conversion of atmospheric N_2 to the plant usable forms. The plant productivity is significantly determined by the ability of plants to fulfill maximum nitrogen requirements from BNF (Vessey and Chemining'wa, 2006). The efficient management of N determines the essentiality of elements for sustainable agriculture. This N management involves BNF that convert atmospheric N_2 to plant usable forms. Furthermore, the effective N management system is also determined by leaching, denitrification, and volatilization (Graham and Vance, 2000). The interaction between rhizobia and legumes contributes to 80% BNF in most of the agricultural systems. The major rhizobium genera developing symbiotic association involve *Sinorhizobium, Rhizobium, Mesorhizobium, Bradyrhizobium, Azorhizobium*, and *Allorhizobium*. The Anabaena–Azolla and Frankia (actinorhizal) symbiotic associations contribute to 20% BNF. Legumes provide major proportion of protein intake (35%) in the world. They serve as an important source for N_2 fixation that fulfills approximately 20% of N for oilseed and grain production worldwide (Sadowsky, 2005).

Nodules are organs developed by rhizobia in legumes under N-deficient environments. Rhizobia transform into endosymbiotic form (bacteroids) in nodules where N_2 is converted to ammonia. Therefore, legume-rhizobia symbiotic association is of significant agronomic and ecological importance. Efficient N_2-fixing rhizobial strains in significant amounts are required to develop symbiotic association. Nearly 70 rhizobial species with variable physiological and symbiotic features have been identified (Vessey and Chemining'wa, 2006). The genetic diversity in rhizobia was also reported in rhizobia. The molecular communications between macrosymbionts and microsymbionts significantly determine infection and nodulation process. The symbiotic association of legumes with rhizobial strains develops indeterminate and determinate nodules. Temperate legumes such as alfalfa, clover, and pea form inderminate nodules with rhizobial strains. Tropical legumes such as common beans and soybean in association with bradyrhizobia form determinate nodules. Initiation of nodulation on stem or roots by rhizobia involve three basic processes such as entry via cavities on primordia of adventitious roots, lesions, cracks, wounds, and root hairs (Zahran, 2010).

A wide range of genes mediating symbiotic associations have been identified as a result of recent advancements in molecular biology during recent decades. Genes driving symbiotic associations are mostly clustered on one or many large plasmids, while these genes are located on chromosomes in bradyrhizobia (Gualtieri and Bisseling, 2000). The coordination in two main groups of genes, namely, fix and nif genes at large determine symbiotic N_2 fixation. Molybdenum-based enzymes are encoded by nif genes and this enzyme displayed functional and structural similarities with N_2-fixing genes in Klebsiella pneumonia. The nif genes reside on plasmids in most rhizobial strains, but in bradyrhizobia these genes are chromosomally located. The nifH and nifDK genes encode nitrogenase enzyme that catalyzes nitrogen fixation in free-living and symbiotic microbes. An iron-bound protein called as component 2 and molybdenum-iron protein as component 1 comprise nitrogenase. Under natural environmental conditions, oxygen and fixed N levels regulate the expression of a nif gene. A variety of other genes in rhizobia may directly or indirectly affect N_2-fixation such as lipopolysaccharides, Β-1,2-glucans, nodulation efficiency, dicarboxylate transport, glutamine synthase, hydrogen uptake, and exopolysaccharide (Sadowsky, 2005). The amount of symbiotic N fixation in wild legumes exceeds those in crops (Zahran, 2006). Legumes species greatly differ for N fixation, which are several folds greater under controlled conditions compared with that in field conditions. The formation of efficient rhizobial strains alongside the breeding of plants with symbiotic ability is the main strategy for improvement of N_2 fixation (Zahran, 2006). Management practices and crop breeding may increase the amount of N_2 fixation by rhizobia-legume association by 300% (Vance, 1998). Food legumes rank third after oilseeds and cereals and serve as essential dietary constituents for animals and humans. Nearly 61.3 million ha area was cultivated with legumes in developing countries while 8.5 million ha was under legume cultivation in developed countries during 2002 (Graham and Vance, 2003). A significant portion of food protein and sustainable agricultural systems come from grain legumes from developing countries (Zahran, 2010).

5.4 Saline and Arid Conditions

Desert regions with severe water shortage due to low rainfall and atmosphere dryness are referred to as arid environments, which significantly deteriorate vegetation alongside the gradual decline in biological diversity in an ecosystem. Huge regions in arid and semiarid conditions stay barren owing to prevailing water scarcity. Soil water potential falls due to soil drying conditions that may also decrease soil hydraulic conductivity. Plants cannot survive under low soil water potential due to their inability to uptake sufficient water. Furthermore, important physiological processes in plants are also significantly affected as a result of low soil water potential. Unavailability of irrigation water in such regions further worsens the scenario because the groundwater is the only source for water and is very deep. Plant growth and productivity significantly fall in aquifers and saline environments. Besides, a variety of plant species are able to grow under such unhydrobiotic conditions often associated with salinity. The rehabilitation of such lands can be done through the use of plants adapted to arid conditions and through efficient utilization of limited water either by preventing water drainage below the root zone or excessive evaporation due to dry environment (Tomar et al., 2003). Temperature and light conditions are optimal for most plants in arid and semiarid regions; however, low rainfall puts even more dependence on irrigation. Plants growing in hot and dry conditions often face combinations of stresses at times that also includes nutrient deficiency (Sánchez-Diaz, 2001).

Rhizobial bacteria can thrive under diverse environmental conditions such as within the root nodules and in the rhizosphere. They exist in soil as free-living saprophytic heterotrophs or in N_2 fixation symbiotic association with legumes. These features bestow rhizobacteria distinct advantages over other bacteria in the soil. Rhizobial bacteria may not require a legume host in the case of saprophytic competence, while diverse rhizobia released as bacteroids enable them to thrive in the soil as free-living heterotrophic saprophytes until the specific host is found (Lindström et al., 1990). Rhizobia are traditionally greatly stress resistant as compared to their host legumes (Vriezen et al., 2007). In this context, *Sinorhizobium* sp. from *Canavalia rosea* is a salt-tolerant rhizobacterial strain when grown under 3.5% NaCl salinity (Chen et al., 2000), *Mesorhizobium* strain, the rhizobial strains tolerant to temperature, pH, and NaCl (Wei et al., 2008) from *Acacia raddiana* and *Acacia gummifera* subjected to 6% NaCl (Essendoubi et al., 2007). In another study, *Bradyrhizobium* sp. from lupine could tolerate 5% NaCl salinity, alkaline (pH 9–10), and acidic pH (4–5) (Raza et al., 2001). Important cellular compounds (e.g., lipopolysaccharides and proteins) synthesis in salinity-tolerant rhizobia is significantly altered due to abiotic stresses such as heat, osmotic, and salt stress (Zahran, 2010). For instance, EBRI 26, a salinity-tolerant rhizobial strain exhibited nearly 49 differentially expressed proteins under 4% NaCl salinity (Shamseldin et al., 2006). Rhizobial bacterial strains maintain osmotic equilibrium by the accumulation of organic and/inorganic solutes that enable them to tolerate salinity (Csonka and Hanson, 1991). In this context, *Rhizobium meliloti* accumulates compatible solutes such as dipeptide, *N*-acetylglutaminyl glutamine amide, trehalose, betaine, proline, glycine betaine, proline, glutamate, and K (Vriezen et al., 2007). Compatible solutes accumulated by rhizobia serve as C or N sources. The amount and type of osmolytes accumulated by rhizobia are related to the intensity of osmotic stress, type of osmolytes, and C source in the growth medium alongside the growth phase of the culture (Smith et al., 1994). Most bacteria contain effective transport systems for osmoprotectants under stressful environments besides other osmoregulated systems (Wood et al., 2001). Salinity tolerance of rhizobia is a multigenic trait (Pandey and Mann, 2000). When *S. meliloti* was exposed to long-term salinity, this resulted in the activation of genes involved in polysaccharide biosynthesis and transport of small molecules such as alcohols, anions, peptides, amines, and amino acids (Rüberg et al., 2003). Uplift in the osmotic stress caused changes in the expression of 137 genes with 85 genes suppressed and 52 genes induced in this rhizobial strain. Besides, osmotic stress induced expression of several genes coding for proteins with unidentified functions (Domínguez-Ferreras et al., 2006).

5.5 Impact of Desiccation, Osmotic and Water Stress on Legume-Rhizobia Symbiosis

Under natural environmental conditions, plants face a combination of environmental stresses that induce substantial negative impact on plant growth and productivity. Salinity and drought are the main limiting factors for plant growth and yield in different regions of the world. Osmotic stress is the condition arisen due to salinity that decreases soil water potential and thereby affects growth and yield (Zhu et al., 1997). The soil enrichment of microbes is significantly affected as a result of the low availability of water that negatively affects exudates, root architecture, nutrient concentration, and plant growth. Tolerance to limited water conditions is greater in microbial cells compared with higher plants. Generally, rhizobia are more tolerant to low soil water potential compared with plants. The impact of drought conditions on microsymbiont associations also affects N_2 fixation (Hungria and Vargas, 2000). Subsequently, the host plant determines the infection by rhizobia and formation of nodules that drives N_2 fixation. In a study, drought-tolerant cultivar of *M. sativa* displayed minimal effect of limited water on N_2 fixation compared with those of drought-sensitive cultivars (Aguirreolea and Sánchez-Díaz, 1989). The degree of drought tolerance is variable among rhizobia species. As an example, fast-growing rhizobacteria are less drought tolerant compared with those of slow-growing rhizobia (Zahran, 2010). Total N_2 fixation, nitrogenase activity, nodule O_2 permeability, and N and C metabolism are significantly affected by water stress influence on symbiosis (Sadowsky, 2005). N_2 fixation is widespread in different legumes. Also, a number of drought-tolerant rhizobial strains have been reported (Njiti and Galiana, 1996). Rhizobial species vary in their tolerance to osmotic stress tolerance, ability for N_2 fixation, and potential to infect plants as reported in *Retama raetam, Albizzia adianthifolia, Sesbania, Gliricidia sepium,* and *Acacia mangium* (Galiana et al., 1998; Mahdhi et al., 2008; Melchor-Marroquín et al., 1999; Rehman and Nautiyal, 2002; Swaine et al., 2007).

5.6 Salinity Stress Affects Rhizobial-Legume Symbiotic Associations

Salinity is the main constraint for plant growth and productivity in arid and semiarid areas of the world. Soil infertility due to the existence of huge quantities of salts and introduction of plants able to tolerate such constraints should be studied (Soussi et al., 1998). Salinity tolerance of plants is at large determined by changes in developmental and morphological changes alongside the physio-biochemical processes. Cell structure and function are significantly hampered by osmotic and specific ion toxicity components of salinity (Munns and Tester, 2008). Specific ion toxicity produces higher levels of reactive oxygen species that arrests plant growth and metabolic activity. Osmotic stress manifested in the form of minimal cell turgor (Munns and Tester, 2008). Salinity is toxic to most organisms excluding halotolerant organisms such as halophytes and halobacteria that possess specific mechanisms to decrease endogenous levels of toxic sodium. Halophytes under salinity tolerance avoid toxic sodium by limiting the uptake of sodium and enhancing sodium sequestration (Ashraf et al., 2018).

Therefore, a better understanding of physiological responses under salt conditions can be of value in programs conducted to breed salt-tolerant crop varieties. In the following section, plant responses to soil salinity are discussed with emphasis on molecular mechanisms of signal transduction and on the physiological consequences of altered gene expression. Understanding the mechanisms by which plants perceive and transduce stress signals to initiate adaptive responses is essential for engineering stress-tolerant crop plants (Xiong and Zhu 2001). Thus, in addition to the existing salt-tolerant crop genotypes, research is needed to develop genotypes with increased tolerance to salinity (Qadir et al. 2000). Genetic variability within a species offers a valuable tool for studying mechanisms of salt tolerance. One of these mechanisms depends on the capacity for osmotic adjustment. A general feature of many plants growing in a saline environment is that they decrease osmotic potential by accumulation of inorganic and/or compatible solutes in their cells. Legume productivity is severely impeded by salinity that affects formation of root nodules and thereby N_2 capacity is limited (Zahran, 2010). Salinity degrades cell ultrastructure, decreases plant biomass, photosynthesis, and endurance of rhizobacteria (Novikova and Gordienko, 1999). Reports are present in the literature on the formation of root nodule symbiotic association under salinity (Zahran, 2010). Rhizobacteria present in saline soils possess special characteristics that facilitate establishment of symbiotic association with legumes (Zahran, 2005). In this context, 15 *S. meliloti* isolated from root nodules of trigonella, melilot, and alfalfa effectively established nodules with salinity-tolerant legumes under both normal and saline conditions (Ibragimova et al., 2006). The ability of isolated bacteria to establish nodules determines the effectiveness of symbiotic interaction (Zahran, 2010).

Legumes are able to develop symbiotic associations with several genera of rhizobia such as *Sinorhizobium, Mesorhizobium,* and *Rhizobium,* which are able to endure arid climate conditions (Nguyen et al., 2004). Grain legumes are salinity sensitive or moderately tolerant. In this context, salinity-sensitive legumes include *Pisum sativum, P. vulgaris, Lens culinaris,* and *Cicer arietinum,* while *Vicia faba* and *Glycine max* are moderately salt tolerant (Phang et al., 2008). Rhizobium-legume association is highly sensitive to salinity, and researchers put significant emphasis on improving rhizobium-legume symbiosis (Zahran, 2005). The breeding and selection of better legume cultivars could circumvent decline in N_2 fixation as a result of environmental constraint. In saline soils, rhizobial population is unable to develop successful symbiotic association with legumes owing to inefficiency of infection process (Bouhmouch et al., 2005). Significantly better N_2 fixation can be achieved when both symbiotic partners (legumes and rhizobia) are salinity tolerant.

5.7 Legume-Rhizobium Relationship Response to pH and Temperature

Plant growth is negatively affected in acidic soils that may decrease 25% crop productivity (Munns and Tester, 2008). The ability of rhizobia to tolerate acidic conditions depends upon the potential of strain to keep internal pH nearly neutral (Graham et al., 1994). Higher tolerance to acidic environment is displayed by bradyrhizobia compared with rhizobia (Sadowsky, 2005). However, some *R. tropici* strains can thrive in acidic environments owing to the production of glutathione (Riccillo et al., 2000). The effect of acidic soils on the symbiotic relationship of rhizobia with legumes is related to the infection process and legume growth (Munns, 1986). The disturbance in signal exchange between microsymbiont and macrosymbiont partly determines the symbiotic relationship (Hungria and Stacey, 1997) alongside the excretion of Nod factors and suppression of nodule genes in rhizobia (Richardson et al., 1988). N_2 fixation, nodulation efficiency, and rhizobial persistence are significantly impeded by stress factors such as acidic soil (Graham and Vance, 2003). Rhizobial strains enable to develop nodulation with *P. vulgaris* in arid environments, which were also evaluated for pH tolerance (Priefer et al., 2001). Rhizome-legume symbiotic relationship is significantly affected due to soil nutrient status. The availability of nutrients is impeded by alterations in soil acidity and matric potential (Sadowsky, 2005). The need of essential nutrients for microbes and plants increased under stressful conditions. The negative effect of low pH on root growth may be offset by the availability of Ca, while nod gene induction increased by ion uptake that in turn affects nodule development (Smit et al., 1992). Phosphorus (P) is also a limiting factor that determines symbiotic interactions and N_2 fixation (Saxena and Rewari, 1991), and nearly 33% of agricultural land is low in P especially in acidic soils (Graham and Vance, 2000). The requirement of P is different for plant and microbes. It is seen that fast-growing rhizobial strain appeared more sensitive to limited P availability compared with slow-growing rhizobia (Beck and Munns, 1985).

The persistence and survival of rhizobial strains are markedly affected by high soil temperature in temperate climate (Boumahdi et al., 2001). It is observed that rhizobial strains from natural-growing legumes are able to thrive under elevated soil temperature (Zahran, 2010). The effect of soil temperature varies with rhizobial strain. For instance, *R. leguminosarum* bv. Trifolii is more tolerant than *Bradyrhizobium* sp. to high soil temperature (Sadowsky, 2005). The infection process of rhizobial strains is hampered by high soil temperature (Segovia et al., 1991). The struggle for nodulation is also influenced by high soil temperature (Triplett and Sadowsky, 1992). In a study, rhizobial strain developed nodules with *P. vulgaris*, *Acacia*, and *Prosopis* under high soil temperature conditions (Zahran, 2010). A temperature range around 30°C–40°C is optimum for efficient rhizobium-symbiotic relationship. However, soil temperature below or above this optimum range damages nodulation, nodulation physiology, and infection process alongside plant growth and yield (Michiels et al., 1994). The inducers of nod genes are directly influenced by high soil temperature in bean and soybean (Hungria and Stacey, 1997). Nodule physiology that includes H_2 evolution, nitrogenase activity, and leghemoglobin synthesis are altered in high soil temperature. Therefore, rhizobial strains with greater tolerance to high soil temperature can be isolated from the indigenous soil microbes that possess the ability to thrive under varied environmental conditions. These rhizobial strains had the potential to improve nutrient uptake and growth of plants in disturbed soils (Hungria and Stacey, 1997).

5.8 Rhizobium-Legume Symbiosis Relationship Under Metal Stress

Heavy metal contamination of soil is the major obstacle for plant growth and productivity. It is seen that most of agricultural soils contain substantial levels of heavy metals. The conventional remediation strategies including ion exchange, soil evacuation, and coagulation-filtration are applied to remediate polluted soils. However, these approaches are expensive and destructive to the soils. Therefore, the reclamation of contaminated soils by in situ strategies is gaining more attraction since these are cost effective (Khan et al., 2009). In this scenario, use of plants for reclamation of polluted soils is an ecologically safe method (Ashraf et al., 2017). Plant exudates are critical for the survival and infection process of rhizobia, which is an important aspect in rhizoremediation (Kuiper et al., 2004). The infection process and spread of microbes relies on root system. Therefore, a plant-microbe relationship emerges as efficient soil reclamation technology. Sustainable agriculture is negatively affected due to the presence of trace metals in soil that significantly damages legume production. In a study, the effect of heavy metals such as chromium (Cr), cadmium (Cd), and copper (Cu) on the symbiotic relationship of rhizobium with pea was examined (Wani et al., 2008b). Among three heavy metals, Cu exerted the most severe effects on growth and seed yield in pea. Likewise, Wani et al. (2008a) also reported rhizobial strains formed nodules in pea under zinc and nickel stress. The study advised the potential of rhizobial strains for improved N_2 fixation, growth, and tolerance to metals in pea. Legumes possess the potential to grow in arsenic (As) contaminated soils, which contain rich diversity in free-living rhizobial strains (Carrasco et al., 2005). Legumes-rhizobial symbiotic relationship is essential for the remediation of arsenic contaminated soils. In this context, contamination decreased the formation of nodules and plant dry weight in soybean (Reichman, 2007). Higher plant dry weight was recorded in soybean plants inoculated with *B. japonicum*, and this increase in growth was ascribed to the production of hormones. Therefore, the use of rhizobial strains for the remediation of trace metal contaminated soils is an efficient and emerging area of research.

Soil contains certain elements that might be beneficial for some organisms but are harmful when present in elevated levels. In this context, soil with 50–200 mg kg^{-1} of trace metals such as zinc, cadmium, copper, and cobalt exert adverse effects on nitrogenase activity, nodulation, and growth in *Lablab perpureus*. Furthermore, increase in soil metal levels results in significant decline in essential nutrients such as Ca, Na, and Ca (Younis, 2007). Despite the available information on the bad effects of metals on legume-rhizobial symbiosis, the influence of greater soil metal levels on this symbiotic association is not fully known. Soil with higher metal concentration significantly abate nutrient and water uptake (Karpiscak et al., 2001), which ultimately suppresses plant growth and development. Metals when taken up by plants induce cellular injuries through the disruption of metabolic activities and cellular components. It is reported that nodulation is significantly impeded in plants grown in soil with even minimal levels of cadmium (Cd). Microsymbionts are sensitive to Cd that disrupts nodule ultrastructure, induces nodule senescence, decreases biomass and number of nodules, and abates nitrogenase activity alongside the inhibited metabolic activities such as photosynthesis (Noriega et al., 2007). Cd leads to oxidative stress with concomitant decrease in soluble protein, leghemoglobin, and antioxidant enzyme activities. Enhancement in thiols and lipid peroxidation is induced by Cd (Benavides et al., 2005). The nodulation index is significantly decreased as a result of trace metals such as lead, zinc, and Cd in white clover (Manier et al., 2009). There are conflicting reports on the influence of heavy metals on legumes and nodulation (Wani et al., 2008a). In this context, Paudyal et al. (2007) reported adverse effects of limited aluminium levels on rhizobial growth kinetics. Wood and Cooper (1988) reported that 50 mM aluminium exerted negative influence on rhizobial strain growth kinetics. By contrast, Kinkle et al. (1987) did not find any influence of trace metals on growth kinetics of *Bradyrhizobium japonicum* and nitrogen fixation and growth of host plant in heavy metal contaminated soils. The symbiotic rhizobium-legume relation is inhibited due to direct influence of heavy metals on photosynthetic pigments, Rubisco activity, and prognosis of nodules. The indirect inhibitory effect of metals on the survival of rhizobial strains also affects rhizobium-legume symbiosis (Wani et al., 2008b). Metals also inhibit the synthesis of molecules that are known as crucial for proliferation and growth of rhizobia (Breen and Murphy, 1995) as a result of enhanced production of reactive oxygen species such as hydroxyl radicals and singlet oxygen. In this context, Cd induced oxidative stress in pea leaves (McCarthy et al., 2001). The effect of abiotic stress on legume interaction with PGPR has also been reviewed in Table 5.1.

It is clear from the literature that integration of PGPR with other defense-related strategies may result in enhancement in plant growth and productivity under changing climate conditions. PGPR strains possess important biological activities including indole acetic acid and siderphore production as well as nitrogen fixation and phosphorus solubilization. These biological activities enable PGPR strains to confer stress tolerance to plants. The literature survey revealed the use of PGPR for sustainable agriculture. Some PGPR strains are stress resistant in addition to their plant growth-promoting characteristics.

TABLE 5.1

Amelioration of Abiotic Stress in Legumes by Plant Growth-Promoting Rhizobacteria

Plant Species	Type of Bacteria	Type of Stress	Plant Response	References
Chickpea	*Bacillus thuringiensis*, *Bacillus subtilis* and *Bacillus megaterium.*	Heavy metal and drought	Treatment of plants with PGPR alone or in combination with plant growth regulators displayed significant increase in the accumulation of heavy metals in chickpea. Nickle (Ni) was more in sensitive cultivar, while lead (Pb) was greater in both cultivars as a result of PGPR treatment. Authors recommended the use of PGPR alongside the application of plant growth regulators for improving phytoremediation by chickpea under drought conditions.	Khan and Bano (2018)
Chickpea	*Mesorhizobium* and Rhizobial strains RHD18	Salinity	Inoculation with rhizobacterial and Mesorhizobial strains in chickpea developed 59 nodules per plant along with concomitant increase in plant biomass compared to un-inoculated plants under salinity stress. Further increase in plant biomass and nodule number was recorded upon co-inoculation with *Mesorhizobium* and rhizobial strains RHD18 in chickpea under salinity.	Chaudhary and Sindhu (2017)
Chickpea	*Pantoea dispersa*	Salinity	Bacterial strains from chickpea rhizosphere were isolated and characterized for salinity tolerance and growth-promoting traits, i.e., ACC (1-aminocyclopropane-1-carboxylate), indole acetic acid (IAA) production, siderphores, and phosphorous (P) solubilization. Among these isolated bacterial strains, one potential isolate was identified as Pantoea dispersa with significant potential for plant growth promotion, i.e., ACC deaminase activity (207.45 nmol/mg/h), P solubilization (3.64 μg/mL), siderphores (60.35%), and IAA (218.3 μg/mL) under 150 mM salinity. Un-inoculated plants under salinity accumulated appreciable amount of Na in leaves along with decrease in K, leaf water content, chlorophyll content, and membrane integrity that impeded growth and yield. However, damaging effects of salinity were minimal in plants inoculated with Pantoea dispersa. Inoculated plants displayed better growth and yield alongside improved membrane integrity, chlorophyll, and K contents. Furthermore, Na accumulation decreased substantially in inoculated plants.	Panwar et al. (2016)
Chickpea	*Pseudomonas pseudoalcaligens* and *Pseudomonas putida*	Salinity	Bacterial strains were isolated from saline regions in Gujarat, India. Among 176 isolated strains, 62 strains were salt tolerant, which were further characterized for P-solublization, siderphores, and IAA production. Fifteen isolates showed better P- solubilization (9–45 μg/mL). These 15 isolates were tested for siderphore production and 13 of them were able to produce substantial siderphores (11–50 mM). IAA production was checked for these 13 isolates, and 10 isolates were able to produce IAA (10-26 μg/mL after 72 h). Five best isolates were selected from 10 isolates for further pot studies where chickpea plants were grown in pots under 300 mM NaCl. Plants inoculated with MSC1 and four other strains showed increase in growth under salinity stress. These isolates were identified as *Pseudomonas pseudoalcaligens* and *Pseudomonas putida* after *16SrRNA* analysis.	Patel et al. (2012)

(*Continued*)

TABLE 5.1 (*Continued*)

Amelioration of Abiotic Stress in Legumes by Plant Growth-Promoting Rhizobacteria

Plant Species	Type of Bacteria	Type of Stress	Plant Response	References
Pea	Not-identified by 16SrRNA analysis	Salinity	Pea had significant sensitivity toward salinity. This study showed that inoculation with PGPR induced salinity tolerance in transgenic peas with overexpression of the *Arabidopsis* Na^+/H^+ gene. The chosen bacterial strain had several plant growth-promoting features such as phosphorous solubilization, proline, and indole acetic acid production. Inoculation with bacterial strain improved growth and biomass in transgenic plants than those of un-inoculated plants. Activities of different antioxidant enzymes also increased in plants inoculated with bacterial strain. Authors suggested that inoculation of peas with rhizobacterial strains resulted in enhanced salinity tolerance.	Ali et al. (2015)
Pea	*Arthrobacter protophormiae*	Salinity	The inoculation of pea with *Arthrobacter protophormiae* (SA3) reduced stress mediated ethylene production through its ACC deaminase activity and thereby improved salinity tolerance of plants.	Barnawal et al. (2014)
Mung beans	*Pseudomonas aeruginosa* GGRJ21	Water stress	*P. aeruginosa* GGRJ21 inoculation significantly improved water stress tolerance in mung beans by improving osmolyte accumulation, antioxidant enzyme activities, and expression of stress related genes.	Sarma and Saikia (2014)
Alfalfa	*Klebsiella* sp.	Salinity	Co-inoculation of alfalfa plants with rhizobial strain *S. meliloti* ARh29 and *Klebsiella* sp. A36, *K. cowanii* A37 significantly increased plant growth under salinity stress.	Noori et al. (2018)

REFERENCES

Aguirreolea, Jone, and Manuel Sanchez-Diaz. "CO_2 evolution by nodulated roots in *Medicago sativa* L. under water stress." *Journal of Plant Physiology* 134, no. 5 (1989): 598–602.

Ali, Zahid, Nasr Ullah, Saadia Naseem, Muhammad Inam Ul Haq, and Hans Joerg Jacobsen. "Soil bacteria conferred a positive relationship and improved salt stress tolerance in transgenic pea (*Pisum sativum* L.) harboring Na^+/H^+ antiporter." *Turkish Journal of Botany* 39, no. 6 (2015): 962–972.

Ashraf, Muhammad Arslan, Ali Akbar, Abida Parveen, Rizwan Rasheed, Iqbal Hussain, and Muhammad Iqbal. "Phenological application of selenium differentially improves growth, oxidative defense and ion homeostasis in maize under salinity stress." *Plant Physiology and Biochemistry* 123 (2018): 268–280.

Ashraf, Muhammad Arslan, Iqbal Hussain, Rizwan Rasheed, Muhammad Iqbal, Muhammad Riaz, and Muhammad Saleem Arif. "Advances in microbe-assisted reclamation of heavy metal contaminated soils over the last decade: A review." *Journal of Environmental Management* 198 (2017): 132–143.

Barnawal, Deepti, Nidhi Bharti, Deepamala Maji, Chandan Singh Chanotiya, and Alok Kalra. "ACC deaminase-containing *Arthrobacter protophormiae* induces NaCl stress tolerance through reduced ACC oxidase activity and ethylene production resulting in improved nodulation and mycorrhization in *Pisum sativum*." *Journal of Plant Physiology* 171, no. 11 (2014): 884–894.

Beck, Douglas P., and Donald N. Munns. "Effect of calcium on the phosphorus nutrition of *Rhizobium meliloti*." *Soil Science Society of America Journal* 49, no. 2 (1985): 334–337.

Begum, Anjuman Ara, Stewart Leibovitch, Pierre Migner, and Feng Zhang. "Specific flavonoids induced nod gene expression and pre-activated nod genes of *Rhizobium leguminosarum* increased pea (*Pisum sativum* L.) and lentil (*Lens culinaris* L.) nodulation in controlled growth chamber environments." *Journal of Experimental Botany* 52, no. 360 (2001): 1537–1543.

Benavides, Maria P., Susana M. Gallego, and María L. Tomaro. "Cadmium toxicity in plants." *Brazilian Journal of Plant Physiology* 17, no. 1 (2005): 21–34.

Bertin, Cecile, Xiaohan Yang, and Leslie A. Weston. "The role of root exudates and allelochemicals in the rhizosphere." *Plant and Soil* 256, no. 1 (2003): 67–83.

Bouhmouch, Ilham, Bouchra Souad-Mouhsine, Fatiha Brhada, and Jamal Aurag. "Influence of host cultivars and *Rhizobium* species on the growth and symbiotic performance of *Phaseolus vulgaris* under salt stress." *Journal of Plant Physiology* 162, no. 10 (2005): 1103–1113.

Boumahdi, Mouad, Patrice Mary, and Jean-Pierre Hornez. "Changes in fatty acid composition and degree of unsaturation of (brady) rhizobia as a response to phases of growth, reduced water activities and mild desiccation." *Antonie van Leeuwenhoek* 79, no. 1 (2001): 73–79.

Breen, Anthony P., and John A. Murphy. "Reactions of oxyl radicals with DNA." *Free Radical Biology and Medicine* 18, no. 6 (1995): 1033–1077.

Brencic, Anja, and Stephen C. Winans. "Detection of and response to signals involved in host-microbe interactions by plant-associated bacteria." *Microbiology and Molecular Biology Reviews* 69, no. 1 (2005): 155–194.

Carrasco, José Antonio, Patricia Armario, Eloísa Pajuelo, Araceli Burgos, Miguel A. Caviedes, Rafael López, Manuel A. Chamber, and Antonio J. Palomares. "Isolation and characterisation of symbiotically effective *Rhizobium* resistant to arsenic and heavy metals after the toxic spill at the Aznalcollar pyrite mine." *Soil Biology and Biochemistry* 37, no. 6 (2005): 1131–1140.

Chaudhary, Deepika, and Satyavir Singh Sindhu. "Amelioration of salt stress in chickpea (*Cicer arietinum* L.) by coinculation of ACC deaminase-containing rhizospheric bacteria with *Mesorhizobium* strains." *Legume Research: An International Journal* 40, no. 1 (2017): 80–86.

Chen, Wen-Ming, Severine Laevens, Tsong-Ming Lee, Tom Coenye, Paul De Vos, Max Mergeay, and Peter Vandamme. "*Ralstonia taiwanensis* sp. nov., isolated from root nodules of *Mimosa* species and sputum of a cystic fibrosis patient." *International Journal of Systematic and Evolutionary Microbiology* 51, no. 5 (2001): 1729–1735.

Chen, Wen-Ming, Tsong-Ming Lee, Chun-Chieh Lan, and Chiu-Ping Cheng. "Characterization of halotolerant rhizobia isolated from root nodules of *Canavalia rosea* from seaside areas." *FEMS Microbiology Ecology* 34, no. 1 (2000): 9–16.

Cooper, Jessica E. "Early interactions between legumes and rhizobia: Disclosing complexity in a molecular dialogue." *Journal of Applied Microbiology* 103, no. 5 (2007): 1355–1365.

Csonka, Laszlo N., and Andrew D. Hanson. "Prokaryotic osmoregulation: Genetics and physiology." *Annual Review of Microbiology* 45, no. 1 (1991): 569–606.

D'haeze, Wim, and Marcelle Holsters. "Nod factor structures, responses, and perception during initiation of nodule development." *Glycobiology* 12, no. 6 (2002): 79R–105R.

Debellé, Frédéric, Charles Rosenberg, Jacques Vasse, Fabienne Maillet, Esperanza Martinez, Jean Dénarié, and Georges Truchet. "Assignment of symbiotic developmental phenotypes to common and specific nodulation (nod) genetic loci of *Rhizobium meliloti*." *Journal of Bacteriology* 168, no. 3 (1986): 1075–1086.

Dharmatilake, Amitha J., and Wolfgang D. Bauer. "Chemotaxis of *Rhizobium meliloti* towards nodulation gene-inducing compounds from alfalfa roots." *Applied and Environmental Microbiology* 58, no. 4 (1992): 1153–1158.

Dixon, Richard A., and Nancy L. Paiva. "Stress-induced phenylpropanoid metabolism." *The Plant Cell* 7, no. 7 (1995): 1085.

Domínguez-Ferreras, Ana, Rebeca Pérez-Arnedo, Anke Becker, José Olivares, María J. Soto, and Juan Sanjuán. "Transcriptome profiling reveals the importance of plasmid pSymB for osmoadaptation of *Sinorhizobium meliloti*." *Journal of Bacteriology* 188, no. 21 (2006): 7617–7625.

Donald N. Munns "Acid soil tolerance in legumes and rhizobia." In: Tinker, B. and Läuchli, A. (Eds.), *Advances in Plant Nutrition*, pp. 63–91. Praeger Publishers, New York, 1986.

Essendoubi, Mohammed, Fatiha Brhada, Jamal Eddine Eljamali, Abdelkarim Filali-Maltouf, Sylvie Bonnassie, Sylvie Georgeault, Carlos Blanco, and Mohamed Jebbar. "Osmoadaptative responses in the rhizobia nodulating Acacia isolated from south-eastern Moroccan Sahara." *Environmental Microbiology* 9, no. 3 (2007): 603–611.

Franche, Claudine, Kristina Lindström, and Claudine Elmerich. "Nitrogen-fixing bacteria associated with leguminous and non-leguminous plants." *Plant and Soil* 321, no. 1–2 (2009): 35–59.

Galiana, Antoine, Guy-Modeste Gnahoua, Jeannine Chaumont, Didier Lesueur, Yves Prin, and Bernard Mallet. "Improvement of nitrogen fixation in *Acacia mangium* through inoculation with rhizobium." *Agroforestry Systems* 40, no. 3 (1998): 297–307.

Ge-Hong, Wei, Xue-ying Yang, Zhi-Xin Zhang, Yang Ya-Zhen, and Kristina Lindsröm. "Strain *Mesorhizobium* sp. CCNWGX035: A stress-tolerant isolate from *Glycyrrhiza glabra* displaying a wide host range of nodulation." *Pedosphere* 18, no. 1 (2008): 102–112.

Gibson, Katherine E., Hajime Kobayashi, and Graham C. Walker. "Molecular determinants of a symbiotic chronic infection." *Annual Review of Genetics* 42 (2008): 413–441.

Giraud, Eric, Lionel Moulin, David Vallenet, Valérie Barbe, Eddie Cytryn, Jean-Christophe Avarre, Marianne Jaubert et al. "Legumes symbioses: Absence of Nod genes in photosynthetic bradyrhizobia." *Science* 316, no. 5829 (2007): 1307–1312.

Graham, Peter H., and Carroll P. Vance. "Nitrogen fixation in perspective: An overview of research and extension needs." *Field Crops Research* 65, no. 2–3 (2000): 93–106.

Graham, Peter H., and Carroll P. Vance. "Legumes: Importance and constraints to greater use." *Plant Physiology* 131, no. 3 (2003): 872–877.

Graham, Peter H., Kathryn J. Draeger, Mark L. Ferrey, Mark J. Conroy, Bruce E. Hammer, Esperanza Martinez, Sharon R. Aarons, and Carmen Quinto. "Acid pH tolerance in strains of *Rhizobium* and *Bradyrhizobium*, and initial studies on the basis for acid tolerance of *Rhizobium tropici* UMR1899. *Canadian Journal of Microbiology* 40, no. 3 (1994): 198–207.

Gualtieri, Gustavo, and Ton Bisseling. "The evolution of nodulation." In: Schilperoort, R., and Dure L. (eds.), *Plant Molecular Evolution*, pp. 181–194. Springer, Dordrecht, the Netherlands, 2000.

Gutjahr, Caroline, and Uta Paszkowski. "Weights in the balance: Jasmonic acid and salicylic acid signaling in root-biotroph interactions." *Molecular Plant-Microbe Interactions* 22, no. 7 (2009): 763–772.

Harborne, Jeffrey B., and Christine A. Williams. "Advances in flavonoid research since 1992." *Phytochemistry* 55, no. 6 (2000): 481–504.

Herman P. Spaink, Guido V. Bloemberg, Antonius Albertus Nicolaas Van Brussel, Ben Jeffery Lichtenberg, Koen M. G. M. Van der Drift, Janine Haverkamp, and Jane Elizabeth Thomas Oates. "Host specificity of *Rhizobium leguminosarum* is determined by the hydrophobicity of highly unsaturated fatty acyl moeities on the nodulation factors." *Molecular Plant-microbe Interactions: MPMI (USA)* 8, no. 1 (1995): 155–164.

Herridge, David F., Mark B. Peoples, and Robert M. Boddey. "Global inputs of biological nitrogen fixation in agricultural systems." *Plant and Soil* 311, no. 1–2 (2008): 1–18.

Hungria, Mariangela, and Donald A. Phillips. "Effects of a seed color mutation on rhizobial nod-gene-inducing flavonoids and nodulation in common bean." *Plant-Microbe Interactions* 6 (1993): 418–422.

Hungria, Mariangela, and Gary Stacey. "Molecular signals exchanged between host plants and rhizobia: Basic aspects and potential application in agriculture." *Soil Biology and Biochemistry* 29, no. 5–6 (1997): 819–830.

Hungria, Mariangela, and Milton A.T. Vargas. "Environmental factors affecting N2 fixation in grain legumes in the tropics, with an emphasis on Brazil." *Field Crops Research* 65, no. 2–3 (2000): 151–164.

Ibragimova, M. V., M. L. Rumyantseva, O. P. Onishchuk, V. S. Belova, O. N. Kurchak, E. E. Andronov, N. I. Dzyubenko, and B. V. Simarov. "Symbiosis between the root-nodule bacterium *Sinorhizobium meliloti* and alfalfa (*Medicago sativa*) under salinization conditions." *Microbiology* 75, no. 1 (2006): 77–81.

Jaeger, Charles H., Steven E. Lindow, Webb Miller, Edward Clark, and Mary K. Firestone. "Mapping of sugar and amino acid availability in soil around roots with bacterial sensors of sucrose and tryptophan." *Applied and Environmental Microbiology* 65, no. 6 (1999): 2685–2690.

Kapulnik, Yoram, Cecillia M. Joseph, and Donald A. Phillips. "Flavone limitations to root nodulation and symbiotic nitrogen fixation in alfalfa." *Plant Physiology* 84, no. 4 (1987): 1193–1196.

Karpiscak, Martin M., L. R. Whiteaker, Janick F. Artiola, and Katie E. Foster. "Nutrient and heavy metal uptake and storage in constructed wetland systems in Arizona." *Water Science and Technology* 44, no. 11–12 (2001): 455–462.

Khan, Mohammad Saghir, Almas Zaidi, Parvaze Ahmad Wani, and Mohammad Oves. "Role of plant growth promoting rhizobacteria in the remediation of metal contaminated soils." *Environmental Chemistry Letters* 7, no. 1 (2009): 1–19.

Khan, Naeem, and Asghari Bano. "Effects of exogenously applied salicylic acid and putrescine alone and in combination with rhizobacteria on the phytoremediation of heavy metals and chickpea growth in sandy soil." *International Journal of Phytoremediation* 20, no. 5 (2018): 405–414.

Kinkle, Brian K., J. Scott Angle, and Harold H. Keyser. "Long-term effects of metal-rich sewage sludge application on soil populations of *Bradyrhizobium japonicum*." *Applied and Environmental Microbiology* 53, no. 2 (1987): 315–319.

Kosslak, Renee M., Rita S. Joshi, Benjamin A. Bowen, Herbert E. Paaren, and Edward R. Appelbaum. "Strain-specific inhibition of nod gene induction in *Bradyrhizobium japonicum* by flavonoid compounds." *Applied and Environmental Microbiology* 56, no. 5 (1990): 1333–1341.

Kosslak, Renee M., Roger Bookland, John Barkei, Herbert E. Paaren, and Edward R. Appelbaum. "Induction of *Bradyrhizobium japonicum* common nod genes by isoflavones isolated from Glycine max." *Proceedings of the National Academy of Sciences* 84, no. 21 (1987): 7428–7432.

Kuiper, Irene, Ellen L. Lagendijk, Guido V. Bloemberg, and Ben J. J. Lugtenberg. "Rhizoremediation: A beneficial plant-microbe interaction." *Molecular Plant-microbe Interactions* 17, no. 1 (2004): 6–15.

Lerouge, Patrice, Philippe Roche, Catherine Faucher, Fabienne Maillet, Georges Truchet, Jean Claude Promé, and Jean Dénarié. "Symbiotic host-specificity of *Rhizobium meliloti* is determined by a sulphated and acylated glucosamine oligosaccharide signal." *Nature* 344, no. 6268 (1990): 781.

Lindström, Kristina, Pertti Lipsanen, and Seppo Kaijalainen. "Stability of markers used for identification of two *Rhizobium galegae* inoculant strains after five years in the field." *Applied and Environmental Microbiology* 56, no. 2 (1990): 444–450.

Mahdhi, Mosbah, Angèle Nzoué, Philippe de Lajudie, and Mohamed Mars. "Characterization of root-nodulating bacteria on *Retama raetam* in arid Tunisian soils." *Progress in Natural Science* 18, no. 1 (2008): 43–49.

Maj, Dominika, Jerzy Wielbo, Monika Marek-Kozaczuk, and Anna Skorupska. "Response to flavonoids as a factor influencing competitiveness and symbiotic activity of *Rhizobium leguminosarum*." *Microbiological Research* 165, no. 1 (2010): 50–60.

Manier, Nicolas, Annabelle Deram, Kris Broos, Franck-Olivier Denayer, and Chantal Van Haluwyn. "White clover nodulation index in heavy metal contaminated soils—A potential bioindicator." *Journal of Environmental Quality* 38, no. 2 (2009): 685–692.

Mark B. Peoples, David F. Herridge, and Jagdish K. Ladha. "Biological nitrogen fixation: An efficient source of nitrogen for sustainable agricultural production?" *Plant and Soil* 174, no. 1–2 (1995): 3–28.

McCarthy, Iva, Maria C. Romero-Puertas, Jose M. Palma, Luisa M. Sandalio, Francisco J. Corpas, Manuel Gómez, and Luis Alfonso Del Rio. "Cadmium induces senescence symptoms in leaf peroxisomes of pea plants." *Plant, Cell & Environment* 24, no. 10 (2001): 1065–1073.

Melchor-Marroquín, J. I., J. Jesus Vargas-Hernandez, Ronald Ferrera-Cerrato, and Laksmi Krishnamurthy. "Screening *Rhizobium* spp. strains associated with *Gliricidia sepium* along an altitudinal transect in Veracruz, Mexico." *Agroforestry Systems* 46, no. 1 (1999): 25–38.

Michiels, Jan, Christel Verreth, and Jos Vanderleyden. "Effects of temperature stress on bean-nodulating *Rhizobium* strains." *Applied and Environmental Microbiology* 60, no. 4 (1994): 1206–1212.

Moulin, Lionel, Antonio Munive, Bernard Dreyfus, and Catherine Boivin-Masson. "Nodulation of legumes by members of the β-subclass of Proteobacteria." *Nature* 411, no. 6840 (2001): 948.

Munns, Rana, and Mark Tester. "Mechanisms of salinity tolerance." *Annual Review of Plant Biology* 59 (2008): 651–681.

Nguyen, Nguyen Tran, Reda EA Moghaieb, Hirofumi Saneoka, and Kounosuke Fujita. "RAPD markers associated with salt tolerance in *Acacia auriculiformis* and *Acacia mangium*." *Plant Science* 167, no. 4 (2004): 797–805.

Njiti, Clément Forkong, and Antoine Galiana. "Symbiotic properties and Rhizobium requirements for effective nodulation of five tropical dry zone acacias." *Agroforestry Systems* 34, no. 3 (1996): 265–275.

Noori, Fatemeh, Hassan Etesami, Hamid Najafi Zarini, Nayer Azam Khoshkholgh-Sima, Ghasem Hosseini Salekdeh, and Farshad Alishahi. "Mining alfalfa (*Medicago sativa* L.) nodules for salinity tolerant non-rhizobial bacteria to improve growth of alfalfa under salinity stress." *Ecotoxicology and Environmental Safety* 162 (2018): 129–138.

Noriega, Guillermo O., Karina B. Balestrasse, Alcira Batlle, and Maria L. Tomaro. "Cadmium induced oxidative stress in soybean plants also by the accumulation of δ-aminolevulinic acid." *Biometals* 20, no. 6 (2007): 841–851.

Novikova Tatyana Ivanovna, and N. Ya Gordienko. "Specific features of functioning of the symbiotic system Rhizobium–*Glycyrrhiza uralensis* under the conditions of chloride salinization." *Sibirskii Ekolog Z* 3 (1999): 295–302.

Pandey, Akhilesh, and Matthias Mann. "Proteomics to study genes and genomes." *Nature* 405, no. 6788 (2000): 837.

Panwar, Meenu, Rupinder Tewari, Arvind Gulati, and Harsh Nayyar. "Indigenous salt-tolerant rhizobacterium *Pantoea dispersa* (PSB3) reduces sodium uptake and mitigates the effects of salt stress on growth and yield of chickpea." *Acta Physiologiae Plantarum* 38, no. 12 (2016): 278.

Patel, Dhara, Chaitanya Kumar Jha, Neelam Tank, and Meenu Saraf. "Growth enhancement of chickpea in saline soils using plant growth-promoting rhizobacteria." *Journal of Plant Growth Regulation* 31, no. 1 (2012): 53–62.

Paudyal, S. P., Rishi R. Aryal, S. V. S. Chauhan, and Dinesh Kumar Maheshwari. "Effect of heavy metals on growth of *Rhizobium* strains and symbiotic efficiency of two species of tropical legumes." *Science World* 5, no. 5 (2007): 27–32.

Peck, Melicent C., Robert F. Fisher, and Sharon R. Long. "Diverse flavonoids stimulate NodD1 binding to nod gene promoters in *Sinorhizobium meliloti*." *Journal of Bacteriology* 188, no. 15 (2006): 5417–5427.

Perret, Xavier, Christian Staehelin, and William J. Broughton. "Molecular basis of symbiotic promiscuity." *Microbiology and Molecular Biology Reviews* 64, no. 1 (2000): 180–201.

Phang, Tsui-Hung, Guihua Shao, and Hon-Ming Lam. "Salt tolerance in soybean." *Journal of Integrative Plant Biology* 50, no. 10 (2008): 1196–1212.

Phillips, Donald A., and Yoram Kapulnik. "Plant isoflavonoids, pathogens and symbionts." *Trends in Microbiology* 3, no. 2 (1995): 58–64.

Phillips, Donald A., Tama C. Fox, Maria D. King, T. V. Bhuvaneswari, and Larry R. Teuber. "Microbial products trigger amino acid exudation from plant roots." *Plant Physiology* 136, no. 1 (2004): 2887–2894.

Priefer, Ursula B., Jamal Aurag, Bert Boesten, Ilham Bouhmouch, Roberto Defez, Abdelkarim Filali-Maltouf, Marco Miklis et al. "Characterisation of Phaseolus symbionts isolated from Mediterranean soils and analysis of genetic factors related to pH tolerance." *Journal of Biotechnology* 91, no. 2–3 (2001): 223–236.

Pueppke, Steven G., and William J. Broughton. "*Rhizobium* sp. strain NGR234 and *R. fredii* USDA257 share exceptionally broad, nested host ranges." *Molecular Plant-Microbe Interactions* 12, no. 4 (1999): 293–318.

Quadir, M., A.Ghafoor, and G.Murtaza. Amelioration strategies for saline soils: a review." Land degradatiom & Development 11, no.6(2000): 501–521.

Raza, Sherow, Bjarne Jørnsgård, H. Abou-Taleb, and Jørgen Lindskrog Christiansen. "Tolerance of *Bradyrhizobium* sp. (Lupini) strains to salinity, pH, $CaCO_3$ and antibiotics." *Letters in Applied Microbiology* 32, no. 6 (2001): 379–383.

Rehman, Ateequr, and Chandra Shekhar Nautiyal. "Effect of drought on the growth and survival of the stress-tolerant bacterium *Rhizobium* sp. NBRI2505 sesbania and its drought-sensitive transposon Tn5 mutant." *Current Microbiology* 45, no. 5 (2002): 368–377.

Reichman, S. M. "The potential use of the legume–rhizobium symbiosis for the remediation of arsenic contaminated sites." *Soil Biology and Biochemistry* 39, no. 10 (2007): 2587–2593.

Riccillo, Pablo M., Cecilia I. Muglia, Frans J. De Bruijn, Andrew J. Roe, Ian R. Booth, and O. Mario Aguilar. "Glutathione is involved in environmental stress responses in *Rhizobium tropici*, including acid tolerance." *Journal of Bacteriology* 182, no. 6 (2000): 1748–1753.

Richardson, Alan E., Richard J. Simpson, Michael A. Djordjevic, and Barry G. Rolfe. "Expression of nodulation genes in *Rhizobium leguminosarum* biovar trifolii is affected by low pH and by Ca and Al ions." *Applied and Environmental Microbiology* 54, no. 10 (1988): 2541–2548.

Rivas, Raul, Encarna Velázquez, Anne Willems, Nieves Vizcaíno, Nanjappa S. Subba-Rao, Pedro F. Mateos, Monique Gillis, Frank B. Dazzo, and Eustoquio Martínez-Molina. "A new species of Devosia that forms a unique nitrogen-fixing root-nodule symbiosis with the aquatic legume *Neptunia natans* (Lf) Druce." *Applied and Environmental Microbiology* 68, no. 11 (2002): 5217–5222.

Rüberg, Silvia, Zhe-Xian Tian, Elizaveta Krol, Burkhard Linke, Folker Meyer, Yiping Wang, Alfred Pühler, Stefan Weidner, and Anke Becker. "Construction and validation of a *Sinorhizobium meliloti* whole genome DNA microarray: Genome-wide profiling of osmoadaptive gene expression." *Journal of Biotechnology* 106, no. 2–3 (2003): 255–268.

Sadowsky, Michael J. "Soil stress factors influencing symbiotic nitrogen fixation." In: Newton, W.E. (ed.), *Nitrogen Fixation in Agriculture, Forestry, Ecology, and the Environment*, pp. 89–112. Springer, Dordrecht, the Netherlands, 2005.

Sánchez-Diaz, M. "Adaptation of legumes to multiple stresses in Mediterranean-type environments." *Options Mediterr* 45 (2001): 145–151.

Sarma, Rupak K., and Ratul Saikia. "Alleviation of drought stress in mung bean by strain *Pseudomonas aeruginosa* GGRJ21." *Plant and Soil* 377, no. 1–2 (2014): 111–126.

Saxena, Anil K., and R. B. Rewari. "Influence of phosphate and zinc on growth, nodulation and mineral composition of chickpea (*Cicer arietinum* L.) under salt stress." *World Journal of Microbiology and Biotechnology* 7, no. 2 (1991): 202–205.

Schwedock, Julie, and Sharon R. Long. "Nucleotide sequence and protein products of two new nodulation genes of *Rhizobium meliloti*, nodP and nodQ." *Mol Plant Microbe Interact* 2, no. 4 (1989): 181–194.

Segovia, Lorenzo, Daniel Pinero, Rafael Palacios, and Esperanza Martinez-Romero. "Genetic structure of a soil population of nonsymbiotic *Rhizobium leguminosarum*." *Applied and Environmental Microbiology* 57, no. 2 (1991): 426–433.

Shamseldin, Abdelaal, Julius Nyalwidhe, and Dietrich Werner. "A proteomic approach towards the analysis of salt tolerance in *Rhizobium etli* and *Sinorhizobium meliloti* strains." *Current Microbiology* 52, no. 5 (2006): 333–339.

Simms, Ellen L., and D. Lee Taylor. "Partner choice in nitrogen-fixation mutualisms of legumes and rhizobia." *Integrative and Comparative Biology* 42, no. 2 (2002): 369–380.

Smit, Gerrit, Saskia Swart, Ben JJ Lugtenberg, and Jan Willem Kijne. "Molecular mechanisms of attachment of Rhizobium bacteria to plant roots." *Molecular Microbiology* 6, no. 20 (1992): 2897–2903.

Smith, Linda Tombras, Gary M. Smith, Marian R. D'souza, Jean-Alain Pocard, Daniel Le Rudulier, and Magdy A. Madkour. "Osmoregulation in *Rhizobium meliloti*: Mechanism and control by other environmental signals." *Journal of Experimental Zoology* 268, no. 2 (1994): 162–165.

Soussi, Meriem, Ameila Ocana, and Carmen Gonzalez Lluch. "Effects of salt stress on growth, photosynthesis and nitrogen fixation in chick-pea (*Cicer arietinum* L.)." *Journal of Experimental Botany* 49, no. 325 (1998): 1329–1337.

Spaink, Herman P. "Root nodulation and infection factors produced by rhizobial bacteria." *Annual Reviews in Microbiology* 54, no. 1 (2000): 257–288.

Subramanian, Senthil, Gary Stacey, and Oliver Yu. "Endogenous isoflavones are essential for the establishment of symbiosis between soybean and *Bradyrhizobium japonicum*." *The Plant Journal* 48, no. 2 (2006): 261–273.

Swaine, Emily K., Michael D. Swaine, and Ken Killham. "Effects of drought on isolates of *Bradyrhizobium elkanii* cultured from *Albizia adianthifolia* seedlings of different provenances." *Agroforestry Systems* 69, no. 2 (2007): 135–145.

Sy, Abdoulaye, Eric Giraud, Philippe Jourand, Nelly Garcia, Anne Willems, Philippe De Lajudie, Yves Prin et al. "Methylotrophic Methylobacterium bacteria nodulate and fix nitrogen in symbiosis with legumes." *Journal of Bacteriology* 183, no. 1 (2001): 214–220.

Teplitski, Max, Jayne B. Robinson, and Wolfgang D. Bauer. "Plants secrete substances that mimic bacterial *N*-acyl homoserine lactone signal activities and affect population density-dependent behaviors in associated bacteria." *Molecular Plant-Microbe Interactions* 13, no. 6 (2000): 637–648.

Tomar, O. S., Paramjit Singh Minhas, Vinod K. Sharma, Yashpal P. Singh, and Raj Kumar Gupta. "Performance of 31 tree species and soil conditions in a plantation established with saline irrigation." *Forest Ecology and Management* 177, no. 1–3 (2003): 333–346.

Triplett, Eric W., and Michael J. Sadowsky. "Genetics of competition for nodulation of legumes." *Annual Review of Microbiology* 46, no. 1 (1992): 399–422.

Valverde, Angel, Encarna Velazquez, Carmen Gutierrez, Emilio Cervantes, Antonio Ventosa, and Jose-Mariano Igual. "Herbaspirillum lusitanum sp. nov., a novel nitrogen-fixing bacterium associated with root nodules of *Phaseolus vulgaris*." *International Journal of Systematic and Evolutionary Microbiology* 53, no. 6 (2003): 1979–1983.

Valverde, Angel, Encarna Velazquez, Felix Fernandez-Santos, Nieves Vizcaíno, Raul Rivas, Pedro F. Mateos, Eustoquio Martínez-Molina, Jose Mariano Igual, and Anne Willems. "*Phyllobacterium trifolii* sp. nov., nodulating *Trifolium* and *Lupinus* in Spanish soils." *International Journal of Systematic and Evolutionary Microbiology* 55, no. 5 (2005): 1985–1989.

Vance, Carroll P. "Legume symbiotic nitrogen fixation: Agronomic aspects." In: Spaink, H.P. Kondorosi, A., and Hoykaas, (Eds.), *The Rhizobiaceae*, pp. 509–530. Springer, Dordrecht, the Netherlands, 1998.

Velázquez, Encarna, Paula García-Fraile, Martha-Helena Ramírez-Bahena, Raúl Rivas, and Eustoquio Martínez-Molina. "Bacteria involved in nitrogen-fixing legume symbiosis: Current taxonomic perspective." In: Khan, M.S., Zaidi, A., and Musarrat, J. (Eds.), *Microbes for Legume Improvement*, pp. 1–25. Springer, Vienna, Austria, 2010.

Vessey, J. Kevin, and George N. Chemining'wa. "The genetic diversity of *Rhizobium leguminosarum* bv. viciae in cultivated soils of the eastern Canadian prairie." *Soil Biology and Biochemistry* 38, no. 1 (2006): 153–163.

Vriezen, Jan AC, Frans J. De Bruijn, and K. Nüsslein. "Responses of rhizobia to desiccation in relation to osmotic stress, oxygen, and temperature." *Applied and Environmental Microbiology* 73, no. 11 (2007): 3451–3459.

Wani, Parvaze Ahmad, Mohammad Saghir Khan, and Almas Zaidi. "Effect of metal-tolerant plant growth-promoting *Rhizobium* on the performance of pea grown in metal-amended soil." *Archives of Environmental Contamination and Toxicology* 55, no. 1 (2008a): 33–42.

Wani, Parvaze Ahmad, Mohammad Saghir Khan, and Almas Zaidi. "Effects of heavy metal toxicity on growth, symbiosis, seed yield and metal uptake in pea grown in metal amended soil." *Bulletin of Environmental Contamination and Toxicology* 81, no. 2 (2008b): 152–158.

Wasson, Anton P., Flavia I. Pellerone, and Ulrike Mathesius. "Silencing the flavonoid pathway in *Medicago truncatula* inhibits root nodule formation and prevents auxin transport regulation by rhizobia." *The Plant Cell* 18, no. 7 (2006): 1617–1629.

Wood, Janet M., Erhard Bremer, Laszlo N. Csonka, Reinhard Kraemer, Bert Poolman, Tiemen van der Heide, and Linda T. Smith. "Osmosensing and osmoregulatory compatible solute accumulation by bacteria." *Comparative Biochemistry and Physiology Part A: Molecular & Integrative Physiology* 130, no. 3 (2001): 437–460.

Wood, Martin A., and John E. Cooper. "Acidity, aluminium and multiplication of *Rhizobium tripolii*: Possible mechanisms of aluminium toxicity." *Soil Biology and Biochemistry* 20, no. 1 (1988): 95–99.

Xiong, Liming, and Jian-Kang Zhu. "Abiotic stress signal transduction in plants: Molecular and genetic perspectives." *Physiologia Plantarum* 112, no. 2 (2001): 152–166.

Younis, M. "Responses of *Lablab purpureus*-Rhizobium symbiosis to heavy metals in pot and field experiments." *World Journal of Agricultural Sciences* 3, no. 1 (2007): 111–122.

Zahran, Hamdi H. "Legumes–Microbes interactions under stressed environments." In: Khan, D.M.S., Musarrat, P.D.J., and Zaidi, D.A. (Eds.), *Microbes for Legume Improvement*, pp. 353–387. Springer, Vienna, Austria, 2010.

Zahran, Hamdi H. "Rhizobial nitrogen fixation in agriculture: Biotechnological perspectives." *Microbial Biotechnology in Agriculture and Aquaculture* 1 (2005): 71–100.

Zahran, Hamdi Hussein. "Wild-legume rhizobia: Biodiversity and potential as biofertilizer." In: Rai, M.K. (Ed.), *Handbook of Microbial Biofertilizers*, pp. 203–222. Haworth Press, New York, 2006.

Zhang, Feng, and Donald L. Smith. "Inoculation of soybean (*Glycine max* (L.) Merr.) with genistein-preincubated *Bradyrhizobium japonicum* or genistein directly applied into soil increases soybean protein and dry matter yield under short season conditions." *Plant and Soil* 179, no. 2 (1996): 233–241.

Zhang, Feng, and Donald L. Smith. "Preincubation of *Bradyrhizobium japonicum* with genistein accelerates nodule development of soybean at suboptimal root zone temperatures." *Plant Physiology* 108, no. 3 (1995): 961–968.

Zhang, Juan, Senthil Subramanian, Gary Stacey, and Oliver Yu. "Flavones and flavonols play distinct critical roles during nodulation of *Medicago truncatula* by *Sinorhizobium meliloti*." *The Plant Journal* 57, no. 1 (2009): 171–183.

Zhu, Jian-Kang, Paul M. Hasegawa, Ray A. Bressan, and Hans J. Bohnert. "Molecular aspects of osmotic stress in plants." *Critical Reviews in Plant Sciences* 16, no. 3 (1997): 253–277.

Zuanazzi, José Angelo Silveira, Pierre Henri Clergeot, Jean-Charles Quirion, Henri-Philippe Husson, Adam Kondorosi, and Pascal Ratet. "Production of *Sinorhizobium meliloti* nod gene activator and repressor flavonoids from *Medicago sativa* roots." *Molecular Plant-Microbe Interactions* 11, no. 8 (1998): 784–794.

Zurdo-Pineiro, Jose Luis, Raul Rivas, Martha E. Trujillo, Nieves Vizcaino, Jose Antonio Carrasco, Manuel Chamber, Antonio Palomares, Pedro F. Mateos, Eustoquio Martinez-Molina, and Encarna Velazquez. "*Ochrobactrum cytisi* sp. nov., isolated from nodules of *Cytisus scoparius* in Spain." *International Journal of Systematic and Evolutionary Microbiology* 57, no. 4 (2007): 784–788.

6

Symbiosis with Fungal Endophytes Conferring Stress Tolerance in Legumes

Chaitanya Ghalagi, D. J. Bagyaraj, and B. Mohan Raju

CONTENTS

6.1 Introduction

Plants are continuously exposed to an enormous amount of external stimulus. Being sessile, they have to deal with various and complex types of interactions involving numerous environmental factors. To survive fluctuating changes in their environments, plants have developed specific mechanisms to perceive, anticipate, and respond to complex stress conditions, minimizing damage while conserving valuable resources for growth and development. Plants do interact with numerous underground beneficial microbes, such as rhizobia, mycorrhiza, endophytes, and plant growth-promoting rhizobacteria (PGPR). These microorganisms abound in the rhizosphere and perform many functions important for plant growth and development. Among the beneficial microbes, fungal symbionts are one of the most important elements in plant microecosystems. Plants form mutualistic association with two major classes of fungal symbionts, namely (a) mycorrhizal fungi that reside only in roots but extend out into the rhizosphere; and (b) fungal endophytes that occur ubiquitously in all plants, often in the intercellular spaces of roots, stems, and/or leaves. In mutualistic association, both the partners are benefited. Fungal symbionts acquire carbohydrates from their host plant and in turn confer host fitness benefits that may enhance host growth, acquisition, and improve the plant's ability to tolerate abiotic stresses such as drought and decrease biotic stresses by enhancing plant's resistance to insects, pathogens, and herbivores. It has been well documented that mycorrhizal fungi facilitate nutrient uptake in plants, particularly diffusion of limited nutrients such as phosphorus, zinc, and copper (Bolan, 1991) and also other nutrients like sulfur, nitrogen, and other micronutrients from the soil. Many researchers recognized that mycorrhizal fungi are multifunctional including conferring protection to host plants against certain pathogens (Sikes, 2010). Similarly, in recent decades, endophytes (both bacterial and fungal) emerged as a potential means to mitigate plant stress under a rapidly changing climate.

Legumes (Leguminosae or Fabaceae) play a very important role in many agricultural systems, providing high protein grains and herbage as well as maintaining and improving soil fertility. Also, it has a significant role in imparting nutritional security to the people. However, the major challenges that tropical countries, like India, face in cultivation of legumes are susceptibility to biotic and abiotic stresses. The yield losses due to biotic and abiotic stresses in legumes vary from 30% to 100% depending on the magnitude of stress severity (Rana et al., 2016). Under the present circumstance of changing climate, it is important to develop strategies to enhance tolerance in food legumes to various stresses. In this context, the present review emphasizes on stress tolerance in legumes mediated by fungal symbionts that may help to ensure sustainable production.

6.2 Endophytic Fungi-Plant Symbiosis and Stress Alleviation

Fungal endophytes are major classes of fungal symbionts that establish different forms of symbiotic interactions with host plants such as commensalism, mutualism, and parasitism (Lewis, 1985). A commensal symbiosis has no beneficial or detrimental effects on hosts. Parasitic fungi, on the other hand, negatively affect host fitness by decreasing growth rates and/or fecundity, or inducing disease symptoms that may result in lethality. However, the mutualistic symbiosis confer host fitness benefits that can result in stress tolerance, increased growth rates, and/or nutrient acquisition (Rodriguez et al., 2004). Hence, it is important for a plant to have mutualistic association with endophytic fungi.

It is evident that all plants are in symbiotic association with fungi. This symbiotic association with mutualistic fungi confers several benefits to plants such as nutrient acquisition (Read, 1999), growth enhancement (Marks and Clay, 1990) tolerance to drought, metals (Read, 1999), temperature (Redman et al., 2002), etc. In a broader context, Jia et al. (2016) have reported two types of beneficial interactions that plant endophytic fungi form with their host plants, namely: (1) enhancement of the growth of host plants, and (2) increase in resistance of the host plants to biotic and abiotic stresses.

A symbiotically conferred abiotic and biotic stress tolerance appears to involve two mechanisms: (i) rapid activation of host stress response systems after symbiotic plants are exposed to stresses (Redman et al., 1999) and (ii) synthesis of anti-stress biochemicals by the fungus (Bacon and Hill, 1996). However, there are no reports on how endophytes in symbiotic association activate host stress response systems. But, there are few reports on anti-stress biochemicals produced by endophytic fungi that confer drought tolerance to host plants (Bacon and Hill, 1996). In essence, endophytic fungi act as a "biological trigger" to confer stress tolerance. This is supported by the fact that the symbiotic plants activate defense systems more quickly than nonsymbiotic plants after pathogen challenge (Rodriguez et al., 2004).

Some fungal endophytes confer host resistance against several fungal pathogens (Freeman and Rodriguez, 1993, Latch, 1993, Duchesne, 1994, Blee and Anderson, 2000, Clay and Schardl, 2002). One of the most dramatic examples of this is represented by the behavior of plant pathogenic fungi, *Colletotrichum* species, which expressed non-mutualistic lifestyles in specific hosts and may establish mutualistic symbiosis with genetically unrelated plant species and confers stress (disease and/or drought) tolerance. With this option, it is possible to use fungal endophytes from hosts thriving in high-stress environments to confer desirable traits for drought, temperature, disease, and salt tolerance to genetically unrelated stress-sensitive plant species. With specific reference to this, Rodriguez et al. (2008) have proposed that the fungal endophytes may confer stress tolerance to plants through a process defined as "habitat-adapted" symbiosis. Therefore, it is important to screen for endophytic fungi, isolated from plants grown under extreme environmental conditions, which can form mutualistic symbiosis and impart tolerance to legumes. This would allow existing legume varieties to generate new capabilities in them for tolerating specific environmental stresses.

6.3 Mycorrhizal Fungi-Legume Symbiosis and Stress Alleviation

In a mycorrhizal association, the fungus colonizes the host plant's root tissues, either intracellularly as in the case of arbuscular mycorrhizal (AM) fungi (AMF), or extracellularly as in ectomycorrhizal fungi. The most common mycorrhizal fungi is the arbuscular type, which has a forging symbiosis with plethora of plant species roots, including many crop plants (Berutti et al., 2016). The widespread occurrence of AM fungi has been found in cultivated legumes (Strzemska, 1975). Many temperate and tropical legumes such as clover, lucerne (alfalfa), beans (*Phaseolus* and *Vicia*), peas, soybean, cowpea, pigeon pea, chickpea, groundnut, stylosanthes, and centrosema form mutualistic symbiosis with AMF. Some arborescent legumes such as *Leucaena* are also associated with AMF, however; the legumes belonging to Caesalpinoideae are associated with ecto-type mycorrhizal fungi. In addition, certain leguminous shrubs such as *Acacia* have both ecto-type mycorrhizal fungi and AM association. Mycorrhizal fungi are found to have multifaceted functionality that they act as biofertilizers, bioprotectants, mycoremediators, and biodegraders. Their association with legumes not only improve nutrient uptake but also protect the plants from abiotic (like drought and nutrient) and biotic (like soil-borne plant pathogenic fungi, bacteria, and nematode) stresses. Therefore, arbuscular mycorrhizal-legume symbiosis is viewed as a better alternative for enhancing plant responses to many kinds of stresses.

6.4 Fungal Endophyte-Abiotic Stress Response in Plants

The most harmful factor concerning the growth and productivity of crops worldwide is abiotic stress. Plants that are symbiotically associated with endophytic fungi rapidly activate host stress response systems when exposed to stress (Redman et al., 1999). The actual mechanism that activates host stress response systems is unknown. But, plants hosting such beneficial fungi undergo significant changes in their key physiological aspects, including phytohormones balance, root exudation, composition of mineral nutrients, and physical modification in soil (Richardson, 2009, Kusari et al., 2013). These changes help host plants to enhance their overall fitness to withstand different abiotic stresses (Chandanie et al., 2006, Khan et al., 2009). Endophytes are found to enhance the nutrient uptake by modifying root morphology, alteration of nitrogen accumulation and metabolism, and enhanced water use efficiency by osmotic adjustment and stomatal regulation. Bayat et al. (2008) have reported that endophytes confer drought tolerance to plants through alteration of various physiological mechanisms such as relative water content (RWC), cell membrane stability (CMS), proline and chlorophyll contents. Waqas et al. (2012) reported that *Phoma glomerata* and *Penicillium* sp. significantly increased plant biomass and related growth parameters, assimilation of essential nutrients such as potassium, calcium, magnesium and reduced the sodium toxicity in cucumber plants under sodium chloride and polyethylene glycol-induced salinity and drought stress when compared with control plants. Similarly, endophyte-infected plants exhibited drought tolerance by enhanced accumulation of solutes in tissues, reduced leaf conductance, and thicker cuticle formation as compared to non-infected plants (Malinowski and Belesky, 2000).

Wu et al. (2019) suggested that endophytic fungus *Serendipita indica* helps host plants (*Cunninghamia lanceolata*) to absorb more nutrients under low phosphate and allocate more nitrogen and phosphate to leaves, thus promoting plant growth. Endophytes promote plant growth by phosphate solubilization and enhanced production of indole acetic acid (IAA), cytokinins, gibberellin (GAs), siderophores, and the supply of essential vitamins (Jha et al., 2013). He et al. (2012) have reported that all endophytes have the ability to decompose organic components including lignin, cellulose, and hemicelluloses, which facilitate in nutrient recycling. In this context, *Piriformospora indica* and *Azotobacter chroococcum* improved the uptake of mineral nutrients (especially Zn) in wheat (Abadi and Sepehri, 2016) and *Pseudomonas* sp. mediated phosphate solubilization in rice and wheat by producing gibberellic acid (Choi et al., 2008). Therefore, the presence of endophytes has a variable impact on plant performance. Under low soil nutrient content, endophyte infected (E+) plants showed higher values for tiller number, plant height, chlorophyll content, shoot dry weight, and agronomic efficiency of water use (WUE) than non-infected (E−) plants (Rahman et al., 2006). Further, the endophytic fungi, *P. indica*, induces salt tolerance in barley by increasing the levels of antioxidants (Baltruschat et al., 2008) and increases heat tolerance in wheat in terms of height, weight of grain, as well as germination of second-generation seeds (Hubbard et al., 2014).

6.5 Mycorrhizal Fungi as Bioprotectants in Legumes to Alleviate Environmental Stresses

The AM symbiosis can alter plant physiology in a way to cope with stresses under stressful conditions (Miransari et al., 2008). A symbiotic association between AMF and host plant regulates the plant nutrient uptake, stomatal conductance, leaf water potential, photosynthesis, and transpiration rate (Yang et al., 2014). It could also induce production of growth-promoting hormones, anti-oxidative enzymes (Zhu et al., 2011); improve plant root architecture (Wu et al., 2012) and soil properties.

Mycorrhizal symbiosis can potentially improve water uptake by plants under stressful environments (Mathimaran et al., 2017). For instance, soybean plants inoculated with AMF showed higher values for relative water content (RWC) and leaf water potential than non-mycorrhizal plants irrespective of soil moisture level (Mathimaran et al., 2017, Ashwin et al., 2018a and b). The nitrogen and potassium content in shoots was also enhanced in AMF inoculated soybean plants, implying AM fungal colonization not only improves plant water status but also nutritional status in plants (Aliasgharzad et al., 2006). Ashwin et al., 2018a while working with soybean reported that the inoculation of soybean cultivars with the selected AM fungi confers tolerance under drought conditions. In another study, Bethlenfalvay et al. (1987) reported that AM fungi under drought stress in soybean stimulated nodule development and activity in soybean besides increasing the CO_2 uptake due to better leaf conductance.

AM symbiosis has been shown to improve tolerance to temperature stress in plants. Both low- and high-temperature stresses affect many physiological and biochemical processes and functions in legumes. Zhu et al. (2017) have reported the possible mechanisms of AM symbiosis in improving temperature stress tolerance of the host plants via enhancing water and nutrient uptake, improving photosynthetic capacity and efficiency, protecting plants against oxidative damage, and increasing accumulation of osmolytes. In maize, Zhu et al. (2010) showed that AM fungus can alleviate the damage caused by temperature stress by reducing membrane lipid peroxidation and membrane permeability and increasing the accumulation of osmotic adjustment compounds and antioxidant enzymes activity. Consequent to AM fungal formation, extreme temperature tolerance of maize plants with increased host plant growth and biomass was achieved.

AM fungi do alleviate salt stress in plants and improve the growth performance of their host plants. Rabie (2005) while working with mung bean suggested that the AM-fungi protected the mung bean (*Vigna radiata*) against the detrimental effects of salt stress. The mycorrhizae-mediated improvement in legumes was due to protection against salt stress provided through better access to nutrients (Zandavalli et al., 2004), osmotic modifications (Rao and Tak, 2002), and improved photosynthesis (Feng et al., 2002).

Bioprotection conferred to plants against many soilborne pathogens such as *Aphanomyces*, *Cylindrocladium*, *Fusarium*, *Macrophomina*, *Phytophthora*, *Pythium*, *Rhizoctonia*, *Sclerotinium*, *Verticillium*, and *Thielaviopsis* and various nematodes by AM fungal colonization of the plant root has been reported in many crop species including legumes. Mycorrhizae have been found to excrete enzymes that are toxic to these soilborne pathogens and thereby confer tolerance against these pathogens. The mechanisms by which AM fungal colonization confers the protective effect are fairly well understood (Bagyaraj, 2018). Studies dealing with AM effects on aboveground pathogens are less abundant compared to soilborne pathogens.

6.6 Mycorrhizal Fungi as Myco-Remediators in Legumes

Several investigations on plants inoculated with mycorrhizae under the impact of bulk metals have demonstrated that the host plants were protected against heavy metal stress and toxicity (Khabou et al., 2014, Yang et al., 2015, Sharma and Kapoor, 2019). The buffering effect of a mycorrhizal fungus, *Glomus intraradices*, on Cd uptake and its subsequent effect in reducing the toxic effect of Cd on growth of pea plants were demonstrated by Rivera et al. (2002). This implies the potential role of AM fungi in mycoremediation, an on-site remediation strategy, which employs fungi to degrade or sequester contaminants from the environment. Mycorrhizal colonization also facilitates phytoremediation/mycoremediation by promoting the growth of hyperaccumulating plants (Firmin et al., 2015, Liu et al., 2015, Nouri et al., 2017). AMF have been reported for dissipation of organic contaminants such as atrazine (Huang et al., 2007), polycyclic aromatic hydrocarbons (PAHs) (Joner et al., 2001, Joner and Leyval, 2003, Xu et al., 2006), DDT (Wu et al., 2008b), etc. In *Vigna radiata* (mung bean), an AM fungus, *Glomus mosseae,* was found to dissipate an organic contaminant, PAHs (Rabie, 2004). AMF also directly help the plants to escape from the buildup of phytotoxic concentrations of certain pollutants by secreting specific detoxifying compounds (organic acids) such as short-chain organic acids, phenolics, and small concentrations of high molecular weight compounds, i.e., enzymes and proteins (Korade and Fulekar, 2009). AM fungi can greatly improve the phytoremediation capacity of some leguminous plants (Palmroth et al., 2002, Smith et al., 2006, Sharma and Kapoor, 2019).

6.7 Mycorrhizal Fungi as Biofertilizer in Legumes

Mycorrhizal fungi being a biofertilizer is either inoculated or supplemented with inorganic fertilizers to crops in order to increase growth and productivity (Zayed et al., 2013, Ashwin and Bagyaraj, 2018, Jyothi and Bagyaraj, 2018). Legumes fix atmospheric nitrogen in their nodules formed as a result of mutualistic symbiosis with compatible *Rhizobium.* Though nodulation is achieved, its effectiveness to fix nitrogen is influenced by environmental factors such as temperature, salinity (Afzal et al., 2015), water availability, and nutrition status (especially phosphorus, potassium, zinc, iron, molybdenum, and cobalt). Even the activity of the enzyme nitrogenase, which is involved in N-fixation, is also dependent on an adequate supply of the trace elements such as Mo, Cu, and Co. AM fungi are considered as natural biofertilizers (Berruti et al., 2016), as they supply adequate nutrients to host plants in exchange for photosynthetic products, and they alleviate the limitation in plant growth and development (Nouri et al., 2014). They not only help plant growth but also aid the bacterial symbionts to fix nitrogen in the nodular tissue by supplying P and other trace elements crucial for nodulation and nitrogen fixation (Berruti et al., 2016). Therefore, the tripartite symbiosis between legumes, rhizobia, and mycorrhizal fungi are generally considered to be beneficial for the nitrogen (N) uptake of legumes (Meng et al., 2015, Desai et al., 2016). The flavonoids such as daidzein, genistein, and coumestrol have been identified as possible signals for regulating the establishment of the tripartite symbiosis (Antunes et al., 2006). However, AM fungal symbiosis with nitrogen-fixing legume is a complex nutrient feedback mechanism; therefore, the plants either promote or limit fungal colonization according to their nutritional need and supply (Nouri et al., 2014). In fact, the suppression of AM colonization due to high Pi levels has been reported in several earlier studies (Breuillin et al., 2010, Balzergue et al., 2013, Bonneau et al., 2013).

Under low-nutrient conditions, AM fungi are capable of significantly improving plant mineral nutrient acquisition, especially phosphorous (Berruti et al., 2016). This was known long back when Asai (1944) suggested that mycorrhizae are indeed necessary precondition for effective nodulation in many legumes. He showed that several legumes grew poorly and failed to nodulate in autoclaved soil unless they were inoculated with mycorrhizal fungi. This effect, however, was not attributed to the improvement of P nutrition by mycorrhizae because it was yet unknown. Later, Crush (1974) showed that AM stimulated the growth and nodulation in several legume species (*Centrosema pubescens, Stylosanthes guyanensis,* and *Trifolium repens*). Thereafter, several literatures state the importance of AM-fungal inoculation in increasing plant growth and yield besides nodulation and nitrogen fixation and P and N content in legumes such as soybean (Carling et al., 1978, Bagyaraj et al., 1979), *Vigna unguiculata* (Islam et al., 1980), *Medicago sativa* (Barea et al., 1980), and *Leucaena leucocephala* (Munns and Mosse, 1980, Manjunath et al., 1984). AM fungi generally obtain and supply extra phosphate from the labile pool rather than by dissolving insoluble phosphates (Raj et al., 1981). Mycorrhiza being a biofertilizer, its inoculation resulted in increased plant growth than when given any amount of P-fertilizers (Bagyaraj and Sreeramulu, 1982). This prompts one to consider the possibility of hormonal effects, which was shown by Allen et al. (1980) in the grass species such as *Boteloua gracilis,* where AM fungi directly affect the levels of plant hormones (cytokinins and gibberellins-like substance).

Dual inoculation of crop legumes with *Rhizobium* and AM fungi has been found to be advantageous. Wu et al. (2009) have reported that dual inoculation of *Glomus mosseae* and *Rhizobium* elicited a synergistic effect resulting in enhanced N, P, and K content in *Medicago sativa* grown on three types of coal mine substrates. They also suggested that inoculation with AM fungi may be a promising approach for revegetation of coal mine substrates. Rhizobia and AM fungi often interact synergistically, resulting in better root nodulation, nutrient uptake, and plant yield compared to plant symbiosis with either of them separately (Manjunath et al., 1984, Javaid, 2010). Similar synergistic effects were also reported for dual inoculation with AM fungi, and phosphate solubilizing bacteria, *Azotobacter* or *Azospirillum* (Domínguez-Núñez et al., 2015). Mycorrhizal stimulation of nitrogenase activity by legume *Rhizobium* symbiosis as measured by the acetylene reduction technique was reported by Crush (1974). In comparison with non-mycorrhizal nodulated plants, mycorrhizal nodulated plants were found to have greater amount of nodular tissues with higher concentrations of leghemoglobin and higher rates of acetylene reduction (Daft and El-Giahmi, 1974, 1975, 1976). Similar observations have been made by Kawai and Yamamoto (1986) in soybean and Lynd et al. (1985) in Siratro. Using ^{15}N tracer technique, Kucey and Paul (1982) confirmed that mycorrhizal nodulated *Vicia faba* plants, fixing more N than those nodulated but non-mycorrhizal.

Smith and Daft (1977) reported that AM inoculation increased in nodulation, nitrogen fixation, and phosphate concentration in *Medicago sativa*. A similar phenomenon was observed in soybean by Asimi et al. (1980), and further, they found that adding soluble P at low level (0.25 g KH_2PO_4/kg soil) eliminated the mycorrhizal effect on plant growth and significantly stimulated nodule formation and nitrogenase activity. This suggests the main role of AM fungi in phosphate transfer to bacteroids by AM fungal specific phosphatases in arbuscules developing inside the root cells adjacent to nodules. In addition to P and N, other elements such as sulfur (Allen and Shachar-Hill, 2009, Sieh et al., 2013), zinc, copper, manganese, iron (Azcón et al., 2003, Ortas and Akpinar, 2006), potassium, aluminum, etc., can also be transferred to plants through AM fungi.

6.8 Fungal Endophyte-Derived Secondary Metabolites in Plant Defense

In their relationship with the host plant, endophytes contribute to the plants by producing a plethora of substances that provide protection and survival benefits to the host plants (Patil et al., 2016). In other words, endophytic fungi are a promising source of bioactive compounds or an alternative source of metabolites produced in higher plants (Kusari and Spiteller, 2011). These compounds have different biological activities like antimycotic, anticancer, antibiotics, insecticidal, and antiviral (Strobel and Daisy, 2003, Verma et al., 2009). All these bioactive compounds belong to different structural groups, such as terpenoids, steroids, xanthones, chinones, phenols, isocoumarins, benzopyranones, tetralones, cytochalasines, and enniatines, indicating that fungal endophytes are able to produce almost all chemical classes of secondary metabolites. Among which, the terpenoids and polypeptides are commonly produced, whereas flavonoids and lignins are rarely produced. These secondary metabolites are particularly active due to metabolic interactions with their hosts (Guo et al., 2007, Lösgen et al., 2008, Lin et al., 2009). However, how each endophytes and their host plants coordinate their metabolic biosynthesis to produce secondary metabolites remains unclear.

Alkaloids are nitrogen-rich compounds. They are one of the largest groups of plant secondary metabolites that act as defense compounds in plants. Therefore, the plant–endophyte symbiosis is considered as a case of "defensive mutualism" (Clay, 1988). Endophytic alkaloids are characterized into four classes: ergot alkaloids (i.e., ergopeptine and ergovaline), indole diterpenes (i.e., lolitrem B and epoxy-janthitrem), pyrrolizidines (i.e., lolines), and peramine (Schardl et al., 2012, 2013, Saikkonen et al., 2013, Gutjahr, 2014, Panaccione et al., 2014, Young et al., 2015). Ergot alkaloids that provide resistance to plants against herbivory are present in various grasses infected by endophytic fungi of the genus *Epichloë* (Guerre, 2015). In addition to producing protective alkaloids, endophytes enhance plant immunity against chewing insects by promoting endogenous defense responses mediated by the jasmonic acid (JA) pathway (Bastías et al., 2017). An anticancer drug camptothecin, isolated from inner bark of plant *Nothapodytes foetida*, produced by fungal endophyte belonging to class Phycomycetes (Puri et al., 2005) is another important compound for human health.

An endophyte derived indole-based compound, loline, has a broad spectrum of anti-insect and anti-aphid activity resulting in increased resistance to host plants against insect herbivory. This compound was found in grass, *Festuca pratensis* (*Lolium pratense*) originating from its fungal endophyte, *Neotyphodium uncinatum* (Blankenship et al., 2001).

Phenols and phenolic acids, on the other hand, have often been isolated from some endophyte cultures originating from a variety of the host plants including legumes (Yu et al., 2010). These are responsible for several bioactivities in plants such as antioxidant, cytotoxic and antimicrobial activity. Pan et al. (2017) have reported that interactions between fungal endophytes and host plants induce the production of phenolic compounds with antioxidant activity. Fungal endophytes produce numerous antioxidant phenolic compounds, confer effective tolerance to ROS under abiotic stress conditions, and also promote growth via biosynthesis of plant hormones and nutrient acquisition (Sati et al., 2014). Not only the endophytic fungi, but some non-endophytic fungi also produce many antioxidant phenolic compounds (Smith et al., 2015). Phenolic compounds, pestalachloride A and B, produced by endophytic fungi were found to have antifungal activity in some plant species. Similarly, Pestalachloride C and D were found to have moderate antibacterial activity in others (Li et al., 2008). In another report, Dai et al. (2006) found that the culture extracts of endophytic fungus *Nodulisporium* sp. contain antimicrobial flavonoids as well.

Many steroids have been shown to be produced by endophytes, but most of the isolated compounds showed moderate antimicrobial and antifungal activities. The liquid extract of fungal endophyte, *Colletotrichum* sp. of *Artemisia annua*, contains various antifungal steroid compounds such as 3β,5α,6β-trihydroxyergosta-7,22-diene; 3β-hydroxyergosta-5ene; 3-oxoergosta-4,6,8,22-tetraene; 3β-hydroxy-5α,8α-epidioxyergosta-6,22-diene; 3β-hydroxy-5α,8α-epidioxyergosta-6,9,22-triene; and 3-oxoergosta-4-ene. Some of these metabolites were shown to be antifungal against some crop pathogens such as *Gaeumannomyces graminis* var. *tritici*, *Rhizoctonia cerealis*, *Helminthosporium sativum*, and *Phytophthora capisici* (Lu et al., 2000, Yu et al., 2010).

Terpenoids, on the other hand, are a larger and diverse class of naturally occurring organic chemicals derived from terpenes. The major terpenoids isolated from endophytes are sesquiterpenes, diterpenoids, and triterpenoids (Yu et al., 2010). *Sesquiterpenes* are a class of terpenes that consist of three isoprene units. Many fungal endophytes, in association with host plants, produce these compounds. One such sesqiterpene is trichodermin, which is obtained from *Trichoderma harzianum*, an endophytic fungus living in an evergreen holly shrub *Ilex cornuta*. It is also found to protect solanaceous plants against the pathogens *Alternaria solani* and *R. solani* in vitro (Chen et al., 2007). Yet another antifungal eremophilane sesquiterpene, phomenone, is produced by *Xylaria* sp., an endophytic fungus associated with *Piper aduncum*. This compound has antifungal activity against *Cladosporium cladosporioides* (wheat pathogen) and *C. sphaerospermum* (common indoor mold) (Silva et al., 2010). Similarly, there are several more sesquiterpenes derived from endophytes that are symbiotically associated with their host plants, having antifungal activity such as Cycloepoxylactone, Cycloepoxytriol B (Hussain et al., 2009), 3,12 Dihydroxycadalene (Silva et al., 2006), Heptelidic acid, Hydroheptelidic acid (Calhoun et al., 1992), and Chokols (Yoshihara et al., 1985).

Diterpenes are a class of chemical compounds composed of two terpene units. Many diterpenes, such as sesquiterpenes, are produced by endophytic fungi in association with the host plants. Stierle et al. (1995) have reported that an endophytic fungus *Taxomyces andreanae* produces antifungal drug paclitaxel (Taxol) when associated with yew tree, *Taxus brevifolia*. Also, there are many other similar compounds such as peroconicins A and B, sordaricin, scoparasin B (antifungal compounds), and guanacastepene (antibacterial compound) that are produced by endophytic fungi, which confer resistance to host plants. There are no reports on endophytic fungi producing bioactive triterpenes as a defense molecule. But, a triterpene, helvolic acid, isolated from yeast, *Pichia guilliermondii*, was found to have antibacterial activity.

Phenolics are aromatic benzene ring compounds with one or more hydroxyl groups produced by plants mainly for protection against biotic and abiotic stresses. Some fungal endophytes do produce phenolic compounds that confer resistance to host plants. 2-Methoxy-4-hydroxy-6-methoxymethylbenzaldehyde, a phenolic compound, was found to be antifungal against the cucumber pathogen *Cladosporium cucumerinum*. The compound is produced by *Pezicula* strain 553, a fungal endophyte colonizing an unknown tree (Schulz et al., 1995). Similarly, tridepside compound, colletotric acid, obtained from the liquid culture of *Colletotrichum gloeosporioides*, a fungus that colonizes the stems of *Artemisia mongolica*, showed protection against insects and pathogens. This compound was also shown to have antimicrobial activity against the bacteria, *Bacillus subtilis*, *Staphylococcus aureus*, and *Sarcina lutea* (*Micrococcus luteus*), and the fungus *Helminthosporium sativum* (Zou et al., 2000).

Endophytic fungus, *Neotyphodium typhnium*, in symbiotic association with Big Bluegrass (*Poa ampla*), produces flavonoids such as tricin and related flavone glycosides. These flavonoids were found to be toxic to the larvae of *Culex pipiens* (Ju et al., 1998). Podophyllotoxin, aryl tetralin lignan, isolated from *Podophyllum* plant species (Stähelin and von Wartburg, 1991) has been shown to have antiviral and insecticidal properties (Sudo et al., 1998; Gao et al., 2004). This compound is produced by both the endophyte (*Phialocephala fortinii*) and its host plant as well (*Podophyllum* sp.).

An aliphatic compound is a hydrocarbon containing carbon and hydrogen joined together in straight chains, branched chains, or non-aromatic rings. Betina (1992) reported that aliphatic compound Brefeldin A has antibacterial, antiviral, anti-nematode, and antifungal activities. Initially, this compound was isolated from the endophytic fungus *Eupenicellium brefeldianum* (Harri et al., 1963). Later, there were some other reports stating that endophytes of different hosts do produce Brefeldin A, including *Paecilomyces* sp. inhabitant of *Taxus mairei* and *Aspergillus clavatus* inhabitant of *Torreya grandis* (Wang et al., 2007). Some other aliphatic compounds such as Pestalofones C and E (obtained from endophytic fungus *Pestalotiopsis fici*) and Gamahonolide A and B (obtained from endophytic fungus *Epichloë typhina*) also have antifungal activity (Mousa and Raizada, 2013).

Antimicrobial peptides are produced as secondary metabolites by endophytic fungi. One such peptide is Leucinostatin A, isolated from cultures of *Acremonium* sp. which colonizes *Taxus baccata*. This compound acts as a fungicide against *Pythium ultimum* (Strobel et al., 1997), the important plant pathogen that causes damping off and root rot diseases. Noble et al. (1991) reported that a lipopeptide purified from cultures of the endophytic fungi *Crytosporiopsis* sp. and *Pezicula* sp. inhibits pathogenic fungi. Yet another lipopeptide, Cryptocandin, was isolated from the fungus *C. quercina*, an endophyte of *Tripterigeum wilfordii.* Cryptocandin inhibited the growth of phytopathogenic fungi including *Sclerotinia sclerotiorum* and *Botrytis cinerea* (Strobel et al., 1999).

Finally, taking together the benefits offered by microbiota, it can be inferred that the plants indeed require the association of useful microbes, which not only help the plants to grow and put on biomass, but also help the plants to mitigate the stress effects. The useful microbiota plays an important role in various biological processes and affects a wide range of traits associated with growth and development of plants including crop yield as well as biotic and abiotic stress tolerance of plants. In fact, a large number of studies have illustrated the importance of useful microbes in crop growth and stress tolerance of crop plants. Therefore, it is highly relevant to have symbiotic association for the benefit of crop plants in general, and food legumes in particular. As the food legumes are one of the major sources of proteins that are generally grown under adverse climatic conditions with reduced soil moisture coupled with high temperature and low soil nutrient status, there is definitely a need for the intervention of microbes to mitigate the stress effects.

6.9 Summary

From time immemorial, plants and microbes have been living together, symbiotically offering a helping hand to each other and in the process, both of them have overcome all odds and survive under extreme adverse conditions. While the plants provide the microbes with necessary carbon and sometimes the shelter, the microbes, especially the soilborne endophytes, offer a number of benefits to the plants—starting from making available nutrient elements to imparting a greater level of stress tolerance to both biotic and abiotic agents besides providing several secondary metabolites including plant defense compounds. The food legumes, like any other plants/crops, also derive lot of benefits from the microbes such as rhizobia, AM fungi, and other endophytes. Being legumes, most often, the food leguminous crops are neglected by farmers and grown generally in less fertile lands under rain-fed conditions where biotic and abiotic stresses are bound to happen. However, by establishing the symbiotic association, the leguminous crops survive and continue to grow even under adverse growing conditions and ensure that their progeny seeds are protected.

REFERENCES

Abadi, V.A.J.M. and Sepehri, M., 2016, Effect of *Piriformospora indica* and *Azotobacter chroococcum* on mitigation of zinc deficiency stress in wheat (*Triticum aestivum* L.). *Symbiosis*, **69**(1):9–19.

Afzal, I., Noor, M.A., Bakhtavar, M.A., Ahmad, A. and Haq, Z., 2015, Improvement of spring maize performance through physical and physiological seed enhancements. *Seed Sci. Technol.*, **43**(2):238–249.

Aliasgharzad, N., Neyshabouri, M.R. and Salimi, G., 2006, Effects of arbuscular mycorrhizal fungi and *Bradyrhizobium japonicum* on drought stress of soybean. *Biol.*, **61**(19):324–328.

Allen, J.W. and Shachar-Hill, Y., 2009, Sulfur transfer through an arbuscular mycorrhiza. *Plant Physiol.*, **149**(1):549–560.

Allen, M.F., Smith, W.K., Moore, T.S.J. and Christensen, M., 1980, Comparative water relations and photosynthesis of mycorrhizal and non-mycorrhizal bouteloua gracilis hbk lag ex steud. *New Phytol.*, **88**(4):683–693.

Antunes, P.M., Rajcan, I. and Goss, M.J., 2006, Specific flavonoids as interconnecting signals in the tripartite symbiosis formed by arbuscular mycorrhizal fungi, *Bradyrhizobium japonicum* (Kirchner) Jordan and soybean (*Glycine max* (L.) Merr.). *Soil Biol. Biochem.*, **38**(3):533–543.

Asai, T., 1944, Uber die Mykorrhizenbildung der Leguminosen Vi\Anzen. *Japanese J. Botauv*, **13**:463.

Ashwin, R., Bagyaraj, D.J. and Mohan Raju, B., 2018a, Evaluation of different arbuscular mycorrhizal fungi for selecting the best for inoculating soybean cultivars MAUS 2 and MAUS 212. *Pertanika J. Trop. Agric. Sci.*, **41**(4):1587–1598.

Ashwin, R., Bagyaraj, D.J. and Mohan Raju, B., 2018b, Symbiotic response of drought tolerant soybean varieties, DSR 2 and DSR12 to different arbuscular mycorrhizal fungi. *Proc. Natl. Acad. Sci.* doi:10.1007/s40011-018-0978-9.

Asimi, S., Gianinazzi-Pearson, V. and Gianinazzi, S., 1980, Influence of increasing soil phosphorus levels on interactions between vesicular–arbuscular mycorrhizae and *Rhizobium* in soybeans. *Can. J. Bot.*, **58**(20):2200–2205.

Azcón, R., Ambrosano, E. and Charest, C., 2003, Nutrient acquisition in mycorrhizal lettuce plants under different phosphorus and nitrogen concentration. *Plant Sci.*, **165**:1137–1145.

Bacon, C.W. and Hill, N.S., 1996, Symptomless grass endophytes: Products of co-evolutionary symbioses and their role in the ecological adaptations of grasses, In *Endophytic Fungi in Grasses and Woody Plants* (Eds.), S.C. Redkin and L.M. Carris, St. Paul, APS Press, pp. 155–178.

Bagyaraj, D.J., 2018, Arbuscular mycorrhizal fungi and biological control of soil-borne plant pathogens. *Kavaka*. **51**:1–6.

Bagyaraj, D.J., Manjunath, A. and Patil, R.B., 1979, Interaction between VA mycorrhiza and *Rhizobium* and their effects on soybean in the field. *New Phytol.*, **82**:141–146.

Baltruschat, H., Fodor, J., Harrach, B.D., Niemczyk, E., Barna, B., Gullner, G., Janeczko, A., Kogel, K.H., Schafer, P., Schwarczinger, I., Zuccaro, A. and Skoczowski, A., 2008, Salt tolerance of barley induced by the root endophyte *Piriformospora indica* is associated with a strong increase in antioxidants. *New Phytol.*, **180**:501–510.

Balzergue, C., Chabaud, M., Barker, D.G., Bécard, G. and Rochange, S.F., 2013, High phosphate reduces host ability to develop arbuscular mycorrhizal symbiosis without affecting root calcium spiking responses to the fungus. *Front. Plant Sci.*, **4**:426.

Barea, J.M., Escudero, J.L. and Azcon, C., 1980, Effects of introduced and indigenous VA mycorrhizal fungi on nodulation, growth and nutrition of *Medicago sativa*. *Plant Soil*, **54**:283–296.

Bayat, F., Mirlohi, A. and Khodambashi, M., 2008, Effects of endophytic fungi on some drought tolerance mechanisms of tall fescue in a hydroponics culture. *Russ. J. Plant Physiol.*, **56**(4):510–516.

Berruti, A., Lumini, E., Balestrini, R. and Bianciotto, V., 2016, Arbuscular mycorrhizal fungi as natural biofertilizers: Let's benefit from past successes. *Front. Microbiol.*, **6**:1–13.

Bethlenfalvay, G.J., Brown, M.S., Mihara, K.L. and Stafford, A.E., 1987, Glycine-Glomus-Rhizobium symbiosis: V. Effects of mycorrhiza on nodule activity and transpiration in soybeans under drought stress. *Plant Physio.*, **85**(1):115–119.

Betina, Y., 1992, Biological effects of the antibiotic Brefeldin-A (Decumbin, Cyanein, Ascotoxin, Synergisidin): A retrospective. *Folia. Microbiol.*, **37**:3–11.

Bagyaraj, D.J. and Sreeramulu, K.R., 1982, Pre-inoculation with VA mycorrhiza improve growth and yield of chilli transplanted in the field and save phosphatic fertilizer. *Plant Soil*, **69**:375–381.

Bastias, D.A., Martínez-Ghersa, M.A., Ballaré, C.L. and Gundel, P.E., 2017, Epichloë fungal endophytes and plant defenses: Not just alkaloids. *Trends Plant Sci.*, **22**(11):939–948.

Blankenship, J.D., Spiering, M.J., Wilkinson, H.H., Fannin, F.F., Bush, L.P. and Schardl, C.L., 2001, Production of loline alkaloids by the grass endophyte, *Neotyphodium uncinatum in defined media. Phytochem.*, **58**(3):395–401.

Blee, K.A. and Anderson, A.J., 2000, Defense responses in plants to arbuscular mycorrhizal fungi. In *Current Advances in Mycorrhizae Research* (Eds.), G.K. Podila and D.D. Douds, Jr. APS Press, St. Paul, MI, pp. 27–44.

Bolan, N.S., 1991, A critical review on the role of mycorrhizal fungi in the uptake of phosphorus by plants. *Plant Soil*, **134**(2):189–207.

Bonneau, L., Huguet, S., Wipf, D., Pauly, N. and Truong, H.N., 2013, Combined phosphate and nitrogen limitation generates a nutrient stress transcriptome favorable for arbuscular mycorrhizal symbiosis in *Medicago truncatula. New Phytol.*, **199**(1):188–202.

Breuillin, F., Schramm, J., Hajirezaei, M., Ahkami, A., Favre, P., Druege, U., Hause, B., Bucher, M., Kretzschmar, T., Bossolini, E. and Kuhlemeier, C., 2010, Phosphate systemically inhibits development of arbuscular mycorrhiza in *Petunia hybrida* and represses genes involved in mycorrhizal functioning. *Plant J.*, **64**(6):1002–1017.

Calhoun, L.A., Findlay, J.A., Miller, J.D. and Whitney, N.J., 1992, Metabolites toxic to spruce budworm from balsam fir needle endophytes. *Mycol. Res.*, **96**(4):281–286.

Carling, D.E., Riehle, W.G., Brown, M.F. and Johnson, D.R., 1978, Effects of a vesicular-arbuscular mycorrhizal fungus on nitrate reductase and nitrogenase activities in nodulating and non-nodulating soybeans. *Phytopathology*, **68**(1590):96.

Chandanie, W.A., Kubota, M. and Hyakumachi, M., 2006, Interactions between plant growth promoting fungi and arbuscular mycorrhizal fungus *Glomus mosseae* and induction of systemic resistance to anthracnose disease in cucumber. *Plant Soil*, **286**(1–2):209–217.

Chen, L.Z., Chen, J.M., Zheng, X.S., Zhang, J.F. and Yu, X.P., 2007, Identification and antifungal activity of the metabolite of endophytic fungi isolated from *Llex cornuta. J. Pest Sci.*, **9**(2):143–8.

Choi, O., Deng, K.K., Kim, N.J., ROSS Jr, L., Surampalli, R.Y. and Hu, Z., 2008, The inhibitory effects of silver nanoparticles, silver ions, and silver chloride colloids on microbial growth. *Water Res.*, **42**(12): 3066–3074.

Clay, K., 1988, Fungal endophytes of grasses: A defensive mutualism between plants and fungi. *Ecol.*, **69**(1):10–16.

Clay, K. and Schardl, C., 2002, Evolutionary origins and ecological consequences of endophyte symbiosis with grasses. *Am. Nat.*, **160**(S4):99–127.

Crush, J.R., 1974, Plant growth responses to vesicular-arbuscular mycorrhiza VII. Growth and modulation of some herbage legumes. *New Phytol.*, **73**(4):743–749.

Daft, M.J. and El-Giahmi, A.A., 1975, Effect of Glomus infection on three legumes. In: *Endomycorrhizas*, (Eds.) Sanders, F.E., Mosse, B. and Tinker, P.B. pp. 581–92.

Daft, M.J. and El-Giahmi, A.A., 1976, Studies on nodulated and mycorrhizal peanuts. *Ann. App. Bio.*, **83**(2):273–276.

Daft, M.J. and Elgiahmi, A., 1974, Effect of endogone mycorrhiza on plant growth VII. Influence of infection on the growth and nodulation in french bean (*Phaseolus vulgaris*). *New Phyto.*, **73**(6):1139–1147.

Dai, J., Krohn, K., Flörke, U., Draeger, S., Schulz, B., KISS-Szikszai, A., Antus, S., Kurtán, T. and Van Ree, T., 2006, Metabolites from the endophytic fungus *Nodulisporium* sp. from *Juniperus cedre. Eur. J. Org. Chem.*, **15**:3498–3506.

Domínguez-Nuñez, J.A., Medina, M., Berrocal-Lobo, M., Anriquez, A. and Albanesi, A., 2015, The combined effects of CECT 844 and the black truffle co-inoculation on *Pinus nigra* seedlings. *IForest*, **8**(5):624.

Duchesne, L.C., 1994, *Role of Ectomycorrhizal Fungi in Biocontrol*. APS Press, St. Paul, MI, pp. 27–46.

Feng, G., Zhang, F., Li, X., Tian, C., Tang, C. and Rengel, Z., 2002, Improved tolerance of maize plants to salt stress by arbuscular mycorrhiza is related to higher accumulation of soluble sugars in roots. *Mycorrhiza*, **12**(4):185–190.

Firmin, S., Labidi, S., Fontaine, J., Laruelle, F., Tisserant, B., Nsanganwimana, F., Pourrut, B., Dalpé, Y., Grandmougin, A., Douay, F. and Shirali, P., 2015, Arbuscular mycorrhizal fungal inoculation protects Miscanthus × giganteus against trace element toxicity in a highly metal-contaminated site. *Sci. Total Environ.*, **527**:91–99.

Freeman, S. and Rodriguez, R.J., 1993, Genetic conversion of a fungal plant pathogen to a non-pathogenic, endophytic mutualist, *Science*, **260**:75–78.

Gao, R., Gao, C., Tian, X., Yu, X., Di, X., Xiao, H. and Zhang, X., 2004, Insecticidal activity of deoxypodophyllotoxin, isolated from *Juniperus sabina* L, and related lignans against larvae of *Pieris rapae* L. *Pest Manag. Sci.*, **60**:1131–1136.

Guerre, P., 2015, Ergot alkaloids produced by endophytic fungi of the genus Epichloë. *Toxins*, **7**(3):773–790.

Guo, Z., She, Z., Shao, C., Wen, L., Liu, F., Zheng, Z. and Lin, Y., 2007, 1 H and 13 C NMR signal assignments of Paecilin A and B, two new chromone derivatives from mangrove endophytic fungus *Paecilomyces* sp. *Magn. Reson. Chem.*, **45**(9):777–780.

Gutjahr, C., 2014, Phytohormone signaling in arbuscular mycorrhiza development. *Curr. Opin. Plant Biol.*, **20**:26–34.

Harri, E., Loemer, W., Singh, H.P., Stahlin, H., Tamm, C., 1963, Die Constitution von Brefeldin A. *Helv. Chem. Acta.*, **46**:1235–1243.

He, X., Han, G., Lin, Y., Tian, X., Xiang, C., Tian, Q., Wang, F. and He, Z., 2012, Diversity and decomposition potential of endophytes in leaves of a *Cinnamomum camphora* plantation in China. *Ecol. Res.*, **27**(2):273–284.

Huang, W.Y., Cai, Y.Z., Xing, J., Corke, H. and Sun, M., 2007, A potential antioxidant resource: Endophytic fungi from medicinal plants. *Eco. Bot.*, **61**(1):14.

Hubbard, M., Germida, J.J. and Vujanovic, V., 2014, Fungal endophytes enhance wheat heat and drought tolerance in terms of grain yield and second generation seed viability. *J. App. Microbiol.*, **116**(1):109–122.

Hussain, H., Akhtar, N., Draeger, S., Schulz, B., Pescitelli, G., Salvadori, P., Antus, S., Kurtán, T. and Krohn, K., 2009, New bioactive 2, 3-epoxycyclohexenes and isocoumarins from the endophytic fungus *Phomopsis* sp. from *Laurus azorica*. *European J. Org. Chem.*, **5**:749–756.

Islam, R., Ayanaba, A. and Sanders, F.E., 1980, Response of cowpea (*Vigna unguiculata*) to inoculation with VA-mycorrhizal fungi and to rock phosphate fertilization in some unsterilized Nigerian soils. *Plant Soil*, **54**(1):107–117.

Javaid, A., 2010, Role of arbuscular mycorrhizal fungi in nitrogen fixation in legumes. In: *Microbes for Legume Improvement*, Vienna, Springer, pp. 409–426.

Jha, P.N., Gupta, G., Jha, P. and Mehrotra, R., 2013, Association of rhizospheric/endophytic bacteria with plants: A potential gateway to sustainable agriculture. *Greener J. Agric. Sci.*, **3**(2):73–84.

Jia, M., Chen, L., Xin, H.L., Zheng, C.J., Rahman, K., Han, T. and Qin, L.P., 2016, A friendly relationship between endophytic fungi and medicinal plants: A systematic review. *Front. Microbiol.*, **7**:906.

Joner, E.J. and Leyval, C., 2003, Rhizosphere gradients of polycyclic aromatic hydrocarbon (PAH) dissipation in two industrial soils and the impact of arbuscular mycorrhiza. *Environ. Sci. Technol.*, **37**(11):2371–2375.

Joner, E.J., Johansen, A., Loibner, A.P., Dela Cruz, M.A., Szolar, O.H., Portal, J.M. and Leyval, C., 2001, Rhizosphere effects on microbial community structure and dissipation and toxicity of polycyclic aromatic hydrocarbons (PAHs) in spiked soil. *Environ. Sci. Tech.*, **35**(13):2773–2777.

Ju, Y., Sacalis, J.N. and Still, C.C., 1998, Bioactive flavonoids from endophyte-infected blue grass (*Poa ampla*). *J. Agric. Food Chem.*, **46**(9):3785–3788.

Jyothi, E. and Bagyaraj, D.J., 2018, Microbial consortia developed for *Ocimum tenuiflorum* reduces application of chemical fertilizers by 50% under field conditions. *Med. Plants—Int. J Phytomed. Related Ind.* **10**:138–144.

Kawai, Y. and Yamamoto, Y., 1986, Increase in the formation and nitrogen fixation of soybean nodules by vesicular-arbuscular mycorrhiza. *Plant Cell Physiol.*, **27**(3):399–405.

Khabou, W., Hajji, B., Zouari, M., Rigane, H. and Abdallah, F.B., 2014, Arbuscular mycorrhizal fungi improve growth and mineral uptake of olive tree under gypsum substrate. *Ecol. Eng.*, **73**:290–296.

Khan, M.S., Zaidi, A., Wani, P.A., Ahemad, M. and Oves, M., 2009, Functional diversity among plant growth-promoting rhizobacteria: Current status. In: *Microbial Strategies for Crop Improvement*, (Eds.) Mohammad S. Khan, Almas Zaidi, and Javed Musarrat, Berlin, Heidelberg, Springer, pp. 105–132.

Korade, D.L. and Fulekar, M.H., 2009, Rhizosphere remediation of chlorpyrifos in mycorrhizospheric soil using ryegrass. *J. Hazard. Mater.*, **172**(2–3):1344–1350.

Kucey, R.M.N. and Paul, E.A., 1982, Carbon flow, photosynthesis, and N2 fixation in mycorrhizal and nodulated faba beans (*Vicia faba* L.). *Soil Boil. Biochem.*, **14**(4):407–412.

Kusari, S. and Spiteller, M., 2011, Are we ready for industrial production of bioactive plant secondary metabolites utilizing endophytes? *Nat. Prod. Rep.*, **28**(7):1203–1207.

Kusari, S., Pandey, S.P. and Spiteller, M., 2013, Untapped mutualistic paradigms linking host plant and endophytic fungal production of similar bioactive secondary metabolites. *Phytochem.*, **91**:81–87.

Latch, G.C., 1993, Physiological interactions of endophytic fungi and their hosts. Biotic stress tolerance imparted to grasses by endophytes. *Agric. Ecosyst. Environ.*, **44**(1–4):143–156.

Lewis, D.H., 1985, Symbiosis and mutualism: Crisp concepts and soggy semantics. In: *The Biology of Mutualism*, (Ed.) D.H. Boucher, London, UK, Croom Helm Ltd, pp. 29–39.

Li, E., Jiang, L., Guo, L., Zhang, H. and Che, Y., 2008, Pestalachlorides A–C, antifungal metabolites from the plant endophytic fungus *Pestalotiopsis adusta*. *Bioorg. Med. Chem.*, **16**(17):7894–7899.

Lin, T., Lin, X., Lu, C., Hu, Z., Huang, W., Huang, Y. and Shen, Y., 2009, Secondary metabolites of *Phomopsis* sp. XZ-26, an endophytic fungus from *Camptotheca acuminate*. *European J. Org. Chem.*, **18**:2975–2982.

Liu, H., Yuan, M., Tan, S., Yang, X., Lan, Z., Jiang, Q., Ye, Z. and Jing, Y., 2015, Enhancement of arbuscular mycorrhizal fungus (*Glomus versiforme*) on the growth and Cd uptake by Cd-hyperaccumulator *Solanum nigrum*. *App. Soil Ecol.*, **89**:44–49.

Lösgen, S., Magull, J., Schulz, B., Draeger, S. and Zeeck, A., 2008. Isofusidienols: Novel chromone-3–oxepines produced by the endophytic fungus *Chalara* sp. *European J. Org. Chem.*, **2008**(4):698–703.

Lu, H., Zou, W.X., Meng, J.C., Hu, J. and Tan, R.X., 2000, New bioactive metabolites produced by *Colletotrichum* sp., an endophytic fungus in *Artemisia annua*. *Plant Sci.*, **151**(1):67–73.

Lynd, J.Q., Tyrl, R.J. and Purcino, A.A.C., 1985, Mycorrhiza-soil fertility effects on regrowth, nodulation and nitrogenase activity of siratro (*Macroptilium atropurpureum* (DC) Urb.). *J. Plant Nutr.*, **8**(11):1047–1059.

Malinowski, D.P. and Belesky, D.P., 2000, Adaptations of endophyte-infected cool-season grasses to environmental stresses: Mechanisms of drought and mineral stress tolerance. *Crop Sci.* **40**:923–940.

Manjunath, A., Bagyaraj, D.J. and GOPALA Gowda, H.S., 1984, Dual inoculation with VA mycorrhiza and *Rhizobium* is beneficial to *Leucaena*. *Plant Soil*, **78**:445–448.

Marks, S. and Clay, K., 1990, Effects of CO_2 enrichment, nutrient addition, and fungal endophyte infection on the growth of two grasses, *Oecologia*, **84**:207–214.

Mathimaran, N., Mahaveer, P.S., Mohan Raju, B. and Bagyaraj, D.J., 2017, Arbuscular mycorrhizal symbiosis and drought tolerance in crop plants. *Mycosphere*, **8**(3):361–376.

Meng, L., Zhang, A., Wang, F., Han, X., Wang, D. and Li, S., 2015, Arbuscular mycorrhizal fungi and rhizobium facilitate nitrogen uptake and transfer in soybean/maize intercropping system. *Front. Plant Sci.*, **6**:339.

Miransari, M., Bahrami, H.A., Rejali, F. and Malakouti, M.J., 2008, Using arbuscular mycorrhiza to alleviate the stress of soil compaction on wheat (*Triticum aestivum* L.) growth. *Soil Biol. Biochem.*, **40**(5): 1197–1206.

Mousa, W.K. and Raizada, M.N., 2013, The diversity of anti-microbial secondary metabolites produced by fungal endophytes: An interdisciplinary perspective. *Front. Microbiol.*, **4**:65.

Munns, D.N. and Mosse, B., 1980, Mineral nutrition of legume crops. In: *Advances in Legume Science*, (Ed.) R. J. Summerfield and A.H. Bunting, *Proceedings of the International Legumes Conference*, Kew. pp. 115–125.

Noble, H.M., Langley, D., Sidebottom, P.J., Lane, S.J. and Fisher, P.J., 1991, An echinocandin from an endophytic *Cryptosporiopsis* sp. and *Pezicula* sp. in *Pinus sylvestris* and *Fagus sylvatica*. *Mycol. Res.*, **95**(12):1439–1440.

Nouri, E., Breuillin-Sessoms, F., Feller, U. and Reinhardt, D., 2014, Phosphorus and nitrogen regulate arbuscular mycorrhizal symbiosis in *Petunia hybrida*. *PLoS One*, **9**(3):908–941.

Nouri, H., Borujeni, S.C., Nirola, R., Hassanli, A., Beecham, S., Alaghmand, S., Saint, C. and Mulcahy, D., 2017, Application of green remediation on soil salinity treatment: A review on halophytoremediation. *Process Saf. Environ.*, **107**:94–107.

Ortas, I. and Akpinar, C., 2006, Response of kidney bean to arbuscular mycorrhizal inoculation and mycorrhizal dependency in P and Zn deficient soils. *Plant Soil Sci.*, **56**:101–109.

Palmroth, M.R., Pichtel, J. and Puhakka, J.A., 2002, Phytoremediation of sub-arctic soil contaminated with diesel fuel. *Biores. Technol.*, **84**(3):221–228.

Pan, F., Su, T.J., Cai, S.M. and Wu, W., 2017, Fungal endophyte-derived *Fritillaria unibracteata* var. wabuensis: Diversity, antioxidant capacities in vitro and relations to phenolic, flavonoid or saponin compounds. *Sci. Rep.*, **7**:42008.

Panaccione, D.G., Beaulieu, W.T. and Cook, D., 2014, Bioactive alkaloids in vertically transmitted fungal endophytes. *Func. Ecol.*, **28**(2): 299–314.

Patil, S.M., Chandanshive, V.V., Rane, N.R., Khandare, R.V., Watharkar, A.D. and Govindwar, S.P., 2016, Bioreactor with *Ipomoea hederifolia* adventitious roots and its endophyte *Cladosporium cladosporioides* for textile dye degradation. *Environ. Res.*, **146**:340–349.

Puri, S.C., Verma, V., Amna, T., Qazi, G.N. and Spiteller, M., 2005, An endophytic fungus from *Nothapodytes foetida* that produces camptothecin. *J. Nat. Prod.*, **68**(12):1717–1719.

Rabie, G.H., 2004, Using wheat-mungbean plant system and arbuscular mycorrhiza to enhance in-situ bioremediation. *J. Food Agric. Environ.*, **2**:381–390.

Rabie, G.H., 2005, Influence of arbuscular mycorrhizal fungi and kinetin on the response of mungbean plants to irrigation with seawater. *Mycorrhiza*, **15**(3):225–230.

Rahman, M.H., Saiga, S., Sabreen, S., Kodama, Y. and Tsuiki, M., 2006, *Neotyphodium* endophyte infection affects the performance of tall fescue in temperate region Andisols. *Grassl. Sci.*, **52**(1):23–28.

Raj, J., Bagyaraj, D.J. and Manjunath, A., 1981, Influence of soil inoculation with vesicular-arbuscular mycorrhiza and a phosphate-dissolving bacterium on plant growth and ^{32}P-uptake. *Soil Biol. Biochem.*, **13**(2):105–108.

Rana, D.S., Dass, A., Rajanna, G.A. and Kaur, R., 2016, Biotic and abiotic stress management in pulses. *Indian J. Agron.*, **61**:238–248.

Rao, A.V. and Tak, R., 2002, Growth of different tree species and their nutrient uptake in limestone mine spoil as influenced by arbuscular mycorrhizal (AM)-fungi in Indian arid zone. *J. Arid Environ.*, **51**(1): 113–119.

Read, D.J., 1999, Mycorrhiza—The state of the art. In: *Mycorrhiza*, (Eds.) A. Varma and B. Hock, Berlin, Germany, Springer-Verlag, pp. 3–34.

Redman, R.S., Ranson, J. and Rodriguez, R.J., 1999, Conversion of the pathogenic fungus *Colletotrichum magna* to a nonpathogenic endophytic mutualist by gene disruption. *Mol. Plant Micr. Int.,* **12**:969–975.

Redman, R.S., Roossinck, M.J., Maher, S., Andrews, Q.C., Schneider, W.L. and Rodriguez, R.J., 2002, Field performance of cucurbit and tomato plants colonized with a nonpathogenic mutant of *Colletotrichum magna* (teleomorph: *Glomerella magna*; Jenkins and Winstead). *Symbiosis*, **32**:55–70.

Richardson, A.E., Barea, J.M., Mcneill, A.M. and PRIGENT-Combaret, C., 2009, Acquisition of phosphorus and nitrogen in the rhizosphere and plant growth promotion by microorganisms. *Plant Soil*, **321**(1–2):305–339.

Rivera-Becerril, F., Calantzis, C., Turnau, K., Caussanel, J.P., Belimov, A.A., Gianinazzi, S., Strasser, R.J. and Gianinazzi-Pearson, V., 2002, Cadmium accumulation and buffering of cadmium-induced stress by arbuscular mycorrhiza in three *Pisum sativum* L. genotypes. *J. Exp. Bot.*, **53**(371):1177–1185.

Rodriguez, R.J., Redman, R.S. and Henson, J.M., 2004, The role of fungal symbioses in the adaptation of plants to high stress environments. *Mitig. Adapt. Strat. GL.*, **9**(3):261–272.

Rodriguez, R.J., White Jr, J.F., Arnold, A.E. and Redman, R.S., 2008, Fungal endophytes: Diversity and functional roles. *New Phytol.*, **182**(2):314–330.

Saikkonen, K., Gundel, P. and Helander, M., 2013, Chemical ecology mediated by fungal endophytes in grasses. *J. Chem. Ecol.*, **39**:962–968.

Sati, S.C. and Joshi, S., 2014, Endophytic fungal associations of plants and antioxidant compounds. In: *Plants as a Source of Natural Antioxidants*, (Ed.) N.K. Dubey, Oxfordshire, UK, CABI, pp. 245–251.

Schardl, C.L., Florea, S., Pan, J., Nagabhyru, P., Bec, S. and Calie, P.J., 2013, The epichloae: Alkaloid diversity and roles in symbiosis with grasses. *Curr. Opin. Plant Biol.*, **16**(4):480–488.

Schardl, C.L., Young, C.A., Faulkner, J.R., Florea, S. and Pan, J., 2012, Chemotypic diversity of epichloae, fungal symbionts of grasses. *Fungal Ecol.*, **5**(3):331–344.

Schulz, B., Sucker, J., Aust, H.J., Krohn, K., Ludewig, K., Jones, P.G. and Döring, D., 1995, Biologically active secondary metabolites of endophytic *Pezicula* species. *Mycol Res.*, **99**(8):1007–1015.

Sharma, S. and Kapoor, R, 2019, Arbuscular mycorrhizal fungi in quenching the detrimental effects of heavy metals for sustainable agriculture of crop plants. In: *Microbes for Restoration of Degraded Ecosystems*, (Eds.) D.J. Bagyaraj and Jamaluddin, Boca Raton, FL, CRC Press, pp. 75–96.

Sieh, D., Watanabe, M., Devers, E.A., Brueckner, F., Hoefgen, R. and Krajinski, F., 2013, The arbuscular mycorrhizal symbiosis influences sulfur starvation responses of *Medicago truncatula*. *New Phytol.*, **197**(2):606–616.

Sikes, B.A., 2010, When do arbuscular mycorrhizal fungi protect plant roots from pathogens? *Plant Signal Behav.*, **5**(6):763–765.

Silva, G.H., DE Oliveira, C.M., Teles, H.L., Pauletti, P.M., CASTRO-Gamboa, I., Silva, D.H., Bolzani, V.S., Young, M.C., Costa-Neto, C.M., Pfenning, L.H. and Berlinck, R.G., 2010, Sesquiterpenes from *Xylaria* sp., an endophytic fungus associated with *Piper aduncum* (Piperaceae). *Phytochem. Lett.*, **3**(3):164–167.

Silva, G.H., Teles, H.L., Zanardi, L.M., Young, M.C.M., Eberlin, M.N., Hadad, R., Pfenning, L.H., Costa-Neto, C.M., Castro-Gamboa, I., Da Silva Bolzani, V. and Araújo, Â.R., 2006, Cadinane sesquiterpenoids of *Phomopsis cassiae*, an endophytic fungus associated with *Cassia spectabilis* (Leguminosae). *Phytochem.*, **67**(17):1964–1969.

Smith, H., Doyle, S. and Murphy, R., 2015. Filamentous fungi as a source of natural antioxidants. *Food Chem.*, **185**:389–397.

Smith, M.J., Flowers, T.H., Duncan, H.J. and Alder, J., 2006, Effects of polycyclic aromatic hydrocarbons on germination and subsequent growth of grasses and legumes in freshly contaminated soil and soil with aged PAHs residues. *Environ. Pollut.*, **141**:519–525.

Smith, S.E. and Daft, M.J., 1977, Interactions between growth, phosphate content and nitrogen fixation in mycorrhizal and non-mycorrhizal *Medicago sativa. Func. Plant Biol.*, **4**(3):403–413.

Stähelin, H.F. and Von Wartburg, A., 1991, The chemical and biological route from podophyllotoxin glucoside to etoposide: Ninth Cain memorial Award lecture. *Cancer Res.*, **51**(1):5–15.

Stierle, A., Strobel, G., Stierle, D., Grothaus, P. and Bignami, G., 1995, The search for a taxol-producing microorganism among the endophytic fungi of the Pacific yew, *Taxus brevifolia. J. Nat. Prod.*, **58**(9):1315–1324.

Strobel, G.A. and Daisy, B., 2003, Bioprospecting for microbial endophytes and their natural products. *Microbiol. Mol. Biol. Rev.*, **67**:491–502.

Strobel, G.A., Miller, R.V., Martinez-Miller, C., Condron, M.M., Teplow, D.B. and Hess, W.M., 1999, Cryptocandin, a potent antimycotic from the endophytic fungus *Cryptosporiopsis quercina. Microbiol.*, **145**(8);1919–1926.

Strobel, G.A., Torczynski, R. and Bollon, A., 1997, *Acremonium* sp.—a leucinostatin A producing endophyte of European yew (*Taxus baccata*). *Plant Sci.*, **128**(1):97–108.

Strzamska, J., 1975, Occurrence and intensity of mycorrhiza and deformation of roots without mycorrhiza in cultivated plants. In: *Endomycorrhizas,* (Eds.), F.E. Sanders, B. Masse, and P.B. Tinker, London, UK, Academic Press, pp. 537–543.

Sudo, K., Konno, K., Shigeta, S. and Yokota, T., 1998, Inhibitory effects of podophyllotoxin derivatives on herpes simplex virus replication. *Antivir. Chem. Chemother.*, **9**(3):263–267.

Verma, V.C., Kharwar, R.N. and Strobel, G.A., 2009, Chemical and functional diversity of natural products from plant associated endophytic fungi. *Nat. Prod. Commun.*, **4**:1511–1532.

Wang, F.W., Jiao, R.H., Cheng, A.B., Tan, S.H. and Song, Y.C., 2007, Antimicrobial potentials of endophytic fungi residing in *Quercus variabilis* and brefeldin A obtained from *Cladosporium* sp. *World J. Microbiol. Biotechnol.*, **23**(1): 79–83.

Waqas, M., Khan, A.L., Kamran, M., Hamayun, M., Kang, S.M., Kim, Y.H. and Lee, I.J., 2012, Endophytic fungi produce gibberellins and indoleacetic acid and promotes host-plant growth during stress. *Molecules*, **17**(9):10754–10773.

Wu, C., Li, B., Wei, Q., Pan, R. and Zhang, W., 2019, Endophytic fungus Serendipita indica increased nutrition absorption and biomass accumulation in *Cunninghamia lanceolata* seedlings under low phosphate. *Acta Ecologica Sinica*, **39**(1):21–29.

Wu, F.Y., Bi, Y. and Wong, M.H., 2009, Dual inoculation with an arbuscular mycorrhizal fungus and Rhizobium to facilitate the growth of alfalfa on coal mine substrates. *J. Plant Nutr.*, **32**(5):755–771.

Wu, N., Zhang, S., Huang, H., Shan, X., Christie, P. and Wang, Y., 2008b, DDT uptake by arbuscular mycorrhizal alfalfa and depletion in soil as influenced by soil application of a non-ionic surfactant. *Environ. Pollut.*, **151**(3):569–575.

Wu, Q.S., Zou, Y.N. and Huang, Y.M., 2012, The arbuscular mycorrhizal fungus *Diversispora spurca* ameliorates effects of waterlogging on growth, root system architecture and antioxidant enzyme activities of citrus seedlings. *Fungal Ecol.*, **6**(1):37–43.

Xu, P., Drewes, J.E., Kim, T.U., Bellona, C. and Amy, G., 2006, Effect of membrane fouling on transport of organic contaminants in NF/RO membrane applications. *J. Membrane Sci.*, **279**(1–2):165–175.

Yang, Y., Liang, Y., Han, X., Chiu, T.Y., Ghosh, A., Chen, H. and Tang, M., 2015, The roles of arbuscular mycorrhizal fungi (AMF) in phytoremediation and tree-herb interactions in Pb contaminated soil. *Sci. Rep.*, **6**:20469.

Yang, Y., Tang, M., Sulpice, R., Chen, H., Tian, S. and Ban, Y., 2014, Arbuscular mycorrhizal fungi alter fractal dimension characteristics of *Robinia pseudoacacia* L. seedlings through regulating plant growth, leaf water status, photosynthesis, and nutrient concentration under drought stress. *J. Plant Growth Regul.*, **33**(3):612–625.

Yoshihara, T., Togiya, S., Koshino, H., Sakamura, S., Shimanuki, T., Sato, T. and Tajimi, A., 1985, Three fungitoxic cyclopentanoid sesquiterpenes from stromata of *Epichloë typhina. Tetrahedron Lett.*, **26**(45):5551–5554.

Young, C.A., Charlton, N.D., Takach, J.E., Swoboda, G.A., Trammell, M.A., Huhman, D.V. and Hopkins, A.A., 2015, Characterization of *Epichloë coenophiala* within the US: Are all tall fescue endophytes created equal? *Front. Chem.*, **2**:95.

Yu, H., Zhang, L., Li, L., Zheng, C., Guo, L., Li, W., Sun, P. and Qin, L., 2010, Recent developments and future prospects of antimicrobial metabolites produced by endophytes. *Microbiol. Res.*, **165**(6):437–449.

Zandavalli, R.B., Dillenburg, L.R. and DE Souza, P.V.D., 2004, Growth responses of *Araucaria angustifolia* (Araucariaceae) to inoculation with the mycorrhizal fungus *Glomus clarum. App. Soil Ecol.*, **25**(3):245–255.

Zayed, M.S., Hassanein, M.K.K., Esa, N.H. and Abdallah, M.M.F., 2013, Productivity of pepper crop (*Capsicum annuum* L.) as affected by organic fertilizer, soil solarization, and endomycorrhizae. *Ann. Agricul. Sci.*, **58**(2):131–137.

Zhu, K.X., Lian, C.X., Guo, X.N., Peng, W. and Zhou, H.M., 2011, Antioxidant activities and total phenolic contents of various extracts from defatted wheat germ. *Food Chem.*, **126**(3):1122–1126.

Zhu, X., Song, F. and Liu, F., 2017, Arbuscular mycorrhizal fungi and tolerance of temperature stress in plants. In: *Arbuscular Mycorrhizas and Stress Tolerance of Plants*, (Ed.) Qiang-Sheng, W.U., Springer Malaysia Representative Office, Springer, Singapore, pp. 163–194.

Zhu, X., Song, F. and Xu, H., 2010, Influence of arbuscular mycorrhiza on lipid peroxidation and antioxidant enzyme activity of maize plants under temperature stress. *Mycorrhiza*, **20**:325–332.

Zou, W.X., Meng, J.C., Lu, H., Chen, G.X., Shi, G.X., Zhang, T.Y. and Tan, R.X., 2000, Metabolites of *Colletotrichum gloeosporioides*, an endophytic fungus in *Artemisia mongolica. J. Nat. Pro.*, **63**(11):1529–1530.

7

Multi-trait Trichoderma *for Improving Plant Health of Grain Legumes*

Sandeep Kumar, Sonika Pandey, Utkarsh Singh Rathore, and Krishna Kumar

CONTENTS

7.1 Introduction

Trichoderma spp. is a symbiotic saprotrophic fungus that is distributed worldwide (Woo et al., 2014; Samuels, 1996), commonly isolated from rhizosphere soil or decaying plant materials (wood bark). Geographical allocation and conditions may limit the distribution of species to strains clusters of this genus (Danielson and Davey, 1973). *Trichoderma*-based biofertilizers have been reported for higher nutrient (N, P, K) uptake, mitigation of soilborne pathogens, increased crop production, and ultimately enhancing fertility and soil health (Saba et al., 2012; Khan et al., 2014; Li et al., 2015; Haqu et al., 2012; Haque et al., 2010). Studies on this particular root-colonizing fungal genera have also demonstrated induced systemic resistance in plant (ISR), a plant defense system (Pieterse et al., 2014). The major mechanisms for promoting plant growth include improvement of nutrient availability (biofertilization), suppression of parasitic and nonparasitic pathogens (biocontrol), and production of plant hormones/and or plant growth-promoting substances (phytostimulation) (Martínez-Viveros et al., 2010; Bhattacharyya and Jha, 2012; Attitalla et al., 2012; Srivastava et al., 2006) *Trichoderma* sp. is ubiquitous in nature and have been reported from almost all parts of the world including Arabic countries (Youssef, 1974; Domsch et al., 1980; Halwagy et al., 1982; Abdel-Kader et al., 1983; Barooah and Borthakur, 1994).

7.2 Soil Types and Distribution of Dominant *Trichoderma* spp.

Soils of India available for agriculture are mainly covered by alluvial black, red, and laterite soils in which alluvial soils cover around 43% of total available agricultural land. Similarly, black, red, laterite, arid, and desert soils cover approx. 15%, 18.5%, 2.62%, and 6.13%, respectively. Whereas, forest and mountain soils contribute 8.67%; soils having high salt concentration and alkaline soils cover 1.68%; and peaty and marshy soils contribute 2.17% of total land area. Some properties of these soils are summarized in Table 7.1. Soil type in an area contributes the microbial composition of that area; this microbial community on a larger scale is crucial for global element and nutrient cycling. Mapping the soil microbial diversity with relation to the flora of soil can establish an understating of plant-microbe interactions. The major problem faced in diversity analysis of soils is that the majority of soil microorganisms cannot be characterized by conventional techniques used for microbial cultivation. Various characterized and well-studied microbes contribute only 1%–10% of total soil microbes, and the rest are yet to be studied by classical methods (Amann et al., 1995). However, newer techniques are being adopted for the purpose, enabling researchers to establish the facts regarding plant-microbe interactions. The fact is now established that one gram of given soil will consist of members belonging to each of the three domains of the biosphere, i.e., the *Bacteria*, the *Eucarya* and the *Archaea* (e.g., Liesack et al., 1997; Gattinger et al., 2002).

TABLE 7.1

Distribution of Dominant Microbes in Different Soil Types

Soils Types	Regions in India	Minerals (Abundant)	Minerals (Lacks)	Dominant Microbial Genera	References
Alluvial	Mainly found in the plains of Gujarat, Punjab, Haryana, UP, Bihar, Jharkhand, etc.	Potash and Lime	Nitrogen and Phosphorous	*Pseudomonas* spp. (49%) *Bacillus* spp. (23%) *Trichoderma* sp.	Gagelidze et al. (2018); Banik and Dey (1981)
Black (Regur soil)	Deccan plateau- Maharashtra, Madhya Pradesh, Gujarat, Andhra Pradesh, Tamil Nadu, Valleys of Krishna and Godavari	Lime, Iron, Magnesia, and Alumina, Potash	Phosphorous, Nitrogen, and Organic matter	*Bacillus* spp. (40%) *Psuedomonas* spp. (24%) *Trichoderma* sp.	Gagelidze et al. (2018); Bhale and Rajkonda (2012)
Red	Eastern and southern part of the Deccan Plateau, Orissa, Chattisgarh, and southern parts of the middle Ganga plain	Iron and Potash	Nitrogen, Phosphorous and Humus	*Rhodococcus* spp. (58%), *Pseudomonas* spp. (22%), *Bacillus* spp. (20%) *Trichoderma* sp.	Gagelidze et al. (2018); Banik and Dey (1981); Bhale and Rajkonda (2012)
Arid and Desert	Western Rajastan, north Gujarat and southern Punjab	Soluble salts, Phosphate	Humus, Nitrogen	*Trichoderma* sp. (1%)	Gagelidze et al. (2018); Idress et al. (2012)
Saline and Alkaline	Western Gujarat, deltas of eastern coast, Sunderban areas of West Bengal, Punjab, and Haryana	Sodium, Potassium, Magnesium	Nitrogen and Calcium	*Bacillus* spp. (58%) *Rhodococcus* spp. (25%), *Pseudomonas* spp. (16%) *Trichoderma harzianum* (abundant)	Gagelidze et al. (2018); Gherbawy et al. (2004)
Peaty and Marshy	Coastal areas like Orissa, West Bengal, Tamil Nadu, and Kerala	Organic matter and Soluble salts	Phosphate and Potash	*Bacillus* spp. (67%) *Rhodococcus* spp. (22%), *Pseudomonas* spp. (11%) *Trichoderma* sp.	Gagelidze et al. (2018); Banik and Dey (1981); Saravanakumar et al. (2016); Wuczkowski et al. (2003)

7.3 Molecular Tools for Monitoring *Trichoderma* in Agricultural Environments

Genus *Trichoderma* came to light in 1974 with four species (Samules, 1996). In recent years, the number of *Trichoderma* species has increased dramatically. Fingerprinting techniques are DNA-based methods relying on PCR, restriction digestion, and southern hybridization banding pattern. Schlick et al. (1994) used fingerprinting techniques to differentiate *Trichoderma harzianum* and their gamma rays-induced mutants. Zimand et al. (1994) designed the first technique to distinguish a commercially used *T. harzianum* T-39 from other *Trichoderma* species. This technique was less time consuming and can be conducted using small amounts of DNA.

Arisan et al. (1995) used wild-type M13 phage DNA for RAPD. They examined 11 strains of *T. viride*, 2 strains of its telomorph *Hypocrea rufa* and 9 of *Trichoderma* species (T. citrinoviride, *T. parceramosum, T. harzianum, T. koningii, T. longibrachiatum, T. piluliferum, T. polysporum, T. reesei, T. saturnisporum*) all these were screened against *Cryphonectria parasitica*. Restriction fragment length polymorphism using HIND III enzyme was designed by Bowen et al., (1996) This method was able to detect the BCA in the commonly isolated mycota. Dodd et al. (2004) combined RFPL marker with dot blot assay and monitored the spread and survival of *T. atroviride*. 65 on kiwi fruit leaves. Bulat et al. (1998) evaluated ITS sequence ability to analyze closely related *Trichoderma* and *Glicodium* strains. Cumagun et al. (2000) analyzed 42 isolates *T. harzzianum* and *T. viridae* using ITS1 (Sadfi-Zouaoui et al., 2009).

Freeman et al. (2002) used arbitrarily primed PCR in an experiment to use *Trichoderma* T39, T161, and T166 for the management of *B. cinerea* and *Colletotrichum acutatumin* strawberry, and to investigate the population survival of these three. Fanti et al. (2003) evaluated finger printing with mini and microsatellite primers for the environmental monitoring of two promising biocontrol *Trichoderma* strains effective against *Cytospora* canker.

The RAPD-PCR method combined with benomyl resistance improved the identification of biocontrol strains from field samples. Tachapattaworakul et al. (2008) used this technique for monitoring of *T. atroviride* strains CBS 470.94, CBS 350.93, L2, and OB. The RAPD procedure with primer OPC8 generated a fragment of 3,000 bp specific to the biocontrol strain L2. This strain showed a tenfold level of resistance to benomyl when compared to the other strains examined, which made its isolation from soil easier. The applied method allowed the recovery and identification of strain L2 from the soil four months after its inoculation. Although the PCR-based fingerprinting tools represent simple and flexible methods, they are not always strain-specific and their reproducibility may be affected by the low annealing temperature and short primer size (Hermosa et al., 2001). However, strain-specific fragments generated by such techniques have the potential to be converted into monitoring markers based on sequence characterized amplified regions (SCARs) (Paran and Michelmore, 1993).

7.4 Species-Specific PCR

The use of species-specific primer is a potent tool for distinguishing individual *Trichoderma* species. Chen et al. (1999) developed a species-specific primer-based test to identify *T. harzianum* biotypes T2 and Th4. The method can be used efficiently for the detection of *T. aggressivum* in green mold-affected mushroom compost. Kredics et al. (2009) describes a multiplex PCR assay allowing the detection of *T. pleuroticola* and *T. pleuroti*. Meena et al. designed SCAR primers for the detection of *T. harzianum* and *T. hamatum*. In order to monitor and quantify, López-Mondéjar et al. (2010) designed a primer set and a TaqMan probe for the ITS region.

Geistlinger et al. (2015) introduced a tool based on the use of simple sequence repeats (SSRs) using Touchdown PCR to track and quantify the species *T. virens*. Primers were designed for 12 different loci. The results confirmed the root-endophytic lifestyle of this species in tomato plants: the fungal biomass in the plant tissues was quantified and co-colonization of the roots by different strains of *T. virens* could also be observed (Figure 7.1).

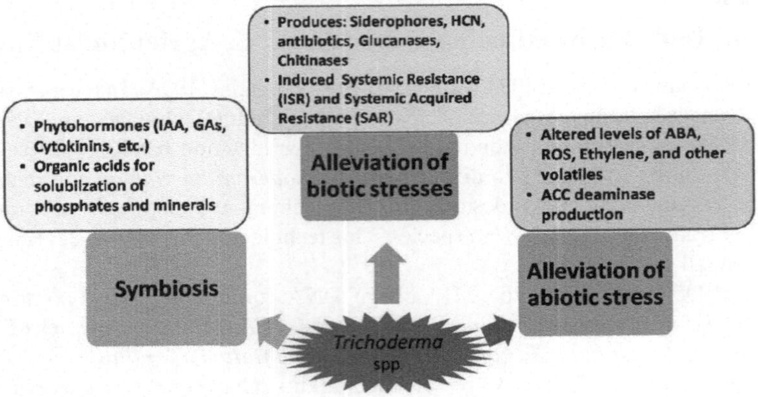

FIGURE 7.1 Potential of *Trichoderma* as a consistent plant growth stimulator.

7.5 Biotic Stress Management

A few endeavors have been made to oversee various diseases of pulses by using bioagents. Among the biocontrol organisms, *Trichoderma* species are the most considered bioagents in pulse biological system (Fig. 7.1). They are effectively used to control diseases caused by various pathogens, e.g., *Fusarium* spp., *Rhizoctonia solani*, *R. bataticola*, *Sclerotium rolfsii*, *Sclerotinia sclerotiorum*, *Ascochyta* spp., *Cercospora* spp., *Alternaria* spp. *Phytophthora* spp. furthermore, and *Pythium* spp., in various pulses and other field crops. *Trichoderma* spp. are effectively used to control various pulse diseases (Table 7.2).

Chickpea wilt is caused by *Fusarium oxysporum* f.sp. *ciceri*, the main constertain for restricting chickpea production worldwide. Numerous *Trichoderma* species, *Trichoderma viride*, *T. harzianum*, and *T. virens* confines from various areas were screened against races of *F. oxysporum* f.sp. *ciceri* (Babychan and Simon, 2017; Kumar et al., 2014). *T. viride* pursued by *T. harzianum*, which hindered the greatest mycelial development of the pathogen. The coordinated use of *T. harzianum* (10g/kg seed) and carboxin (2 g/kg seed) as a seed treatment diminished wilt frequency upto 44.1%–60.3% and expanded seed germination by 12.0%–14.0% and grain yield by 42.6%–72.9% (Dubey et al., 2007). *T. virens*-based seed dressing, *viz*., Pusa 5SD and soil application details, *viz*., Pusa Biopellet 16G (PBP 16G) and Pusa Biogranule 6 (PBG 6) were utilized for the board of wet root decay brought about by *R. solani* (Dubey et al., 2012; Dubey et al., 2009). Different illnesses such as damping off root as well as stem rot can be controlled by T. viride (detach TVM2) (Dubey and Patel B., 2012; Rehman et al., 2012). Patole et al., 2017 used *Trichoderma viride* and *Trichoderma harzianum* for the management of chickpea wilt.

Wilt of pigeon pea brought about by *Fusarium udum* is an important disease, causing generous yield decrease all through the nation. *Trichoderma* is transcendently utilized against pigeon pea wilt (Shukla and Chaudhary, 2007). Mandhare and Suryawanshi (2008) revealed that 63.25% wilt decrease was accomplished by seed and soil application of *Trichoderma* spp. Roy et al. (2005) utilized mutant strains of *T. harzianum viz*., 50Th3I1, 75Th4IV, and 125Th4l for the management of wilt illness. Shukla and Chaudhary (2007) examined the potential of 30 distinctive *Trichoderma* spp. against *F. udum* and discovered variety in the antagonistic property of these isolates' percentage. *Phytophthora* stem scourge (PSB), brought about by *Phytophthora drechsleri* f. sp. *cajani*, is another significant disease of pigeon pea (*Cajanus cajan*). A few detaches of *Trichoderma* species are assessed against *Phytophthora* of pigeon pea.

Fusarium oxysporum f.sp. *lentis* causes lentil wilt worldwide and is a major cause of yield loss. *T. hamatum* was reported to reduce up to 33% of lentil wilt damage (El-Hasan et al., 2013).

TABLE 7.2
Biocontrol Efficacy of *Trichoderma* in Pulses Against Soilborne Pathogens

Plant Name	Disease Causing Agent	*Trichoderm* Species Used	References
Vigna mungo (Black gram)	*Macrophomina phaseolina, Alternaria alternate*	*T. viride, T. harzianum*	Raguchander et al. (1997); Dubey et al. (2012); Mishra et al. (2011)
Cicer arietinum (Chickpea)	*Fusarium oxysporum, Rhictonia solani, Aspergillus niger, Chaetomium* spp., *Sclerotium rolfsii, Penicillum* spp., *Macropho phaseolina*	*T. harzianum, T. viride*	Mukherjee et al. (1997); Haware et al. (1999); Pandey et al. (2003); Poddar et al. (2004)
Vigna sinensis (Cowpea)	*Rhictonia solani*	*T. harzianum*	Pan and Das (2011)
Cajanus cajan (Pigeon pea)	*Fusarium udum, Phytophthora drechsleri* f.sp., *cajani*	*T. viride, T. harzianum, T. hamatum* and *T. koningi*	Hukma and Pandey (2011)
Lentil	*Fusarium oxysporum* f.sp. *lentis, Macrophomina phaseolina*	*T. harzianum, T. viride*	El-Hassan et al. (2013)
Lentil	*Sclerotium roflsi*	*Trichoderma viride, T. virens* and *T. harzianum*	Kushwaha et al. (2018)
Vigna mungo	*Fusarium oxysporum* and *Alternaria alternata*	*Trichoderma viride*	Surekha et al. (2014)
Peanut	*Aspergillus niger*	*Trichoderma viride*	Sayiprathap et al. (2017)
Cowpea	*Erysiphe flexuosa*	*Glomus versiforme* and *Trichoderma harzianum*	Omomowo et al. (2018)
Bean (*Phaseolus vulgaris* L.)	*Rhizoctonia solani*	*T. harzianum* T019	Mayo et al. (2015)
Bean	*Rhizoctonia solani*	*T. asperellum, T. harzianum* and *Trichoderma* spp.	Asad et al. (2014)
Faba bean	*Fusarium solani*	*Trichoderma harzianum*	Belay and Anteneh (2016)
Bean	*Fusarium oxysporum* f.sp. *phaseoli*	*T. reesei* and *T. koningii*	Otadoh et al. (2011)
Faba bean	*Black Root Rot (Fusarium solani*	*Trichoderma harzianum, T. koningii* and *T. viride*	Belete et al. (2015)
Chickpea (*Cicer arietinum* L.)	*Fusarium oxysporum* f.sp. *ciceri w*	*Trichoderma harzianum*	Nirmalkar et al. (2017)
Chickpea	*Fusarium oxysporum* f.sp. *ciceri and Rhizoctonia solani*	*T. harzianum, T. hamatum* and *T. viride.*	Khan et al. (2014)
Soybean (*Glycine max* L.)	*Myrothecium, Anthracnose and Rhizoctonia*	*Trichoderma viride*	Kuchlan et al. (2017)
Chickpea (*Cicer arietinum*)	*Fusarium oxysporum* f.sp. *ciceri*	*Trichoderma viride* and *Trichoderma harzianum*	Shabir-U-Rehman et al. (2013)
Chickpea (*Cicer arietinum*)	*Fusarium oxysporum* f.sp. *ciceri*	*Trichoderma viride* and *Trichoderma harzianum*	Patole et al. (2017)
Lentil (*Lens culinaris* Medikus subsp. *culinaris*) L.	*Fusarium oxysporum* f.sp. *lentis:Fol*	*Trichoderma viride, Trichoderma koningi* and *Trichoderma harzianum*	Tiwari et al. (2017)

Root rots are caused by various fungal pathogens *viz., Rhizoctonia bataticola, R. solani, S. rolfsii, M. phaseolina*, and *F. solani*. Soil application of *Trichoderma virens*-based formulations, Pusa Biogranule 6 and Pusa Biopellet16G (PBP 16G) and seed dressing formulation, Pusa 5SD (PBG6), were found to be superior in reducing wet rot disease incidence and increasing seed germination and shoot and root lengths in mung bean (Dubey et al., 2009).

Root rot disease complex is considered the most destructive in pea and bean, as it affects its initial plant stand and more than 20 different pathogens have been reported to be associated with the disease from different parts of the world (Idowu et al., 2016). Among these, *Fusarium oxysporum* f.sp. *pisi*, and *Pythium* spp. are highly destructive. Asad et al., 2014, demonstrated the use of *T. asperellum*, *T. harzianum*, and *Trichoderma* spp in beans against *Rhizoctonia solani*, *T. reesei* and *T. koningii* were used efficiently in beans against *Fusarium oxysporum* f.sp. *phaseoli*. In peanut, *Trichoderma viride* is employed for the management of disease caused by *Aspergillus niger*.

7.6 Mechanisms Involved in Biocontrol Activity

There are various mechanisms through which *Trichoderma* inhibit the growth of pathogens. These mechanisms include mycoparasitism, antibiosis, starvation, and induced resistance.

7.6.1 Mycoparasitism

This process involves the direct attack of one fungus on the other fungi. This is the main mechanism that is used by *Trichoderma* to control phytopathogens. The process of mycoparasitism is a cascade of following events—chemotrophic growth of *Trichoderma*, recognition of the host, coiling and appresoria formation, cell wall degrading enzyme secretion, hyphal penetration, and lysis of host cell. The process of recognition involves binding of carbohydrates in the *Trichoderma* cell wall to lectins on the target fungus; coiling of hyphae around the pathogen involves higher amounts of osmotic solutes such as glycerol; and attacking the hyphae of pathogen is done through the secretion of cell wall-degrading enzymes (Harman et al., 2004a). This synergistic mechanism involves the host cell lysis and death. There are 20–30 genes, proteins, and other metabolites that are directly involved in mycoparasitism.

7.6.2 Antibiosis

This process involves the production of low-molecular-weight diffusible secondary metabolites and antibiotics that are harmful to the pathogens. A variety of antibiotics such as gliovirin, gliotoxin, viridin, viridol, koninginins, pyrones, and peptaibols against fungal phytopathogens are secreted by *Trichoderma* (Howell et al., 2004a; Mukhtar, 2008; Harman et al., 2004b; Vinale et al., 2014). Trichokonins via peptabiol is secreted by *Trichoderma pseudokoningi* SMF, it exerts antibiotic action through extensive apoptotic programmed cell death in the pathogen (Shi et al., 2016). *Trichoderma* secondary metabolites gliotoxin and gliovirin, which belong to P and Q family, are the main secondary metabolites that play a role in phytopathoneic action. The P group is secreted by *T. virens* and is active against *P. ultimum* but not against *R. solani*, and the Q group is active against *R. soleni* but not against *P. ultimum* (Howell, 2003), "Koninginin D also inhibited the growth of soilborne plant pathogens, such as *R. solani*, *P. cinnamomi*, *Pythium middletonii*, *Fusarium oxysporum* and *Bipolaris sorokiniana*" (Dunlop et al., 1989). Viridins isolated from diverse *Trichoderma* spp. (*T. koningii, T. viride, T. virens*) prevent spore germination of *Botrytis alli*, *Colletotrichum lini*, *Fusarium caeruleum*, *Penicillium expansum*, *Aspergillus niger*, and *Stachybotrys atra* (Singh et al., 2005). Harzianic acid isolated from a *T. harzianum* strain showed in vitro antibiotic activity against *Pythium irregulare*, *Sclerotinia sclerotiorum*, and *R. solani* (Vinale et al., 2009). The isolation and overexpression of tri5 (trichodiene synthase) gene in *T. brevibacterium Tb4ltri5, a* transformant, has been found to increase the production of trichodermin and antifungal activity against *Aspergillus fumigates* and *Fusarium* spp. (Tijerino et al., 2011). *T. asperellum* strain produces two asperelines (A and E) and five trichotoxins (T5D2, T5E, T5F, T5G, and 1717A), which can be associated with antibiosis (Brotman et al., 2013).

7.6.3 Starvation

It is the most common reason for the death of a microorganism, and thus competition for nutrients results in the biological control of phytopathogens. Rhizosphere is a rich source of sugars, amino acids, iron, vitamins, organic acids, etc. Iron is essential for many filamentous fungi, and for the iron uptake from the environment, many fungi secrete low-molecular-weight ferric ion chelators termed as siderophore;

siderophore helps in the mobilization of environmental iron. *Trichoderma* species secrete siderophore, which binds to the insoluble iron and converts it to the soluble form for plant absorption and thus inhibits the growth of pathogens by Depriving them (Doni et al., 2014; Leong, 1986). Through this mechanism, *Trichoderma* decreased the growth of *Pythium*. *T. harzianum* T35 controls *Fusarium oxysporum* by competing for both rhizospheric colonization and nutrients. Competition has proved very efficient for the control of various phytopathogens such as *Botrytis, Fusarium, Phytium*, etc. *Trichoderma* is more efficient than any other fungi to mobilize and take soil nutrients. Promoter gene analysis related to antagonism revealed the presence of consensus sequences for transcription factors responsible for carbon (CreaA), nitrogen (AreA), Stress (Msn2/Msn4), pH (PacC), and mycoparasitism (MYC) regulation. Appropriate manipulation of these regulators will give an alternative way to isolate more competitive BCAs.

7.7 Induced Resistance

Induced resistance is the most indirect form of antagonism. It can be local or systemic. Salicylic acid (SA) and non-expressor of pathogenesis-related genes (NPR1) are key players in systemic acquired resistance. *T. harzianum* when inoculated on to roots or on to leaves of grapes provides control of diseases caused by *Botrytis cinerea* on leaves spatially separated from the site of application of the BCA (Deshmukh et al., 2006). Many classes of compounds are released by the *Trichoderma* sp. into the zone of interaction that induces resistance in plants. The first class is proteins with enzymatic or other activity. Fungal proteins such as xylanase, cellulases and swollenins are secreted by *Trichoderma* species' (Martínze et al., 2001). *Trichoderma* endochitinase can also enhance defense, probably through induction of plant defense-related proteins. Active rhizosphere colonizers *T. harzianum* and *T. virens* produce some cell wall-degrading enzymes, antibiotics such as gliotoxin and viridin (Tronsmo and Harman, 1992), and also certain biologically active heat-stable metabolites such as ethyl acetate (Claydown et al., 1987), which inhibit various pathogens present in the soil. In another study, *T. hamatum* was found to be an effective biocontrol agent for management of lentil wilt caused By *F. oxysporum* f.sp. *lentis* (Chowdappa et al., 2013). It inhibits growth of wilt pathogens by production of antifungal enzymes, competition for key nutrients and/or ecological niches, complex mechanisms of mycoparasitism, growth promotion, and a combination of these effects. During mycoparasitism, *Trichoderma* produces appressoria-like structures to penetrate the target fungus cell wall (Chet, 1987).

7.8 Biocontrol Genes from *Trichoderma*

Whole genome sequencing of *Trichoderma* species has provided many valuable data for the understanding of various genes. To date, whole genome sequence of seven *Trichoderma* species is available *viz. Trichoderma harzianum, Trichoderma asperellum, Trichoderma reesei, Trichoderma virens, Trichoderma atroviride, Trichoderma longibrachiatum*, and *Trichoderma citrinoviride* are available (Sharma et al., 2009). These studies have revealed that *Trichoderma* species contain many valuable genes that help in the control of growth of phytopathogens and plant growth promotional activities.

Cell wall-degradating genes-A gene named tvspl, responsible for encoding serine protease from *Trichoderma* virens, has been cloned. Serine protease helps in the degradation of the fungal cell wall. It has an important role in pathogenesis against *Rhizoctonia solani* (Pozo et al., 2004). tri5 gene (trichodiene synthase gene) isolated and characterized from *Trichoderma harzianum* is responsible for the secretion of enzyme trichothecene; this enzyme stops the DNA and protein synthesis in the cells of pathogens and hence stops their growth and ultimately kills them. This gene has shown phytotoxicity against *Fusarium* species. *Trichoderma asperellum* expresses the gene tag83, which is responsible for the secretion of exo-β-glucanase; this gene with *R. soleni* has shown inhibition action (Marcello et al., 2010). β-1,3 and β-1,6 are the glucanase enzyme, which are coded by coding TvBgn2 and TvBgn3 genes. These enzymes help in cell wall degradation of phytopathogens. From *Trichoderma atroviridae*, a gene named gluc 78, which encodes 1,3-β-glucosidase, was isolated, cloned, and sequenced; this gene is responsible for pathogen cell wall degradation (Donzelli et al., 2001). Crel, it is a glucose repressor gene was isolated form *Trichoderma harzianum*. This gene causes the repression of cellulase and xylanase-encoding genes; these two are the key cell wall-degrading enzymes. β-tubulin genes are isolated and characterized from *T. harzianum*; β-tubulin are structural components of most cell walls. A 3-D model of β-tubulin gene was done by a Swiss model automated comparative protein modeling server (Li and Yang, 2007). Sm1, a cysteine-rich protein, was isolated from *T. virens*; it is responsible for the disease suppression in dicots and monocots plants (Buensanteai et al., 2010). From *T. harzianum*, a gene named SL41 has been cloned and expressed; serine proteases play a key role in fungal biology and biocontrol activity. ThPG1 encodes endopolygalacturonase, isolated from *T. harzianum*, which is responsible for cell wall degradation of pathogens such as *R. soleni* and *P. ultimum*. G protein α subunit genes TgaA and TgaB were isolated from *T. virens* and cloned and characterized. These genes have antagonistic activity against *R. soleni* and *Sclerotium rolfsi*.

7.9 Genes Involved in Biotic Stress Management

From *T. harzianum*, a gene named thkel1 has been isolated and characterized. This gene codes putative kelch-repeat protein, which helps in regulating the glucosidase activity and enhances tolerance to salt and osmotic stresses in *Arabidopsis thaliana* plants (Hermosa et al., 2011). TvGST (glutathione transferase gene) has been cloned from *T. virens*; this acts as a cadmium-tolerant gene (Dixit et al., 2011). From *T. virens*, an adenalyte cyclase-encoding gene, tac1, had been isolated and cloned; this gene has its role in mycoparasitism against *R. soleni* and *P. ultimum* (Mukherjee et al., 2007). ThPTR2 gene obtained from *T. harzianum* has a significant role in mycoparasitism against *Botyrytis cinerea*. Taabc2 obtained from *Trichoderma atroviride* has a significant role in ATP binding cassette transporter in cell membrane pump that helps in mycoparasitic activity. ech42 and prb1 genes have been isolated and characterized from *Trichoderma harzianum*; these genes are very effective against *R. solani* and *S. rolfsii*. egl1, which is isolated from *Trichoderma longibrachiatum*, encodes β-1,4-endoglucanase. This gene shows good biocontrol activity (Migheli et al., 1998). Mitogen activated protein kinase-encoding gene TmkA isolated from *T. virens* has been found to cause mycoparasitic action against *R. soleni* and *S. rolfsi* (Mukherjee et al., 2003).

A transcription factor gene named Thctf1 was isolated from *T. harzianum*; this gene is involved in the function production of 6-pentyl-2h-pyran (6PP), which has antifungal activity against *R. solani*, *B. cinerea*, and *S. rolfsii*. Trip5, a trichodiene synthase gene encoded by *T. brevicomactum*, is responsible for the production of trichodermin, which has antifungal activity against *S. cerevisiae*, *Kluyveromyces marxianus*, *Candida albicans*, *C. glabrata*, *C. tropicalis*; Th-chit encoded by *Trichoderma harzianum* has antifungal activity against transgenic tobacco— this gene confers antifungal activity against *A. alternata* (Saiprasad et al., 2009). *T. harzianum* encodes ergl gene, which regulates the synthesis of enzyme squalene epoxidase, helps in the synthesis of ergosterol; silencing of this gene provides resistance to antifungal compound terbinafine. The monooxygenase gene, from *T. hamatum*, has been isolated; this gene has antifungal activity against *S. sclerotiorum*, *Sclerotinia minor*, and *Sclerotium cepivorum* (Carpenter et al., 2008). *T. ressei* encodes a gene named TrCCD1, which is involved in carotenoid metabolism and helps in conidiospore and hyphal growth (Zhong et al., 2009).

7.10 Abiotic Stress Management

Abiotic stresses exerted by crops include temperature extremities compared to optimal, salinity, drought, flooding, and heavy metals concentrations, reducing the yield (Mahajan and Tuteja, 2005). The type, variety, and abiotic stress period exerted by the crop may lead to total crop loss. Leavening behind a non-fertile or contaminated soil which is of no use for farmers. Abiotic stresses account for about 70% loss in total crop yield and cropping area.

7.11 *Trichoderma* sp. in Mitigation of Salt Stresses

Among all this, salinity is the most typical abiotic stress (Mahajan and Tuteja, 2005). Salt stress is an important growth-limiting factor for most non-halophyte plants. Plant growth is ultimately reduced due to saline conditions, although plant species differ in their tolerance toward salinity levels (Munns and Termaat, 1986). Saline soil affects many aspects of growth and metabolism in plants; as a result, yields are reduced. Excess salt in the soil may adversely affect plant growth through either osmotic inhibition of water uptake by roots or specific ion effects.

The area affected by salinity in the world covers about 400 million hectares, of which 54 million are found in South and Southeast Asia (Akbar and Ponnamperuma, 1982). The coastal areas of Andaman and Nicobar Islands affected by the *Tsunami* are related to salt accumulation in the soil. This concern reflects an important fear within the agricultural sector and rural communities, of facing a strong decrease in soil fertility and productivity due to high salinity resulting from sea intrusion. Due to the scarcity of fresh water (usually with an electric conductivity [EC] of 2 dS m^{-1} or less), especially in arid and semiarid areas, some agricultural crops are irrigated with saline water (EC up to 5 dS m^{-1}). Crops tolerant to salt—such as maize, cotton, and barley—can withstand EC levels about 2 to 4 dS m^{-1}, and seawater has an EC of 52 dS m^{-1}.

However, incorporation of *Trichoderma* during seed biopriming treatments in many cereal and vegetable crops has resulted in increased levels of plant growth hormones and improved seed performance (Howell, 2003). Biopriming is a process of biological seed treatment that refers to a combination of seed hydration and seed inoculation with beneficial organisms to protect seed. The technique helps seeds to evenly germinate even under adverse Soil conditions (Naznin et al., 2015; Singh et al., 2003). A biocontrol agent, *Trichoderma*, releases a variety of compounds that induce resistance responses to biotic and abiotic stresses (Harman et al., 2004b; Cardona and Rodriguez, 2006). Several studies have shown that root colonization by *Trichoderma harzianum* results in increased levels of plant enzymes, including various peroxidase, chitinases, β-1,3-glucanases, lipoxygenase-pathway hydroperoxidelyase accompanied with phenols and phytoalexins to provide durable resistance against such stress (Harman, 2006; Hoitink et al., 2006; Gachomo and Kotchoni, 2008).

Trichoderma spp. impose numerous mechanisms that influence rhizospheric surroundings and plant growth (Harman et al., 2004a), which mainly include enhancing the solubilization of soil nutrients (Altomare et al., 1999; Yedidia et al., 2001; Yadav et al., 2009; Kapri and Tewari, 2010), increasing root length and number of root hairs to explore larger spaces of soil to absorb nutrients (Bjorkman et al., 1998; Bjorkman, 2004; Samolski et al., 2012; Tucci et al. 2011) and improving the production of plant stimulatory compounds, such as growth hormones (indole acetic acids, cytokinin, gibberellins, and zeatin) (Windham et al., 1986; Roco and Perez, 2001; Gravel et al., 2007; Contreras-Cornejo et al., 2009). Promotion of plant growth by *Trichoderma* spp. may also result from their secretion of antifungal substances, such as iron chelators (siderophore) and hydrogen cyanide (Anke et al., 1991; Samuels et al., 2002; Dutta et al., 2006; Ushamailini et al., 2008), that are together undertaking front against the phytopathogens, protecting plants from infection (Benítez et al., 2000; Whipps, 2001).

Accumulation of some compatible solutes has been observed under salt stress conditions and has been suggested as part of mechanisms that control salt tolerance in plants. Compatible solutes such as proline and glycine betaine presumably have another role in stabilizing the tertiary structure of proteins and function as osmoprotectants in salt-stressed cells (Munns and Tester, 2008). Proline is one of the best-known solutes; however, its relative importance for tolerance and precise protective function during stress require investigations (Bohnert and Jensen, 1996). Lipid peroxidation determined by malondialdehyde content is an effective indicator of cellular oxidative damage (Heath and Packer, 1968), and salinity tolerance has been reported to correlate with lower levels of lipid peroxidation associated with up-regulation of antioxidative enzymes (Demiral and Türkan, 2005; Moradi and Ismail, 2007).

7.12 *Trichoderma* sp. in Mitigation of Drought Stresses

In arid and semiarid regions of India, crop yield is limited due to increased rate of drought. Researchers are putting efforts in developing varieties by considering stage and intensity of stress in relation to plant growth (Haggag et al., 2015). In pulse crops, flower induction and flowering period are directly affected, causing shedding of flowers and leaves. Along this, drought stress during pod filling reduces the pod number, seed number, and seed weight (Tahir et al., 2019).

Reactive oxygen species (ROS) are commonly produced by the plant tissue but are quenched by the plant cell defenses. Conversely, ROS production under a drought-stressed scenario is induced and elevated (Farooq et al., 2009). However, enzymatic *viz.*, superoxide dismutase (SOD), ascorbate peroxidase (APX), catalase (CAT), and glutathione reductase (GR) as well as non-enzymatic components such as ascorbic acid (ASA) and glutathione (GSH) hostile to ROS are posed by plant. *Trichoderma* employ primarily three mechanisms against drought stress *viz.*, secretion of phytohormonal analogues, alleviation of damage by ROS, and water-use efficiency in enhancing plant growth (Chepsergon et al., 2014; Mastouri et al., 2010). Study reports from Mastouri et al. (2012) suggest that root colonization by *Trichoderma* sp. under water stress conditions enhances tolerance and induces expression of antioxidant enzymes coding genes in plants (Dixit et al., 2015). Tomato seedling bioprimed with *Trichoderma atroviride* ID20G demonstrated augmented antioxidant activity rather than the control under drought conditions (Guler et al., 2016). Studies have also revealed that root colonization by *Trichoderma* sp. resulted in elevated levels of peroxidase, chitinases, glucanases, lipoxygenases, and phytoalexins-like compounds that impart consistent shielding effect against oxidative stress (Harman, 2006; Shukla et al., 2012; Shukla et al., 2015).

Bae and companion (2009) reported that cacao plant cells get loaded with high concentration of amino acids under drought stress, but *T. Hamatum* DIS2196 inoculated seedlings showed delayed drought–induced changes and also detracted the negative effects, allowing the growth in the cacao plants.

Protein contents from the roots and leaves of the plants inoculated with *Trichoderma* sp. have shown increased levels of antioxidative enzymes such as SOD, the chief scavenger of superoxide radicals and other detoxifying enzymes, respectively, accompanied with elevated scales of peroxidase, glutathione-reductase, and Glutathione-S-transferase (GST) (Shoresh and Harman, 2008; Dixit et al., 2015). SOD acts as the primary cell defense against the ROS under stress-like drought, deploying dismutation reaction, which converts the toxic superoxide (O2–) to hydrogen peroxide and oxygen (Chepsergon et al., 2014).

A study reported by Gusain et al. (2014) revealed that enhanced production of SOD, CAT, and POD antioxidative enzymes in *T. harzianium* (T-35)-treated rice plants augmented tolerance under severe drought stress through the reduction of oxidative stress. In a similar study, Shukla et al. (2014) showed that bioprimed seed of *Triticum aestivum* with *Trichoderma harzianum* not only promoted the plant growth but also increased level PAL and ROS scavenging enzymes. The study also reported enhanced root vigor along with the osmoregulatory processes in the plants (Shukla et al., 2014).

Increase in proline and total soluble protein content was observed in tomato plants inoculated with *T. harzianum*. Simultaneous increase in phenol and flavonoid contents was also observed. *T. harzianum* also employed the growth-promoting traits under the drought stress, maintaining the plant health (Mona et al., 2017; Haqu et al., 2012; Lalitha et al., 2012).

7.13 Genes Involved in Abiotic Stress Management

Trichoderma has several genes that help the plant to survive against abiotic and biotic stress. A gene that is responsible for the resistance against heat, salt tolerance, osmotic and oxidative tolerance, named hsp70, has been isolated from *T. harzianum* and has been cloned and characterized (Mantero Barrientos et al., 2008). From *Trichoderma harzianum*, a gene named thkel1 has been isolated and characterized. This gene codes putative kelch-repeat protein, which helps in regulating the glucosidase activity and enhances tolerance to salt and osmotic stresses in Arabidopsis thaliana plants (Hermosa et al., 2011). TvGST (glutathione transferase gene) has been cloned from *T. virens*; this acts as a cadmium-tolerant gene (Dixit et al., 2011).

7.14 Conclusion

Chemical fungicides that are normally used to control plant disease are very harmful for the environment. They pollute air, water, and soil, causing harm to human health. Use of biocontrol agents makes the environment safe. *Trichoderma* is a well-known biocontrol agent and it is effective against a variety of pathogens. There are many biocontrol genes in *Trichoderma*, which help the plant in developing resistance against plant pathogens. *Trichoderma*-based bioformulations are cheap, safe, and environmental friendly. However, there is a need for more work to be done to develop stable, improved, and cost-effective Trichoderma formulations that can control pathogens more effectively and efficiently. In present agricultural scenarios, use of biocontrol agents (BCAs) is of utmost importance in pulses for biotic stresses management, but their potential has yet to be exploited fully, mainly because the research in this area is still confined to the laboratory and very little attention has been paid to produce the commercial formulations of biocontrol agents. Moreover, whatever has been commercially produced has not been used efficiently by the farmers, owing to the lack of information regarding its application and importance. So, the concept of biological control and the different formulations need to be popularized at the field level.

REFERENCES

Abdel-Kader MIA, Abdel-Hafez AII and Abdel-Hafez SII. (1983). Composition of the fungal flora of Syrian soils. Ii. Cellulose decomposing fungi. *Mycopathologia* 81: 167–171.

Akbar M, Ponnamperuma FN. (1982). Saline Soils of South and South East Asia as potential rice lands. *Rice Research Strategies for the Future*. IRRI, Manila, Philippines, pp. 265–281.

Altomare C, Norvell WA, Björkman T, Harman GE. (1999). Solubilization of phosphates and micronutrients by the plant-growth promoting and biocontrol fungus *Trichoderma harzianum* rifai 1295–1222. *Appl. Environ. Microbiol.* 65, 2926–2933.

Amann RI, Ludwig W, Schleifer KH. (1995). Phylogenetic identification and in situ detection of individual microbial cells without cultivation. *Microbiol. Rev.* 59:143–169.

Anke H, Kinn J, Bergquist K, Olov S. (1991). Production of siderophores by strains of the genus *Trichoderma*. *Bio. Metals.*, 4:176–180.

Arisan-Atac I, Heidenreich E, Kubicek CP. (1995). Randomly amplified polymorphic DNA fingerprinting identifies subgroups of *Trichoderma viride* and other *Trichoderma* sp. capable of chestnut blight biocontrol. *FEMS Microbiol. Lett.* 126:249–255. doi:10.1016/0378-1097(95)00017-Y.

Asad SA, Ali N, Hameed A, Khan SA, Rafiq Ahmad, Muhammad Bilal, Shahzad M, Tabassum A. (2014). Biocontrol efficacy of different isolates of *Trichoderma* against soil borne pathogen *Rhizoctonia solani*. *Pol. J. Microbiol.* 63(1):95–103.

Babychan M, Simon S. (2017). Efficacy of *Trichoderma* spp. against *Fusarium oxysporum* f.sp. *lycopersici*. (FOL) infecting pre-and post-seedling of tomato. *J. Pharmacogn. Phytochem.* 6:616–619.

Bae H, Sicher RC, Kim MS, Kim SH, Strem MD. (2009). The beneficial endophyte *Trichoderma hamatum* isolate DIS 219b promotes growth and delays the onset of the drought response in *Theobroma cacao*. *J. Exp. Bot.* 60:3279–3295.

Barooah M, Borthakur HP. (1994). Isolation of hydrocarbon degrading microorganisms from oil fields of upper Assam. *J. Ind. Pollut. Control* 10:31–34.

Belay H, Anteneh B. (2016). Biocontrol of faba bean black root rot caused by *Fusarium solani* using seed dressing and soil application of *Trichoderma harzianum*. *J. Biol. Control.* 3(30):169–176.

Belete E, Ayalew A, Ahmed S. (2015). Evaluation of local isolates of *Trichoderma* spp. against black root rot (*Fusarium solani*) on faba bean. *J. Plant Pathol. Microb.* 6(6):279.

Benítez T, Rey M, Delgado-Jarana J, Rincon M, Limón MC. (2000). Improvement of *Trichoderma* strains for biocontrol. *Revista iberoamericana de micología*. 17:S31–S36. *Biology*, 59:651–668.

Bhattacharyya PN, Jhan DK. (2012). Plant growth-promoting rhizobacteria (PGPR): Emergence in agriculture. *World J. Microbiol. Biotechnol.* 28:1327–1350.

Bjorkman T, Blanchard LM, Harman GE. (1998). Growth enhancement of shrunken-2 sweet corn when colonized with *Trichoderma harzianum* 1295-22 effect of environmental stress. *J. Am. Soc. Hortic Sci.* 123:35–40.

Bohnert HJ, Jensen RG. (1996). Strategies for engineering water-stress tolerance in plants. *Trends Biotechnol.* 14:89–97.

Bowen JK, Franicevic SC, Crowhurst RN, Templeton MD, Stewart A. (1996). Differentiation of a specific Trichoderma biological control agent by restriction fragment length polymorphism (RFLP) analysis. *N. Z. J. Crop Hort. Sci.* 24:201–211. doi:10.1080/01140671.1996.9513955

Brotman Y, Landau U, Cuadros-Inostoza A, Takayuki T, Fernie AR et al. (2013). *Trichoderma*-Plant Root Colonization: Escaping early plant defense responses and activation of the antioxidant machinery for saline stress tolerance. *PLoS Pathog.* 9(3):e1003221.

Buensanteai N, Mukherjee PM, Horwitz BA, Cheng C. (2010). Expression and purification of biologically active *Trichoderma virens* proteinaceous elicitor Sm1 in *Pichia pastoris. Protein Express Purif.* 72:131–138.

Bulat SA, Lübeck M, Mironenko NV, Jensen DF, Lübeck PS. (1998). UP-PCR analysis and ITS1 ribotyping of *Trichoderma* and *Gliocladium* fungi. *Mycol. Res.* 102:933–943. doi:10.1017/S0953756297005686.

Cardona R, Rodríguez H. (2006). Effects of *Trichoderma harzianum* fungus on the incidence of the charcoal rot disease on sesame. *Rev. Fac. Agron. (LUZ).* 23:42–47.

Carpenter MA, Ridgway HJ, Stringer AM, Hay AJ, Stewart A. (2008). Characterization of a *Trichoderma hamatum* monooxygenase gene involved in antagonistic activity against fungal plant pathogens. *Curr. Genet.* 53, 193–205.

Chepsergon J, Mwamburi L, Kassim MK. (2014). Mechanism of drought tolerance in plants using *Trichoderma* spp. *Int. J. Sci. Res.* 3(11):1592–1595.

Chet I. (1987). *Trichoderma* application, mode of action, and potential as biocontrol agent of soil-borne pathogenic fungi. In: *Innovative Approaches to Plant Disease Control.* I. Chet (Ed.), pp. 137–160, John & Wiley Son, New York.

Chen X, Romaine CP, Tan Q, Schlagnhaufer B, Ospina-Giraldo MD, Royse DJ, Huff DR. (1999). PCR-based genotyping of epidemic and preepidemic Trichoderma isolates associated with green mold of Agaricus bisporus. *Appl. Environ. Microbiol.* 65:2674–2678.

Claydown KL, Emerson OH, Sauthwell RJ. (1987). The isolation of a toxic substance from the culture filtrates of *Trichoderma. Phytopathol.* 36:1068.

Contreras-Cornejo HA, Macías-Rodríguez L, Cortés-Penagos C. (2009). *Trichoderma* virens, a plant beneficial fungus enhances biomass production and promotes lateral root growth through an auxin dependent mechanism in Arabidopsis. *Plant Physiol.* 149:1579–1592.

Cumagun JCR, Hockenhull J, Lübeck M. (2000). Characterization of *Trichoderma* isolates from Philippine rice fields by UP-PCR and rDNA-ITS1 analysis: Identification of UP-PCR Markers. *J. Phytopathol.* 148:109–115. doi:10.1046/j.1439-0434.2000.0 0467.x.

Danielson RM, Davey CB. (1973). Nonnutritional factors affecting growth of *Trichoderma* in culture. *Soil Biol. Biochem.* 5, 495–504.

Demiral T, Turkan I. (2005). Comparative lipid peroxidation, antioxidant defense systems and proline content in roots of two rice cultivars differing in salt tolerance. *Environ. Exp. Bot.* 53:247–257.

Deshmukh S, Hüeckelhoven R, Schäefer P, Imani J, Sharma M. et al. (2006). The root endophytic fungus *Piriformospora indica* requires host cell death for proliferation during mutualistic symbiosis with barley. *Proc. Nat. Acad. Sci. USA.* 103:18450–18457.

Dixit P, Mukherjee PK, Ramachandran V, Eapen S. (2011). Glutathione transferase from *Trichoderma virens* enhances cadmium tolerance without enhancing its accumulation in transgenic *Nicotiana tabacum. PLoS One.* 6(1):e16360.

Dodd SL, Hill RA, Stewart, A. (2004). Monitoring the survival and spread of the biocontrol fungus *Trichoderma atroviride* (C65) on kiwifruit using a molecular marker. *Austr. Plant Pathol.* 33:189–196. doi:10.1071/AP03070.

Domsch KH, Gams, W, Anderson, TH. (1980). *Compendium of Soil Fungi.* Vol. 1 and 2, Academic Press, London UK, p. 859.

Donzelli BGG, Lorito M, Scala F, Harman GE. (2001). Cloning, sequence and structure of a gene encoding an antifungal glucan 1,3-β-glucosidase from *Trichoderma atroviride* (*T. harzianum*). *Gene* 277:199–208.

Dubey SC, Bhavani R and Singh B. (2009). Development of Pusa 5SD for seed dressing and Pusa Biopellet 10G for soil application formulation of *T. harzianum* and their evaluation for integrated management of dry root rot of mungbean (*Vigna radiata*). *Bio. Control* 50:231–242.

Dubey SC, Patel B. (2012). Evaluation of fungal antagonists against *Thanatephorus cucumeris* causing web blight of urd and mung bean. *Indian Phytopathol.* 54(2):206–209.

Dubey SC, Tripathi A, Singh B. (2012). Combination of soil application and seed treatment formulations of *Trichoderma* species for integrated management of wet root rot caused by *Rhizoctonia solani* in chickpea (*Cicer arietinum*). *Indian J. Agril. Sci.* 82:356–362.

Dubey SS, Suresh M, Singh B. (2007). Evaluation of *Trichoderma* species against *Fusarium oxysporum* f.sp. *ciceri* for integrated management of chickpea wilt. *Biol. Control.* 40(1):118–127.

Dunlop RW, Simon A, Siwasithamparam D, Ghisalberti EL. (1989). An antibiotic from *Trichoderma koningii* active against soil-borne plant pathogens. *J. Nat. Prod.* 52:67–74.

Dutta T, Jana M, Pahari P, Bhattacharya T. (2006). The effect of temperature, pH, and salt on *Amylase* in *Heliodiaptomus viduus* (Gurney) (*Crustacea*: *Copepoda*: Calanoida). *Turk. J. Zool.* 30.

El-Hassan SA, Gowen SR, Pembroke B. (2013). Use of *Trichoderma hamatum* for biocontrol of lentil vascular wilt disease: Efficacy, mechanisms of interaction and future prospects. *J. Plant Prot. Res.* 53(1):12–26.

El-Katatney MH and Idris MM. (2014). Effects of single and combined inoculations with *Azospirillum brasilense* and *Trichoderma harzianum* on seedling growth or yield parameters of wheat (*Triticum vulgaris* L., Giza 168) and Corn (*Zea mays* L., Hybrid 310). *J. Plant Nutri.* 37:1913–1936.

Fanti S, Barbieri A, Vannacci, G. (2003). Environmental monitoring of antagonistic *Trichoderma*, In *Abstracts of the 1st Federation of European Microbiological Societies (FEMS) Congress of European Microbiologists, FEMS*, Ljubljana, Slovenia, p. 428.

Farooq M, Wahid A, Kobayashi N, Fujita D, Basra SMA. (2009). Plant drought stress: Effects, mechanisms and management. *Agro. Sust. Dev.* 29:185–212.

Freeman S, Kolesnik I, Barbul O, Zveibil A, Maymon M, Nitzani Y. et al. (2002). Use of *Trichoderma* spp. for biocontrol of *Colletotrichum acutatum* (anthracnose) and *Botrytis cinerea* (grey mould) in strawberry, and study of biocontrol population survival by PCR. *IOBCWPRS Bull.* 25:167–170.

Gachomo EW, Kotchoni SO. (2008). The Use of *Trichoderma harzianum* and *T. viride* as potential biocontrol agents against peanut microflora and their effectiveness in reducing *Aflatoxin* contamination of infected kernels. *Biotechnology* 7:439–447.

Gattinger A, Ruser R, Schloter M, Munch JC. (2002) Microbial community structure varies in different soil zones of a potato field. *Plant Nutr. Soil Sci.* 165:421–428.

Geistlinger J, Zwanzig J, Heckendorff S, Schellenberg I. (2015). SSR markers for *Trichoderma virens*: Their evaluation and application to identify and quantify root-endophytic strains. *Diversity* 7:360–384. doi:10.3390/d70 40360.

Gherbawy Y, Druzhinina I, Shaban MI, Wuczkowsky M, Yaser M, El-Naghy MA, Prillinger H-J, Kubicek CP. (2004). *Trichoderma* populations from alkaline agricultural soil in the Nile valley, Egypt, consist of only two species. *Mycol. Prog.* 3(3):211–218.

Gravel V, Antoun V, Tweddell RJ. (2007). Growth stimulation and fruit yield improvement of greenhouse tomato plants by inoculation with *Pseudomonas putida* or *Trichoderma atroviride*: Possible role of indoleacetic acid (IAA). *Soil Biol. Biochem.* 39:1968–1977.

Guler NS, Pehlivan N, Karaoglu SA, Guzel S, Bozdeveci A. (2016). *Trichoderma atroviride* ID20G inoculation ameliorates drought stress-induced damages by improving antioxidant defence in maize seedlings. *Acta Physiol. Plant* 38:132.

Gusain YS, Singh US, Sharma AK. (2014). Enhance activity of stress related enzymes in rice (*Oryza sativa* L.) induced by plant growth promoting fungi under drought stress. *Afri. J. Agri. Res.* 9:1430–1434. doi:10.5897/AJAR2014.8575.

Haggag WM, Abouziena HF, Abd-El-Kreem F, Habbasha S. (2015). Agriculture biotechnology for management of multiple biotic and abiotic environmental stress in crops. *J. Chem. Pharm.* 7(10):882–889.

Halwagy R, Moustafa AE, Kame SM. (1982). Ecology of the soil mycoflora in the desert of Kuwait. *J. Arid. Environ.* 5:109–125.21.

Haque MM, Haque MA, Ilias GNM, Molla AH. (2010). *Trichoderma*-enriched biofertilizer: A prospective substitute of inorganic fertilizer for mustard (*Brassica campestris*) production. *The Agri.* 8:66–73.

Harman GE. (2006). Overview of mechanisms and uses of *Trichoderma* spp. *Phytopathology* 96:190–194.

Harman GE, Björkman T. (1998). Potential and existing uses of *Trichoderma* and *Gliocladium* for plant disease control and plant growth enhancement. In: *Trichoderma and Gliocladium. Vol. 2. Enzymes, Biological Control, and Commercial Applications*, Harman G, Kubicek C. (Eds), Taylor & Francis Group, London, UK, pp. 229–265.

Harman GE, Howell CR, Viterbo A, Chet I, Lorito M. (2004a). *Trichoderma* species opportunistic, avirulent plant symbionts. *Nat. Rev. Microbiol.* 2:43–56.

Harman GE, Lorito M, Lynch JM. (2004b). Uses of *Trichoderma* spp. to alleviate or remediate soil and water pollution. *Adv. Appl. Microbiol.* 56:313–330.

Harman GE, Tronsmo A. (1992). Methods of genetic manipulation for the production of improved bioprotectant fungi. In: *New Approaches in Biological Control of Soil-Borne Diseases*, Jensen DF, Hockenhull J, Fokkema NJ. (Eds), IOBC/WPRS Bulletin XV/1:181–187.

Haware MP, Mukherjee PK, Lenne IM, Jayanthi S, Tripathi HS, Rathi YPS. (1999). Integrated biological-chemical control of Botrytis gray mould of chickpea. *Indian Phytopathol.* 52(2):174–176.

Heath RL, Packer L. (1968). Photoperoxidation in isolated chloroplast I. Kinetics and stoichiometry of fatty acid peroxidation. *Arch. Biochem. Biophys.* 125:189–198.

Hermosa MR, Grondona I, Díaz-Mínguez JM, Iturriaga EA, Monte E. (2001). Development of a strain-specific SCAR marker for the detection of the *Trichoderma atroviride*, a biological control agent against soil borne fungal plant pathogens. *Curr. Genet.* 38:343–350. doi:10.1007/s002940000173.

Hermosa R, Botella L, Keck E, Jiménez JA, Montero-Barrientos M, Arbona V, Gómez-Cadenas A, Monte E, Nicolás C. (2011). The overexpression in *Arabidopsis thaliana* of a *Trichoderma harzianum* gene that modulates glucosidase activity, and enhances tolerance to salt and osmotic stresses. *J. Plant Physiol.* 168:1295–1302.

Hoitink HAJ, Madden LV, Dorrance AE. (2006). Systemic resistance induced by *Trichoderma* spp.: Interactions between the host, the pathogen, the biocontrol agent, and soil organic matter quality. *Phytopathology* 96:186–189.

Howell CR. (2003). Mechanisms employed by *Trichoderma* species in the biological control of plant diseases: The history and evolution of current concepts. *Plant Dis.* 87:4–10.

Howell CR, Hanson LE, Stipanovic RD, Puckhaber LS. (2000). Induction of terpenoid synthesis in cotton roots and control of *Rhizoctonia solani* by seed treatment with *Trichoderma virens*. *Phytopathology* 90:248–252.

Hukma R, Pandey RN. (2011). Efficacy of biocontrol agents and fungicides in the management of wilt of pigeonpea. *Indian Phytopath.* 64(3):269–271.

Kapri A, Tewari L. (2010). Phosphate solubilization potential and phosphatase activity of rhizospheric *Trichoderma* spp. *Braz. J. Microbiol.* 41(3):787–795.

Khan MR, Ashraf S, Rasool F. (2014). Field performance of *Trichoderma* species against wilt disease complex of chickpea caused by *Fusarium oxysporum* f.sp. *ciceri* and *Rhizoctonia solani*. *Turk. J. Agric. For.* 38(4):447–454.

Kredics L, Kocsubé, S, Nagy L, Komon-Zelazowska, M, Manczinger L, Sajben E. et al. (2009). Molecular identification of *Trichoderma* species associated with *Pleurotus ostreatus* and natural substrates of the oyster mushroom. *FEMS Microbiol. Lett.* 300:58–67. doi:10.1111/j.1574-6968.2009.01765.x.

Kuchlan P, Kuchlan MK, Husain SM. (2017). Effect of foliar application of growth activator, promoter and antioxidant on seed quality of soybean. *Legume Res.* 40(2):313–318.

Kushwaha SK, Kumar S, Chaudhary B. (2018). Efficacy of *Trichoderma* against *Sclerotium rolfsii* causing collar rot disease of lentil under *in-vitro* conditions. *J. Nat. Appl. Sci.* 10(1):307–312.

Leong J. (1986). Siderophores: Their biochemistry and possible role in biocontrol of plant pathogen. *Annu. Rev. Phytopathol.* 24:187–209.

Li M, Yang Q. (2007). Isolation and characterization of a β-tubulin gene from *Trichoderma harzianum*. *Biochem. Genet.* 45:529–534.

Li RX, Cai F, Pang G, Shen QR, Rong Li., Chen W. (2015). Solubilisation of phosphate and micronutrients by *Trichoderma harzianum* and its relationship with the promotion of tomato plant growth. *PLoS One.* 10(6):e0130081.

Liesack W, Janssen PH, Rainey FA, Ward-Rainey N, Stackebrandt E. (1997). Microbial diversity in soil: The need for a combined approach using molecular and cultivation techniques. In: *Modern Soil Microbiology*, Van Elsas JD, Trevors JT, Wellington EMH. (Eds), Marcel Dekker, New York, pp. 375–439.

López-Mondéjar R, Antón A, Raidl S, Ros M, Pascual JA. (2010). Quantification of the biocontrol agent *Trichoderma harzianum* with real-time TaqMan PCR and its potential extrapolation to the hyphal biomass. *Biores. Technol.* 101:2888–2891. doi:10.1016/j.biortech.2009.10.019.

Mahajan S, Tuteja N. (2005). Cold, salinity and drought stresses: An overview. *Arch. Biochem. Biophys.* 444(2):139–158.

Mandhare VK, Suryawanshi AV. 2008. Efficacy of some botanicals and *Trichoderma* species against soil borne pathogens infecting chickpea. *J. Food Leg.* 21(2):122–124.

Mantero Barrientos M, Hermosa R, Nicolas C, Cardoza RE, Gutierrez S, Monte E. (2008). Over expression of a *Trichoderma* HSP70 gene increases fungal resistance to heat and other abiotic stresses. *Fungal Genet. Biol.* 45:1506–1513.

Marcello CM, Steindorff AS, Silva SP, Silva RN, Bataus LAM. (2010). Expression analysis of the exo-β-1,3-glucanase from the mycoparasitic fungus *Trichoderma asperellum*. *Microbiol. Res.* 165:75–81.

Martínez C, Blanc F, Le CE, Besnard O, Nicole M, Baccou JC. (2001). Salicylic acid and ethylene pathways are differentially activated in melon cotyledons by active or heat-denatured cellulase from *Trichoderma longibrachiatum*. *Plant Physiol.* 127:334–344.

Martínez-Viveros O, Jorquera MA, Crowley DE, Gajardo G, Mora ML. (2010). Mechanisms and practical considerations involved in plant growth promotion by rhizobacteria. *J. Soil. Sci. Plant Nutr.* 10(3):293–319.

Mastouri F, Bjorkman T, Harman GE. (2012). *Trichoderma harzianum* enhances antioxidant defense of tomato seedlings and resistance to water deficit. *Mol. Plant Microbe Interact. J.* 25:1264–1271.

Mayo S, Gutiérrez S, Malmierca MG, Lorenzana A, Campelo MP, Hermosa R, Casquero PA. (2015). Influence of *Rhizoctonia solani* and *Trichoderma* spp. in growth of bean (*Phaseolus vulgaris* L.) and in the induction of plant defense-related genes. *Plant Sci.* 6:685.

Migheli Q, Gonza´lez-Candelas L, Dealessi L, Camponogara A, Ramo´n-Vidal D. (1998). Transformants of *Trichoderma longibrachiatum* overexpressing the beta-1,4-endoglucanase gene egl1 show enhanced biocontrol of *Pythium ultimum* on cucumber. *Phytopathol.* 88:673–677.

Mishra DS, Gupta AK, Prajapati CR, Singh US. (2011). Combination of fungal and bacterial antagonists for management of root and stem rot disease of soybean. *Pak. J. Bot.* 43(5):2569–2574.

Mona SA, Abeer H, Elsayed FA, Abdulaziz AA, Dina WKS, Stephan W, Dilfuza E. (2017). Increased resistance of drought by *Trichoderma harzianum* fungal treatment correlates with increased secondary metabolites and proline content. *J. Integr. Agr.* 16(8):1751–1757.

Moradi F, Ismail AM. (2007). Responses of photosynthesis, chlorophyll fluorescence and ROS-scavenging systems to salt stress during seedling and reproductive stages in rice. *Ann. Bot.* 99:1161–1173.

Mukherjee M, Mukherjee PK, Kale PS. (2007). CAMP signalling is involved in growth, germination, mycoparasitism and secondary metabolism in *Trichoderma virens*. *Microbiol.* 153:1734–1742.

Mukherjee PK, Haware MP, Raghu K. (1997). Induction and evaluation of benomyl-tolerant mutants of *Trichoderma viride* for biological control of Botrytis grey mould of chickpea. *Indian Phytopathol.* 50(4):485–489.

Mukherjee PM, Latha J, Hardar R, Horwitz BA. (2003). TMKA, Mitogen activated protein kinase of *Trichoderma virens* is involved in biocontrol properties and repression of conidiation in the dark. *Eukaryot. Cell* 2: 446–455.

Munns R, Termaat A. (1986). Whole-plant responses to salinity. *Aust. J. Plant Physiol.* 13: 143–160.

Munns R, Tester M. (2008). Mechanisms of salinity tolerance. *Annu. Rev. Plant Biol.* 59:651–681.

Nirmalkar VK, Said PP and Kaushik DK. (2017). Efficacy of Fungicides and Bio-Agents against *Pyricularia gresia* in Paddy and Yield Gap Analysis Thought Frontline Demonstration. *Int. J. Curr. Microbiol. App. Sci.* 6(4):2338–2346.

Omomowo, IO, Fadijia AE, Omomowo OI. (2018). Assessment of bio-efficacy of *Glomus versiforme* and *Trichoderma harzianum* in inhibiting powdery mildew disease and enhancing the growth of cowpea. *Ann. Agric. Sci.* 63(1):9–17.

Otadoh J, Okoth S, Ochanda J, Kahindi J. (2011). Assessment of *Trichoderma* isolates for virulence efficacy on *Fusarium oxysporum* f.sp. *Phaseoli*. *Trop. Subtrop. Agroecosystems* 13:99–107.

Pan S, Das A. (2011). Control of cowpea (*Vigna sinensis*) root and collar rot (*Rhizoctonia solani*) with some organic formulations of *Trichoderma harzianum* under field condition. *J. Plant Prot. Sci.* 3(2):20–25.

Pandey G, Pandey RK and Pant H. (2003). Efficacy of different levels of *Trichoderma viride* against root-knot nematode in chickpea (*Cicer arietinum* L.). *An. Plant Prot. Sci.* 11:101–103.

Paran I, Michelmore RW. (1993). Development of reliable PCR-based markers linked to downy mildew resistance genes in lettuce. *Theor. Appl. Genet.* 85:985–993. doi:10.1007/BF00215038.

Patole SP, Shankara K, Pradhan RS, Dhore SB. (2017). Isolation and characterization of toxin from *Alternaria helianthi* inciting blight in sunflower. *Int. J. Curr. Microbiol. App. Sci.* 6(10):2892–2896.

Pieterse CM, Leon-Reyes A, Vander Ent S, Van Wees SC. (2009). Networking by small-molecule hormones in plant immunity. *Nat. Chem. Biol.* 5:308–316. *PLoS One* 6(1):1–15.

Pieterse CM, Zamioudis C, Berendsen RL, Weller DM, Saskia Van Wees CM, Bakke PA. (2014). Induced systemic resistance by beneficial microbes. *Annu. Rev. Phytopathol.* 52:347–375.

Poddar RK, Singh DV, Dubey SC. (2004). Integrated application of *Trichoderma harzianum* mutants and carbendazim to manage chickpea wilt (*Fusarium oxysporum* f.sp. *ciceris*). *Indian J. Agric. Sci.* 74(6):346–348.

Pozo MJ, JongMin B, Garcia JM, Kenerley CM. (2004). Functional analysis of tvsp1, a serine protease-encoding gene in the biocontrol agent *Trichoderma virens*. *Fungal Genet. Biol.* 41:336–348.

Raguchander T, Rajappan K, Samiappan R. (1997). Evaluating methods of application of biocontrol agents in the control of Mungbean root rot. *Ind. Phytopathol.* 50:229–234.

Roco A, Perez LM. (2001). In vitro biocontrol activity of *Trichoderma harzianum* on *Alternaria alternata* in the presence of growth regulators. *Electronic J. Biotechnol.* 4, Available from www.ejbiotechnology.info/ ntent/vol 4/issue 2/full/1/1.

Roy N, Bose S, Jash S, Khalko S, Pan S. (2005). Biological control of *Aiacrophomina phaseolina* by stable mutants of *Trichoderma harzianum*. *J. MStcopath. Res.* 43:239–243.

Sadfi-Zouaoui N, Châabani S, Rouaissi M, Hedi A, Hajlaoui MR, Boudabous A. (2009). Analysis of the diversity of *Trichoderma* spp. in soil horizons using digested ITS regions. *Ann. Microbiol.* 59(3):459–463.

Saiprasad GVS, Mythili JB, Anand L, Sunnetha C, Rashmi HJ, Naveena C. Ganeshan, G. (2009). Development of *Trichoderma harzianum* gene construct conferring antifungal activity in transgenic tobacco. *Indian J. Biotechnol.* 8:199–206.

Samolski I, Rincon AM, Pinzon LM, Viterbo A, Monte E. (2012). The qid74 gene from *Trichoderma harzianum* has a role in root architecture and plant biofertilization. *Microbiol.* 158:129–138.

Samuels GJ, Dodd SL, Gams W, Castlebury L, Petrini O. (2002). *Trichoderma* species associated with the green mold epidemic of commercially grown agaricus bisporus. *Mycologia.* 94:146–170. doi:10.2307/3761854.

Samuels GJ. (1996). *Trichoderma*: A review of biology and systematics of the genus. *Mycol. Res.* 100:923–935.

Saravanakumar K, Yu C, Dou K, Wang M, Li Y. et al. (2016). Biodiversity of *Trichoderma* Community in the Tidal Flats and Wetland of Southeastern China. *PLoS One.* 11(12):e0168020.

Sayiprathap B, Sarathbabu K, Ganesh T, Mahendra T. (2017). Efficacy of *Trichoderma viride* isolates against *Aspergillus niger* inciting collar rot in peanut (*Arachis hypogaea* L.). *Bull. Env. Pharmacol. Life Sci.* 6(1): 230–233.

Schlick A, Kuhls K, Meyer W, Lieckfeldt E, Börner T, Messner K. (1994). Fingerprinting reveals gamma-ray induced mutations in fungal DNA: Implications for identification of patent strains of *Trichoderma harzianum*. *Current Genetic.* 26:74–78.

Shabir-U-Rehman WA, Dar SA, Ganie Javid A, Bhat G, Hassan Mir Rubina Lawrence, Narayan S, Singh PK. (2013). Comparative efficacy of *Trichoderma viride* and *Trichoderma harzianum* against *Fusarium oxysporum* f.sp. *ciceri* causing wilt of chickpea. *Afr. J. Microbiol. Res.* 7:5731–5736. doi:10.5897/AJMR2013.

Sharma K, Mishra AK, Misra RJ. (2009). Morphological, biochemical and molecular characterization of *Trichoderma harzianum* isolates for their efficacy as biocontrol agents. *J. Phytopathol.* 157:51–56. doi:10.1111/j.1439-0434.2008.01451.x.

Shi WL, Chen XL, Wang LX et al. (2016). Cellular and molecular insight into the inhibition of primary root growth of Arabidopsis induced by peptaibols, a class of linear peptide antibiotics mainly produced by *Trichoderma* spp. *J. Exp. Bot.* doi:10.1093/jxb/erw023.

Shoresh M, Harman GE. (2008). Genome-wide identification, expression and chromosomal location of the genes encoding chitinolytic enzymes in *Zea mays*. *Mol. Genet. Genomics* 280(2):173–185. doi:10.1007/s00438-008-0354-1.

Shukla N, Awasthi RP, Rawat L. (2015). Seed biopriming with drought tolerant isolates of *Trichoderma harzianum* promote growth and drought tolerance in *Triticum aestivum*. *Annals Appl. Biol*. 66:171–182.

Shukla N, Awasthi RP, Rawat L, Kumar J. (2012). Biochemical and physiological responses of rice (*Oryza sativa* L.) as influenced by *Trichoderma harzianum* under drought stress. *Plant Physiol. Biochem*. 54:78–88.

Shukla N, Awasthi RP, Rawat L, Kumar J. (2014). Seed biopriming with drought tolerant isolates of *Trichoderma harzianum* promote growth and drought tolerance in *Triticum aestivum*. *Ann. Appl. Biol*. 166(2):171–182.

Singh S, Dureja P, Tanwar RS, Singh A. (2005). Production and antifungal activity of secondary metabolites of *Trichoderma virens*. *Pestic. Sci. J*. 17: 26–29.

Singh US, Zaidi NW, Joshi D, Varshney S, Khan T. (2003). Current status of *Trichoderma* as biocontrol agent. In: *Current Status of Biological Control of Plant Diseases Using Antagonistic Organism in India*. (Eds.), Project Directorate of Biological Control, Bangalore, pp. 13–48.

Stepanova AN, Yun J, Likhacheva AV, Alonso JM. (2007). Multilevel interactions between ethylene and auxin in Arabidopsis roots. *Plant Cell*. 19:2169–2185.

Surekha CH, Neelapu NRR, Siva Prasad B, Sankar Ganesh P. (2014). Induction of defense enzymes and phenolic content by *Trichoderma viride* in *Vigna mungo* infested with *Fusarium oxysporum* and *Alternaria alternata*. *Int. J. Agri. Sci. Res*. 4(4):31–40.

Tachapattaworakul T, Fontana-Tachon A, Guiraud JP, Ratomahenina R. (2008). Application of benomyl resistance and RAPD-PCR for detecting a *Trichoderma atroviride* strain used as biocontrol agent. *Biol. Agric. Hortic*. 25:369–377. doi:10.1080/01448765.2008.9755062.

Tahir A, Saira S, Saif R, Tahir S, Qadir M. (2019). Morphological and yield response of pulses against drought stress: A review. *TURJAF*. 7:202. doi:10.24925/turjaf.v7i2.202-208.2178.

Thakur AK, Uphoff N, Antony E. (2010). An assessment of physiological effects of system of rice intensification (sri) practices compared with recommended rice cultivation practices in India. *Exp. Agric*. 46:77–98.

Thakur M, Sohal BS. (2013). Role of elicitors in inducing resistance in plants against pathogen infection: A review. *ISRN Biochemistry*: 1–10. doi:10.1155/2013/762412.

Tijerino A, Cardoza RE, Moraga J, Malmierca MG, Vicente F, Aleu J, Collado IG, Gutierrez S, Monte E, Hermosa R. (2011). Overexpression of the trichodiene synthase gene tri5 increases trichodermin production and antimicrobial activity in *Trichoderma brevicompactum*. *Fungal Genet. Biol*. 48:285–296.

Tiwari N, Ahmed S, Kumar S and Sarker A. (2017). Fusarium wilt: A killer disease of lentil. *Fusarium-Plant Diseases, pathogen diversity, genetic diversity, resistance and molecular markers. IntechOpen, Rijeka*, 119-138.

Tucci M, Ruocco M, De Masi L, De Palma M, Lorito M. (2011). The beneficial effect of *Trichoderma* spp. on tomato is modulated by the plant genotype. *Mol. Plant Pathol*. 12:341–354. doi:10.1111/j. 1364-3703.2010.00674.

Ushamailini C, Nakkeeran S, Marimuthu T. (2008). Development of biomanure for the management of turmeric rhizome rot. *Archiv. Phytopathol. Plant Protec*.. 41:365–376. doi:10.1080/03235400600796711.

Vinale F, Flematti G, Sivasithamparam K et al.(2009). Harzianic acid, an antifungal and plant growth promoting metabolite from *Trichoderma harzianum*. *J. Nat. Prod*. 72:2032–2035.

Whipps JM. (2001). Microbial interactions and biocontrol in the rhizosphere. *J. Exp. Bot*. 52:487–511.

Windham MT, Elad Y, Baker R. (1986). A mechanism for increased plant growth induced by *Trichoderma* spp. *Phytopathol*. 6:518–521.

Woo SL, Ruocco M, Vinale F, Nigro M, Marra R, Nadia L, Pascale A, Lanzuise S, Manganiello G, Lorito M. (2014). *Trichoderma* based products and their widespread use in agriculture. *Open Mycol. J*. 8:71–126.

Wuczkowski M, Druzhinina I, Gherbawy Y, Klug B, Prillinger H, Kubicek CP. (2003). Species pattern and genetic diversity of *Trichoderma* in a mid-European, primeval floodplain-forest. *Microbiol. Res*. 158:125–133.

Yadav G, Gokhale R, Mohanty D. (2009). Towards prediction of metabolic products of polyketide synthases: An in silico analysis. *PLoS Comput. Biol*. 5:1–14.

Yedidia I, Srivastva AK, Kapulnik Y, Chet I. (2001). Effect of *Trichoderma harzianum* on microelement concentrations and increased growth of cucumber plants. *Plant Soil* 235, 235–242.

Yoshioka Y, Ichikawa H, Naznin HA, Kogure A and Hyakumachi M. (2011). Systemic resistance induced in *Arabidopsis thaliana* by *Trichoderma asperellum* SKT-1, a microbial pesticide of seedborne diseases of rice. *Pest Manag. Sci*. doi:10.1002/ps.2220 (Epub ahead of print).

Youssef YA. (1974). On the fungal flora of libyan soils. *Arch. Microbiol*. 99: 167–171.

Zhong YH, Wang TH, Wang XL, Zhang GT, Yu HN. (2009). Identification and characterization of a novel gene, TrCCD1, and its possible function in hyphal growth and conidiospore development of *Trichoderma reesei*. *Fungal Genet. Biol*. 46: 255–263.

Zimand G, Valinsky L, Elad Y, Chet I, Manulis S. (1994). Use of the RAPD procedure for the identification of *Trichoderma* strains. *Mycol. Res*. 98, 531–543. doi:10.1016/S0953-7562(09)80474-7.

8

Effects of Pesticides and Heavy Metals on the Growth and Yield of Food Legume Plants

A. Sankaranarayanan, N. Amaresan, and Abhishek Sharma

CONTENTS

8.1 Introduction

Needless to mention, agriculture is an important sector that plays a vital role in the world's economy, and its vitality has to be maintained for sustainability in agriculture to meet the nutritional demands of the world's population. Food legume plants yield diversified benefits such as high nutrition, biofuel, medicinal potential, farmer's livelihood, poverty alleviation, eradication of malnutrition, and development of sustainable agriculture productivity (Figure 8.1), though they face many stress factors from biotic and abiotic sources. These stresses are major challenges to the yield and growth of food legume plants. The biotic stresses are rendered from bacteria, fungi, virus, nematode, weeds, and insects; meanwhile, the abiotic stress is sourced from drought, salinity, temperature, cold, frost, light, and unfertile soil. There is a need in this alarming situation for the various cellular and molecular mechanisms to be studied (Ahuja et al., 2010) and adapted by the plant to overcome the stress response. In this juncture, the biotic stresses are controlled by various strategic approaches such as biological control, plant extracts, development of transgenic plants, and molecular markers; however, the farmers are highly reliant on pesticides to control the biotic stress-causing agents. On the other hand, it has not been ruled out, based on the available reports, that the long-term use of agrochemicals is also among the important sources for heavy metals in plants due to its presence in agricultural soils (Mishra et al., 2017). The presence of cadmium in agricultural soils through the application of phosphate fertilizers has also been reported (Mann et al., 2002). The abiotic stresses are managed by the plant by its own inherent mechanisms and with the association of microbial mediated mitigation concepts.

Plant diseases which lowered the productivity of agriculture (Adejumo, 2005) and insect pests played a pivotal role in this process (Nampala et al., 2002) (Table 8.1).

Though various pest management strategies have been employed to control the pests (Abate et al., 2000; Adati et al., 2008) in legume plants (Table 8.2), chemicals have been used as an effective agent to control the pests in various crops including food legume plants, and due to swift action (Adejumo, 2005),

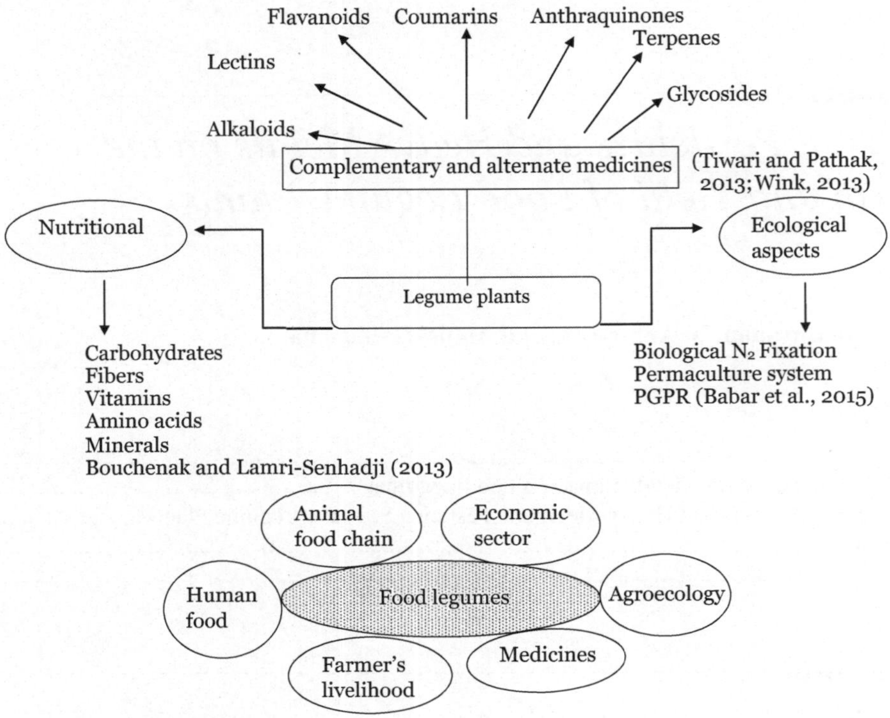

FIGURE 8.1 Diversified benefits of food legume plants.

TABLE 8.1

Details of Some Insect Pests Affected the Food Legume Plants

S.No.	Name of the Insect Pest	Common Name	References
1.	*Helicoverpa armigera*	Pod borer	Pawar et al. (1986); Margam et al. (2011); Chhangani et al. (2018)
2.	*Helicoverpa punctigera*	Pod borer	Murray and Zalucki (1990)
3.	*Maruca vitrata*	Spotted pod borer	Sharma et al. (1999); Margam et al. (2011)
4.	*Etiella zinckenella*	Spiny pod borer	Meena and Sharma (2016); Helalia et al. (2011)
5.	*Etiella behrii*	Seed web moth/Etiella moth	Austin et al. (1993)
6.	*Melanagromyza obtuse*	Pod fly	Sharma et al. (2003)
7.	*Liriomyza cicerina*	Leaf miner	Weigand et al. (1994)
8.	*Ophiomyia phaseoli*	Stem fly/bean fly	Moorthy and Srinivasan (1991)
9.	*Aphis craccivora*	Aphids	Abdallah (2012)
10.	*Aphis fabae*	Aphids/black bean aphids	Hansen et al. (2008)
11.	*Bemisia tabaci*	White fly	Lakshminarayana and Mishra (1992)
12.	*Mylabris* spp.	Blister beetles	Lal and Sastawa (2000)
13.	*Mylabris calida*	Beetle	Khelil (1994)
14.	*Collasobruchus chinensis*	Bruchids	Aslam (2004)
15.	*Bruchus pisorum*	Bruchids	Clement et al. (1999)
16.	*Megaleurothrips dorsalis*	Flower thrips	Abtew et al. (2015)
17.	*Megaleurothrips sjostedti*	Bud thrips	Abudulai et al. (2006)
18.	*Caliothrips indicus*	Thrips	Dixit et al. (2016)
19.	*Empoasca* spp.	Leafhoppers	Murray et al. (2000)
20.	*Tanaostigmodes cajaninae*	Pod wasp	Sharma et al. (2003)

TABLE 8.2
Integrated Pest Management (IPM) Strategy to Control the Plant Legume Pests

S.No.	Strategy Methods	Name of the Pest	References
1.	Biological control	*Thrips palmi* *Frankiliniella occidentalis* *Helicoverpa armigera*	Cox et al. (2006); Ugine et al. (2007); Pawar et al. (1986); Bhatnagar et al. (1983)
2.	Plant extracts	Flower bud thrips	Abtew et al. (2015); Khan et al. (2010)
3.	Transgenic plant	*Helicoverpa armigera*	Gopalaswamy et al. (2003)
4.	Molecular markers	Pod borer	Lawlor et al. (1998)
5.	Resistant variety	*Melanagromyza phaseoli*; *Chromatomyia horticola*; *Helicoverpa armigera*	Mittal and Ujagir (2005)
6.	Wild variety	*Melanagromyza obtuse* Malloch; *Tanaostigmodes cajaninae* La Salle	Sharma et al. (2003)
7.	Partially resistant and intercropping	*Aphis fabae*	Hansen et al. (2008)
8.	Biopesticides	*Helicoverpa armigera*	Chhangani et al. (2018)

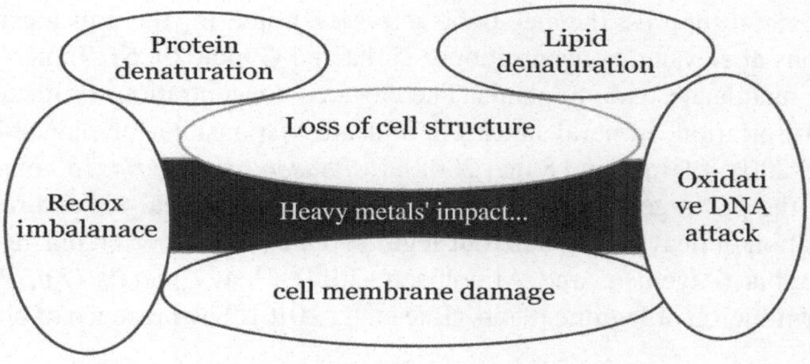

FIGURE 8.2 Heavy metal impact on various processes in food legume plants.

they are reliable and rendered immediate effect against pests. On the other hand, they pose a danger to the environment, health risk to humans and animals, and provide a platform to develop the risk of resistance.

Heavy metal toxicity in food legume plants suppresses the plant growth and deteriorates the development by dwindling the various essential processes in plants by multiple ways (Figure 8.2). Heavy metals impaired various essential mechanisms in food legume plants and it leads to the process of programmed cell death (Emamverdian et al., 2015) (Figure 8.2). In general, heavy metals from different sources (Alloway, 1990; Raskin et al., 1994; Shen et al., 2002), including fertilizers and pesticides, posed a major threat to agriculture. Another noteworthy point mentioned here about the presence of arsenic, chromium, nickel, and cobalt in many formulations of pesticides—and their limit was also alarming (Defarge et al., 2018; Rahman et al., 2006). Recent scientific reports revealed that the presence of heavy metals—especially Zn, Fe, Cu, Cr, and Pb—in beans and cowpea from Ibadan of Nigeria (Olutona and Aderemi, 2019) indicated the seriousness of this issue. The accumulation of heavy metals in agricultural crops leads to various health-related consequences in human beings (Huang et al., 2007), and the disorders due to heavy metal in various human systems were listed by many researchers (Waisberg et al., 2003; Bocca et al., 2004; Li et al., 2013; Wang, 2013). Heavy metal disturbed the number and diversity of soil microorganisms and impaired their activity in soil. However, the toxicity of heavy metals on soil microbes depends on various parameters such as pH, soil temperature, organic matter, and chemical nature of metals (Baath, 1989; Giller et al., 1998; Friedlova, 2010).

8.2 Pesticides and Heavy Metals Impact in Food Legumes

Environment poses a threat to legume plants by increasing the rate of attack by biological agents. By various biotic agents attacking the food legume plants, the physical constraints also played a major role in attracting the insect pests on food legume plants (Lipiec et al., 2013; Jha et al., 2014) (Figure 8.3).

Hence, the usage of pesticide or other ways to control the pests thus reduces the crop loss by it and increases the yield. But the problem is coming because of the intensive usage of pesticides. The intensive and large-scale usage of pesticides poses serious threats to the environment by the deposition of remains in the soils, its own effect on target and non-target species (Braganga et al., 2018), and deposition in plant parts and their yields, too (Olutona and Aderemi, 2019). Soil is an integral component of the environment, and the toxicity studies in soil, which render the details about the pollutants in soil by pesticides, ultimately reveals that the environment affects the health of human beings (Kapanen and Itävaara, 2001) and other biotic species in the universe.

The list of pesticides used to control the pests in legume plants has already been reported by Lama (2008). Once the pesticides taken up by the plants, they are translocated into the shoot part by xylem or phloem or may be stored in an absorption site near the translocation pathway and then further transferred to other plant parts (Kaur et al., 2015). Afterwards, the pesticides may be detoxified or activated by three phases and that reaction was mediated through the cytochrome P45o enzymes, conjugation, and hydrolysis reactions (Karthikeyan et al., 2003; Singh, 2012). The long half-life period and persistence of the pesticides in various food legume plants are reported by several researchers (Table 8.3).

Heavy metals in general suppress the metabolic activities, hindering the translocation of reserve material to growing regions at elevated concentrations (Sinha and Gupta, 2005) (Table 8.4). The excess concentrations of heavy metals are toxic to plants. The elevated concentrations of nickel showed its impact on photosynthesis, respiration, mineral nutrition, and are responsible for induced oxidative stress in plants (Llamas et al., 2008; Llamas and Sanz, 2008). In *Phaseolus vulgaris*, 150 mg/kg of nickel though despite of growth of the plant, genotoxicity was assessed by RAPD analysis and reported (Al-Qurainy, 2009). The accumulation of heavy metals in four legume plants was reported during summer and winter seasons in Saudi Arabia (Alyemeni and Almohisen, 2014). Heavy metals (Zn, Pb, Ni, Cu, and Cd) affected the seed germination of legume plants (Bae et al., 2016). The presence of chromium (250 mg/L)

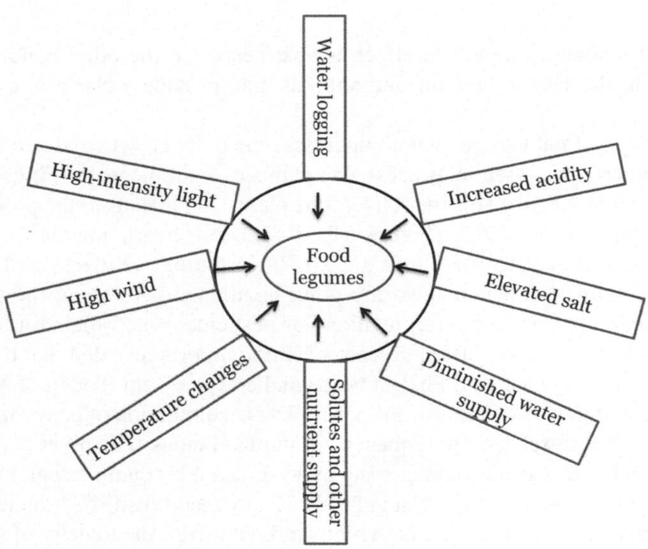

FIGURE 8.3 Environmental threats affect the various activities in food legumes.

TABLE 8.3
Pesticide Residues in Various Food Legume Plants

Name of the Pesticide/ Residue	Name of the Food Legume Plant	Country	Reference
DDT	Pigeon pea, green gram, black gram		Neetu (2013)
Pirimiphos-methyl	Chickpea	Canary Islands	Gonzalez-Curbelo et al. (2012)
Isocarbophos and triazophos	Kidney pea; cowpea; legumes	China	Ping-Sheng and Qi (2010); Chen et al. (2011)
Isofetamid	Legumes	Canada and USA	FAO and WHO (2019)
Dimethoate	Legumes	Nordic countries	Skretteberg et al. (2014)
Cyclodiene	Cowpea	Nigeria	Olutona and Aderemi (2019)
Triazole	Pea	Thailand	EFSA (2009); Skretteberg et al. (2014)
Methamidophos	Legumes	China	Chen et al. (2011)
Triazophos	Legumes	China	Chen et al. (2011)

TABLE 8.4
Effects of Heavy Metals in Food Legume Plants

Name of the Plant	Name of the Heavy Metals	Effects	References
Vigna mungo	Copper; lead	Photosynthesis; chlorophyll pigmentation	Bibi and Hussain (2005)
Lathyrus sativus	Lead	Root	Brunet et al. (2008)
Vicia faba	Lead	Chlorophyll; protein	Bouziani and Yssaad (2015)
Vigna unguiculata	Copper	Nodulation	Kopittke et al. (2007)
Chick pea	Chromium	Seed germination; reduction of biomass; root cell division and elongation	Velez et al. (2016)
Soybean	Arsenic	Root and shoot growth; nodule production; necrosis and lesions in leaves	Reichman (2007); Neumann et al. (1998)

decreased the synthesis of β amylase and enhanced the enzyme involved in starch hydrolysis, leading to the weakening of sugar supply—a vital need for seed germination in chickpea (Velez et al., 2016). Some of the heavy metals and their impact are given in Table 8.4.

8.3 Microbial Mitigation of Heavy Metal and Pesticide Stress in Legume Plants

In general, a plant has its own inherent mechanisms to mitigate various abiotic and biotic stresses. For example, the chickpea germination rate tolerated 100 mg/L of chromium (Valez et al., 2016). Apart from these inherent mechanisms, "rhizomediation" of the combination of plants and microbes, especially rhizhospheric bacteria, played a major role in the mitigation of heavy metals (Velez et al., 2016), particularly nodulating bacteria. Nodulating bacteria reducing Cr (VI) to Cr (III) mediated through reductase enzyme (Ontañon et al., 2015). In another study, inoculation of chickpea with *Mesorhizobium amorphae*, *Mesorhizobium* RC3 reduced chromium (VI) to (III) mediated through chromate reductase enzyme by *M. ciceri* (Wani et al., 2008; Ontañon et al., 2015). *Rhizobium* bacteria associated with the root nodules of legume reduced the arsenic (V) into (III), and then released it into the surrounding environment (Fitz and Wenzel, 2002). In a cowpea plants *S. acidiscabies* E13 showed heavy metal (copper, ferrous, nickel, uranium, and manganese) tolerant PGPR activities by producing siderophores (Dimkpa et al., 2009) and protected the plant from metal toxicity and promoted the plant growth by binding ferrous and nickel (Dimkpa et al., 2008).

In general, the microorganisms residing in the soil degrade the pollutants, which includes partitioning (Pignatello and Xing, 1996), adsorption and absorption (Weber and Huang, 1996), chemisorption (Maruya et al., 1996), diffusion (Ehlers and Luthy, 2003), and covalent binding (Bollag, 1992) or may tolerate the pesticides that are sprayed in soil samples/antagonistic activities (Kyei-Boahen et al., 2017). *Bacillus cereus* and *B. safensis* isolated from methomyl, carbendazim, and imidacloprid (up to 0.4%) sprayed pesticide soil samples showed tolerance to pesticide concentrations, and produced PGPR activities in an *in vitro* study to observe its antagonistic activities against *Alternaria* sp in *Lens culnaris* (Roy et al., 2018). In the fertilizer, the phosphorous (P) contents supported the symbiosis performance (Bargaz et al., 2018) in cowpea plants (Kyei-Boahen et al., 2017). Further, the influential role of P on the efficiency of *Rhizobium* has been reported (Kyei-Boahen et al., 2017). This mechanism has revealed the efficient utilization of P by the *Rhizobium* bacteria. Co-inoculation studies with *Rhizobium* sp. and *Penicillium* sp. showed the same impact on the growth of soybean plants (Wang et al., 2011). In an *in vitro* study, PGPR strains showed the efficient degradation of acibenzolar-S-methyl and the growth of PGPR organisms observed as high, compared to other tested pesticides (Myresiotis et al., 2012).

8.4 Conclusion

Pesticides and heavy metals persist inside the plants for prolonged duration due to their long half-life. However, both the pesticides and heavy metals are needed for the growth and yield of food legume plants. In the meantime, the present conditions are highly alarming. The presence of pesticide residues in food legume plants directly affect humanity by causing various deleterious diseases, and the presence of excess heavy metal in soil affects the growth and yield of food legume plants. Deeper insights are needed into the stress-mitigating mechanisms by microbes to maintain agricultural sustainability.

REFERENCES

Abate T, van-Huis A, Ampofo JKO (2000) Pest management strategies in traditional agriculture: An African perspective. *Annual Review of Entomology* 45, 631–659.

Abdallah YEY (2012) Effect of plant traps and sowing dates on population density of major soybean pests. *The Journal of Basic and Applied Zoology* 65, 37–46.

Abtew A, Subramanian S, Chesto X, Kreiter S, Garzia GT, Martin T (2015) Repellency of plant extracts against the legume flower thrips *Megalurothrips sjostedti* (Thysanoptera: Thripidae). *Insects* 6(3), 608–625.

Abudulai M, Salifu AB, Mohammed H (2006) Screening of cowpea for resistance to the flower bud thrips, *Megalurothrips sjostedti* Trybom (Thysanoptera: Thripidae). *Journal of Applied Sciences* 7, 1621–1624.

Adati T, Tamo M, Yusuf R, Downham M, Singh B, Hammond W (2008) Integrated pest management for cowpea-cereal cropping systems in the West African savannah. *International Journal of Tropical Insect Science* 27, 123–137.

Adejumo TO (2005) Crop protecting strategies for major diseases of cocoa, coffee and cashew nut in Nigeria. *African Journal of Biotechnology* 4, 143–150.

Ahuja I, de Vos RC, Bones AM, Hall RD (2010) Plant molecular stress responses face climate change. *Trends in Plant Science* 15, 664–674.

Al-Qurainy F (2009) Toxicity of heavy metals and their molecular detection on *Phaseolus vulgaris* (L.). *Australian Journal of Basic and Applied Sciences* 3(3), 3025–3035.

Alloway BJ (1990) *Heavy Metal in Soils*, John Wiley & Sons, New York.

Alyemeni MN, Almohisen I (2014) Effect of anthropogenic activities on heavy metals in legume crops, Riyadh, Saudi Arabia. *APCBEE Procedia* 10, 275–280.

Aslam M (2004) Pest status of stored chickpea beetle, *Collasobruchus chinenesis* Linnaeus on chickpea. *Journal of Entomology* 1, 28–33.

Austin AD, White TCR, Maelzer DA, Taylor DG (1993) Biology of *Etiellabehrii zeller* (Lepidoptera: Pyralidae): A pest of seed lucerne in South Australia. *Transactions of the Royal Society of South Australia Incorporated* 117(2), 67–76.

Baath E (1989) Effects of heavy metals in soil on microbial processes and populations (a review). *Water, Air, & Soil Pollution* 47(3–4), 335–379.

Babar MM, Zaidi NSS, Azooz MM, Kazi AG (2015) Biotechnology approaches to overcome biotic and abiotic stress constraints in legumes. In: *Legumes under Environmental Stress: Yield, Improvement and Adaptations*. John Wiley & Sons, Chichester, UK, pp. 247–264.

Bae J, Benoit DL, Watson AK (2016) Effect of heavy metals on seed germination and seedling growth of common ragweed and road side ground cover legumes. *Environmental Pollution* 113, 112–118.

Bargaz A, Lyamlouli K, Chtouki M, Zeroual Y, Dhiba D (2018) Soil Microbial resources for improving fertilizers efficiency in an Integrated plant nutrient management system. *Frontiers in Microbiology*. doi:10.3389/fmicb.2018.01606.

Bhatnagar VS, Sithanantham S, Pawar CS, Jadhav DS, Rao VK, Reed W (1983) Conservation and augmentation of natural enemies with reference to IPM in chickpea and pigeonpea. In: *Proceedings of the International Workshop on Integrated Pest Control in Grain Legumes*, 4–9 April 1983. Empresa Braseleira Pesquisa Agropocuria (EMRAPA), Goiania, Brazil, pp. 157–180.

Bibi M, Hussain M (2005) Effect of copper and lead on photosynthesis and plant pigments in black gram [*Vigna mungo* (L.) Hepper]. *Bulletin of Environmental Contamination and Toxicology* 74, 1126–1133. doi:10.1007/s00128-005-0698-8.

Bocca B, Alimonti A, Petrucci F, Violante N, Sancesario G, Forte G, Senofonte O (2004) Quantification of trace elements by sector field inductively coupled plasma mass spectrometry in urine, serum, blood and cerebrospinal fluid of patients with Parkinson's disease. *Spectrochimica Acta Part B: Atomic Spectroscopy* 59, 559–566.

Bollag JM (1992) Decontaminating soil with enzymes. *Environmental Science and Technology* 26, 1876–1881.

Bouchenak M, Lamri-Senhadji M (2013) Nutritional quality of legumes, and their role in cardio-metabolic risk prevention: A review. *Journal of Medicinal Food* 16, 185–198.

Bouziani EH, Yssaad R (2015) Phytotoxicity of lead on the Physiological parameters of two varieties of Broad bean (*Vicia faba*). *International Journal of Agricultural and Biosystems Engineering* 9(8), 942–945.

Braganga I, Grosso C, Rede D, Sousa SR, Lemos PC, Domnigues VF, Delerue-Matos C (2018) Ecotoxicological effects of insecticides in plants assessed by germination and other phytotoxicity tools. In: S Vats (Ed.) *Biotic and Abiotic Stress Tolerance in Plants*. doi:10.1007/978-981-10-9029-5.

Brunet J, Repellin A, Varrault G, Terryn N, Zuily-Fodil Y (2008) Lead accumulation in the roots of grass pea (*Lathyrus sativus* L.): A novel plant for phytoremediation systems? *Comptes Rendus Biologies* 331(11), 859–864.

Chen C, Qian Y, Chen Q, Tao C, Li C, Li Y (2011) Evaluation of pesticide residues in fruits and vegetables from Xiamen, China. *Food Control* 22(7), 1114–1120.

Chhangani G, Tali MK, Sing B (2018) Management of gram pod borer *Helicoverpa armigera* (Hubner) by bio-pesticides in farmscaping system. *Journal of Entomology and Zoology Studies* 6(4), 1052–1054.

Clement SL, Cristofaro M, Cowgill SB, Weigand S (1999) Germplasm resources, insect resistance, and grain legume improvement. In: SL Clement and SS Quisenberry (Ed.) *Global Plant Genetic Resources for Insect Resistant Crops*. CRC Press, Boca Raton, FL, pp. 131–148.

Cox PD, Matthews L, Jacobson RJ, Cannon R, MacLeod A, Walters KFA (2006) Potential for the use of biological agents for the control of *Thripspalmi* (Thysanoptera: Thripidae) outbreaks. *Biocontrol Science and Technology* 16, 871–891.

Defarge N, Vendomois JS, Seralini GE (2018) Toxicity of formulations and heavy metals in glyphosate based herbicides and other pesticides. *Toxicology Reports* 5, 156–163.

Dimkpa C, Svatoš A, Merten D, Büchel G, Kothe E (2008) Hydroxamate siderophores produced by *Streptomyces acidiscabies* E13 bind nickel and promote growth in cowpea (*Vigna unguiculata* L.) under nickel stress. *Canadian Journal of Microbiology* 54, 163–172.

Dimkpa CO, Merten D, Svatoš A, Büchel G, Kothe E (2009) Metal-induced oxidative stress impacting plant growth in contaminated soil is alleviated by microbial siderophores. *Soil Biology and Biochemistry* 41, 154–162.

Dixit GP, Parihar AK, Bohra A, Singh NP (2016) Achievements and prospects of grass pea (*Lathyrus sativus* L.) improvement for sustainable food production. *The Crop Journal* 4(5) 407–416.

EFSA (2009) Scientific opinion on risk assessment for a selected group of pesticides from the triazole group to test possible methodologies to assess cumulative effects from exposure through food from these pesticides on human health. *EFSA Journal* 7(9), 1167.

Ehlers LJ, Luthy RG (2003) Contaminant bioavailability in soil and sediment: Improving risk assessment and remediation rests on better understanding bioavailability. *Environmental Science and Technology* 37, 295A–302A.

Emamverdian A, Ding Y, Mokhberdoran F, Xie Y (2015) Heavy metal stress and some mechanisms of plant defense response. *The Scientific World Journal* 2015, 1–18. doi:10.1155/2015/756120.

FAO and WHO (2019) Pesticide residues in food 2018–Report 2018–Joint FAO/WHO Meeting on Pesticide Residues. FAO Plant Production and Protection Paper no. 234. Rome, Italy, 668 pp.

Fitz WJ, Wenzel WW (2002) Arsenic transformations in the soilrhizosphere-plant system: Fundamentals and potential application to phytoremediation. *Journal of Biotechnology* 99, 259–278.

Friedlova M (2010) The influence of heavy metals on soil biological and chemical properties. *Soil and Water Research* 5(1), 21–27.

Giller KE, Witter E, Mcgrath SP (1998) Toxicity of heavy metals to microorganisms and microbial processes in agricultural soils. *Soil Biology and Biochemistry* 30(10–11), 1389–1414.

Gonzalez-Curbelo MA, Dionis-Delgado S, Asensio-Ramos M, Hernandez-Borges J (2012) Pesticide analysis in toasted barley and chickpea flours. *Journal of Separation Science* 35, 299–307.

Gopalaswamy SVS, Kumar S, Subaratnam GV, Sharma HE, Sharma KK (2003) Transgenic pigeonpea: A new tool to manage *Helicoverpa armigera*. In: *National Symposium on Bioresources, Biotechnology and Bioenterprise*, 19–20 November 2003. Osmania University, Hyderabad, India.

Hansen LM, Lorentsen L, Boelt B (2008) How to reduce the incidence of black bean aphids (*Aphis fabae* Scop.) attacking organic growing field beans (*Vicia faba* L.) by growing partially resistant bean varieties and by intercropping field beans with cereal. *Acta Agriculturae Scandinavica Section B–Soil and Plant Science* 58(4), 359–364.

Helalia AAR, Ali FAF, Hegab MFA, Kamal KHA (2011) Effect of sowing dates of three cowpea cultivars on their infestation rate with cowpea pod borer, *Etiella zinckenella*. *Arab Universities Journal of Agricultural Sciences* 19(1), 247–259.

Huang SS, Liao QL, Hua M, Wu XM, Bi KS, Yan CY, Chen B, Zhang XY (2007) Survey of heavy metal pollution and assessment of agricultural soil in Yangzhong district, Jiangsu Province, China. *Chemosphere* 67, 2148–2155.

Jha UC, Chaturvedi SK, Bohra A, Basu PS, Khan MS, Barh D (2014) Abiotic stresses, constraints and improvement strategies in chickpea. *Plant Breed.* doi:10.1111/pbr.12150.

Kapanen A, Itävaara M (2001) Ecotoxicity tests for compost applications. *Ecotoxicology and Environmental Safety* 49, 1–16.

Karthikeyan R, Davis LC, Erickson LE et al. (2003) Studies on responses of non-target plants to pesticides: A review. Hazardous Substance Research Centre, Kansas State University. Manhattan, KS.

Kaur H, Bhardwaj R, Kumar V, Sharma A, Singh R, Thukral AK (2015) Effect of pesticides on leguminous plants: An overview. In: MM Azooz and P Ahmed (Eds.) *Legumes Under Environmental Stress: Yield, Improvement and Adaptations*, 1st edition. John Wiley & Sons, Chichester, UK, pp. 91–101.

Khan ZR, Midega CA, Bruce TJ, Hooper AM, Pickett JA (2010) Exploiting phytochemicals for developing a 'push pull' crop protection strategy for cereal farmers in Africa. *Journal of Experimental Botany* 61, 4185–4196.

Khelil MA (1994) Spatial and temporal structures of the entomofauna in the alfa (*Stipa tenacissima* L.) stepp in the high steppic plain of Tlemcen (Algeria). Application to two insects *Mylabris oleae* and *Mylabris calida* (Coleoptera, Meloidae). *Conference 46th International Symposium on Crop Protection, Gent (Belgium)*, 3 May 1994. Mededelingen-Faculteit Landbouwkundige en Toegepaste Biologische Wetenschappen Universiteit Gent, 59(2a), 429–438. (In French).

Kopittke PM, Dart PJ, Menzies NW (2007) Toxic effects of low concentrations of Cu on nodulation of cowpea (*Vigna unguiculata*). *Environmental Pollution* 145, 309–315.

Kyei-Boahen S, Savala CE, Chikoye D, Abaidoo R (2017). Growth and Yield responses of cowpea to inoculation and Phosphorous fertilization in different environments. *Frontiers in Plant Science* 8, 1–15.

Lakshminarayana S, Misra DS (1992) Management of yellow mosaic virus through whitefly (*Bemisia tabaci*) resistant green gram varieties. In: *National Seminar on Changing Scenario in Pests and Pest Management in India*, 31 January–2 February 1992. Plant Protection Association of India, Rajendranagar, Hyderabad, India.

Lal NES, Sastawa BM (2000) Evaluation of host plant resistance, sowing date modification and intercropping as methods for the control of *Mylabris* and *Coryna* species (Coleoptera: Meloidae) infesting pearl millet in the Nigerian Sudan savanna. *Journal of Arid Environments* 46(3), 263–280.

Lama JP (2008) Standard setting on pesticide residues to ensure food safety. *Journal of Agriculture and Environments* 9, 46–53.

Lawlor HI, Siddique KHM, Sedgley RH, Thurling N (1998) Improving cold tolerance and insect resistance in chickpea and the use of AFLPs for the identification of molecular markers for these traits. *Acta Horticulturae* 461, 185–192.

Li H, Qian X, Hu W, Wang Y, Gao H (2013) Chemical speciation and human health risk of trace metals in urban street dusts from a metropolitan city, Nanjing, SE China. *Science of the Total Environment* 456, 212–221.

Lipiec J, Doussan C, Nosalewicz A, Kondracka K (2013) Effect of drought and heat stresses on plant growth and yield: A review. *International Agrophysics* 27, 463–477.

Llamas A, Sanz A (2008) Organ-distinctive changes in respiration rates of rice plant sunder nickel stress. *Plant Growth Regulation* 54, 63–69.

Llamas A, Ullrich CI, Sanz A (2008) Ni2+ toxicity in rice: Effect on membrane functionality and plant water content. *Plant Physiology and Biochemistry* 46, 905–910.

Mann, SS, Rate AW, Gilkes RJ (2002) Cadmium accumulation in agricultural soils in Western Australia. *Water, Air, Soil Pollution* 141, 281–297.

Margam VM, Coates BS, Ba MN, Sun W, Binso-Dabire CL, Baoua I, Ishiyaku MF et al. (2011) Geographic distribution of phylogenetically-distinct legume pod borer *Maruca vitrata* (Lepidoptera: Pyraloidea: Crambidae). *Molecular Biology Reports* 38, 893–903.

Maruya KA, Risebrough RW, Horne AJ (1996) Partitioning of polynuclear aromatic hydrocarbons between sediments from San Francisco and their porewaters. *Environmental Science and Technology* 30, 2942–2947.

Meena NL, Sharma US (2006) Effect of sowing date and rowspacing on incidence of major insect pests of soybean, *Glycine max* (L.) Merrill. *Soybean Research* 4, 73–76.

Mishra J, Singh R, Arora NK (2017) Alleviation of heavy metals stress in plants and remediation of soil by rhizosphere microorganisms. *Frontiers in Microbiology* 8. doi:10.3389/fmicb.2017.01706.

Mittal V, Ujagir U (2005) Field screening of pea, *Pisum sativum* L. germplasm for resistance against major insect pests. *Journal of Plant Protection and Environment* 2(1), 50–58.

Moorthy PNK, Srinivasan K (1991) Diagnosis and management of bean flies in beans. In: J Prakash and RLM Pierik RLM (Eds.) *Horticulture—New Technologies and Applications. Current Plant Science and Biotechnology in Agriculture*, vol 12. Springer, Dordrecht, the Netherlands.

Murray DAH, Zalucki MP (1990) Survival of *Helicoverpa punctigera* (Wallengren) and *H. armigera* (Hubner) (LepIdoptera: Noctuidae) pupae submerged in water. *Australian Journal of Entomology* 29, 91–192.

Murray J, Wylde A, Pauls KP, Michaels T, Schaafsma A (2000) Detection of genetic loci for resistance to potato leafhopper (*Empoasca fabae*) in the common bean (*Phaseolus vulgaris*). *Annual Report, Bean Improvement Cooperative* 43, 27–28.

Myresiotis CK, Vryzas Z, Papadopoulou-Mourkidou E (2012) Biodegradation of soil applied pesticides by selected strains of plant growth promoting rhizobacteria (PGPR) and their effects on bacterial growth. *Biodegradation* 23, 297–310. doi:10.1007/s10532-011-9509-6.

Nampala, P, Ogenga-Latigo MW, Kyamanywa S, Adipala E, Oyabo N, Jackai LEN (2002) Potential impact of intercropping on major cowpea field pests in Uganda. *African Crop Science Journal* 10, 335–344.

Neetu T (2013) Determination of chlorinated pesticide in vegetables, cereals and pulses by gas chromatography in East National Capital Region, Delhi, India. *Journal of Agriculture and Forestry Sciences* 1, 27–28.

Neumann H, Bode-Kirchhoff A, Madeheim A, Wetzel A (1998) Toxicity testing of heavy metals with the Rhizobium-legume symbiosis: High sensitivity to cadmium and arsenic compounds. *Environmental Science and Pollution Research* 5, 28–36.

Olutona GO, Aderemi MA (2019) Organochlorine pesticide residue and heavy metals in leguminous food crops from selected markers in Ibadan, Nigeria. *Legume Science*. doi:10.1002/leg3.3.

Ontañon OM, González PS. Agostini E (2015) Biochemical and molecular mechanisms involved in simultaneous phenol and Cr (VI) removal by *Acinetobacter guillouiae* SFC 500-1A. *Environmental Science and Pollution* 22(17), 13014–13023.

Pawar CS, Bhatnagar VS, Jadhav DR (1986) *Heliothis* species and their natural enemies, with their potential for biological control. *Proceedings of the Indian Academy of Science (Animal Science)* 95, 695–703.

Pignatello JJ, Xing B (1996) Mechanisms of slow sorption of organic chemicals to natural particles. *Environmental Science and Technology* 30, 1–11.

Ping-Sheng Z, Qi L (2010) Investigation on organophosphorous pesticide residue in fresh melon and legume vegetables in Huizhou. *Guangxi Agricultural Sciences* 41, 338–340.

Rahman MM, Sengupta MK, Chowdhury UK, Lodh D, Das B, Ahamed S, Mandal D et al. (2006) Arsenic contamination incidents around the world. In: R Naidu, E Smith, G Owens, P Bhattacharya, and P Nadebaum (Eds.) *Managing Arsenic in the Environment: From Soil to Human Health.* CSIRO, Collingwood, Victoria, pp. 3–30.

Raskin I, Kumar PBAN, Dushenkov S, Salt DE (1994) Bio concentration of heavy metals by plants. *Current Opinion in Biotechnology* 5(3), 285–290.

Reichman SM (2007) The potential use of the legume-rhizobium symbiosis for the remediation of arsenic contaminated sites. *Soil Biology and Biochemistry* 39, 2587–2593.

Roy T, Bandopadhyay A, Sonawane PJ, Majumdar S, Mahapatra NR, Alam S, Das N (2018) Bio-effective disease control and plant growth promotion in lentil by two pesticide degrading strains of *Bacillus* sp. *Biological Control.* doi:10.1016/j.biocontrol.2018.08.018.

Sharma HC, Pampapathy G, Reddy LJ (2003) Wild relatives of pigeon pea as a source of resistance to the pod fly (*Melanagromyza obtusa* Malloch) and pod wasp (*Tanaostigmodes cajaninae* La Salle). *Genetic Resources and Crop Evolution* 50(8), 817–824.

Sharma HC, Saxena KB, Bhagwat VR (1999) Legume pod borer, *Maruca vitrata*: Bionomics and management. Information Bulletin 55. *International Crops Research Institute for the Semi-Arid Tropics (ICRISAT)*, Patancheru, India, 35 pp.

Shen Z, Li X, Wang C, Chen H, Chua H (2002) Lead phytoextraction from contaminated soil with high-biomass plant species. *Journal of Environmental Quality* 31(6), 1893–1900.

Singh DK (2012) *Toxicology: Agriculture and Environment-Pesticide Chemistry and Toxicology.* Bentham Science Publishers, Oak Park, IL.

Sinha S, Gupta AK (2005) Assessment of metals in leguminous green manuring plant of *Sesbania cannabina* L. grown on Xy ash amended soil: Effect on antioxidants. *Chemosphere* 61, 1204–1214.

Skretteberg LG, Lyrån B, Holen B, Jansson A, Fohgelberg P, Siivinen K, Andersen JH, Jensen BH (2014) Pesticide residues in food of plan origin from Southeast Asia–A Nordic project. *Food Control.* doi:10.1016/j.foodcont.2014.11.008.

Tiwari P, Pathak M (2013) Screening of phytocontent of different plant parts of a medicinal legume herb. *Asian Journal of Biotechnology and BioResource* 4, 49–52.

Ugine TA, Wraight SP, Sanderson JP (2007) Effects of manipulating spray application parameters on efficacy of the entomopathogenic fungus *Beauveria bassiana* against Western flower thrips *Frankiliniella occidentalis*, infesting green house impatiens crops. *Biocontrol Science and Technology* 17, 193–219.

Velez PA, Talano MA, Paisio CE, Agostini E, Gonsalez PS (2016) Synergistic effect of chickpea plants and *Mesorhizobium* as a natural system for chromium phytoremediation. *Environmental Technology.* doi:10.1080/09593330.2016.1247198.

Waisberg M, Joseph P, Hale B, Beyersmann D (2003) Molecular and cellular mechanisms of cadmium carcinogenesis. *Toxicology* 192, 95–117.

Wang WX (2013) Dietary toxicity of metals in aquatic animals: Recent studies and perspectives. *Chinese Science Bulletin* 58, 203–213.

Wang X, Pan Q, Chen F, Yan X, Liao H (2011) Effects of co-inoculation with arbuscular mycorrhizal fungi and rhizobia on soybean growth as related to root architecture and availability of N and P. *Mycorrhiza* 21, 173–181. doi:10.1007/s00572-010-0319-1.

Wani PA, Khan MS, Zaidi A (2008) Chromium-reducing and plant growth promoting Mesorhizobium improves chickpea growth in chromium amended soil. *Biotechnology Letters* 30, 159–163.

Weber WJ, Huang W (1996) A distributed reactivity model for sorption by soils and sediments. 4. Intraparticle heterogeneity and phase-distribution relationships under non equilibrium conditions. *Environmental Science and Technology* 30, 881–888.

Weigand S, Lateef SS, El Din Sharaf N, Mahmoud SF, Ahmed K, Ali K (1994) Integrated control of insect pests of cool season food legumes. In: EJ Muehlbauer and WJ Kaiser (Eds.) *Expanding the Production and Use of Cool Season Food Legumes.* Kluwer Academic Publishers, Dordrecht, the Netherlands, pp. 679–694.

Wink M (2013) Evolution of secondary metabolites in legumes (Fabaceae). *South African Journal of Botany* 89, 164–175.

Section II

Crop Responses to Stress-Soybean

9

*Biotic and Abiotic Stress Responses of Hyacinth Bean (*Lablab purpureus*) and Soybean (*Glycine max*): A Mini-review*

Difo Voukang Harouna and Elmugheira M. Ibrahim Mohammed

CONTENTS

9.1 Introduction

Hyacinth bean (*Lablab purpureus*) belongs to the family Leguminosae (Fabaceae) and is also known as field beans, Indian bean, and lablab bean (Kalloo & Pandey, 2012). It is a herbaceous perennial but is grown as an annual crop in many places. It is also known as one of the most diverse domesticated legume species with multiple uses and more than 150 common names (Maass, Knox, Venkatesha, & Pengelly, 2010). Although its largest agro-morphological diversity occurs in South Asia, its origin appears to be Africa where it is known as a "lost crop" and better adapted to drought as compared with common beans (*Phaseolus vulgaris*) or cowpea (*Vigna unguiculata*), both of which have been preferred to lablab in African agricultural production systems (Maass et al., 2010).

Similarly, soybean, *Glycine max*, is also a herbaceous annual plant in the family Fabaceae grown for its edible seeds. The soybean plant is usually an erect bush with woody stems and alternately arranged leaves. The leaves possess three individual leaflets that are oval or lance-like in shape, growing to a length of 3–10 cm (1.2–4.0 in). The soybean plant produces small white or purple flowers and curved seed pods that are 3–15 cm (1.2–6 in) in length and can contain 1–5 seeds. The seeds can be a variety of colors including yellow, green, brown, black, or a mottled combination.

Soybean [*Glycine max* (L.) Merrill] is an important leguminous crop, which is grown since age-old times and one of the oldest grown crops known to humans. It is grown in tropical, subtropical, and temperate regions of the world. It harbors more than 300 insects, among them only a few attain major pest status (Gaur & Mogalapu, 2018). These insects damage the crop from seedling stage to maturity and cause nearly 25% reduction of yield.

Just like many other food legumes, hyacinth beans and soybeans are being affected by many biotic and abiotic factors worldwide. This review proposes to highlight the scientific efforts so far done on investigating the biotic and abiotic stresses of these two important legumes.

9.2 Biotic Stress Responses of Hyacinth Bean (*Lablab purpureus*) and Soybean (*Glycine max*)

9.2.1 Hyacinth Bean (*Lablab purpureus*)

Dolichos (*Lablab purpureus*) is an important, highly nutritious human food and animal feed as mentioned earlier. It is used as a cover crop, green manure, and also known to have medicinal value. Despite its diverse uses, it also possesses common production constraints that include low yields, poor cooking and flavor qualities, and susceptibility to diseases and pests, which lower the economic value of the crop.

The biotic pressure from insect pests and diseases seriously affects lablab's maximum production in terms of yield and quality (Boit, Kinyua, Kiplagat, & Chepkoech, 2018). Legume pod borer (*Maruca vitrata*) has been reported as the most serious pest of lablab worldwide (Liao & Lin, 2000).

Among the insects that infest the plant, lepidopterous and dipterous pod borers are of considerable importance. The insect pests so far recorded on dolichos include *Adisura atkinsoni* Moore (Lep: Noctuidae); Heliothis *armigera* Hubner (Noctuidae); *Maruca testulalis* Geyer (Pyralidae); *Exelastis atomosa* Walsingham; *Spenarches caffer* Zeller (Pterophoridae); *Lampides boeticus* Linnaeus (Lycaenidae); *Etiella zinkenella* Treitsctike (Pyralidae); *Marasmarcha liophanes* MEYRICK (Pterophoridae); *Melanagromyza obtusa* Malloch (Dipt: Agromyzidae); Mylabris beetles; Leaf beetle, *Ceretoma ruficormis*; Cock-shaver larvae, *Schizonycha* species; *Callosobruchus* species; Chrysomelid beetle, *Sagra femorata Durray*; *Aphis craccivora*; *Lablab bug, Coptosoma cribraria*; and Mites, *Eutetranychus orientalis* L. (Byre Gowda, 2019). A number of experiments have been carried out to test the damaging effects of these insects on hyacinth bean. The first study focused on many pod borers in 1936 and proposed solutions such as applying a solution of bleaching powder (Byre Gowda, 2019).

In a recent study, some accessions of Dolichos were identified as promising resistant genotypes and showed potentialities to be exploited in a Dolichos breeding program (Boit et al., 2018). A first report of leaf blight disease of hyacinth bean caused by *C. cucurbitarum* in India was also published (Das, Dutta, & Mandal, 2017).

Studies on the biotic responses of lablab beans are mainly focused on the identification of resistant varieties to the pest and diseases identified. Molecular evidences of the responses to biotic stress need to be addressed also.

9.2.2 Soybean (*Glycine max*)

Biotic factors that mainly affect soybean growth and development are pests and diseases. Several diseases, including *Phytophthora* root and stem rot, pod and stem blight, frogeye leaf spot, brown spot, downy mildew, *Cercopsora* leaf blight and purple seed stain, and *Sclerotinia* stem rot (white mold), are known to affect soybeans (Jenn Thomas-Murphy, 2019).

Seed and seedling diseases of soybean are common and significant problems, as they can decrease plant populations that result in replanting and production losses. Several different pathogens can cause these diseases, and the most common tend to be *Fusarium*, *Rhizoctonia*, *Phytophthora*, and *Pythium*. They can kill and rot seeds before germination or cause seedling death. They are most common when soil is very wet in the first few weeks after planting and in heavy, poorly drained soils (Malvick, 2018).

Most of the insects associated with soybean are harmful to the crop and can cause yield losses. Reports have shown that about 15%–20% of total soybean production is lost directly and indirectly due to attacks by insect pests (Asala, Olubusuyi, & Abdulrazaq, 2016). The prevalence of insect pests and the damage that they cause vary from one year to another and also from one location to the other. The classes of insects that cause the greatest potential damage to soybean crop are the defoliators (Asala et al., 2016). Volunteer researchers at Penn State University have carefully established a list of pests and diseases of soybeans that can be easily accessed by the public at the following website: https://plantvillage.psu.edu/topics/soybean/infos.

The discovery of the soybean aphid, *Aphis glycines*, in North America in 2000, has developed a great deal of interest in the biology, ecology, and control of this insect in its native range of eastern Asia (Wu, Schenk-Hamlin, Zhan, Ragsdale, & Heimpel, 2004).

Experiments aimed at investigation of biotic responses of soybean plants at the molecular level have been published, though not enough.

Structural variants in the soybean genome to localize clusters of biotic stress-response genes have been reported (McHale et al., 2012). The study revealed that the most abundant classes of gene families associated with these regions were the nucleotide-binding and receptor-like protein classes, both of which are important for plant biotic defense. Novel soybean microRNAs (miRNAs) involved in abiotic and biotic stresses have been identified. It was found that differentially expressed plant miRNAs provide molecular evidence for their possible involvement in the process of water deficit- and rust-stress responses (Kulcheski et al., 2011).

9.3 Abiotic Stress Responses of Hyacinth Bean (*Lablab purpureus*) and Soybean (*Glycine max*)

9.3.1 Hyacinth Bean (*Lablab purpureus*)

Plants are constantly exposed to environmental challenges such as drought, salinity, flooding, and heavy metal stresses; in single or in combination these can affect plants.

Lablab purpureus (hyacinth bean) is an important pulse crop in Southeast Asia and Eastern Africa. It has been used for several purposes such as nitrogen fixer, cover crop or for livestock fodder, intercrop, weed suppressor, and soil erosion retardant (D'Souza & Devaraj, 2010). The plant is known to be tolerant to mild salinity and drought.

Except for the general information about its mild tolerance to salt and drought, there are few reports indicating biochemical and physiological bases of its response to salinity and other abiotic stresses.

The effect of salinity on antioxidants, antioxidant enzymes, and other markers of abiotic stress in Hyacinth bean have been evaluated (D'Souza & Devaraj, 2010). The study proved that salinity enhanced antioxidant enzymes, peroxidase, and glutathione reductase activities and reduced that of catalase in concentration and time-dependent manner in leaves. Antioxidant enzymes in roots showed inverse relationship with concentration and time of exposure. The Metabolic enzyme β-amylase activity increased in both leaves and roots.

In a more recent study, stress indicators such as H_2O_2, lipid peroxidation, antioxidants (glutathione, proline, ascorbic acid), antioxidant enzymes (guaiacol peroxidase, catalase, and glutathione reductase), and metabolic enzymes (acid phosphatase and amylase) have also been monitored in hyacinth beans in response to drought stress (Kokila, Myrene, & Devaraj, 2014). The study revealed that some cultivars showed an increase in glutathione reductase, guaiacol peroxidase concentration, and reduced catalase activity. An elevated level of amylase activity was also observed in some lablab cultivars while in some, acid phosphatase showed an increase in activity under stress. A comparable study earlier showed similar results when ten-days-old seedlings were tested under the effect of drought (D'souza & Devaraj, 2011).

Drought stress-related genes in the hyacinth bean were evaluated, and the study revealed that 17 functionally relevant genes were found to be overrepresented under drought stress using enrichment analysis (Yao, Wang, Cheng, & Wu, 2013). The study further revealed through BLASTX analysis that the majority top-matched genes from hyacinth beans were proteins from soya beans (*Glycine max* [L.]).

9.3.2 Soybean (*Glycine max*)

Abiotic stress is the negative impact of nonliving factors on the organism. Plants are subjected to different stresses including that of heavy metals, drought, high salinity, and low temperature.

As mentioned earlier, it was found that differentially expressed plant microRNAs (miRNAs) provide molecular evidence for their possible involvement in the process of biotic and abiotic-stress responses (Kulcheski et al., 2011).

Recent reports have demonstrated that hundreds of miRNAs have been discovered or identified in soybean (Khan, Tyagi, Singh, & Magadum, 2010).

During various abiotic stresses such as drought, salinity, temperature fluctuation, and oxidative environment, several stress-regulated miRNAs have been identified in different plants. Molecular techniques were used to conduct genome-wide analysis of microRNAs in temperature-stressed soybeans—belonging to different developmental stages—in drought, cold, salinity, and water lodging conditions.

Salinity is also a major problem in agriculture. Several salt-regulated microRNAs have been identified that play either a direct or indirect role in salt stress alleviation. In a study involving *Arabidopsis* and legumes plant crops, it was revealed that miRNAs under specific condition can control or regulate the expression of specific genes associated with abiotic stress tolerances (Khan et al., 2010). Experiments have shown that 133 conserved miRNAs representing 95 miRNA families were expressed in soybean under three treatments. In addition, it has been proved that many miRNAs are inducible and are differentially expressed in response to certain stress. Thus, the miRNAs act as a master regulator modulating various gene expressions in plants in response to abiotic stress and plant growth. They act by cleavage or through translational repression and play a role in feedback regulation of small RNA pathways and in the biogenesis of trans-acting siRNAs. They play a major role in stress responses, enabling plants to survive under adverse conditions such as drought, salinity, and high/low temperature.

To understand the regulatory networks of stress tolerance, quantitative and qualitative analyses of gene expression are necessary. In order to restore the cellular function and make plants more tolerant to stress, transferring a single gene may not be sufficient to reach the required tolerance levels. To overcome such constraints, enhancing expression of a stress inducible transcription factor that regulates a number of other genes is a promising approach.

Scientists have experimented with eight libraries of small RNAs, which were constructed to identify new miRNA in soybean to verify those that are possibly drought regulated and submitted to Solexa sequencing. After sequencing and subsequent analyses of the library, 269 miRNAs were detected. The libraries were developed from drought-sensitive and drought-tolerant seedlings of soybean with or without stressors (Khan et al., 2010). In a recent study (Kunert, Vorster, Fenta, Kibido, & Davis, 2016), researchers concluded that the direct screening of root and nodule traits in the field as well as identification of genes, proteins, and metabolites involved in such traits will be essential in order to gain a better understanding of the regulation of root architecture, bacteroid development, and life span in relation to drought tolerance in soybean.

Salt stress (abiotic stresses) is also a serious problem for crop plants worldwide. High salt stress leads to disruption of homeostasis in water potential and ion distribution, which occurs at both the cellular and the whole plant levels. Drastic changes in ion and water homeostasis lead to molecular damage, growth arrest, and even death of plants. Salt stress furthermore induces the abscisic acid synthesis that closes stomata when transported to guard cells, which ultimately decreases photosynthesis activity and leads to oxidative stress. A huge number of salt-regulated microRNAs have been identified in soybean.

9.4 Biotic and Abiotic Management Strategies in Hyacinth Bean (*Lablab purpureus*) and Soybean (*Glycine max*)

Biotic and abiotic stresses cause significant yield losses in legumes and can significantly affect their productivity. Biotechnology tools such as marker-assisted breeding, tissue culture, *in vitro* mutagenesis, and genetic transformation can contribute to solve or reduce some of these constraints. However, only limited success has been achieved so far. The emergence of "omic" technologies and the establishment of model legume plants such as *Medicago truncatula* and *Lotus japonicus* are promising strategies for understanding the molecular genetic basis of stress resistance, which is an important bottleneck for molecular breeding. Understanding the mechanisms that regulate the expression of stress-related genes is a fundamental issue in plant biology and will be necessary for the genetic improvement of legumes.

Most biotic stress-causing agents are usually handled by simpler and less costly techniques. For example, seed and seedling diseases of soybean are difficult to manage, and the different pathogens and diseases require different management approaches. Thus, accurate diagnosis can help in managing them. In general, these diseases can be reduced by planting good-quality seed in well-drained, noncompacted fields. Delaying planting until soils are warmer than 55 degrees F and relatively dry to allow for rapid emergence and growth can be beneficial. Crop rotation and tillage may be of some benefit. Genetic resistance in soybean varieties is only clearly available for managing *Phytophthora* infections. Fungicidal seed treatments may reduce seed and seedling diseases. Seed treatments with two or more active ingredients should typically be used. For example, products containing mefenoxam (ApronXL®) or metalayl (Allegiance®) can be effective against *Pythium* and *Phytophthora*, and products containing fludioxonil (Maxim®) or a strobilurin product (azoxystrobin, trifloxystrobin, or pyraclostrobin) may help to reduce damage from true fungi such as *Fusarium* and *Rhizoctonia* (Malvick, 2018).

9.5 Conclusion

Biotic and abiotic stresses are the major factors that affect the growth and development of both hyacinth beans and soybeans and therefore hinder their production and yield. However, much work is being done to assess or understand the mechanism of abiotic stress resistance in both crops as compared with biotic stress responses, though much more in soybean than hyacinth bean.

REFERENCES

Asala, S. W., Olubusuyi, P. T., & Abdulrazaq, S. (2016). Evaluation of insect pests associated with growth of soybean [*Glycine max* (L.) Merr] in Nigerian Guinea Savanna. *International Journal of Advance Agricultural Research*, *4*, 51–56.

Boit, L. C., Kinyua, M., Kiplagat, O., & Chepkoech, E. (2018). *Evaluating Kenyan Dolichos (Lablab purpureus L.) Genotypes for Resistance to Legume Pod Borers (Maruca vitrata and Helicoverpa armigera) Using Morphological Markers*. *5*(1), 344–351.

Byre Gowda, M. (2019). Dolichos bean, Lablab bean, Hyacinth bean, Field bean, Common bean, Indian bean, Pole bean, Kidney bean, Bonavist bean, Waby bean, Lubia bean, Tonga bean, Kikuyu bean, Seim, Avare, Avarai, Anumulu, Wal, Mochai, Bata. Retrieved April 18, 2019, from http://www.lablablab.org/html/insects-management.html.

D'souza, M. R., & Devaraj, R. V. (2011). Specific and non-specific responses of Hyacinth bean (*Dolichos lablab*) to drought stress. *Indian Journal of Biotechnology*, *10*, 130–139.

D'Souza, M. R., & Devaraj, V. R. (2010). Biochemical responses of Hyacinth bean (*Lablab purpureus*) to salinity stress. *Acta Physiologiae Plantarum*, *32*(2), 341–353. https://doi.org/10.1007/s11738-009-0412-2.

Das, S., Dutta, S., & Mandal, B. (2017). First report of *Choanephora cucurbitarum*, causing leaf blight of hyacinth bean in India. *Journal of Plant Pathology*, *99*(2), 541.

Gaur, N., & Mogalapu, S. (2018). Pests of soybean. In *Pests and Their Management* (pp. 137–162). https://doi.org/10.1007/978-981-10-8687-8_6.

Jenn Thomas-Murphy. (2019). Diseases of Soybeans | Field Crops. Retrieved April 20, 2019, from https://fieldcrops.cals.cornell.edu/soybeans/diseases-soybeans/.

Kalloo, G., & Pandey, S. C. (2012). Hyacinth bean: *Lablab purpureus* (L) Sweet. In *Genetic Improvement of Vegetable Crops* (pp. 387–389). https://doi.org/https://doi.org/10.1016/B978-0-08-040826-2.50030-8.

Khan, Y. J., Tyagi, H., Singh, A. K., & Magadum, S. (2010). Abiotic stress tolerance in soybean: Regulated by ncRNA. *Journal of AgriSearch*, *3*(1), 1–6.

Kokila, S., Myrene, R. D., & Devaraj, V. R. (2014). Response of *Lablab purpureus* (Hyacinth bean) cultivars to drought stress. *Asian Journal of Plant Science and Research*, *4*(5), 48–55.

Kulcheski, F. R., de Oliveira, L. F. V., Molina, L. G., Almerão, M. P., Rodrigues, F. A., Marcolino, J., … Margis, R. (2011). Identification of novel soybean microRNAs involved in abiotic and biotic stresses. *BMC Genomics*, *12*(1), 307. https://doi.org/10.1186/1471-2164-12-307.

Kunert, K. J., Vorster, B. J., Fenta, B. A., Kibido, T., & Davis, T. S. (2016). *Drought Stress Responses in Soybean Roots and Nodules*, *7*, 1–7. https://doi.org/10.3389/fpls.2016.01015.

Liao, C. T., & Lin, C. S. (2000). Occurrence of legume pod borer *Maruca testulalis* (*Lepidoptera*) on cowpea (*Vigna unguiculata*) an its insecticides application trial. *Plant Protection Bulletin*, *42*, 213–222.

Maass, B. L., Knox, M. R., Venkatesha, S. C., & Pengelly, B. C. (2010). *Lablab purpureus*—A Crop Lost for Africa? *Tropical Plant Biology*, 123–135. https://doi.org/10.1007/s12042-010-9046-1.

McHale, L. K., Haun, W. J., Xu, W. W., Bhaskar, P. B., Anderson, J. E., Hyten, D. L., … Stupar, R. M. (2012). Structural variants in the soybean genome localize to clusters of biotic stress-response genes. *Plant Physiology*, *159*(4), 1295–1308. https://doi.org/10.1104/pp.112.194605.

Malvick, D. (2018). Soybean seed and seedling diseases|UMN Extension. Retrieved April 20, 2019, from https://extension.umn.edu/pest-management/soybean-seed-and-seedling-diseases.

Wu, Z., Schenk-Hamlin, D., Zhan, W., Ragsdale, D. W., & Heimpel, G. E. (2004). The soybean aphid in China: A historical review. *Annals of the Entomological Society of America*, *97*(2), 209–218. https://doi.org/10.1093/aesa/97.2.209.

Yao, L., Wang, B., Cheng, L., & Wu, T. (2013). Identification of key drought stress-related genes in the hyacinth bean. *PLoS One*, *8*(3). https://doi.org/10.1371/journal.pone.0058108.

Section III

Crop Responses and Microbe-Mediated Stress Management

Groundnut

10

Biological Control of Soilborne Diseases and Water Stress Management in Groundnut

Dhruti Amin, Abhishek Sharma, Sanket Ray, and Siva Satya Mohan Jampala

CONTENTS

10.1 Introduction

Groundnut (*Arachis hypogaea* L.) is a multipurpose oil seed legume, cultivated in tropical, subtropical agroclimatic areas of Asia, Africa, and the Americas (FAO 2010). Groundnut seeds are nutritious, a rich source of edible oil (43%–55%) and protein (25%–28%) as well as medicinally and economically important. Moreover, groundnut contains bioactive components such as phenolics, flavonoids, polyphenols, and resveratrol and is also involved in soil improvement through biological nitrogen fixation.

The major groundnut-producing countries in the world are India, China, Nigeria, Senegal, Sudan, Burma, and the United States of America. Asia recorded around 50% of area and 60% of world production of groundnut, with the largest share of coverage in India and China (>40%), and the highest share in China (40%) for the total production in the world. India ranks second in groundnut production. Despite the fact that India is the largest grower of groundnut crop in terms of acreage, low yields kept the country ranked second in terms of output. Groundnut yield and prices are low in most Asian countries due to several factors such as cultivation on marginal lands; water stress; disease and pest attacks; low input use; poor storage conditions; seed treatment with chemicals (reduces seed quality); increase in price of other competitive oils especially in India, viz., soy oil, palm oil, cottonseed oil, sunflower oil, and mustard oil; trade policies; festivals and celebrations (which also increase consumption and the prices), etc. Lack of adoption of improved technologies is one of the main factors limiting improvement in groundnut yield. A majority of farmers grow traditional varieties that are adapted to local agroecological conditions, but have low genetic yield potential.

Mainly, environmental factors, i.e., biotic and abiotic stress, potentially reduce the overall production of the groundnut. Moreover, these factors are increasingly important due to climate changes. Therefore, it is very important to know how these two factors negatively influence the groundnut yield and how we can overcome their negative impact.

The scope to raise groundnut production in Asia through area expansion is extremely limited. While enhancing production, the main challenges for research and development are to:

1. Bridge the gap between actual and attainable yields by enhancing farmers' access to quality inputs, improved technologies, and information
2. Incentivize farmers to improve the competitiveness of oilseed crops through domestic and border protection measures in relation to their substitutes and competing crops
3. Achieve a technological breakthrough that not only overcomes yield barriers but also provides effective protection against insect pests and diseases and resistance to moisture stress (Birthal et al. 2010).

Herein, we briefly discussed the effect of biotic and abiotic (especially water) stresses on groundnut crop and various environmentally friendly protection mechanisms toward soilborne pathogens and water deficit conditions.

10.2 Biotic Stress

Groundnut is a cash crop providing income and livelihoods to farmers. It also contributes nutrition to farm families through consumption of energy and protein-rich groundnut kernels and provides nutritious fodder (haulms) to livestock. Thus, groundnut cultivation contributes to the sustainability of mixed crop-livestock production systems, the most predominant system of the semiarid areas. In India, groundnut is cultivated during *kharif* and *rabi*-summer. The major groundnut-growing states in India are Gujarat, Andhra Pradesh, Tamil Nadu, Karnataka, Rajasthan, and Maharashtra—contributing about 80% area and production. The average yield of *rabi*-summer groundnut is around 1600 kg/ha, whereas *kharif*-groundnut is around 1000 kg/ha, which is lower than major groundnut-growing countries. Disease and pest attack is considered one of the major factors for low groundnut yield (Singh and Oswalt 1992). Groundnut crop is susceptible to more than 55 disease-causing pathogens including viruses, though all of them may not be economically important (Roy and Shiyani 2000). The following Table 10.1 shows the most common disease-causing pathogens that affect groundnut at various stages, i.e., seed germination, vegetative, and reproductive stages.

Among fungal foliar diseases, very few are economically important in India such as leaf spots (early and late) and rust, which are widely distributed and can cause losses in susceptible genotypes to the extent of 70%–80% when both of them occur together. The leaf spot caused by *Alternaria alternata* is increasingly becoming a major threat to rabi/summer crop and also to kharif groundnut crop. Other fungal foliar diseases such as anthracnose, leaf scorch, *Phomosis* leaf spot, *Phyllosticta* leaf spot, *Pestalotiopsis* leaf spot, *Phoma* leaf diseases, *Drechslera* leaf blight, Zonate leaf spot, and *Cylindrocladium* leaf spot are not economically important at the present time. Aflatoxin contamination caused by *Aspergillus flavus* and *Aspergillus parasiticus* fungi is one of the other important constraints affecting groundnut quality in most Asian countries. Similarly, seed and soilborne diseases such as collar rot, stem rot, and dry root rot have been realized as major limitations in crop production. These diseases cause severe seedling mortality, resulting in patchy crop stands in sandy loam soils, and reduce the yields by 30%–40%. In recent years in India, groundnut production declination has been observed due to several viral diseases such as bud necrosis, peanut (groundnut) mottle, and peanut clump disease and is becoming economically important. Bud necrosis and peanut mottle virus (PMV) is a serious, widespread disease of groundnut,

TABLE 10.1
List of Common Groundnut Diseases in India

Disease	Pathogen	Parts of Plant Affected
Fungal Diseases		
Crown rot or collar rot	*Aspergillus niger*	Seedling
Stem rot	*Sclerotium rolfsii*	Stem, peg, pod
Dry root rot	*Macrophomina phaseolina*	Seedling
Aflaroot/yellow mold	*Aspergillus flavus*	Seed
Early leaf spot	*Cercospora arachidicola*	Leaf
Late leaf spot	*Cercosporidium personatum*	Leaf
Rust	*Puccinia arachidis*	Leaf
Sclerotinia blight	*Sclerotinia minor*	Stem, peg, pod
Alternaria leaf blight	*Alternaria* spp.	Leaf
Rust	*Puccinia arachidis*	Leaf
Aflatoxin contamination	*Aspergillus flavus* and *Aspergillus parasiticus*	Seed
Bacterial Diseases		
Bacterial wilt	*Pseudomonas solanacearum*	Leaf
Nematodes		
Root-knot	*Meloidogyne arenaria, M. hapla,* and *M. javanica*	Root
Kalahasti malady	*Tylenchorhynchus brevelineatus*	Peg and pod
Virus Diseases		
Groundnut bud necrosis	Groundnut bud necrosis tospovirus	Whole plant
Groundnut stem necrosis	Tobacco streak ilarvirus	Leaf
Peanut mottle	Peanut mottle potyvirus	Leaf
Peanut clump	Peanut clump furovirus	Whole plant

with a wide host range; it can cause severe yield losses ranging up to 60% (http://agropedia.iitk.ac.in/node/25554). Herein, this chapter mainly emphasizes types of soil-borne disease-causing pathogens, their symptoms, and disease management approaches such as cultural methods, chemical methods, and biological control.

10.2.1 Influence of Soilborne Disease on Groundnut

India is a country with immense variations in seasons and other ecological aspects. With these remarkable climatic changes from region to region, groundnut plants are susceptible to diseases caused by soilborne fungal pathogens. Soilborne plant pathogenic fungi cause heavy crop losses all over the world. Soilborne diseases such as stem rot, collar rot, and aflaroot are potential threats to groundnut cultivation and yield.

10.2.1.1 Stem Rot/Sclerotium Wilt

- **Disease causing pathogen:** *Sclerotium rolfsii*
  - *Sclerotium rolfsii* Sacc. is a soilborne plant pathogen that causes diseases on a wide range of agricultural and horticultural crops. It has wide geographic diversity and is commonly found in the tropics, subtropics, and other warm temperate regions, especially the Southern United States, Central and South America, West Indies, southern European countries bordering the Mediterranean, Africa, India, Japan, Philippines, and Hawaii (Aycock 1966; Domsch et al. 1980; Farr et al. 1989). Rolfs (1892) first reported this pathogen, and then in 1911 the pathogen was named as *Sclerotium rolfsii* by Saccardo (1911).

- **Disease cycle**
  - *Sclerotium rolfsii* survives in the form of sclerotia that can remain viable in the soil for 2–3 years. They serve as the principle overwintering structures and primary inoculums for disease and are capable of initiating infection with or without an additional food base (Aycock 1966; Punja 1985). High soil moisture, dense planting, and frequent irrigation promote infection. Generally, growth is observed at optimum pH 3–5 and temperature 27°C–30°C. Sclerotia may exist free in the soil or near the soil surface or in association with plant debris. Sclerotia (0.5–2.0 mm diameter) begin to develop after 4–7 days of mycelial growth. The fungus can survive as mycelium in the soil only for short periods (1–2 months). The mycelium can be seed-borne. Eruptive germination is characterized by aggregates of mycelium bursting through the sclerotial rind in soil and is greatest at 21°C–30°C. Initially, sclerotia appear in white, then quickly melanize into a dark brown coloration. The fungus becomes active primarily at the soil surface, and a mat of hyphae is formed over the basal portions of groundnut plants. It penetrates the plant tissues by forming appressoria and grows into the host cells, killed in advance by the fungus. The fungus kills the host tissue by producing large amounts of oxalic acid, pectolytic, and cellulolytic enzymes.
- **Symptoms**
  - The first sign of infection is the development of white fungal threads surrounding the plant tissues, mainly on the stem, which results in sudden wilting of a branch that is completely or partially in contact with the soil. Slowly the leaves turn brown and wilt but remain attached to the plant. Gradually a white mycelium web spreads over the soil and basal canopy of the plant. The white sclerotia of small seed size, produced in the infected tissues, later turn to brown as the disease develops and spreads. The entire plant may be killed, or only two or three branches may be affected. Infections of pegs can take place independently, together with the stem. Lesions on the developing pegs can retard pod development, and seeds in the infected pods show a characteristic bluish-gray discoloration.

10.2.1.2 Collar Rot/Seedling Blight/Crown Rot

- **Disease-causing pathogen:** *Aspergillus niger*, *A. pulverulentum*
  - The mycelium of *Aspergillus niger* fungus is hyaline to subhyaline. Conidiophores start growing directly from the substrate, and their features are septate, thick walled, hyaline, or olive brown in color. The vesicles are mostly globose and have two rows of hyaline phialides, viz. primary and secondary phialides. The conidial heads are dark brown to black. The conidia are globose, dark brown in color, and produce in long chains.
- **Disease cycle**
  - The pathogen survives in plant debris in the soil, not necessarily from a groundnut crop. Soilborne conidia cause disease carryover from season to season. Chances of disease occurrence increase when soil temperature is high (i.e., 30°C–35°C) and low moisture content and seeds are sown deeper in soil. Moreover, infected seeds are also considered a primary source of disease occurrence and are also known as seed-borne in nature.
- **Symptoms**
  - Generally, the disease occurrence is seen in three different stages of groundnut life cycle—seed, young seedling, and adult plants.

1. **Preemergence rot**

   Seeds are infected by soilborne conidia and caused rotting of seeds. The seeds are covered with black masses of spores that make internal tissues of the seed soft and watery; consequently, the seed fails to germinate.

2. **Post-emergence rot**

   The pathogen attacks the emerging young seedling and causes circular brown spots on the cotyledons. Lately, this symptom has also spread to the hypocotyl and stem. Brown discolored spots appear on the collar region. The affected portion becomes soft and rotten, resulting in the collapse of the seedling. The collar region is covered by profuse growth of fungus and conidia, and the affected stem also shows a shredding symptom.

3. **Crown rot**

   When the infection occurs in adult plants, it shows crown rot symptoms. Large lesions develop on the stem below the soil and spread upwards along the branches causing drooping of leaves and wilting of the plant.

10.2.1.3 Aflaroot/Yellow Mold

- **Disease-causing pathogen:** *Aspergillus flavus*
  - Colorless, septate, and branched hyphae are produced by *Aspergillus flavus.* A vesicle is borne at the end of each of the long conidiophores, followed by development of sterigmata that bear chains of yellow-green to blue-green conidia. The conidial heads, each containing a mass of conidial chains, appear as masses up to 600 μm in length and 100 μm in diameter. Sterigmata of *A. flavus* are borne in single or double series on an elongate to subglobose vesicle. The radiate heads split with maturity. *A. flavus* may also produce sclerotia.
- **Disease cycle**
  - The level of yellow mold damage and aflatoxin production is dependent on the environmental conditions and production, harvesting, and storage practices. This pathogen is seedborne as well as soilborne. It becomes active in high humidity (90%–98%), low soil moisture. Favorable growth is observed at 17°C–42°C temperatures in soilborne cases while aflatoxin production in seeds occurs between 25°C and 35°C.
- **Symptoms**
  - The fungus attacks the seed and un-emerged seedlings and rapidly reduces to a shriveled, dried, brown or black mass covered by yellow or greenish spores. Mostly fast decaying is observed when infected seeds are sown. Moreover, the more chances of infection and decaying are with emerging radicals and hypocotyls. The fungus-infected emergence cotyledons lead to the display of necrotic lesions with reddish brown margins. This necrosis terminates at or near the cotyledonary axis. Also, under field conditions the diseased plants are stunted and are often chlorotic. The leaflets are reduced in size with pointed tips, widely varied in shape and sometimes with veinal clearing. Furthermore, the affected plants lack secondary root development.

10.2.2 Management of Soilborne Disease

Disease management requires sensible and acceptable management tools that fulfill the current production system scenario. It demands for cost-effective, easily adaptable, and eco-friendly tools for the effective soilborne diseases management. The vital among them are discussed below:

10.2.2.1 Culture Practices

Heat: In some large nurseries or greenhouses, it may be possible to treat beds or bulk soil with aerated steam. All areas must be brought to a temperature of 160°F–180°F for 30 minutes even after steam treatment; some sclerotia may survive and losses may occur.

Solarization: Clear plastic sheeting of 0.025–0.4 mm thick must be applied to the area for 4–8 weeks, depending upon the time of year. Treated areas should receive direct and full sunlight. Soil solarization may significantly reduce viable sclerotia. But it is expensive and not practical for use in large field situations (Mullen 2001).

Crop sanitation: Crop sanitation by burning of crop residues, deep plowing before sowing is essential for soilborne disease management. Such practices remove the causal fungus of its food base, which makes it difficult to proliferate pathogen and disease development.

Crop rotation: Crop rotation is a deliberate order of crops planted in the same field over several seasons. It is well established that rotation of groundnut with cotton for 3–4 years can greatly reduce soilborne disease incidence and severity. A gap of at least 4–5 years between groundnut crops is advised for most effective control. Groundnut-cotton rotation systems are successfully used in vertisols in central and western India, where stem and pod rots can be severe in the post rainy season. Besides this, it also includes limiting the buildup of fungal inoculum, weed control, and promoting good soil fertility (Liebman and Dyck 1993; Peters et al. 2003).

Irrigation: Maintaining adequate moisture without overwatering peanuts may limit disease incidence by various soilborne pathogens; however, drastic reductions in irrigation can limit yield potential. The dense canopy produced by groundnut imposes a microclimate with increased humidity that may influence disease incidence (Thiessen and Woodward 2012).

Soil amendments: Soil fertility refers to the nutrient availability in a soil, and the fertility of a soil may be affected by crop rotations, soil pH, and soil moisture levels. Improving soil fertility may directly affect pathogens, improve plant health by limiting its susceptibility, or increase antagonistic microbial growth. Soil application of urea, ammonium nitrate, and ammonium bicarbonate is likely to inhibit fungal germination and retard mycelial growth. The cultural practices of controlling disease have limited efficacy and are time consuming. Moreover, some of them are season dependent and not widely accepted and used.

10.2.2.2 Chemical Control

Chemical fungicides are extensively used in current agriculture for controlling disease. They possess inherent limitations such as high cost, nonavailability, toxicity, development of resistant strains, environmental pollution, deteriorating human health, and adverse effects on beneficial soil microflora and fauna, which have compounded the problem in India and all other countries producing groundnut. Due to these reasons, it becomes important to identify alternative methods for plant protection that are less dependent on chemicals and are more environmentally friendly. Nowadays the green revolution has led to intensified agriculture to meet the ever-increasing demands for food and fiber, which is practiced at great cost to the environment, resulting in continuous damage of natural ecosystems, groundwater and food-stuff pollution, and other environmental degradation. Keeping this in view, studies have been initiated to include crop protection and growth promotion using the native microorganisms as a component of integrated pest management (IPM).

Microbial antagonists are widely used for the biocontrol of fungal plant diseases. Mehan et al. (1995) reported that progress has been made in developing screening techniques, identifying sources of resistance, and breeding resistant cultivars. Certain well-studied chemical pesticide management strategies are available for reducing damage by soilborne pathogens, but increasing awareness about the health hazards and environmental problems due to the use of chemical pesticides resulted in the development of IPM. Therefore, in the present situation, greater emphasis has been placed on utilization of biological pesticides or biocontrol agents for management of diseases in groundnut.

10.2.2.3 Biological Control

Biological control has proved to be a promising disease management technology especially against soilborne plant pathogens. Various types of compounds (elicitor molecules) including oligosaccharides, glycoproteins, glycopeptides, and lipids have been shown to induce defense responses in plant cells, and their involvement in the detection of (potential) pathogens in plant has been discussed (Boller 1995; Ebel and Cosio 1994; Hattori and Ohta 1985). Oligosaccharides derived from fungal and plant cell wall polysaccharides are one class of well-characterized elicitors that, in some cases, can induce defense responses at a very low concentration and hence can be used as potent biocontrol agents. Higher plants have the ability to initiate various defense reactions such as the production of phytoalexins, antimicrobial proteins, reactive oxygen species, and reinforcement of cell walls when they are infected by pathogens such as fungi, bacteria, and viruses (Somssica and Hahlbroack 1998). Southern stem rot on peanut has not been successfully managed with fungicides. Pentachloronitrobenzene (PCNB), a widely used fungicide, can reduce disease incidence up to 50% and gives modest yield increases when applied at 11.2 kg ha (Hagan et al. 1986; Hagan et al. 1988; Csinos 1989). Fungicides superior to PCNB are now being evaluated. Several researchers demonstrated that either diniconazole or tebuconazole gave excellent control over southern stem rot when applied on a 14-day schedule throughout the growing season (Backman and Crawford 1985; Csinos 1987; Kvien et al. 1987) whereas Diniconazole and flutolanil provide excellent disease control when applied at the flowering to mid-peg stage of development (Csinos 1987, 1989). Also, *in vitro* antagonism of *Trichoderma harzianum* against *S. rolfsii* has been reported long back (Wells et al. 1972; Chet et al. 1978). Successful control of diseases caused by *S. rolfsii* by application of *T. harzianum* into the soil has also been reported for several crops such as tomato (Elad et al.1980; Charmswarng 1992), groundnut (Grienstein et al. 1979; Maiti and Sen 1985; Muthamilan and Jeyrajan 1996), sugar beet (Upadhyay and Mukhopadhyay 1986), bean (Elad et al. 1980), and snap bean (Papavizas and Lewis 1989). Jadon et al. (2015) reported that seeds treated with tebuconazole 2 DS @ 1.5 g kg^{-1}, mancozeb 75% WP @ 3 g kg^{-1}, carbendazim 12% + mancozeb 63% WP @ 3 g kg^{-1}, respectively, were very effective in the management of soilborne diseases with apparent yield advantage over untreated plots. A commercially available biocontrol agent, i.e., Headline 20% WG, Opera 183 g/L SE, BASF India Ltd., as well as fungal culture filtrate derived from *Sclerotium rolfsii* as elicitors, was used in controlling soilborne diseases and induction of systemic acquired resistance in groundnut crop (Nandini et al. 2010; Amin et al. 2015; Amin and Jampala 2018). Table 10.2 shows various elicitors or beneficial organisms used to induce resistance/tolerance to biotic stress on groundnut plant by various researchers from India and overseas.

TABLE 10.2

List of Beneficial Organisms Used to Induce Resistance in Groundnut against Soilborne Disease

Elicitor	Results	Author/Year
Trichoderma harzianum stain kd	Application of Tkd to groundnut seeds may reduce infection of the groundnut seeds by *Aspergillus flavus*, and hence, it may reduce the contamination of the seed by aflatoxin, especially under drought stress condition. Concurrently, Tkd treatment may result in yields being enhanced by more than 35%.	Kifle et al. (2017)
Trichoderma viride as biocontroller	Seed treatment of *T. viride* JAU60 reduced 51%–58% collar rot disease incidence in different groundnut varieties under pathogen-infected soil culture. *Trichoderma* treatment remarkably increased the 2.3-fold superoxide dismutase, 5-fold guaiacol peroxidase, and 2.5-fold ascorbate peroxidase activities during disease development in tolerant varieties and the same was found about 1.2-, 1.5-, and 2.0-fold, respectively, in susceptible varieties. It possibly suggests the induction of antioxidant defense response by *Trichoderma* biocontroller to combat oxidative burst produced by invading pathogens.	Gajera et al. (2016)
Trichoderma viride	Collar rot disease observed between 31.0% (J-11, GG-2) and 67.4% (GG-20) in different groundnut varieties under pathogen infested without any treatment whereas treatment with *Trichoderma* significantly reduced collar rot disease from 58.1% (J-11, GG-2) to 51.6% (GG-20). The specific activities of polyphenol oxidase and β-1,3-glucanase elevated 3.5- and 2.3-fold, respectively, at 3 days; phenylalanine ammonia lyase evident 1.6-fold higher at 6 days; and chitinase 2.3–2.8 folds up to 9 days in *Trichoderma*-treated + pathogen-infested seedlings of tolerant varieties (J-11, GG-2) compared with moderate and susceptible varieties (GAUG-10, GG-13, GG-20). This study indicates the synergism activation of defense enzymes under the pathogenic conditions or induced resistance by *T. viride* JAU60 in different groundnut varieties susceptible to collar rot disease.	Gajera et al. (2015)
Six *Trichoderma* strains	*Tricoderma viride* (NBAII Tv 23) inhibited 61% growth of phytopathogenic fungi *Sclerotium rolfsii* whereas *T. harzianum* (NBAII Th1) inhibited 55% growth of pathogen and also elevated levels of chitinase; β-1,3-glucanase and protease were observed in biological control of plant pathogen *S. rolfsii*.	Parmar et al. (2015)

(*Continued*)

TABLE 10.2 (*Continued*)

List of Beneficial Organisms Used to Induce Resistance in Groundnut against Soilborne Disease

Elicitor	Results	Author/Year
Aspergillus flavus strain as biocontrol agent	Significant reductions in aflatoxin levels averaging 71% were detected in treated plots with different inoculation treatments.	Alaniz Zanon et al. (2013)
PGPR cells, viz., *Methylobacterium* (MB-5) and *Rhizobium* sp. (RM-5)	The application of PGPR cells as co-aggregates was found to augment the total phenol content and defense enzyme activities, such as peroxidase and polyphenol oxidase content of groundnut plant, to a higher level whereas a reduction in reducing and non-reducing sugar level was recorded, which ultimately led to a reduction in *Sclerotium rolfsii* incidence in rainfed groundnut.	Srinivasan et al. (2013)
Trichoderma viride	*Trichoderma* seed treatment reduced 51.6% of the disease incidence in susceptible varieties (GG20) and 58.1% in tolerant varieties (J11 and GG2) at 15 DAS, under *A. niger* infection in a pot culture study.	Gajera et al. (2011)
Sclerotium rolfsii derived elicitors	There was an increase in POX, PAL, SA, β-1,3-glucanase and lignin content in fungal culture filtrate (FCF) and mycelial cell wall (MCW)-treated plants of four varieties (J-11, GG-20, TG-26, and TPG41) of groundnut when compared to that of control plants.	Nandini et al. (2010)
	Results indicate that the treatment of fungal culture filtrate successfully induce higher accumulation of phenols thereby inducing systemic resistance in the four different varieties of groundnut against *S. rolfsii.*	Nandini et al. (2009)
Rhizobium and *Trichoderma harzianum* (ITCC- 4572)	The results indicated that the application of these native microorganisms successfully decreases the stem rot incidence and also increases the growth of the groundnut plants.	Ganesan et al. (2007)
Pseudomonas aeruginosa GSE 18	Bacterial seed treatment induced the rapid accumulation of defense-related enzymes such as chitinase, β-1,3-glucanase, peroxidase, and phenylalanine ammonia lyase (PAL) in groundnut seedlings when compared with the control. Thus, activation of host defense responses and inhibition of fungal cell wall-degrading enzymes appear to play a significant role in the biocontrol activity of GSE 18 in collar rot of groundnut.	Kishore et al. (2006)
Aqueous leaf extracts (20% w/v) of *Prosopis juliflora* and chlorothalonil	Foliar application of *P. juliflora* extract at 45, 75, and 90 days after sowing (DAS) and chlorothalonil at 60 DAS effectively reduced foliar diseases severity and increased the pod yields by 81%–98%. This study identified *P. juliflora* extract as a significant component for the integrated management of groundnut foliar diseases.	Kishore and Pande (2005)
Chitin-supplemented foliar application of *Serratia marcescens* GPS 5	Thirty-day-old groundnut (cv. TMV 2) plants pretreated with GPS 5 and GSE 18 (with and without supplementation of 1% colloidal chitin) were challenge inoculated after 24 h with *Phaeoisariopsis personata*, the causal agent of late leaf spot (LLS) disease of groundnut. GPS 5 and GSE 18, applied as a prophylactic spray, reduced the lesion frequency by 23% and 67%, respectively, compared with control. Moreover, an enhanced activity of chitinase, β-1,3-glucanase, peroxidase, and phenylalanine ammonia lyase enzymes with Chitin-supplemented foliar application of *Serratia marcescens* GPS 5 when compared with phosphate buffer and colloidal chitin-treated controls. These observations suggest the application of chitinolytic biocontrol agents for improved biocontrol through activation of host plant resistance.	Kishore et al. (2005)
Pseudomonas aeruginosa	Cell-free culture filtrates of *P. aeruginosa* GSE 18 and GSE 19 inhibited the activity *in vitro* of the cell wall-degrading enzymes (CWDE) polygalacturonase and cellulase by *S. rolfsii* up to a maximum of 55% and 50%, respectively, when measured 6 days after inoculation. *Pseudomonas aeruginosa* GSE 18 and GSE 19 with a known tolerance to thiram, a commonly used seed dressing fungicide, suppressed the growth of *S. rolfsii*, inhibited the activity of CWDE, and reduced the incidence of stem rot, suggesting the usefulness of these biocontrol strains as components in the integrated management of groundnut stem rot.	Kishore et al. (2005)

(*Continued*)

TABLE 10.2 (*Continued*)

List of Beneficial Organisms Used to Induce Resistance in Groundnut against Soilborne Disease

Elicitor	Results	Author/Year
Isolates of plant growth-promoting rhizobacteria (PGPR)	Seed inoculation of these three isolates, viz. PGPR1, PGPR2, and PGPR4, resulted in a significantly higher pod yield than the control, in pots, during rainy and post-rainy seasons. The contents of nitrogen and phosphorus in soil, shoot, and kernel were also enhanced significantly in treatments inoculated with these rhizobacterial isolates in pots during both the seasons as well, as all three isolates suppressed the soilborne fungal diseases such as collar rot of peanut caused by *A. niger* while PGPR4 also suppressed stem rot caused by *S. rolfsii*.	Dey et al. (2004)
Water-soluble glucan purified from the culture filtrate of *Acremonium obclavatum*	Treatment of groundnut leaves with glucan showed enhanced levels of endogenous salicylic acid, chitinase, and β-1,3 glucanase activities, which reached its peak on day 8 after treatment. Enhanced activities of chitinases and β-1,3-glucanases may provide a hostile environment to *P. arachidis*, which is known to colonize intercellularly. Results indicate a dual role for *A. obclavatum* glucan in suppressing the development of rust disease on groundnut leaves.	Sathiyabama and Balasubramanian (1998)

Existing spray regimes are targeted to reduce yield losses. Optimizing time for initial application of biocontrol agents is important in reducing yield losses to disease as well as minimizing the costs of utilizing biocontrol agents. In general, application timing of biocontrol agents is based on the disease model and typically 60 days after planting (DAP), i.e., prior to disease incidence and followed by more than one or two sequential applications with biocontrol agents made between 90 and 120 DAP to inhibit complete growth of soilborne pathogens (Langston et al. 2002; Rideout et al. 2008).

10.3 Abiotic Stress

Groundnut is a significant source of plant-based proteins in the diet for humans and livestock, and there is a need to improve abiotic stress tolerance under the changing climate (Araujo et al. 2015). Maintaining the higher nutritional, medicinal, and commercial values of groundnut and its products and improving the response of peanut varieties to environmental stresses are an important goal (Sarkar et al. 2016).

10.3.1 Influence of Water Stress on Groundnut

Water stress is the most important factor limiting crop productivity throughout the world and has been the focus of much research. Groundnut is grown on 35.5 million ha across 82 countries in the world. More than half of the production area, which accounts for 70% of the peanut-growing area, falls under arid and semiarid regions, where groundnut is frequently subjected to drought stresses for different duration and intensities (Reddy et al. 2003). The water stress affects the crop at different growth stages during the growing season. The flowering and pod filling stage is critical for yield and agronomic characters under drought stress. Not only the yield but also the quality of product decreases under drought stresses (Rucker et al. 1995). Drought stress has adverse influence on water relations (Ramesh Babu and Rao 1983), photosynthesis (Bhagsari et al. 1976), mineral nutrition, metabolism, growth, and yield of groundnut (Suther and Patel 1992); moreover, it influences the growth of weeds, agronomic management and nature, and intensity of insects, pests, and diseases (Wightman and Wightman 1994; Wheatley et al. 1989).

In the tropical and subtropical regions, groundnut is cultivated under rainfed conditions by resource-poor farmers where it is exposed to abiotic stresses such as drought and salinity (Bhauso et al. 2014; Sarkar et al. 2014). A major part of global groundnut cultivation takes place in areas exposed to water shortage. Due to water shortage, over six million tons in annual loss of groundnut production occur (Bhatnagar-Mathur et al. 2013; Gautami et al. 2011). The drought-mediated losses in the groundnut production are due to various reasons. Groundnut plants bear aerial flowers and, after pollination, the fertilized ovaries develop into gynophores or pegs (Haro et al. 2011). The peg is a positively geotropic stalklike structure containing embryos in which basal cells of the fertilized ovary elongate (Jacobs 1947). Proembryo development is required for the elongation of pegs, which initiate penetration into the topsoil layer (Periasamy and Sampoornam 1984). This phenomenon is termed as pegging and occurs when the initial pegs attain a length of 3–4 mm (Ziv 1981). Under sufficient water availability, pegging is a fast process (Haro et al. 2008), and the pod growth initiates after this phase resulting in subterranean fruits (Smith 1950; Periasamy and Sampoornam 1984). But due to drought condition, topsoil becomes hard and it restricts or delays peg penetration into the soil. Hence, failure in the pegging stage results in limited pod set and subsequent seed numbers (Haro et al. 2010, 2011). However, in contrast, if the drought-affected pegs are rewatered, viable pegs start penetrating the soil again (Haro et al. 2008). The resumption of pegging upon rewatering is an adaptive trait in peanut plants that ensures long-term survival of the fertilized embryo compared with other grain crops such as maize and soybean where embryo viability is rapidly lost (Westgate and Boyer 1986). Overall, drought is the major abiotic stress that limits peanut crop yield (Boote et al. 1976; Haro et al. 2008). Furthermore, owing to a lack of genotypic variability for better water use efficiency among groundnut plants (Gautami et al. 2011), the development of drought-tolerant peanut genotypes is a major focus of plant breeders (Sarkar et al. 2016).

10.3.2 Water Stress Management

There are several farming practices, operations, and actions that could be taken to repay a successful production of groundnut crop under arid and semiarid conditions.

10.3.2.1 Chemical Treatment and Agronomical Crop Management Practices

Chemical treatment and agronomical crop management practices have been tried to improve the water deficit effects (Manivannan et al. 2007). Triazole compounds, such as triadimefon (TDM), ketoconazole (KCZ), paclobutrazol (PBZ), propiconazole (PCZ), etc., are widely used as fungicides; moreover, they possess varying degrees of plant growth-regulating properties (Fletcher et al. 2000). It is important to understand how fungicides cum plant growth regulator respond to changes in the global environmental conditions and how they interact with the crops for which they are used. Triazoles have been called plant multi-protectants because of their ability to induce tolerance by a reduction in free radical damage and increase in antioxidant potential in plants to environmental and chemical stresses (Jaleel et al. 2007). Triazole-treated plants have a more efficient free radical scavenging system that enables them to detoxify active oxygen (Kopyra and Gwozdz 2003). Triazoles affect the isoprenoid pathway and alter the levels of certain plant hormones by inhibiting gibberellin synthesis, reducing ethylene evolution and elevating cytokinin levels (Kamountsis and Chronopoulon 1999). Paclobutrazol has been used to induce water stress tolerance in some plants such as *Arachis hypogaea* L. and *Phillyrea angustifolia* L. (Fernandez et al. 2006; Sankar et al. 2007). Tekalign and Hammes (2004) reported induction of chlorophyll a and chlorophyll b contents in potato leaf, which resulted in water stress tolerance. The proline content also increased with increasing water stress tolerance (Jaleel et al. 2007; Srivastav et al. 2010). When plants are subjected to water stress, proline is synthesized from glutamic acid to act as osmoprotectant for keeping the water balance in cells and the outer environment (Delauney and Verma 1993). Higher water stress tolerance is also caused from an increase of antioxidants, a decrease of electrolyte leakage and lipid peroxidation, which is a well-known indicator to evaluate cell membrane damage (Gopi et al. 2007; Jaleel et al. 2007; Srivastav et al. 2010). Some physiological changes, such as reduction of plant height and leaf size, increase of leaf thickness and stomatal density are also known to occur (Sankhla et al. 1985; Burrows et al. 1992). Sankar et al. (2014) reported that Paclobutrazol (PBZ) treatment increased the root length when compared to control, while it decreased the stem length and leaf area. Abscisic acid (ABA)-treated plants decreased root length, while increasing the stem length and leaf area when compared to control. Fresh and dry weight decreased in drought-stressed plants when compared to control. PBZ and ABA applied to the drought-stressed plants increased the fresh and dry weight. PBZ and ABA treatments decreased the fresh and dry weight when compared to the unstressed plants. Growth regulator treatments with paclobutrazol and abscisic acid to the drought stressed plants have great impact on the anatomy of *A. hypogaea* plants (Sankar et al. 2013). One possible approach to reduce the effect of water deficit on plant productivity is through the addition of calcium supplements to irrigation water, as previously done in the case of salt stress (Singh and Jain 1982).

10.3.2.2 Biotechnological Approaches

Biotechnological approaches have made it possible for researchers to identify genetic resources and the manipulation of these resources for the development of transgenic stress-tolerant crops. Some comprehensive detailed reviews reported advances in genomics and genetic engineering for cultivated groundnut improvement (Pandey et al. 2012; Krishna et al. 2015). Some of the important components of these approaches are discussed here.

Drought tolerance AtDREB1A transgenic groundnut (cv. JL24) plants exhibited considerable drought tolerance (Bhatnagar et al. 2009). Manjulatha et al. (2014) found that overexpression of a *PDH45* gene (pea DNA helicase) improved drought tolerance in groundnut and so its productivity. Also, the regulation of isopentenyl transferase (IPT) gene expression resulted in drought tolerance in groundnut and improved its yield attributes (Qin et al. 2011). Bhatnagar et al. (2008) found that the transfer of a single gene may not fully confer plant tolerance to multiple stresses and achieve the required level of tolerance. Rather, it is more sensible to transfer a gene-encoding transcription factor (Sarkar et al. 2014). By binding specific DNA sequences, TFs offer a secondary level of control over the expression of genes under different environmental conditions including biotic and abiotic stresses to regulate growth and development accordingly (Bartels and Sunkar 2005; Mitsuda and Ohme-Takagi 2009; Nakashima et al. 2009; Yamaguchi-Shinozaki and Shinozaki 2006). Overexpressing *DREB1A* transcription factor by transgenic groundnut plants showed drought tolerance and exhibited enhanced yield (Bhatnagar et al. 2013). Similarly, constitutive over-expression of *AtDREB2A, AtHB7,* and *AtABF3TFs* genes in groundnut plants resulted in higher biomass under abiotic stresses (Pruthvi et al. 2014). Thus, transgenic groundnut plants encoding stress-responsive genes and transcription factors (TFs) can acquire improved abiotic stress tolerance. With this perspective, many TF genes such as *DREB, MYB,* and *bZIP* have been identified that modulate plant molecular stress responses under various stresses (Ahuja et al. 2010). *MYB* TFs are reported as the largest family of TFs in eukaryotes (Du et al. 2012). Up until now, 30 *MYB* TF genes have been identified in groundnut plants with their significant involvement in the biological processes (Chen et al. 2014, 2016). Several researchers reported that *MYB* proteins regulate plant primary and secondary metabolisms, cellular development, cell cycle regulation, hormone and flavonoid biosynthesis, and (the most significant) signal transduction process as well as plant responses to biotic and abiotic stress conditions (Dubos et al. 2008; Feller et al. 2011; Ma et al. 2009; Stracke et al. 2001). Genome-wide analysis has shown considerable diversity of TFs and their involvement in *Arachis* abiotic stress tolerance (Wang et al. 2017). Pandurangaiah et al. (2014) reported overexpression of horse gram (*Macrotyloma uniflorum* Lam. Verdc.) NAC transcriptional factor (MuNAC4) in groundnut confers enhanced drought tolerance by reducing the damage to membrane structures and enhancing osmotic adjustment and antioxidative enzyme regulation under stress. This study validates MuNAC4 as an important candidate gene for future phytoengineering approaches for drought tolerance in crop plants. Recently, transgenic groundnut plants overexpressing *MuWRKY3* transcription factor conferred drought tolerance linked with better antioxidant capacity (Kiranmai et al. 2018). Similarly, transgenic groundnut plants expressing an *Arabidopsis* TF (*AtHDG11*) exhibited enhanced drought and salinity tolerance (Banavath et al. 2018). Still, fewer genes or TFs have been targeted to develop transgenic groundnut plants for optimum yield and this requires further investigation.

REFERENCES

Agropedia. Integrated disease management in groundnut. http://agropedia.iitk.ac.in/node/25554.

Ahuja, I., de Vos, R.C., Bones, A. M., Hall, R.D. (2010). Plant molecular stress responses face climate change. *Trends in Plant Science*, 15: 664–674.

Alaniz Zanon, M.S., Chiotta, M.L., Giaj-Merlera, G., Barros, G., Chulze, S. (2013). Evaluation of potential biocontrol agent for aflatoxin in Argentinean peanuts. *International Journal of Food Microbiology*, 162: 220–225.

Amin, D., Jampala, S.S.M., Patel, D. (2015). Induced systemic resistance in groundnut by foliar application of Headline (Pyraclostrobin 20%WG). *Archives of Phytopathology and Plant Protection*, 48(7): 578–587.

Amin, D., Jampala, S.S.M. (2018). Evaluating the physiological effect of strobilurin-based fungicide (Opera 183 g/L SE) against *Sclerotium rolfsii* and *Aspergillus niger* pathogens of *Arachis hypogaea* L. *Archives of Phytopathology and Plant Protection*, 51(9–10): 485–504.

Araujo, S.S., Beebe, S., Crespi, M., Delbreil, B., González, E.M., Gruber, V. et al. (2015). Abiotic stress responses in legumes, strategies used to cope with environmental challenges. *Critical Reviews in Plant Science*, 34(1–3): 237–280.

Aycock, R. (1966). Stem rot and other diseases cause by *Sclerotium rolfsii. Agriculture Experimental Station*, 174: 202.

Backman, P.A., Crawford, M.A. (1985). Effects of triazole fungicides on soil-borne diseases of peanuts. *Proceedings of American Peanut Research and Education Association*, 17: 42.

Banavath, J.N., Chakradhar, T., Pandit, V., Konduru, S., Guduru, K.K., Akila, C.S. et al. (2018). Stress inducible overexpression of *AtHDG11* leads to improved drought and salt stress tolerance in peanut (*Arachis hypogaea* L.). *Frontiers in Chemistry*, 6. https://doi.org/10.3389/fchem.2018.00034.

Bartels, D., Sunkar, R. (2005). Drought and salt tolerance in plants. *Critical Reviews in Plant Sciences*, 24: 23–58.

Bhagsari, A.S., Brown, R.H., Schepers, J.S. (1976). Effect of moisture stress on photosynthesis and some related physiological characteristics in peanuts. *Crop Science*, 16: 712–715.

Bhatnagar-Mathur, P., Devi, M.J., Vadez, V., Sharma, K.K. (2009). Differential antioxidative responses in transgenic peanut bear no relationship to their superior transpiration efficiency under drought stress. *Journal of Plant Physiology*, 166: 1207–1217.

Bhatnagar-Mathur, P., Rao, J.S., Vadez, V., Dumbala, S.R., Rathore, A., Yamaguchi-Shinozaki, K. et al. (2013). Transgenic peanut overexpressing the *DREB1A* transcription factor has higher yields under drought stress. *Molecular Breeding*, 33: 327–340.

Bhatnagar-Mathur, P., Vadez, V., Sharma, K.K. (2008). Transgenic approaches for abiotic stress tolerance in plants: Retrospect and prospects. *Plant Cell Reports*, 27(3): 411–424.

Bhauso, T.D., Radhakrishnan, T., Kumar, A., Mishra, G.P., Dobaria, J.R., Rajam, M.V. (2014). Over-expression of bacterial *mtlD* gene confers enhanced tolerance to salt-stress and water-deficit stress in transgenic peanut (*Arachis hypogaea*) through accumulation of mannitol. *Australian Journal of Crop Science*, 8: 413–421.

Birthal, P.S., Rao, P.P., Nigam, S.N., Bantilan, C.S., Bhagavatulu, S. (2010). *Groundnut and Soybean economies in Asia: Facts, trends and outlook.* Andhra Pradesh, India: International Crops Research Institute for the Semi-Arid Tropics, p. 92.

Boller, T. (1995). Chemoperception of microbial signals in plant cells. *Annual Review of Plant Physiology and Plant Molecular Biology*, 46: 189–214.

Boote, K.J., Varnell, R.J., Duncan, W.G. (1976). Relationships of size, osmotic concentration and sugar concentration of peanut pods to soil water. *Proceedings of the Soil and Crop Science Society of Florida*, 35: 47–50.

Burrows, G.E., Boag, T.S., Stewart, W.P. (1992). Changes in leaf, stem, and root anatomy of *Chrysanthemum* cv. Lillian Hoek following paclobutrazol application. *Journal of Plant Growth Regulation*, 11: 189–194.

Charmswarng, C. (1992). Biological control of *Sclerotium* stem rot of tomato in Thailand. *Journal of Plant Proto Tropics*, 9: 77–83.

Chen, N., Su, M., Chi, X., Zhang, Z., Pan, L., Chen, M. et al. (2016). Transcriptome analysis reveals salt-stress-regulated biological processes and key pathways in roots of peanut (*Arachis hypogaea* L.). *Genes & Genomics*, 38(6): 493–507.

Chen, N., Yang, Q., Pan, L., Chi, X., Chen, M., Hu, D., Yang, Z. et al. (2014). Identification of 30 MYB transcription factor genes and analysis of their expression during abiotic stress in peanut (*Arachis hypogaea* L.). *Gene*, 533: 332–345.

Chet, L., Hadar, Y., Elad, Y., Henis, Y. (1978). Biological control of soil-borne plant pathogens by *Trichoderma harzianum. Proceedings 3rd International Congress of Plant Pathology*, 185.

Csinos, A.S. (1987). Control of southern stem rot and Rhizoctonia limb rot of peanut with flutolanil. *Peanut Science*, 14: 55–58.

Csinos, A.S. (1989.) Targeting fungicides for control of southern stem rot on peanut. *Plant Disease*, 73: 723–726.

Delauney, A.J., Verma, D.S. (1993). Proline biosynthesis and osmoregulation in plants. *Plant Journal*, 4: 215–223.

Dey, R., Pal, K., Bhatt, D., Chauhan, S. (2004). Growth promotion and yield enhancement of peanut (*Arachis hypogaea* L.) by application of plant growth-promoting rhizobacteria. *Microbiological Research*, 159: 371–394.

Domsch, K.H., Gams, W., Anderson, T.H. (1980). *Compendium of Soil Fungi*. London, New York: Academic Press.

Du, H., Yang, S.S., Liang, Z., Feng, B.R., Liu, L., Huang, Y.B. et al. (2012). Genome-wide analysis of the MYB transcription factor superfamily in soybean. *BMC Plant Biology*, 12: 106.

Dubos, C., Le Gourrierec, J., Baudry, A., Huep, G., Lanet, E., Debeaujon, I. et al. (2008). *MYBL2* is a new regulator of flavonoid biosynthesis in *Arabidopsis thaliana*. *Plant Journal*, 55: 940–953.

Ebel, J., Cosio, E.G. (1994). Elicitors of plant defense responses. *International Review of Cytology*, 148: 1–36.

Elad, Y., Chet, I., Katan, J. (1980). *Trichoderma harzianum*, a biological agent effective against *Sclerotium rolfsii* and *Rhizoctonia solani*. *Phytopathology*, 70: 119–121.

FAO. (2010). Retrieved from http://faostat.fao.org/default.aspx. Accessed August 2010.

Farr, D.F., Bills, G.F., Chamuris, G.P., Rossman, A.Y. (1989). Fungi on Plants and Plant Products in the United States. VIII, 1252 S. The American Phytopathological Society (APS) Press, St. Paul (Minnesota), ISBN 0-89054-099-3.

Feller, A., Machemer, K., Braun, E.L., Grotewold, E. (2011). Evolutionary and comparative analysis of MYB and bHLH plant transcription factors. *Plant Journal*, 66: 94–116.

Fernandez, J.A., Balenzategui, L., Banon, S., Franco, J.A. (2006). Induction of drought tolerance by paclobutrazol and irrigation deficit in *Phillyrea angustifolia* during the nursery period. *Scientia Horticulturae*, 107: 277–283.

Fletcher, R.A., Gilley, A., Davis, T.D., Sankhla, N. (2000). Triazoles as plant growth regulators and stress protectants. *Horticulture Review*, 24: 55–138.

Gajera, H., Rakholiya, K., Vakharia, D. (2011). Bioefficacy of *Trichoderma* isolates against *Aspergillus niger* Van Tieghem inciting collar rot in groundnut (*Arachis hypogaea* L.). *Journal of Plant Protection Research*, 51(3): 240–247.

Gajera, H.P., Katakpara, Z.A., Patel, S.V., Golakiya, B.A. (2016). Antioxidant defense response induced by *Trichoderma viride* against *Aspergillus niger* Van Tieghem causing collar rot in groundnut (*Arachis hypogaea* L.). *Microbial Pathogenesis*, 91: 26–34.

Gajera, H.P., Savaliya, D.D., Patel, S.V., Golakiya, B.A. (2015). *Trichoderma viride* induces pathogenesis related defense response against rot pathogen infection in groundnut (*Arachis hypogaea* L.). *Infection, Genetics and Evolution*, 34: 314–325.

Ganesan, S., Kuppusamy, R., Sekar, R. (2007). Integrated management of stem rot disease (*Sclerotium rolfsii*) of groundnut (*Arachis hypogaea* L.) using *Rhizobium* and *Trichoderma harzianum* (ITCC-4572). *Turkish Journal of Agriculture and Forestry*, 31: 103–108.

Gautami, B., Pandey, M.K., Vadez, V., Nigam, S.N., Ratnakumar, P., Krishnamurthy, L. et al. (2011). Quantitative trait locus analysis and construction of consensus genetic map for drought tolerance traits based on three recombinant inbred line populations in cultivated groundnut (*Arachis hypogaea* L.). *Molecular Breeding*, 30: 757–772.

Gopi, R., Jaleel, C.A., Sairam, R., Lakshmanan, G.A., Gomathinayagam, M., Panneerselvam, R. (2007). Differential effects of hexaconazole and paclobutrazol on biomass, electrolyte leakage, lipid peroxidation and antioxidant potential *of Daucus carota* L. *Colloids and Surfaces B: Biointerfaces*, 60: 180–186.

Grienstein, A., Elad, Y., Katan, Y., Chet, I. (1979). Control of *Sclerotium rolfsii* by means of herbicides and *Trichoderma harzianum*. *Plant Disease Report*, 63: 823–826.

Hagan, A.K., Weeks, J.R., McGuire, J.A. (1988). Comparison of soil insecticides alone and in combination with PCNB for suppression of southern stem rot. *Peanut Science*, 15: 35–38.

Hagan, A.K., Weeks, J.R., Reed, R.B. (1986). Southern stem rot suppression on peanut with the insecticide chlorpyrifos. *Peanut Science*, 13: 36–37.

Haro, R.J., Dardanelli, J.L., Otegui, M.E., Collino, D.J. (2008). Seed yield determination of peanut crops under water deficit, soil strength effects on pod set, the source-sink ratio and radiation use efficiency. *Field Crops Research*, 109: 24–33.

Haro, R.J., Dardanelli, J.L., Otegui, M.E., Collino, D.J. (2010). Water deficit and impaired pegging effects on peanut seed yield, links with water and photosynthetically active radiation use efficiencies. *Crop and Pasture Science*, 61: 343–352.

Haro, R.J., Mantese, A., Otegui, M.E. (2011). Peg viability and pod set in peanut, response to impaired pegging and water deficit. *Flora*, 206: 865–871.

Hattori, T., Ohta, Y. (1985). Induction of phenylalanine ammonia-lyase activation and iso-avone glucosiden accumulation in suspension-cultured cells of red bean, *Vigna angularis*, by phytoalexin elicitors, vanadate, and elevation of medium pH Plant. *Cell Physiology*, 26: 1101–1110.

Jacobs, W.P. (1947). The development of the gynophore of the peanut plant (*Arachis hypogaea* L.). The distribution of mitosis, the region of greatest elongation, and the maintenance of vascular continuity in the intercalary meristem. *American Journal of Botany*, 34: 361–371.

Jadon, K.S., Thirumalaisamy, P.P., Kumar, V., Koradia, V.G., Padavi, R.D. (2015). Management of soil borne diseases of groundnut through seed dressing fungicides. *Crop Protection*, 78: 198–203.

Jaleel, C.A., Manivannan, P., Sankar, B., Kishorekumar, A., Sankari, S., Panneerselvam, R. (2007). Paclobutrazol enhances photosynthesis and ajmalicine production in *Catharanthus roseus. Process Biochemistry*, 42: 1566–1570.

Kamountsis, A.P., Chronopoulon-Sereli, A.G. (1999). Paclobutrazol affects growth and flower bud production in gardenia under different light regimes. *Horticultural Science*, 34: 674–675.

Kifle, M.H., Yobo, K.S., Laing, M.D. (2017). Biocontrol of Aspergillus flavus in groundnut using *Trichoderma harzianum* stain kd. *Journal of Plant Diseases and Protection*, 124(1), 51–56.

Kiranmai, K., Rao, L.G., Pandurangaiah, M., Nareshkumar, A., Reddy, A.V., Lokesh, U. (2018). A novel WRKY transcription factor, MuWRKY3 (*Macrotyloma uniflorum* Lam. Verdc.) enhances drought stress tolerance in transgenic groundnut (*Arachis hypogaea* L.) plants. *Frontiers in Plant Science*, 9. https://doi//org/10.3389/fpls.2018.00346.

Kishore, G., Pande, S., Rao, J., Podile, A. (2005). *Pseudomonas aeruginosa* inhibits the plant cell wall degrading enzymes of *Sclerotium rolfsii* and reduces the severity of groundnut stem rot. *European Journal of Plant Pathology*, 113: 315–320.

Kishore, G., Pande, S. (2005). Integrated management of late leaf spot and rust diseases of groundnut (*Arachis hypogaea* L.) with *Prosopis juliflora* leaf extract and chlorothalonil. *International Journal of Pest Management*, 51(4): 325–332.

Kishore, G.K., Pande, S., Podile, A.R. (2006). *Pseudomonas aeruginosa* GSE 18 inhibits the cell wall degrading enzymes of *Aspergillus niger* and activates defence-related enzymes of groundnut in control of collar rot disease. *Australasian Plant Pathology*, 35: 259–263.

Kopyra, M., Gwozdz, E.A. (2003). Antioxidant enzymes in paraquat and cadmium resistant cell lines of horseradish. *Biology Letters*, 40: 61–69.

Krishna, G., Singh, B.K., Kim, E.K., Morya, V.K., Ramteke, P.W. (2015). Progress in genetic engineering of peanut (*Arachis hypogaea* L.): A review. *Plant Biotechnology Journal*, 13(2), 147–162.

Kvien, C.S., Csinos, A.S., Ross, L.F., Conkerton, E.J., Styer, C. (1987). Diniconazole's effect on peanut (*Arachis hypogaea* L.) growth and development. *Journal of Plant Growth Regulation*, 6: 233–244.

Langston Jr, D.B., Phipps, P.M., Stipes, R.J. (2002). An algorithm for predicting outbreaks of Sclerotinia blight of peanut and improving the timing of fungicide sprays. *Plant Disease*, 86(2), 118–126.

Liebman, M., Dyck, E. (1993). Crop rotation and intercropping strategies for weed management. *Ecological Applications*, 3(1), 92–122.

Ma, Q., Dai, X., Xu, Y., Guo, J., Liu, Y., Chen, N. et al. (2009). Enhanced tolerance to chilling stress in *OsMYB3R-2* transgenic rice is mediated by alteration in cell cycle and ectopic expression of stress genes. *Plant Physiology*, 150(1), 244–256.

Maiti, D., Sen, C. (1985). Integrated biocontrol of *Sclerotium rolfsii* with nitrogenous fertilizer and *Trichoderma harzianum. Indian Journal of Agricultural Science*, 55: 464–467.

Manivannan, P., Jaleel, C.A., Kishorekumar, A., Sankar, B., Somasundaram, R., Sridhara, R., Panneerselvam, R. (2007). Propiconazole induced changes in antioxidant metabolism and drought stress amelioration in *Vigna unguiculata* (L.) Walp. by propiconazole under water deficit stress. *Colloids and Surfaces B: Biointerfaces*, 57: 69–74.

Manjulatha, M., Sreevathsa, R., Kumar, A.M., Sudhakar, C., Prasad, T.G., Tuteja, N. et al. (2014). Overexpression of a pea DNA helicase (*PDH45*) in peanut (*Arachis hypogaea* L.) confers improvement of cellular level tolerance and productivity under drought stress. *Molecular Biotechnology*, 56(2), 111–125.

Mehan, V.K., Mayee, C.D., Brenneman, T.B., McDonald, D. (1995). Stem and pod rots of groundnut. Information Bulletin No. 44, *International Crops Research Institute for the Semi-Arid Tropics*, Andhra Pradesh, India: Peanut Collaborative Research Support Program, p. 28.

Mitsuda, N., Ohme-Takagi, M. (2009). Functional analysis of transcription factors in Arabidopsis. *Plant Cell Physiology*, 50: 1232–1248.

Mullen J. (2001). Southern blight, Southern stem blight, White mold. *The Plant Health Instructor*. doi:10.1094/PHI-I-2001-0104-01.

Muthamilan M, Jeyrajan R. (1996). Integrated management of *Sclerotium* root rot of groundnut involving *Trichoderma harzianum, Rhizobium* and carbendazim. *Indian Journal of Mycology and Plant Pathology*, 26: 204–209.

Nakashima, K., Ito, Y., Yamaguchi-Shinozaki, K. (2009). Transcriptional regulatory networks in response to abiotic stresses in Arabidopsis and grasses. *Plant Physiology*, 149: 88–95.

Nandini, D., Mohan, J.S.S., Gunjan, S. (2010). Induction of systemic acquired resistance in *Arachis hypogaea* L. by *Sclerotium rolfsii* derived elicitors. *Journal of Phytopathology*, 158: 594–600.

Nandini, D., Mohan, J.S.S. (2009). Qualitative profiling of induced phenols in different varieties of *Arachis hypogaea* L. by using fungal culture filtrate of *Sclerotium rolfsii*. *PRAJNA-Journal of Pure and Applied Sciences*, 17: 17–22.

Pandey, M.K., Monyo, E., Ozias-Akins, P., Liang, X., Guimarães, P., Nigam, S.N. et al. (2012). Advances in *Arachis* genomics for peanut improvement. *Biotechnology Advances*, 30: 639–651.

Pandurangaiah, M., Rao, G.L., Sudhakarbabu, O., Nareshkumar, A., Kiranmai, K., Lokesh, U., Thapa, G., Sudhakar, C., 2014. Overexpression of horsegram (*Macrotyloma uniflorum* Lam. Verdc.) NAC transcriptional factor (MuNAC4) in groundnut confers enhanced drought tolerance. *Molecular Biotechnology*, 56(8), 758–769.

Papavizas, G.S., Lewis, J.S. (1989). Effect of *Gliocladium* and *Trichoderma* on damping off and blight of snapbean caused by *Sclerotium rolfsii* Sacc. in green house. *Plant Pathology*, 38: 278–286.

Parmar, H.J., Bodar, N.P., Lakhani, H.N., Patel, S.V., Umrania, V.V., Hassan, M.M. (2015). Production of lytic enzymes by *Trichoderma* strains during in vitro antagonism with *Sclerotium rolfsii*, the causal agent of stem rot in groundnut. *African Journal of Microbiology Research*, 9(6): 365–372.

Periasamy, K., Sampoornam, C. (1984). The morphology and anatomy of ovule and fruit development in *Arachis hypogaea* L. *Annals of Botany*, 53: 399–411.

Peters, R.D., Sturz, A.V., Carter, M.R., Sanderson, J.B. (2003). Developing disease-suppressive soils through crop rotation and tillage management practices. *Soil and Tillage Research*, 72(2): 181–192.

Pruthvi, V., Narasimhan, R., Nataraja, K.N. (2014). Simultaneous expression of abiotic stress responsive transcription factors, *AtDREB2A, AtHB7* and *AtABF3* improves salinity and drought tolerance in peanut (*Arachis hypogaea* L.). *PLoS One*, 9: e111152.

Punja, Z.K. (1988). *Sclerotium* (Athelia) *rolfsii*, a pathogen of many plant species. In *Advances in Plant Pathology*, Vol. 6, G.S. Sindhu (Ed.), San Diego, CA: Academic Press, pp. 523–534.

Qin, H., Gu, Q., Zhang, J., Sun, L., Kuppu, S., Zhang, Y. et al. (2011b). Regulated expression of an isopentenyl transferase gene (*IPT*) in peanut significantly improves drought tolerance and increases yield under field conditions. *Plant Cell Physiology*, 52(11), 1904–1914.

Ramesh Babu, V., Rao, D.V.M. (1983). Water stress adaptations in the groundnut (*Arachis hypogaea* L): foliar characteristics and adaptations to moisture stress. *Plant Physiology & Biochemistry*, 10(1): 64–80.

Reddy, T.Y., Reddy, V.R., Anbumozhi, V. (2003). Physiological responses of groundnut (*Arachis hypogaea* L.) to drought stress and its amelioration: A critical review. *Plant Growth Regulation*, 41: 75–88.

Rideout, S.L., Brenneman, T.B., Culbreath, A.K., Langston Jr, D.B. (2008). Evaluation of weather-based spray advisories for improved control of peanut stem rot. *Plant Disease*, 92(3), 392–400.

Rolfs, P.H. (1892). The tomato and some of its disease Florida University of Agriculture experimental station. *Bulletin*, 21: 1–38.

Roy, B.C., Shiyani, R.L. (2000). Rainfed groundnut in India: Prioritizing production constraints and implication for future research. *Bangladesh Journal of Agricultural Economics*, 33: 19–34.

Rucker, K.S., Kvien, C.K., Holbrook, C.C., Hook, J.E. (1995). Identification of peanut genotypes with improved drought avoidance traits. *Peanut Science*, 22: 14–18.

Saccardo, P.A. (1911). Notae Mycologicae. *Annals Mycologici*, 9: 249–257.

Sankar, B., Gopinathan, P., Karthishwaran, K., Somasundaram, R. (2014). Variation in growth of peanut plants under drought stress condition and in combination with paclobutrazol and abscisic acid. *Current Botany*, 5: 14–21.

Sankar, B., Jaleel, A., Manivannan, P., Kishorekumar, A., Somasundaram, R., Panneerselvam, R. (2007). Effect of paclobutrazol on water stress amelioration through antioxidants and free radical scavenging enzymes in *Arachis hypogaea* L. *Colloids and Surfaces B: Biointerfaces*, 60: 229–235.

Sankar, B., Karthishwaran, K., Somasundaram, R. (2013). Leaf anatomical changes in peanut plants in relation to drought stress with or without paclobutrazol and abscisic acid. *Journal of Phytology*, 5: 25–29.

Sankhla, N., Davis, T.D., Upadhyaya, A., Sankhla, D., Walser, R.H., Smith, B.N. (1985). Growth and metabolism of soybean as affected by paclobutrazol. *Plant and Cell Physiology*, 26: 913–921.

Sarkar, T., Thankappan, R., Kumar, A., Mishra, G.P., Dobaria, J.R. (2014). Heterologous expression of the *AtDREB1A* gene in transgenic peanut conferred tolerance to drought and salinity stresses. *PLoS One*, 9: e110507.

Sarkar, T., Thankappan, R., Kumar, A., Mishra, G.P., Dobaria, J.R. (2016). Stress Inducible Expression of *AtDREB1A* transcription factor in transgenic peanut (*Arachis hypogaea* L.) conferred tolerance to soil-moisture deficit stress. *Frontiers in Plant Science*, 7: 935.

Sathiyabama, M., Balasubramanian, R. (1998). Effect of *Acremonium obclavatum* glucan on groundnut leaf rust disease caused by *Puccinia arachidis* Speg. *World Journal of Microbiology & Biotechnology*, 14: 783–784.

Singh, G., Jain, S. (1982). Effect of some growth regulators on certain biochemical parameters during seed development in chickpea under salinity. *Indian Journal of Plant Physiology*, 25: 167–179.

Singh, F., Oswalt, D.L. (1992). Major diseases of groundnut. Report from ICRISAT Human Resource Development Program-skill development series No. 6, pp. 36–50.

Smith, B.W. (1950). *Arachis hypogaea* L. Aerial flower and subterranean fruit. *American Journal of Botany*, 37: 802–815.

Somssica, I.E., Hahlbrock, K. (1998). Pathogen defenses in plants a paradigm of biological complexity. *Trends in Plant Science*, 3(3): 86–90.

Srinivasan, T., Chitra, A.M., Sekar, C. (2013). Application Effect Of Exopolysaccharide (Eps) Rich, Pgpr Coaggregates On The Enhancement Of Isr Mediated Biocontrol In Groundnut *Sclerotium Rolfsii* Pathosystem Under Rainfed Condition. *Indian Streams Research Journal*, 3: 3–9.

Srivastav, M., Kishor, A., Dahuja, A., Sharma, R. (2010). Effect of paclobutrazol and salinity on ion leakage, proline content and activities of antioxidant enzymes in mango (*Mangifera indica* L.). *Scientia Horticulture*, 125: 785–788.

Stracke, R., Werber, M., Weisshaar, B. (2001). The *R2R3-MYB* gene family in *Arabidopsis thaliana*. *Current Opinion in Plant Biology*, 4: 447–456.

Suther, D.M., Patel, M.S. (1992). Yield and nutrient absorption by groundnut and iron availability in soil as influenced by lime and soil water. *Journal of the Indian Society of Soil Science*, 40(3): 594–596.

Tekalign, T., Hammes, P.S. (2004). Response of potato grown under non-inductive condition to paclobutrazol: shoot growth, chlorophyll content, net photosynthesis, assimilate partitioning, tuber yield, quality, and dormancy. *Plant Growth Regulation*, 43: 227–236.

Thiessen, L.D., Woodward, J.E. (2012). Diseases of peanut caused by soilborne pathogens in the southwestern United States. *International Scholarly Research Network ISRN Agronomy*, 2012: 1–9.

Upadhyay, J.P., Mukhopadhyay, A.N. (1986). Biological control of *Sclerolium rolfsii* by *Trichoderma harzianum* in sugarbeet. *Tropical Pest Management*, 32: 215–220.

Wang, P., Song, H., Li, C., Li, P., Li, A., Guan, H. et al. (2017). Genome-wide dissection of the heat shock transcription factor family genes in *Arachis*. *Frontiers in Plant Science*, 8: 106.

Wells, H.D., Bell, D.K., Jarowskii, C.A. (1972). Efficacy of *Trichoderma harzianum* as biocontrol agent for *Sclerolium rolfsii*. *Phytopathology*, 62: 442–447.

Westgate, M.E., Boyer, J.S. (1986). Reproduction at low silk and pollen water potentials in maize. *Crop Science*, 26: 951–956.

Wheatley, A.D., Whiteman, J.A., Williams, J.H., Wheatly, S.J. (1989). The influence of drought stress on the distribution of insects on four groundnut genotypes grown near Hyderabad. *India Bulletin of Entomological Research*, 79: 567–577.

Wightman, J.A., Wightman, A.S. (1994). An insect, agronomic and sociological survey of groundnut fields in southern Africa. *Agriculture Ecosystems & Environment*, 51(3): 311–331.

Yamaguchi-Shinozaki, K., Shinozaki, K. (2006). Transcriptional regulatory networks in cellular responses and tolerance to dehydration and cold stresses. *Annual Review of Plant Biology*, 57(1), 781–803.

Ziv, M. (1981). Photomorphogenesis of the gynophore, pod and embryo in peanut, *Arachis hypogaea* L. *Annals of Botany*, 48: 353–359.

Section IV

Crop Responses to Microbe-Mediated Abiotic Stress Management

11

Microbial Mitigation of Abiotic Stress Responses in Legumes

Utkarsh M. Bitla, Akash L. Shinde, Ajay M. Sorty, Kamlesh K. Meena, and Narendra P. Singh

CONTENTS

11.1 Introduction

The foremost contest for the agricultural industry is ensuring the food security for approximately eight to nine billion people, while global agricultural land is affected by abiotic stresses including salinity, drought, sodicity, high temperature, and heavy metal toxicity-related problems. Abiotic stress(es) have been thought to be the most predominant cause of deteriorating crop productivity—up to as much as 50% (Rodríguez et al., 2006). Anthropogenic activities, along with the changing agroclimatic circumstances, have been estimated to worsen the influence of abiotic stress(es) on productivity, thus necessitating the efforts to design crop-specific adaptation strategies for effective attenuation of the stress-induced detrimental responses (Beebe et al., 2011; Sorty et al., 2016). Legumes (Leguminosae or Fabaceae) comprise the second important plant family in agriculture following the Poaceae (grass family). They provide the largest single source of vegetable protein in human diets and livestock forage; additionally, they also provide oil, fiber, and micronutrients, and they contribute to agriculture by playing a vital role in cropping cycles due to their ability to form symbiotic associations with atmospheric nitrogen-fixing rhizobia (El-Enany et al., 2013; Mantri et al., 2013). Like other crops, legume production also suffers the negative influence of abiotic stresses. Around 64% of the total land area has been affected by drought, 13% by flood, 15% by acidic soil, and 6% by salinity (Cramer et al., 2011). The majority of the observations relating to the abiotic stress tolerance seem acquired under controlled conditions in the laboratory trials that could hardly perform under the actual field conditions. Therefore, a significant gap might exist among the information added by these studies and that essential to develop the crop with enhanced tolerance under field environments. Among the several strategies proposed to develop abiotic stress tolerance in legume crops, the rhizobium inoculation has gained significant attention, owing to the crop-specific colonization ability of the *Rhizobium* strains, and the role contributed by these characteristic microorganisms toward mounting the stress tolerance mechanisms in the host. Literature is available to endorse the association of *Rhizobium* with wild leguminous plants, growing under harsh environmental circumstances. The successful association of the *Rhizobium* strains isolated from wild legumes has been demonstrated. This chapter mainly focuses on challenges in legume growth under abiotic stress and microbial strategies to alleviate abiotic stress—particularly drought and salinity.

11.2 Abiotic Stress Responses in Legumes

Exposure to abiotic stress(es) induces a sequence of morphological, physiological, biochemical, and molecular alterations, which deleteriously influence plant development, productivity, and yield (Wang et al., 2001; Bita and Gerats, 2013). Plants suffer multiple effects of these stresses including altered physiological functions such as germination, photosynthesis, respiration, nitrogen fixation, reproduction, and oxidative metabolism (Iba, 2002; Farooq et al., 2008; Meena et al., 2012). Reactive oxygen species (ROS) are constantly produced by plants as a consequence of aerobic metabolism during their development or in response to abiotic and biotic stresses. ROS include singlet oxygen ($O^{\cdot\cdot}$), superoxide anion (O^{2-}), hydrogen peroxide (H_2O_2), and hydroxyl radical (OH^{-}.). These molecules are highly toxic since they are able to modify all the primary constituents of the cell such as lipids, DNA, carbohydrates, and proteins (Moller et al., 2007; Meena et al., 2017).

At the reproductive stage, high temperature becomes a severe constraint for the production of grain legumes with their cultivation expanding to warmer environments; moreover, the temperature variability has also increased due to climate change. A primary process such as photosynthesis, along with cell growth, is affected by drought (Chaves, 1991). Similarly, high temperature adversely affects flower initiation, pollen viability (germination and tube growth), stigma receptivity, ovule viability, ovule size, fertilization, seed/fruit set, seed composition, grain filling, and seed quality (Barnabás et al., 2008).

Soil salinity leads to the accumulation of high concentration of sodium (Na^+), chloride (Cl^-), and sulfate (SO_4^{2-}) ions in plant cells, inactivating enzymes and inhibiting protein synthesis. The resultant condition reduces the rate of photosynthesis and in turn plant growth (Shannon, 1997). Significant genetic variability in salt tolerance among legume species and cultivars has been described. Some tree legumes such as *Prosopis* and *Acacia* spp. are highly tolerant to salinity, while grain legumes have long been recognized as either sensitive or only moderately tolerant to saline conditions. *Cicer arietinum, Phaseolus vulgaris*, and *Pisum sativum* are known to be highly sensitive to salt stress, while *Glycine max* and *Vicia faba* are considered particularly salt-tolerant grain legumes (Abdel-Wahab and Zahran, 1981; Delgado et al., 1994). Flooding is another important abiotic stress that affects many plants by developing oxygen deficiency (anoxia). This results in various symptoms, including the excess of ethylene synthesis, and leads to yield reduction (Barnawal et al., 2012).

Some studies have reported a reduction in protein content under drought conditions, 30; however, the responses differ with the type of protein. For instance, proteins including albumins, globulins, and protamine decreased in fava bean under drought while glutamines increased; yet, the total protein content was enhanced (Alghamdi, 2009). In legumes under extreme drought, reduction in N uptake was reported, which could be linked to reduced protein content in grain (Liu et al., 2004). Among the fatty acids, the oleic acid content is enhanced while linolenic acid and/or linoleic acid decreased in food legumes under drought (Dwivedi et al., 1996; Kirnak et al., 2010). The protein content in soybean, however, was less affected under abiotic stress particularly due to improved mobilization of amino acids (Rotundo and Westgate, 2009).

Under thermal stress conditions, total oil content increased whereas grain protein content decreased in a variety of food legumes with few exceptions. In soybean and peanut, for instance, oil content increased by 37% and 20%, respectively (Dwivedi et al., 1996; Kirnak et al., 2010). Excess heat also causes a change in fatty acid composition. For instance, an increase in temperature caused a significant increase in the oleic acid contents whereas the heat stress caused a decrease in the linoleic acid content (Carver et al., 1986; Gibson et al., 1996). The concentrations of N and P in soybean grain increased with temperature up to 40°C/30°C (day/night), following which a decline was reported (Thomas et al., 2003).

Higher accrual of heavy metals in arable soils not only pollutes the environment but uptake of these heavy metals by crops—including food legumes—permits the entry of heavy metals into the food chain, causing serious environmental and health hazards (Muchuweti et al. 2006; Wani et al., 2007). Moreover, food legumes grown in soils contaminated with heavy metals such as cadmium (Cd), lead (Pb), copper (Cu), nickel (Ni), and arsenic (As) may significantly reduce grain protein content due to reduced N uptake and supply to developing grains (Wani et al., 2007).

11.3 Influence of Abiotic Stressor(s) on Legume-Rhizobium Interactions

Legume-rhizobium interaction is a well-known symbiotic association. Nodule bacteria belonging to genera *Rhizobium, Bradyrhizobium, Mesorhizobium, Sinorhizobium,* and *Azorhizobium* establish a beneficial relationship with a variety of leguminous plants. They are thus involved in direct promotion of plant growth by fulfilling the nitrogen requirement of legumes through the process known as biological nitrogen fixation (BNF), which occurs in the root nodules (Relic et al., 1994). These bacteria generally belong to the alpha-proteobacteria and are capable of inducing nitrogen-fixing nodules on the roots of several hundreds of leguminous plants (Lorenzo et al., 2000). Successful establishment of the symbiotic interaction involves chemotaxis of the bacteria toward the root, root colonization, root hair deformation, and rapid division of root cortex cells (van Rhijn and Vanderleyden, 1995) (Figure 11.1).

Extreme environmental conditions adversely affect rhizobia-legume interactions wherein the rhizobial partner utilizes endogenous or exogenous osmolytes and secretes specific proteins to alleviate the problem of aridity, salinity, and toxicity (Karmakar et al., 2015). Nodulation and nitrogen fixation in

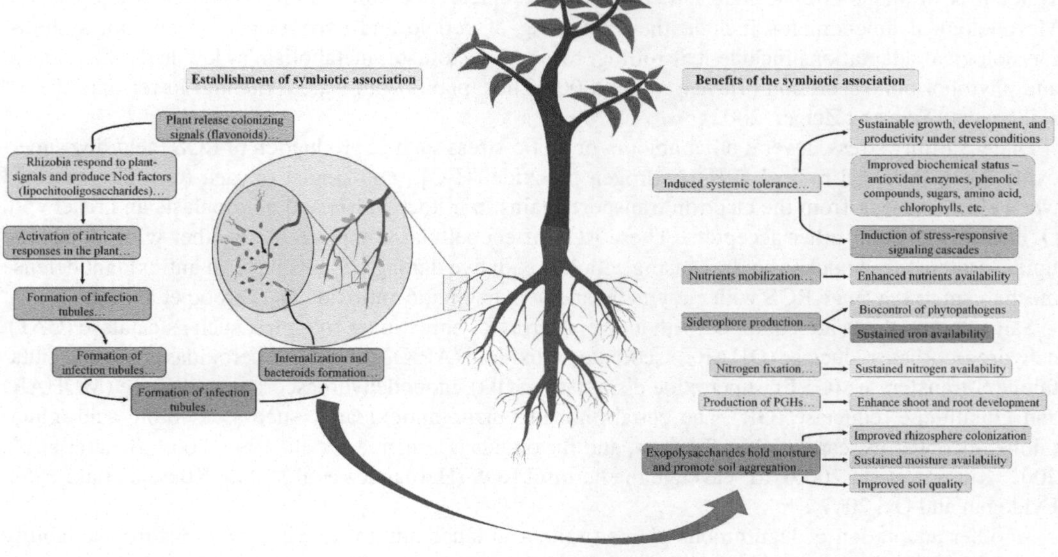

FIGURE 11.1 Stress-tolerant rhizobia (alone/in combination with other beneficial microbes) can actively establish symbiotic association under abiotic stress-prone circumstances. A successful association can benefit the host plant by a variety of mechanisms including direct and indirect plant growth promotion. Rhizobia produce multiple biomolecules that help the plant tolerate abiotic stress and achieve sustainable growth even under stress conditions.

legume-*Rhizobium* symbiosis are detrimentally affected by salinity and drought, which can influence the legume establishment and growth. High temperature is one of the most important abiotic stresses that can affect the legume-rhizobium interaction. In legumes, it also inhibits the process of symbiotic nitrogen fixation. Several hypotheses have been advanced to explain the negative effects of salt on nitrogen fixation in legumes, diminished photosynthate supply to the nodule, reduced supply of respiratory substrates to the bacteroids, and alterations in the oxygen diffusion barrier (Soussi et al., 1998). Drought stress induces oxidative damage in legumes and this has a detrimental effect on nodule performance and BNF (Arrese-Igor et al., 2011). Temperature critically affects root hair infection, bacteroid differentiation, nodule structure, and the working of the legume root nodule. Root temperature influences nodulation and dinitrogen fixation by peanut nodulating *Bradyrhizobium* sp.; for instance, at 25°C, at 30°C and 35°C, the nitrogenase activity of peanut root nodules was reduced with increase in temperature (Nambiar and Dart, 1983).

Most of the leguminous plants require neutral or slightly acidic soil for growth and also for symbiotic nitrogen fixation. Acidity affects several steps in the development of symbiosis, including the exchange of molecular signals between the legume and the microsymbiont (Hungria and Vargas, 2000). Soil acidity can adversely affect the formation of the symbiotic association between *Rhizobium* species and their host legume plants. Munns and Keyser (1981) reported that the acid-sensitive strains of *R. meliloti* failed to maintain a pH gradient in acid conditions and were thus unable to control intracellular pH.

11.4 Abiotic Stress Responses in Legumes

Accumulation of osmoprotectants, production of superoxide radical scavenging mechanisms, exclusion or compartmentation of ions by efficient transporter and symporter systems, and production of specific enzymes involved in the regulation of plant hormones are some of the mechanisms that plants have evolved for adaptation to abiotic stresses (Mahajan and Tuteja, 2005; Parida and Das, 2005; Santner et al., 2009; Des Marais and Juenger, 2010; Tiwari et al., 2011).

The principal mechanism of plants' adaptation to drought conditions is the regulation of water loss. Water loss in plants is controlled either by morphological amendments or physiological adaptations. Morphological amendments include the thickening of cuticle and expansion of deep root systems. Physiological adaptations include leaf rolling, responsive stomata, metabolism at low leaf water status, and phytohormone secretion (Turner et al., 2000). Thus, plants with water potential lower than that of soil persist (Tiaz and Zeiger, 2002).

Under salinity stress, several other abiotic- or biotic-stress such as production of ROS including superoxide (O_2^-), hydroxyl radical (OH), hydrogen peroxide (H_2O_2), and singlet oxygen (O_2) are generated when electrons leak from the electron transport chains in mitochondria and chloroplasts and react with O_2 in the absence of other acceptors. These ROS affect cellular components, together with membrane lipids, proteins, and nucleic acids, causing cellular oxidative damage. Plants have an antioxidant defense mechanism to scavenge ROS with enzymatic and nonenzymatic antioxidants (Farooq et al., 2015).

Salt-tolerant plants are equipped with a diverse array of antioxidant enzymes, such as catalase (CAT), dehydroascorbate reductase (DHAR), ascorbate peroxidase (APX), glutathione peroxidases (GPX), glutathione S-transferases (GST), superoxide dismutase (SOD), monodehydroascorbate reductase (MDHAR) and glutathione reductase (GR), and certain nonenzymatic antioxidants, such as ascorbic acid, glutathione, tocopherols, carotenoids, flavones, and flavonoids (Hernandez et al., 1999, 2000; Schafer et al., 2002; Kukreja et al., 2005), to scavenge the harmful ROS (Hernandez et al., 2000; Kukreja et al., 2005; Eyidogan and Oz, 2007).

Another adaptation of leguminous plants to the arid lands and those with low-moisture availability is their infection by mycorrhizal fungi in addition to *Rhizobium*. Mycorrhizal inoculation has alleviated the harmful impacts of drought stress on *Acacia* and *Leucaena* under arid conditions (Robson and Bottomley, 1991).

11.5 Mechanisms Incorporated through Different Strategies

Several mechanisms are involved in the mitigation of abiotic stress. A positive interaction has been found between efficient rhizobial symbiosis and osmotic stress tolerance in drought-stressed legumes, indicating the role of efficient nodulation in conferring drought tolerance in terms of growth, enhancement of plant water status, and alleviation of oxidative stress (Aroca et al., 2009). Legumes such as *Vigna unguiculata* and *Cicer arietinum* are considered tolerant to water deficit, and important mechanisms were developed by these species to tolerate inadequate water supply. For instance, biochemical modifications in carbon metabolism, such as increase in sucrose (Lobato et al., 2009), as well as significant interference in nitrogen metabolism, like reduction of soluble proteins (Costa et al., 2011) and increase in total amino acids (Lobato et al., 2008), contribute to osmotic adjustment of these plants. Under high-temperature conditions, plants activate the lipid-based signaling cascades that enhance the Ca^{2+} influx (Saidi et al., 2011), heat shock proteins (HSPs) working as molecular chaperones to inhibit denaturation/aggregation of target proteins and enabling protein refolding (Lohar and Peat, 1998).

Several phytohormones (ABA, brassinosteroids, etc.) as well as many signaling molecules (nitric oxide, etc.) purportedly play an important role under heat stress to confer heat tolerance (Hasanuzzaman et al., 2013; Asthir, 2015). Brassinosteroids (BRs) improve the growth and biomass of french beans under heat stress (Upreti and Murti, 2004) by stimulating cell elongation (Salchert et al., 1998). Salicylic acid (SA) is a natural derivative of phenols formed by phenylpropanoid metabolism. It is an important signaling molecule under stress conditions and an effective protectant under heat stress (Yuan et al., 2008; Hasanuzzaman et al., 2013). SA modifies the activity of many enzymes and enhances chlorophyll and carotenoid levels along with photosynthetic rates. In addition, SA has improved plant growth, flower induction, ion uptake, and thermogenesis, and it can affect stomatal movement and ethylene biosynthesis (Hayat et al., 2009). Nitric oxide (NO) is an important concentration-dependent and redox-related signaling molecule (Fancy et al., 2017). NO regulates various physiological processes and has a vital role in conferring tolerance to plants under abiotic stress including heat stress (Hasanuzzaman et al., 2010, 2011, 2012, 2013; Waraich et al., 2012).

The water supply under the flowering and seed development stage has a significant impact on the pods and seed yield and also legume grain quality. Under long-standing drought periods (above 14 days), the growth of plants is retarded and the plants try to overcome the water deficiency by decreasing water loss and leaf area or by increasing water use efficiency. During flowering, a longer period of water deficiency results in the loss of leaves and flowers that was much more intensive in large-leaved bean varieties than in those with smaller leaves (Nemeskéri et al., 2010).

11.6 Microbe-Mediated Abiotic Stress Mitigation in Legumes

Nodulation and the associated legume-rhizobium symbiosis are complex processes involving the expression of both bacterial and plant genes that were initiated by the secretion of a cocktail of phenolic molecules termed flavonoids, which can passively diffuse through the bacterial membrane (Smith and Wollum, 1989; Wang et al., 2012). Rhizobial strains with the ability to colonize rhizosphere competitively catabolize secondary metabolites present in root exudates (Boivin et al., 1991; Howieson et al., 1995), synthesize osmolytes under stress de novo, and produce high quantities of EPS under the water and nutrient-limiting conditions; such strains must be selected through keen screening from the natural populations. The performance of selected strains under stress can be further enhanced by genetically modifying the strains for traits such as rhizosphere competence, rhizopine catabolism, and toxin production. For example, the introduction of NolA (a repressor of *nod* gene) prevents colonization of legume roots by native bacteria (Vlassak and Vanderleyden, 1997).

Salinity influences the existence and distribution of rhizobia in bulk soil and the rhizosphere (Jenkins et al., 1989; Tate, 1995). Salt-tolerant rhizobia have been isolated from various crop and wild legumes (Sorty et al., 2016). Additionally, salt-tolerant legume genotypes should be selected for better performance of both the plant and associating rhizobial strain. During the past few years, rhizobia forming successful N_2-fixing *Rhizobium*-legume symbioses under salt stress have been selected (El-Mokadem et al., 1991; El-Sheikh and Wood, 1995). *R. meliloti* formed successful symbiosis with *Medicago sativa* under saline conditions (100 mM NaCl). These rhizobia are local strains isolated from Saudi Arabian soil in arid lands (Mashhady et al., 1998). Genetic modification of a *Mesorhizobium* strain by expression of an exogenous *acd*S gene can improve its symbiotic performance under salt stress; it has also been proved that under saline stress conditions ACC deaminase can enhance the plant–rhizobium symbiosis (Brígido et al. 2013).

Rhizobium-legume symbioses are presently the most significant nitrogen-fixing systems. They have been shown to potentially enhance N input in arid lands. The leguminous plants include species or varieties that are extremely adapted to the drastic conditions of arid lands. *Medicago sativa, Arachis hypogaea, Cyamposis tetragonoloba*, and *Melilotus* spp. are known to be adapted to environments prevailing in arid regions. In addition, a drought-tolerant cultivar of *Phaseolus vulgaris* has recently been identified (Ramos et al. 1999). The dry weight of this legume remained unaffected by water stress, although the number and weight of nodules, as well as N_2 fixation (acetylene reduction), were obviously reduced. However, these legume species require drought-tolerant rhizobia to form effective symbiosis under arid climates. Rhizobia with survival ability, which showed effective symbiotic characteristics with their host legumes (e.g., Prosopis rhizobia) in desert soils and arid regions, have been identified (Fuhrmann et al., 1986, Jenkins et al., 1987, 1989); such strains could be efficiently implemented for sustained productivity in arid soil.

11.7 Microbial Responses to Abiotic Stress

Plant-microbe interactions are important to the adaptation and existence of both the associates in any abiotic environment. Induced systemic tolerance (IST) is the term being used for microbe-mediated induction of abiotic stress responses. PGP microbes can be used as stress relievers for developing stress-tolerant crops with minimized production costs and environmental hazards (Sorty et al., 2018). Among them, indigenous and native microbes were more effective and competitive, as they are well adapted to the native environment (Mrabet et al., 2005). Co-inoculation of *Rhizobium* and *Azospirillum* in legumes significantly increases several plant-growth variables when compared with single inoculations (Tchebotar et al., 1998; Itzigsohn et al., 2000; Hamaoui et al., 2001; Remans et al., 2007; Remans et al., 2008b). *Azospirillum* is considered a *Rhizobium*-helper that stimulates nodulation, nodule function, and possibly plant metabolism (Molla et al., 2001; Verma et al., 2010). Phytohormones produced by *Azospirillum* promote epidermal-cell differentiation in root hairs that increase the number of potential sites for rhizobial infection leading to the formation of more nodules. Morphological and physiological changes in the root system are also stimulated (Bashan and Levanony, 1990; Pacovsky, 1990; Sarig et al., 1992; Volpin and Kapulnik, 1994). An increase in the number of lateral roots and root hairs significantly adds to the root surface available for nutrients and water uptake. Higher water and nutrient uptake by inoculated roots causes an improved water status of plant, which in turn could be the main factor enhancing plant growth (Boddey et al., 1986; Dalla Santa et al., 2004; Fallik and Okon, 1996; Mostajeran et al., 2002).

11.8 Drought

Drought stress disturbs plant hormonal stability by reducing the endogenous cytokinin level and increasing abscisic acid (ABA) content in the leaves. This effect leads to stomatal closure. Most of the N_2-fixing species suffer harmful physiological effects of water scarcity and show reduced nitrogen fixation when exposed to drought conditions (Zahran, 1999; Sharma et al., 2010). Bacteria-producing extracellular materials were reported for their beneficial effects. Exopolysaccharides (EPS) act as adhering material that helps bacteria to attach to soil particles and by binding soil particles together leads to soil aggregate formation. EPS contributes to hold water within soil aggregate and thus lead to ameliorating plant growth under drought conditions. Among EPS-producing bacteria, rhizobia were applied by Hartel and Alexander (1986) where they observed a significant correlation between the levels of EPS produced by cowpea-associated *Bradyrhizobium* strains and their drought tolerance. Legumes, when inoculated with mixed cultures of rhizobium and ACC deaminase-positive PGPR, stimulate nodulation through inhibition of ethylene biosynthesis thereby enhancing nodulation and nitrogen fixation. A few examples include early growth and enhancement of nodulation in *Glycine max* by ACC deaminase-producing rhizobacteria and improved nodulation in *Pisum sativum* by ACC deaminase-producing *Rhizobium leguminosarum* bv. Viciae 128C53K (Cattelan et al., 1999). Phytohormones-producing microbes also enhance legume growth under drought conditions (Sanget-Mo et al., 2014).

11.9 Salinity

Salinity generates the deleterious osmotic potential that significantly lowers the soil water potential. Similar to plants, rhizobial strains also secrete an array of metabolites called compatible solutes [trehalose, N-acetylglutaminylglutamine amide (NAGGN) and glutamate], osmoprotectants [betaine, glycine-betaine, proline-betaine, glucans, trehalose, sucrose, ectoine, 3-dimethylsulfoniopropionate (3-dimethylpropiothetin or DMSP), 2-dimethylsulfonioacetate (2-dimethylthetin or DMSA)], and pipecolic acid and certain cations [calcium, potassium] as one of the tolerance mechanisms (Streeter, 2003; Sugawara et al., 2010; Chen, 2011). Salt tolerance mechanisms involve several gene families that have been reported largely in *S. melilotti* followed by *R. etli*, *R. tropici*, *Rhizobium* sp., *S. fredii*, and *B. japonicum*. Some of the identified gene families include betaine (betS/betABCI/hutWXV) (Boscari et al., 2002), glycine-betaine (AraC) and proline-betaine (betS/prb) (Boscari et al., 2004; Payakapong et al., 2006; Alloing et al., 2006), glucans (ndvABCD), sucrose (zwf) and trehalose (zwf) (Chen et al., 2002; Jenson et al., 2002; Barra et al., 2003), cation efflux (phaA2/phaD2/phaF2/phaG2) (Jiang et al., 2004) and rpoH2 (Tittabutr et al., 2006), ntrY, ntrX, greA, alaS, dnaJ, nifS, noeJ, kup (Nogales et al., 2002), and omp10, relA, greA, and nuoL (Wei et al., 2004). Leguminous plants are able to detoxify ROS by two mechanisms: nonenzymatic and enzymatic. Nonenzymatic management of ROs is achieved through the accumulation of flavanones, carotenoids, and ascorbic acid, whereas the enzymatic mechanism involves activity of antioxidant enzymes including superoxide dismutases, peroxidases, and catalases. In bacteroids, peroxidases are not produced; hence, an alternative ascorbate-glutathione (ASC-GSH) cycle enzymes become operative (Jebara et al., 2010). Saline conditions considerably decrease plant nutrient uptake capability, especially phosphorus, since under salt-stressed conditions phosphate ions are precipitated (Grattan and Grieve, 1999). Some PGPRs actively solubilize the insoluble phosphates thus enhancing P availability to plants; similarly, some nitrogen fixers also solubilize P and other nutrients and enhance the overall nutrition (Gyaneshwar et al., 2002).

11.10 Co-inoculation of Rhizobia

Co-inoculation techniques could be a new approach to enhance the abiotic stress tolerance and productivity of legumes. Non-rhizobial PGPRs such as *Azospirillum* have been shown to increase the competitiveness of the rhizobial strains and to create additional infection sites that can be later occupied by the rhizobia (Perez-Montano et al., 2014). Providing a supply of biologically fixed N. Salt tolerance of *Galega officinalis* markedly improved when the plant was co-inoculated with its specific symbiont *R. galegae* sv. *officinalis*, with either of the two PGPR strains, *P. extremorientalis* TSAU20 or *P. trivialis* 3Re27; colonization of *G. officinalis* root tips by *Rhizobium* cells almost doubled under saline conditions when the plants were co-inoculated with *Rhizobium* and *Pseudomonas* strains (Egamberdieva et al., 2013). Also, under drought stress, *R. tropici* co-inoculated with *Paenibacillus polymyxa* in kidney enhanced plant height, shoot dry weight, and nodule number (Figueiredo et al., 2008). Therefore, a co-inoculation strategy can be a potential option for sustainable farming of legume crops in abiotic stress-prone areas.

Colonization signal molecules characterized by root exudates of leguminous plants are being added to the end product. The molecules being used in this refined technology include flavonoids, sugars, organic acids, amino acids, amines, and secondary metabolites that function as elicitor or signaling molecules for successful microbe-plant interaction (Skorupska et al., 2010; Bitla et al., 2017). Formulations containing biosignals are available for improvement in microbial colonization development of the symbiotic relationship, triggering mechanisms to combat abiotic stress conditions and induction of defense response in plants.

11.11 Future Perspectives

PGPR strains development is hindered mostly by the fact that the microorganisms are vulnerable to harsh environmental conditions, including high concentrations of environmental contaminants salts, extreme pH and temperature, etc. Genetic engineering can be used to develop improved PGPR strains capable of performing effectively at low inoculum doses and under a variety of environmental circumstances. It is crucial to develop more effective PGPR strains with longer shelf lives to achieve adequate longevity during transport and storage under ambient conditions. Recent advances in the areas of microbiology, biotechnology, molecular biology, and bioinformatics have opened up new gateways to identify novel genes involved in abiotic stress tolerance. Development of microbial inoculant could be achieved by exploring the diverse mechanisms of actions in disease suppression and plant growth promotion, revealing important genes and traits associated with abiotic stress alleviation, bacterial host plant interaction, as well as relationships between the bacteria and various environmental factors. Concepts of microbiotechnology application in agriculture should be employed to isolate indigenous PGPR strains from the stress-affected soils, and stress-based screening may permit rapid, efficient selection of candidate strains that could be used as bio-inoculants for crops grown in dryland areas. Selection of competitive rhizobial strains, breeding for competitive legume host, and using elicitor-loaded biofertilizers can be beneficial in sustaining green revolution to evergreen revolution.

ACKNOWLEDGMENTS

The authors are grateful to the Indian Council of Agricultural Research (ICAR) for financial support through the Network Project on Application of Microorganisms in Agriculture and Allied Sectors (AMAAS).

REFERENCES

Abdel-Wahab, A.M.A., Zahran, H.H. "Effects of salt stress on nitrogenase activity and growth of four legumes." *Biol Plants* 23(1981):16–23.

Alghamdi, S.S. "Chemical composition of faba bean (*Vicia faba* L.) genotypes under various water regimes." *Pak J Nutri* 8(2009):477–482.

Alloing, G., Travers, I., Sagot, B., Le Rudulier, D., Dupont, L. "Proline betaine uptake in *Sinorhizobium meliloti*: Characterization of Prb, an opp-like ABC transporter regulated by both proline betaine and salinity stress." *J Bacteriol* 188(2006):6308–6317.

Aroca, R. and Ruiz-Lozano, J.M. "Induction of plant tolerance to semi-arid environments by beneficial soil microorganisms—A review." *Sustain Agric Rev* 2(2009):121–136.

Arrese-Igor, C., Gordon, C., Gonzalez, E.M. et al. "Physiological response of legume nodules to drought." *Plant Stress* 5(special issue 1)(2011):24–31.

Asthir, B. "Protective mechanisms of heat tolerance in crop plants." *J Plant Interact* 10(2015):202–210. doi:10.1080/17429145.2015.1067726.

Barnabás, B., Jäger, K., Fehér, A. "The effect of drought and heat stress on reproductive processes in cereals." *Plant Cell Environ* 31(2008):11–38. doi:10.1111/j.1365-3040.2007.01727.x.

Barnawal, D., Bharti, N., Maji, D., Chanotiya, C.S., Kalra, A. "1-Aminocyclopropane-1-carboxylic acid (ACC) deaminase-containing rhizobacteria protect *Ocimum sanctum* plants during waterlogging stress via reduced ethylene generation." *Plant Physiol Biochem* 58(2012):227–235.

Barra, L., Pica, N., Gouffi, K., Walker, G.C., Blanco, C., Trautwetter, A. "Glucose 6-phosphate dehydrogenase is required for sucrose and trehalose to be efficient osmoprotectants in *Sinorhizobium meliloti*." *FEMS Microbiol Lett* 229(2003):183–188.

Bashan, Y., Levanony, H. "Current status of Azospirillum inoculation technology: Azospirillum as a challenge for agriculture." *Can J Microbiol* 36(1990):591–608.

Beebe, S., Ramirez, J., Jarvis, A., Rao, I., Mosquera, G., Bueno, G. "Genetic improvement of common beans and the challenges of climate change," in *Crop Adaptation of Climate Change*, 1st edn, eds., S.S. Yadav, R.J. Redden, J.L. Hatfield, H. Lotze-Campen, and A.E. Hall (New York: Wiley) (2011):356–369.

Bita, C.E., Gerats, T. "Plant tolerance to high temperature in a changing environment: Scientific fundamentals and production of heat tolerance crops." *Front Plant Sci* 4(2013):273. doi:10.3389/fpls.2013.00273.

Bitla, U.M., Sorty, A.M., Meena, K.K., Singh, N.P. "Rhizosphere signaling cascades: Fundamentals and determinants," in *Plant-Microbe Interactions in Agro-Ecological Perspectives*, Vol. I, eds., D.P. Singh, H.B. Singh, and R. Prabha (Singapore: Springer Nature) (2017):211–226.

Boddey, R.M., Baldani, V.L.D., Baldani, J.I., Döbereiner, J. "Effect of inoculation of Azospirillum spp. on nitrogen accumulation by field-grown wheat." *Plant Soil*, 95(1986):109–121.

Boivin, C., Barran, L.R., Malpica, C.A., Rosenberg, C. "Genetic analysis of a region of the *Rhizobium meliloti* pSym plasmid specifying catabolism of trigonelline, a secondary metabolite present in legumes." *J Bacteriol* 173(1991):2809–2817.

Boscari, A., Mandon, K., Dupont, L., Poggi, M.C., Le-Rudulier, D. "BetS is a major glycine betaine/praline betaine transporter required for early osmotic adjustment in *Sinorhizobium meliloti*." *J Bacteriol* 184(2002):2654–2663.

Boscari, A., Mandon, K., Poggi, M.C., Le Rudulier, D. "Functional expression of *Sinorhizobium meliloti* BetS, a high-affinity betaine transporter, in *Bradyrhizobium japonicum* USDA110." *Appl Environ Microbiol* 70(2004):5916–5922.

Brígido, C., Nascimento, F.X., Duan, J., Glick, B.R., Oliveira, S. "Expression of an exogenous 1-aminocyclopropane-1-carboxylate deaminase gene in *Mesorhizobium* spp. reduces the negative effects of salt stress in chickpea." *FEMS Microbiol Lett* 349(2013):46–53.

Carver, B.F., Burton, J.W., Carter, T.E., Jr., Wilson, R.F. "Response to environmental variation of soybean lines selected for altered unsaturated fatty acid composition." *Crop Sci* 26(1986):1176–1180.

Cattelan, A.J., Hartel, P.G., Fuhrman, J.J. "Screening for plant growth promoting rhizobacteria to promote early soybean growth." *Soil Sci Soc Am J* 63(1999):1670–1680.

Chaves, M.M. "Effects of water deficits on carbon assimilation." *J Exp Bot* 42(1991):1–16.

Chen, P. "Symbiotic effectiveness, competitiveness and salt tolerance of lucerne rhizobia." RMIT University, Australia (2011).

Chen, R.J., Bhagwat, A.A., Yaklich, R., Keister, D.L. "Characterization of ndvD, the third gene involved in the synthesis of cyclic beta-(1–[3)(1–[6)-D-glucans in *Bradyrhizobium japonicum*." *Can J Microbiol* 48(2002):1008–1016.

Chua, N.H., Hetherington, A.M., Hooley, R. et al. "Control of cell elongation and stress responses by steroid hormones and carbon catabolic repression in plants." *Phil Trans R Soc Lond B* 353(1998). doi:10.1098/rstb.1998.0307.

Costa, R.C.L., Lobato, A.K.S., Silveira, J.A.G., Laughinghouse, H.D. "ABA-mediated proline synthesis in cowpea leaves exposed to water deficiency and rehydration." *Turk J Agric For* 35(2011):309–317.

Cramer, G.R., Urano, K., Delrot, S., Pezzotti, M., Shinozaki, K. "Effects of abiotic stress on plants: A systems biology perspective." *BMC Plant Biol* 11(2011):163. doi:10.1186/1471-2229-11-163.

Dalla Santa, O.R., Hernández, R.F., Alvarez, G.L.M., Ronzelli Junior, P., Soccol, C.R. "Azospirillum sp. inoculation in wheat, barley and oats seeds greenhouse experiments." *Brazi Arch Biolo Tech* 47(2004):843–850. doi:10.1590/s1516-89132004000600002.

Delgado, M.J., Ligero, F., Lluch, C. "Effects of salt stress on growth and nitrogen fixation by pea, faba-bean, common bean and soybean plants." *Soil Biol Biochem* 26(1994):371–376.

Des Marais, D.L., Juenger, T.E. "Pleiotropy, plasticity, and the evolution of plant abiotic stress tolerance." *Ann N Y Acad Sci* 1206(2010):56–79.

Dwivedi, S.L., Nigam, S.N., Nageshwara, R.R.C., Singh, U., Rao, K.V.S. "Effect of drought on oil, fatty acids and protein contents of groundnut (*Arachis hypogaea* L.) seeds." *Field Crops Res* 48(1996):125–133.

Egamberdieva, D., Berg, G., Lindström, K., Räsänen, L.A. "Alleviation of salt stress of symbiotic *Galega officinalis* L. (Goat's Rue) by co-inoculation of *Rhizobium* with root colonizing *Pseudomonas*." *Plant Soil* (2013). doi:10.1007/s11104-013-1586-3.

El-Mokadem, M.T., Helemish, F.A., Abdel-Wahab, S.M., Abou-El-Nour, M.M. "Salt response of clover and alfalfa inoculated with salt tolerant strains of *Rhizobium*." *Ain Shams Sci Bull* 28B(1991):441–468.

El-Sheikh, E.A.E., Wood, M. "Nodulation and N_2 fixation by soybean inoculated with salt-tolerant rhizobia or salt-sensitive *bradyrhizobia* in saline soil." *Soil Biol Biochem* 27(1995):657–661.

Eyidogan, F., Oz, M.T. "Effect of salinity on antioxidant responses of chickpea seedlings." *Acta Physiol Plant* 29(2007):485e493. doi:10.1007/s11738- 007-0059-9.

Fallik, E., Okon, Y. "Inoculants of *Azospirillum brasilense*: Biomass production, survival and growth promotion of Setaria italica and Zea mays." *Soil Biol Biochem* 28(1996):123–126. doi:10.1016/0038-0717(95)00084-4.

Fancy, N.N., Bahlmann, A.K., Loake, G.J. "Nitric oxide function in plant abiotic stress." *Plant Cell Environ* 40(2017):462–472. doi:10.1111/pce.12707.

Farooq, M., Basra, S.M.A., Wahid, A., Cheema, Z.A., Cheema, M.A., Khaliq, A. "Physiological role of exogenously applied glycine betaine to improve drought tolerance in fine grain aromatic rice (*Oryza sativa* L.)." *J Agron Crop Sci* 194(2008):325–333. doi:10.1111/j.1439-037X.2008.00323.x

Farooq, M., Hussain, M., Wakeel, A., Siddique, K.H.M. "Salt stress in maize effects resistance mechanisms and management: A review." *Agron Sustain Dev* 35(2015):461e481. doi:10.1007/s13593-015-0287-0.

Figueiredo, M.V.B., Burity, H.A., Martinez, C.R., Chanway, C.P. "Alleviation of drought stress in common bean (*Phaseolus vulgaris* L.) by co-inoculation with *Paenibacillus polymyxa* and *Rhizobium tropici*." *Appl Soil Ecol* 40(2008):182–188.

Fuhrmann, J., Davey, C.B., Wollum, A.G. "Desiccation tolerance in clover rhizobia in sterile soils." *Soil Sci Soc Am J* 50(1986):639–644.

Gibson, L.R., Mullen, R.E. "Soybean seed composition under high day and night growth temperatures." *J Am Oil Chem Soc* 73(1996):733–737.

Grattan, S.R., Grieve, C.M. "Salinity mineral nutrient relations in horticultural crops." *Sci Hortic (Amst)* 78(1999):127–157.

Gyaneshwar, P., Naresh, K.G., Parekh, L.J. "Effect of buffering on the phosphate solubilizing ability of microorganisms." *World J Microbiol Biotechnol* 14(2002):669–673.

Hamaoula, B., Abbadi, J.M., Burdman, S., Rashid, A., Sarig S., Okon Y. "Effects of inoculation with *Azospirillum brasilense* on chickpeas (*Cicer arietinum*) and faba beans (*Vicia faba*) under different growth conditions." *Agronomie* (2001):553–560.

Hartel, P.G., Alexander, M. "Role of extracellular polysaccharide production and clays in the desiccation tolerance of cowpea *Bradyrhizobia*." *Soil Sci Soc Am J* 50(1986):1193–1198.

Hasanuzzaman, M., Hossain, M.A., Fujita, M. "Nitric oxide modulates antioxidant defense and the methylglyoxal detoxification system and reduces salinity-induced damage of wheat seedlings." *Plant Biotechnol Rep* 5(2011):353–365. doi:10.1007/s11816-011-0189-9.

Hasanuzzaman, M., Hossain, M.A., Fujita, M. "Physiological and biochemical mechanisms of nitric oxide induced abiotic stress tolerance in plants." *Am J Plant Sci* 5(2010):295–324. doi:10.1007/978-1-4614-4747-42.

Hasanuzzaman, M., Hossain, M.A., Teixeira, da Silva, J.A., Fujita, M. "Plant responses and tolerance to abiotic oxidative stress: Antioxidant defenses is a key factors," in *Crop Stress and Its Management: Perspectives and Strategies*, eds., V. Bandi, A.K. Shanker, C. Shanker, and M. Mandapaka (Berlin, Germany: Springer) (2012):261–316.

Hasanuzzaman, M., Nahar, K., Alam, M.M., Roychowdhury, R., Fujita, M. "Physiological, biochemical, and molecular mechanisms of heat stress tolerance in plants." *Int J Mol Sci* 14(2013):9643–9684. doi:10.3390/ijms14059643.

Hayat, S., Masood, A., Yusuf, M., Fariduddin, Q., Ahmad, A. "Growth of Indian mustard (*Brassica juncea* L.) in response to salicylic acid under high temperature stress." *Braz J Plant Physiol* 21(2009):187–195. doi:10.1590/S1677-04202009000300003.

Hernandez, J.A., Campillo, A., Jimenez, A., Alarcon, J.J., Sevilla, F. "Responses of antioxidant systems and leaf water relations to NaCl stress in pea plants." *New Phytol* 141(1999):241e251. doi:10.1046/j.1469-8137.1999.00341.x.

Hernandez, J.A., Jimenez, A., Mullineaux, P., Sevilia, F. "Tolerance of pea (*Pisum sativum* L.) to long-term salt stress is associated with induction of antioxidant defenses." *Plant Cell Environ* 23(2000):853e862.

Howieson, J.G., Loi, A., Carr, S.J. "*Biserrula pelecinus* L.–a legume pasture species with potential for acid, duplex soils which is nodulated by unique root-nodule bacteria." *Aust J Agric Res* 46(1995):997–1009.

Hungria, M., Vargas, M.A. "Environmental factors affecting N_2 fixation in grain legumes in the tropics, with an emphasis on Brazil." *Field Crop Res* 65(2000):151–164.

Iba, K. "Acclimative response to temperature stress in higher plants: Approaches of gene engineering for temperature tolerance." *Annu Rev Plant Biol* 53(2002):225–245. doi:10.1146/annurev.arplant.53.100201.160729.

Itzigsohn, R., Burdman, S., Okon, Y. et al. "Plant-growth promotion in natural pastures by inoculation with *Azospirillum brasilense* under suboptimal growth conditions." *Arid Soil Res Rehabil* 14(2000):151–158. doi:10.1080/089030600263076.

Jebara, S., Drevon, J.J., Jebara, M. "Modulation of symbiotic efficiency and nodular antioxidant enzyme activities in two *Phaseolus vulgaris* genotypes under salinity." *Acta Physiol Plant* 32(2010):925–932.

Jenkins, M.B., Virginia, R.A., Jarrel, W.M. "Ecology of fastgrowing and slow-growing mesquite-nodulating rhizobia in Chihuahua and Sonoran desert ecosystems." *Soil Sci Soc Am J* 53(1989):543–549.

Jenkins, M.B., Virginia, R.A., Jarrel, W.M. "Rhizobial ecology of the woody legume mesquite (*Prosopis glandulosa*) in the Sonoran desert." *Appl Environ Microbiol* 33(1987):36–40.

Jenson, J.B., Peters, N.K., Bhuvaneswari, T.V. "Redundancy in periplasmic binding protein-dependent transport systems for trehalose, sucrose, and maltose in *Sinorhizobium meliloti*." *J Bacteriol* 184(2002):2978–2986.

Jiang, J.Q., Wei, W., Du, B.H., Li, X.H., Wang, L., Yang, S.S. "Salt tolerance genes involved in cation efflux and osmoregulation of *Sinorhizobium fredii* RT19 detected by isolation and characterization of Tn5 mutants." *FEMS Microbiol Lett* 239(2004):139–146.

Karmakar, K., Rana, A., Rajwar, A., Sahgal, M. "Legume-rhizobia symbiosis under stress." Arora, N.K. (Ed.), *Plant Microbes Symbiosis: Applied Facets*, 241(2015). doi:10.1007/978-81-322-2068-8_12.

Kirnak, H., Dogan, E., Turkoglu, H. "Effect of drip irrigation intensity on soybean seed yield and quality in the semi-arid Harran plain Turkey." *Spanish J Agri Res* 8(2010):1208–1217.

Kukreja, S., Nandwal, A.S., Kumar, N., Sharma, S.K., Sharma, S.K., Unvi, V., Sharma, P.K. "Plant water status, H_2O_2 scavenging enzymes, ethylene evolution and membrane integrity of *Cicer arietinum* roots as affected by salinity." *Biol Plant* 49(2005):305e308. doi:10.1007/s10535-005-5308-4.

Liu, F., Jensen, C.R., Andersen, M.N. "Drought stress effect on carbohydrate concentration in soybean leaves and pods during early reproductive development: Its implication in altering pod set." *Field Crops Res* 86(2004):1–13.

Lobato, A.K.S., Costa, R.C.L., Oliveira Neto, C.F. "Responses of the pigments and carbon metabolism in *Vigna unguiculata* cultivars submitted to water deficit." *Res J Biol Sci* 4(2009):593–598.

Lobato, A.K.S., Oliveira Neto, C.F., Costa, R.C.L., Santos Filho, B.G., Cruz, F.J.R. "Laughing house IV HD. Biochemical and physiological behavior of *Vigna unguiculata* (L.) Walp. under water stress during the vegetative phase." *Asian J Plant Sci* 7(2008):44–49.

Lohar, D., Peat, W. "Floral characteristics of heat tolerant and heat sensitive tomato (*Lycopersicon esculentum* Mill.) cultivars at high temperature." *Sci Hortic* 73(1998):53–60. doi:10.1016/S0304-4238(97) 00056-3.

Lorenzo, A.J., Velazquez, E., Perez-Galdona, R. "RFLP analysis of 16S rDNA and low molecular weight RNA profiling of rhizobial isolates from shrubby legumes endemic to Canary islands." *Syst Appl Microbiol* 23(2000):418–425.

Mahajan, S., Tuteja, N. "Cold, salinity and drought stresses: An overview." *Arch Biochem Biophys* 444(2005):139–158.

Mashhady, A.S., Salem, S.H., Barakh, F.N., Heggo, A.M. "Effect of salinity on survival and symbiotic performance between *Rhizobium meliloti* and *Medicago sativa* in Saudi Arabian soils." *Arid Soil Res Rehabil* 12(1998):3–14.

Meena, K.K., Kumar, M., Kalyuzhnaya, M.G., Yandigeri, M.S., Singh, D.P., Saxena, A.K., Arora, D.K. "Epiphytic pink-pigmented methylotrophic bacteria enhance germination and seedling growth of wheat (*Triticum aestivum*) by producing phytohormones" *Antonie Van Leeuwenhoek* 101(2012):777–786.

Meena, K.K., Sorty, A.M., Bitla, U.M., Choudhary, K., Gupta, P., Pareek, A., Singh, D.P. et al. "Abiotic stress responses and microbe-mediated mitigation in plants: The omics strategies." *Front Plant Sci* 8(2017):172.

Molla, A.H., Shamsuddin, Z.H., Halimi, M.S., Morziah, M., Puteh, A.B. "Potential for enhancement of root growth and nodulation of soybean co-inoculated with Azospirillum and Bradyrhizobium in laboratory systems." *Soil Biol Biochem* 33(2001):457–463.

Moller, I.M., Jensen, P.E., Hansson, A. "Oxidative modifications to cellular components in plants." *Annu Rev Plant Biol* 58(2007):459–481.

Mostajeran, A., Amooaghaie, R., Emtiazi G. "Root hair density and deformation of inoculated roots of wheat cultivars by *Azospirillum brasilense* Azospirillum/Trichoderma: The effects on dry bean and role of IAA in this phenomenon." *Iranian Biol J* 13(2002):18–28.

Mrabet, M., Mhamdi, R., Tajini, F., Tiwari, R., Trabelsi, M., Aouani, M.E. "Competitiveness and symbiotic effectiveness of a *R. gallicum* strain isolated from root nodules of *Phaseolus vulgaris*." *Europ J Agron* 22(2005):209–216.

Muchuweti, M., Birkett, J.W., Chinyanga, E., Zvauya, R., Scrimshaw, M.D., Lester, J.N. "Heavy metal content of vegetables irrigated with a mixture of wastewater and sewage sludge in Zimbabwe: Implications for human health." *Agric Agric Ecosyst Environ* 112(2006):41–48.

Munns, D.N., H.H. Keyser. "Tolerance of rhizobia to acidity and aluminum and phosphate." *Soil Sci Soc Am J* 34(1981):519–523.

Nambiar, P.T.C., Dart, P.J. "Factors influencing nitrogenase activity by root nodules of groundnut (*Arachis hypogea* L.)." *Peanut Sci* 10(1983):26–29.

Nemeskéri, E., Sárdi, É., Remenyik, J., Kőszegi, B., Nagy, P. "Study of defensive mechanisms against drought of French bean (*Phaseolus vulgaris* L.) varieties." *Acta Physiol Plant* 32(2010):1125–1134. doi:10.1007/s11738-010-0504-z.

Nogales, J., Campos, R., Ben-Abdelkhalek, H., Olivares, J., Lluch, C., Sanjuan, J. "*Rhizobium tropici* genes involved in free-living salt tolerance are required for the establishing of efficient nitrogen-fixing symbiosis with *Phaseolus vulgaris*." *Mol Plant Microbe Interact* 15(2002):225–232.

Pacovsky, R.S. "Development and growth effects in the *Sorghum-Azospirillum* association." *J Appl Bacterio* 68(1990):555–563. doi:10.1111/j.1365-2672.1990.tb05220.x.

Parida, A.K., Das, A.B. "Salt tolerance and salinity effects on plants: A review." *Ecotox Environ Safe* 60(2005):324–349.

Payakapong, W., Tittabutr, P., Teaumroong, N., Boonkerd, N., Singleton, P.W., Borthakur, D. "Identification of two clusters of genes involved in salt tolerance in *Sinorhizobium* sp. strain BL3." *Symbiosis* 41(2006):47–53.

Perez-Montano, F., Villegas, A., Belligin, R.A., del Cerro, P., Espany, M.R., Jimenez-Guerrero, I., Lopez-Baena, F.J., Ollero, F.J., Cubo, T. "Plant growth promotion in cereal and leguminous agricultural important plants: From microorganism capacities to crop production." *Microbiol Res* 169(2014):325–336.

Ramos, M.L.G., Gordon, A.J., Minchin, F.R., Sprent, J.I., Parsons, R. "Effect of water stress on nodule physiology and biochemistry of a drought tolerant cultivar of common bean (*Phaseolus vulgaris* L.)." *Ann Bot* 83(1999):57–63.

Relic, B., Perret, X., Estrada-García, M.T., Kopcinska, J. et al. "Nod factors of *Rhizobium* are a key to the legume door." *Mol Microbiol* 13(1994):171–178.

Remans, R., Beebe, S., Blair, M. et al. "Physiological and genetic analysis of root responsiveness to auxinproducing plant growth-promoting bacteria in common bean (Phaseolus vulgaris L.)" Plant Soil 302(2008):149–161.

Remans, R., Croonenborghs, A., Torres Gutierrez, R., Michiels, J., Vanderleyden, J. "Effects of plant growth promoting rhizobacteria on nodulation of *Phaseolus vulgaris* L. are dependent on plant P nutrition." *Eur J Plant Pathol* 119(2007):341–351.

Robson, A.D., Bottomley, P.J. "Limitations in the use of legumes in agriculture and forestry," in *Biology and Biochemistry of Nitrogen Fixation*, eds., M.J. Dilworth and A.R. Glenn (Elsevier Publishers: Amsterdam, the Netherlands) (1991):320–349.

Rodríguez, M., Canales, E., Borroto, C.J., Carmona, E., Lopez, J., Pujol, M. "Identification of genes induced upon water-deficit stress in a drought tolerant rice cultivar." *J Plant Physiol* 163(2006):577–584. doi:10.1016/j.jplph.2005. 07.005.

Rotundo, J.L., Westgate, M.E. "Meta-analysis of environmental effects on soybean seed composition." *Field Crops Res* 110(2009):147–156.

Saidi, Y., Finka, A., Goloubinoff, P. "Heat perception and signaling in plants: A tortuous path to thermos tolerance." *New Phytol* 190(2011):556–565. doi:10.1111/j.1469-8137.2010.03571.x.

Sang-Mo, K., Radhakrishnan, R., Khan, A.L., Min-Ji, K., Jae-Man, P., Bo-Ra, K., Dong-Hyun, S., In-Jung, L. "Gibberellin secreting rhizobacterium, *Pseudomonas putida* H-2-3 modulates the hormonal and stress physiology of soybean to improve the plant growth under saline and drought conditions." *Plant Physiol Biochem* 84(2014):115–124.

Santner, A., Calderon-Villalobos, L.I.A, Estelle, M. "Plant hormones are versatile chemical regulators of plant growth." *Nat Chem Biol* 5(2009):301–307.

Sarig, S., Okon, Y., Blum, A. "Effect of *Azospirillum brasilense* inoculation on growth dynamics and hydraulic conductivity of sorghum bicolor roots." *J Plant Nutr* 15(1992):805–819. doi:10.1080/01904169209364364.

Schafer, R.Q., Wang, H.P., Kelley, E.E., Cueno, K.L., Martin, S.M., Buettner, G.R. "Comparing beta-carotene, vitamin E and nitric oxide as membrane antioxidants." *Biol Chem* 383(2002):671e681. doi:10.1515/BC.2002.069.

Shannon, M.C. "Adaptation of plants to salinity." *Adv Agron* 60(1997):75–120.

Sharma, M.P., Srivastava, K., Sharma, S.K. "Biochemical characterization and metabolic diversity of soybean rhizobia isolated from Malwa region of Central India." *Plant Soil Environ* 56(2010):375–383.

Skorupska, A., Wielbo, J., Kidaj, D., Marek-Kozaczuk, M. "Enhancing *Rhizobium*-legume symbiosis using signaling factors," in *Microbes for Legume Improvement*, eds., M.S. Khan, A. Zaidi, J. Musarrat (Springer: New York) (2010):27–44.

Smith, G.B., Wullum, A.G. "Nodulation of *Glycine max* by six *Bradyrhizobium japonicum* strains with different competitive abilities." *Appl Environ Microbiol* 55(1989):1957–1962.

Sorty, A.M., Bitla, U.M., Meena, K.K., Singh, N.P. "Role of microorganisms in alleviating abiotic stresses," in *Microorganisms for Green Revolution*, eds., D.G. Panpatte et al. (Springer Nature: Singapore) (2018):115–128.

Sorty, A.M., Meena, K.K., Choudhary, K., Bitla, U.M., Minhas, P.S., Krishnani, K.K. "Effect of plant growth promoting bacteria associated with halophytic weed (*Psoralea corylifolia* L.) on germination and seedling growth of wheat under saline conditions." *Appl Biochem Biotechnol*, 180(2016):872–882.

Soussi, M., Ocana, A., Lluch, C. "Effect of salt stress on growth, photosynthesis and nitrogen fixation in chickpea (*Cicer arietinum* L.)." *J Exp Bot* 325(1998):1329–1337.

Streeter, J.G. "Effect of trehalose on survival of *Bradyrhizobium japonicum* during desiccation." *J Appl Microbiol* 95(2003):484–491.

Sugawara, M., Cytryn, E.J., Sadowsky, M.J. "Functional role of *Bradyrhizobium japonicum* trehalose biosynthesis and metabolism genes during physiological stress and nodulation." *Appl Environ Microbiol* 76(2010):1071–1081.

Tate, R.L. *Soil microbiology (symbiotic nitrogen fixation)* (John Wiley & Sons: New York) (1995):307–333.

Tchebotar, V.K., Kang, U.G., Asis, C.A., Akao, J.S. "The use of GUS-reporter gene to study the effect of *Azospirillum–Rhizobium* coinoculation on nodulation of white clover." *Biol Ferti Soils* 27(1998):349–352.

Thomas, J.M.G., Boote, K.J., Allen, L.H., Jr., Gallo-Meagher, M., Davis, J.M. "Elevated temperature and carbon dioxide effects on soybean seed composition and transcript abundance." *Crop Sci* 43(2003):1548–1557.

Tiaz, L., Zeiger, E. *Plant Physiology* (Sinauer Associates: Sunderland, MA) (2002):591–621.

Tittabutr, P., Payakapoig, W., Teaumroong, N., Boonkerd, N., Singleton, P.W., Borthakur, D. "The alternative sigma factor rpoH2 is required for salt tolerance in *Sinorhizobium* sp. strain BL3." *Res Microbiol* 157(2006):811–818.

Tiwari, S., Singh, P., Tiwari, R., Meena, K.K., Yandigeri, M., Singh, D.P., Arora, D.K. "Salt-tolerant rhizobacteria-mediated induced tolerance in wheat (*Triticum aestivum*) and chemical diversity in rhizosphere enhance plant growth." *Biol Fertil Soils* 47(2011):907.

Turner, N.C., Wright, G.C., Siddique, K.H.M. "Adaptation of grain legumes (pulses) to water-limited environments." *Adv Agron* 71(2000):193–231.

Upreti, K.K., Murti, G.S.R. "Effects of brassinosteroids on growth, nodulation, phytohormone content and nitrogenase activity in French bean under water stress." *Biol Plant* 48(2004):407–411. doi:10.1023/B:BIOP.0000041094.13342.1b.

van Rhijn, P., Vanderleyden, J. "The Rhizobium-plant symbiosis." *Microbiol Mol Biol Rev* 59(1995):124–142.

Verma, J.P., Yadav J., Tiwari K.N. "Application of Rhizobium sp. BHURC01 and plant growth promoting rhizobacteria on nodulation, plant biomass and yields of chickpea (*Cicer arietinum* L.)." *J Agric Res* 5(2010):148–156.

Vlassak, K.M., Vanderleyden, J. "Factors influencing nodule occupancy by inoculant rhizobia." *Crit Rev Plant Sci* 16(1997):163–229.

Volpin, H., Kapulnik, Y. "Interaction of *Azospirillum* with beneficial soil microorganisms," in *Azospirillum/plant Associations*, eds., Y. Okon (CRC Press, Boca Raton, FL) (1994):111–118.

Wang, D., Yang, S., Tang, F., Zhu, H. "Symbiosis specificity in the legume rhizobial mutualism." *Cell Microbiol* 14(2012):334–342.

Wang, W.X., Vinocur, B., Shoseyov, O., Altman, A. "Biotechnology of plant osmotic stress tolerance: Physiological and molecular considerations." *Acta Hortic* 560(2001):285–292. doi:10.17660/ActaHortic.2001.560.54.

Wani, P.A., Khan, M.S., Zaidi, A. "Cadmium, chromium and copper in green gram plants." *Agron Sustain Dev* 27(2007):145–153.

Waraich, E.A., Ahmad, R., Halim, A., Aziz, T. "Alleviation of temperature stress by nutrient management in crop plants: A review." *J Plant Nutr Soil Sci* 12(2012):221–244. doi:10.4067/S0718-95162012000200003.

Wei, W., Jiang, J., Li, X., Wang, L., Yang, S.S. "Isolation of salt sensitive mutants from *Sinorhizobium meliloti* and characterization of genes involved in salt tolerance." *Lett Appl Microbiol* 39(2004):278–283.

Yuan, Z.C., Haudecoeur, E., Faure, D., Kerr, K.F., Nester, E.W. "Comparative transcriptome analysis of *Agrobacterium tumefaciens* in response to plant signal salicylic acid, indole-3-acetic acid, and gamma-aminobutyric acid reveals signaling cross-talk and *Agrobacterium*–plant co-evolution." *Cell Microbiol* 10(2008):2339–2354.

Zahran, H.H. "Rhizobium-Legume symbiosis and nitrogen fixation under severe conditions and in an arid climate." *Microbiol Mol Biol Rev* 63(1999):968–989.

12

Microbial Bioinoculants for Salt Stress Tolerance in Plants

Mohammed Shariq Iqbal and Mohammad Israil Ansari

CONTENTS

12.1 Introduction

Starvation remains a global distress because of continuous escalating world population. Therefore, an amplified and supportable food production is essential. Thus, foraging continuously escalates the world's population; therefore, an amplified and supportable food production is essential. As a source of food as well as income, about 36% of the world's population is dependent upon agriculture for survival (OECD 2016; Hussain and Sulaimon 2018). At present, climate changes are causing antagonistic effects on agriculture, resulting in susceptibility to food security (Schmidhuber and Tubiello 2007). The consequences of climatic variations on agriculture are governed by several factors, such as concentration of atmospheric CO_2, temperature variation, weather conditions, percent precipitation, etc., resulting in lower yields or forfeiture of plants (Thornton et al. 2014). Temperature variations cause effects on growth and development of specific plants; change in concentration of CO_2 affects the yield of many crops while variation of weather conditions also affects the productivity of various crops. A slight rise in temperature prompts evaporation of water from soil, thus leaving behind augmented concentration of salts in soil. Surplus deposition of salt within the soil with inadequate water resource or facilities converts productive lands to infertile fields (Meritxell et al. 2009). It was reported that about 6% (approx.) of the terrestrial land is extremely altered by salt; however, it was also noted that about 20% of land used in irrigation purpose and drylands of about 2% have been influenced by soil salinity (Ilangumaran and Smith 2017; Machado and Serralheiro 2017). Soil salinity decreases crop yield, especially cereals, legumes, and many vegetable and herbal crops. To some extent the direct or indirect interruptions in cultivating crops might perhaps disturb food sanctuary as well as nourishment for humans (Shrivastava and Kumar 2014). Osmotic and ionic stresses are caused by salinity in soil, which imparts hinderance in the growth and development of plants. Reduced turgor pressure of the cell and changes in cell elongation as well as the cell division rate are generally due to change in water equilibrium within the cell created by osmotic stress (Feng et al. 2016). However, ion homeostasis is controlled by ionic stress within the cell, imparting alterations in rate of transpiration, hormonal modifications, photosynthesis, translocation of nutrients, and various additional metabolic progressions (Munns 2002). Several studies on breeding have been led on various crops to explore genotypes that are salt tolerant, but very little accomplishment has been achieved to date. The identified genotypes that are of definite region have futile to subsist at different situations having stress conditions (Baye et al. 2011). Consequently, conformist breeding techniques are at present overlooked by researchers because of the time-consuming procedure and abridged output. However, in this regard genetic engineering techniques would be better, but it's a costly affair. With this concern, application of microorganisms in plants is advantageous with respect to enhancing salt forbearance, which is a viable and alternate methodology to retrieve the cultivable fields that are susceptible to salinity (Allam et al. 2018). Beneficial microorganisms boost water-soil-plant association by persuading the signaling of various phytochemicals and activate other phyto-mechanisms that work in a cohesive manner to improve drought and salt stress forbearance in plants (Forni et al. 2017) (Figure 12.1).

With this consideration there are various microorganisms that are useful, such as arbuscular mycorrhizal fungi (AMF), plant growth-promoting bacteria (PGPB), and various other endophytes. However, there are many reports clarifying that tolerance of a plant to stress conditions is directly linked to its concomitant microbes (Qin et al. 2016; Jalil and Ansari 2018). Endophytic bioinoculants endure plant aptness in extreme environmental surroundings. There were various investigations that support different endophytic mechanisms to enhance growth of plants under extreme and diverse abiotic conditions (Souza et al. 2015). However, voluminous amounts of beneficial microorganisms fail to utilize their possessions on growth and development of plants under stress conditions such as salinity or drought, heavy metal toxicity, and various other extreme circumstances (Egamberdieva et al. 2014; Ansari and Khan 2018). As many microorganisms are not capable to stay alive in extreme environmental conditions, this causes a perplexing job for scientists to promote bioinoculants. Thus, newer methodologies are desirable to discover microbes that are stress-tolerant and their applicability in present agricultural scenario. In this chapter, we specifically focus on endophytic bioinoculants and their role in tolerance of plants contrary to salinity stress, and its mechanism. Hence, this report would be helpful to promote consciousness in the scientific community about the utilization of various endophytes to reinforce plants alongside salinity stress that could enhance agricultural growth and production under brine condition.

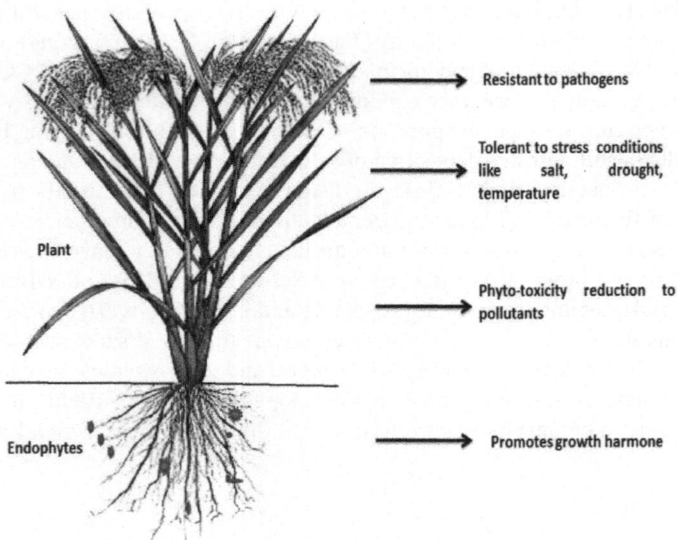

FIGURE 12.1 Plant-endophytic bioinoculants interaction leads to various beneficial effects.

12.2 Plants Mechanism for Salt Tolerance

Since plants are immobile, they possess extraordinary capability to survive under diverse environmental conditions (Flowers and Colmer 2015). Depending upon the ability of plants in resisting against salinity like stress conditions, they are categorized as: halophytes (can tolerate high levels of salinity) and glycophytes (can tolerate only low levels of salinity. Maximum crop plants are sensitive to salinity and have their place in the glycophyte group. The level of forbearance in plants depends upon the plants' genetic characters and allied biotic elements (Choudhary 2012). Plants obligate their particular immune system for various physiological mechanisms, which helps in a signaling chute and prompts forbearance against stress conditions (Andersen et al. 2018). Additionally, plant-microbes concomitant plays a significant role in endurance against stress environments by the plants (Choudhary 2012).

12.3 Plant Immune System in Contrast to Tolerance against Salinity

Plasma membrane receptors of plant cells recognize stress signals that lead to altered signal transduction pathways that encompass signaling of response from the apex of the root system to the apex of the shoot system of the plant (Tuteja and Mahajan 2007). With respect to saline conditions, roots of the plant can primarily sense together ionic as well as osmotic strains and retort swiftly to modify signaling. Generally phytohormones, calcium ions, reactive oxygen species (ROS), and inositol phosphate are responsible for signaling as second messengers (Xiong et al. 2002). As it is observed in Figure 12.2 the cell receives external stimuli (salt, draught or temperature stress); the receptors through the plasma membrane receive signals and pass to second messengers within the cytoplasm, causing the cell to respond accordingly, leading to activate signal transduction pathways.

An escalation in concentration of cytosolic calcium is recognized by calcineurin B and calcium dependent kinases, which initiate various ion transporters found on the plasma membrane of the plant cell. These transporters play vital roles in maintenance of Na^+ ion noxiousness within the cell (Tuteja and Mahajan 2007). Due to stress conditions caused by salinity, plants respond by altering morphophysiological, molecular, and biochemical mechanisms. Plants respond to osmotic stress by maintaining their water content, which is attained by minimizing cell division and decreasing cell elongation. Furthermore, plants sustain a shoot/root ratio for persistence under salinity stress condition (Moya et al. 1999). These modifications are controlled by various hormones formed by plant, viz. abscisic acid, auxin, cytokinin, ethylene, and gibberellin (Wáskiewicz et al. 2016). Apart from the commencement of osmotic stress, a decline in the photosynthetic rate, further closure of stomata and reduced leaf area were observed to inhibit loss of water from the plant cell. This stomatal closure adjusts the CO_2 fixation rate within the

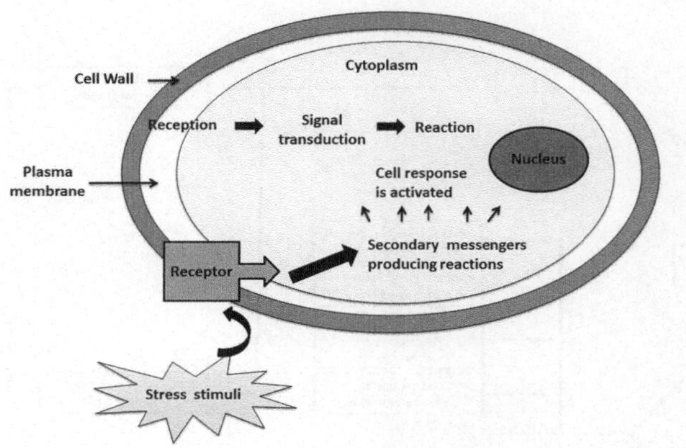

FIGURE 12.2 Salt stress stimuli signals received by plasma membrane receptors lead to signal transduction pathways within the plant cell.

cell and eventually prevents light-independent responses. Further, leaf expansion declines that affects the buildup of deposited carbohydrate, which causes response signals to decrease the rate of photosynthesis and enhances metabolic progressions (Pérez-Alfocea et al. 2010). Under such conditions caused due to stress, the enriched activities of enzymatic antioxidants are noted as a forbearance mechanism within the plant body (Christou et al. 2013). As a consequence to salt stress, enhanced content of glutathione and ascorbate are also reported (Choe et al. 2013). Moreover, secondary phytochemicals such as carotenoids, flavonoids, polyamines, phenolic compounds, and tocopherols play a defensive role in contrast to stress due to oxidative damage (Khan et al. 2014). Due to sustained susceptibility of salt stress, plants are exaggerated by Na^+ ionic stress and this leads to cytotoxicity. To elude Na^+ toxicity, two diverse mechanisms are implemented by plants, which include amplified compartmentation of intracellular space and vacuolar confiscation and amplified Na^+ deposition (Sun et al. 2009). Thus, the effectiveness of these physiognomies deliberates salinity forbearance within the plant.

12.4 Effect of Endophytic Bioinoculants against Plant Salt Tolerance

Under stress conditions, plants enhance their nutrition by making symbiotic relations with microorganisms, and this characteristic of plant and microbe interaction is regarded as habitat-adapted symbiosis (Rodriguez and Redman 2008). As an illustration, a halophytic plant does not work alone as a self-governing individual, but it functions with an association of assorted populations of symbiotic microorganisms that are interdependent for their sustained survival (Ruppel et al. 2013). This type of symbiotic association of plants with a vast group of microorganisms such as bacteria, fungi, and archaea existing in the soil, having potential to acclimatize themselves against stress conditions and linked with the host plant, leads to sustained survival of both (Ruppel et al. 2013). The mechanistic action of these microorganisms to stimulate plant growth takes place in direct as well as indirect ways (Vaishnav et al. 2014; Choudhary et al. 2015) (Figure 12.3).

As a direct mechanism, microorganisms enrich the acceptance efficacy of plant nutrient through solubilization of elements such as phosphorus, potassium, zinc, fixation of nitrogen, and the fabrication of siderophore (Fe chelation) that activates the inflection in plant hormones within the plant tissues (Shen et al. 2016). To withstand stress conditions, plants synthesize hormones with interaction to microorganism, viz. abscisic acid, auxin, cytokinin, and gibberellins, which swiftly alters the physiology of the plant (Waadt et al. 2015). When it comes to indirect mechanistic action, apposite microorganisms hinder the growth of harmful microorganisms, which causes infection in the plant. In this course of action, endophytic bioinoculants function as a biocontrol agent, thus constraining the progression of plant pathogens (fungal and bacterial) by the limiting action of production of various phytocompounds such as antibodies and enzymes that can degrade the microbial cell wall (Hardoim et al. 2015; Choudhary et al. 2015).

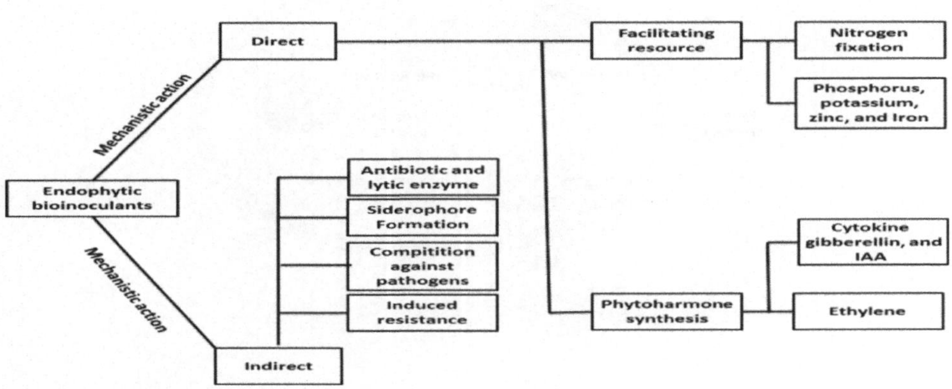

FIGURE 12.3 The mechanistic action of endophytic bioinoculants stimulates plant growth.

The plant growth-promoting bacteria arbitrated organic compounds are reported to escalate forbearance levels and enhance plant growth under stress environment (Vaishnav et al. 2017). With consideration to habitat-adapted symbiosis, endophytic bioinoculants are considered to be extra active because these microorganisms live inside the tissues of the plant and set forth an uninterrupted advantageous influence (Reiter et al. 2002; Weyens et al. 2013). Some of the endophytes with their association to various plants are listed in Table 12.1. In an analysis it was observed that endophytic bacterial linked with *Salicornia* plants delivered confrontation with counter to various abiotic stresses and stimulated plant growth under drought as well as salinity surroundings (Mapelli et al. 2013). Endophytes are good sources of plant growth promoters that may be exploited in agricultural methods needed for generating crops that are salinity tolerant (Sharma et al. 2016). It was reported that an endophyte *Enterobacter* sp. linked with a halophytic weed known as *Psoralea corylifolia* L. improved seed potency in *Triticum aestivum* plant grown under the condition of salinity stress (Sorty et al. 2016). In *Triticum aestivum*, biofortification was reported where *Arthobacter* spp. played an important role by mobilizing nutrients within the plants (Singh et al. 2017).

TABLE 12.1

List of Some of the Endophytic Bioinoculants and Its Associated Plants

S. No	Endophytic Bioinoculants	Associated Plant	References
1.	*Streptomyces* sp.	*Zingiber officinale*	Taechowisan et al. (2013)
2.	*Streptomyces* sp.	Tomato	Dias et al. (2017); Tan et al. (2006)
3.	*Micromonospora* and *Streptomyces*	Chinese cabbage	Sun et al. (2008)
4.	*Potentilla discolor, Senecio declouxii, Juncus effuses, Stellera* sp., *Vaccinium bracteatum*	*S. albogrisedus, S.griseocameus, S. viridis, Micromonospora, Nonomuraea, Oerskovia*	Zhao et al. (2011)
5.	*Streptomyces flavoviridis*	*Withania somnifera, Ocimum basillicum,* and *Rauvolfia tetraphylla*	Waheeda and Shyam (2017)
6.	*Streptomyces, Streptosporangium, Streptoverticillium,* and *Saccharomonospora* sp.	*Azadirachta indica*	Verma et al. (2009)
7.	*Micromonospora, Actinopolyspora,* and *Saccharopolyspora*	*Ocimum Sanctum*, Aloe vera, *Mentha arvensis*	Gangwar et al. (2014)
8.	*Actinomycetes*	*Syzygium cumini*	Saini et al. (2016)
9.	*Pseudomonas, Acinetobacter, Enterobacte, Pantoea,* and *Ralstonia*	Soybean	Kuklinsky-Sobral et al. (2004)
10.	*Burkholderia, Erwinia, Bacillus,* and *Pseudomonas*	*Paphiopedilum*	Tsavkelova et al. (2007)
11.	*Azospirillum* spp.	Maize	Roesch et al. (2007)
12.	*Achromobacter xylosoxidans*	Wheat	Jha and Kumar (2009)
13.	*Bacillus* and *Sphingopyxis*	Strawberry	Dias et al. (2009)
14.	*Ralstonia* sp., *Pantoea agglomerans,* and *Pseudomonas thivervalensis*	*Brassica napus*	Zhang et al. (2011)
15.	*Burkholderia kururiensis*	Rice	Mattos et al. (2008)
16.	*Serratia*	Banana	Ting et al. (2008)
17.	*Talaromyces Funiculosus, Aspergillus fumigatus,* and *Cladosporium sphaerospermum*	Soybean	Hamayun et al. (2009)
18.	*Penicillium* sp. and *Phoma glomerata*	Rice	Waqas et al. (2012)
19.	*Alternaria alternata* and *Fusarium tricinctum*	*Solanum nigrum*	Khan et al. (2015)
20.	*Aspergillus terreus* and *Penicillium citrinum*	*Helianthus annuus* L.	Waqas et al. (2015)
21.	*Trichoderma pseudokoningii*	Tomato	Chadha et al. (2015)
22.	*Fusarium proliferatum*	*Physalis alkekengi* var. *franchetii*	Rim et al. (2005)

12.5 Conclusion with Future Perspectives

It has been observed now a days that, fluctuating ecological conditions has effected harshly on the growth and development of the crop plants resulting in deprived yields. To overcome such an issue, improvement in throughput and sustainability in the field of agriculture is required by working on naturally safe and worthwhile efforts; targeting on deprived usage of fertilizers and pesticides is obligatory. Excessive use of fertilizers and pesticides causes soil salinity that leads to salt stress conditions in plants. However, it is recognized that stress tolerance of plants is associated with their accompanying microbes (endophytic bioinoculants). These endophytic bioinoculants' arbitrated stress forbearance in plants is a sustainable methodology to attain improved growth and crop production. Beneficial endophytic bioinoculants could be a virtuous alternative to substitute fertilizers and pesticides that are chemically produced and used in cultivation. An enormous variety of plant-accompanying endophytic bioinoculants makes it challenging to comprehend the specific mechanisms mode inside the precise plant–microorganism symbiotic collaboration. Hence, endophytes have turned out to be a potential technique for the betterment of plant progression and production under various stress conditions. It is obligatory now to concentrate our investigations on identifying the varieties of endophytic bioinoculants at several phases of plants and their association with plant growth factors and stress forbearance conditions. Host-accompanying microorganisms were observed to have genes accountable for salt stress adaptation, nutrient solubility, and root establishment. Additionally, these microorganisms also enhance nonhost growth of plant and forbearance toward salinity conditions. Therefore, comprehensive exploration of the mechanistic action of microbes with plants under stress conditions is necessary. An ethical understanding of data generated by investigational research on endophytic bioinoculants will permit us to exploit it efficiently, which can result in sustainable food yields under adverse stress conditions.

REFERENCES

Allam NG, Kinany R, El-Refai E, Ali WY (2018) Potential use of beneficial salt tolerant bacteria for improving wheat productivity grown in salinized soil. *J Microbiol Res* 8(2):43–53. doi:10.5923/j.microbiology.20180802.03.

Andersen EJ, Ali S, Byamukama E, Yen Y, Nepal MP (2018) Disease resistance mechanisms in plants. *Genes (Basel)* 9(7):339. doi: 10.3390/genes9070339.

Ansari MI, Khan MMAA (2018) Plant and heavy metal toxicity. In: Khan MMAA, Abid M, Ansari MI, Chharang H (eds.), *Perspectives in Ecotoxicology*. Lenin Media Private Limited, New Delhi, India. pp. 59–71.

Baye TM, Abebe T, Wilke RA (2011) Genotype-environment interactions and their translational implications. *Per Med* 8(1):59–70.

Chadha N, Prasad R, Varma A (2015) Plant promoting activities of fungal endophytes associated with tomato roots from central Himalaya, India and their interaction with *Piriformospora Indica*. *IJPBS* 6(1):333–343.

Choe YH, Kim YS, Kim IS, Bae MJ, Lee EJ, Kim YH, Park HM, Yoon HS (2013) Homologous expression of gamma-glutamylcysteine synthetase increases grain yield and tolerance of transgenic rice plants to environmental stresses. *J Plant Physiol* 170:610–618.

Choudhary DK (2012) Microbial rescue to plant under habitat imposed abiotic and biotic stresses. *Appl Microbiol Biotechnol* 96:1137–1155.

Choudhary DK, Kasotia A, Jain S, Vaishnav A, Kumari S, Sharma KP, Varma A (2015) Bacterial mediated tolerance and resistance to plants under abiotic and biotic stresses. *J Plant Growth Regul* 35:276–300.

Christou A, Manganaris GA, Papadopoulos I, Fotopoulos V (2013) Hydrogen sulfide induces systemic tolerance to salinity and nonionic osmotic stress in strawberry plants through modification of reactive species biosynthesis and transcriptional regulation of multiple defence pathways. *J Exp Bot* 64:1953–1966.

Dias AC, Costa FE, Andreote FD, Lacava PT, Teixeira MA, Assumpção LC, Araújo WL, Azevedo JL, Melo IS (2009) Isolation of micropropagated strawberry endophytic bacteria and assessment of their potential for plant growth promotion. *World J Microbiol Biotechnol* 25(2):189–195.

Dias MP, Bastos MS, Xavier VB, Cassel E, Astarita LV, Santarém ER (2017) Plant growth and resistance promoted by *Streptomyces* spp. in tomato. *Plant Physiol Biochem* 118:479–493. doi: 10.1016/j.plaphy.2017.07.017.

Egamberdieva D, Shurigin V, Gopalakrishnan S, Sharma R (2014) Growth and symbiotic performance of chickpea (*Cicer arietinum*) cultivars under saline soil conditions. *J Biol Chem Res* 31:333–341.

Feng W, Lindner H, Robbins NE, Dinneny JR (2016) Growing out of stress: The role of cell- and organ-scale growth control in plant water-stress responses. *Plant Cell* 28(8):1769–1782.

Flowers TJ, Colmer TD (2015) Plant salt tolerance: Adaptations in halophytes. *Ann Bot* 115(3):327–331.

Forni C, Duca D, Glick BR (2017) Mechanisms of plant response to salt and drought stress and their alteration by rhizobacteria. *Plant Soil* 410:335–356.

Gangwar M, Dogra S, Gupta UP, Kharwar RN (2014) Diversity and biopotential of endophytic actinomycetes from three medicinal plants in India. *Afr J Microbiol Res* 8(2):184–191.

Hamayun M, Khan SA, Ahmad N, et al. (2009) *Cladosporium sphaerospermum* as a new plant growth-promoting endophyte from the roots of *Glycine max* (L.) Merr. *World J Microbiol Biotechnol* 25(4):627–632.

Hardoim PR, van Overbeek LS, Berg G, et al. (2015) The hidden world within plants: Ecological and evolutionary considerations for defining functioning of microbial endophytes. *Microbiol Mol Biol Rev* 79:293–320.

Hussain AO, Sulaimon HO (2018) A comparative study of food security in Africa amid growing population. *Inter J Sci Res Publ* 8(10):6–12. doi: 10.29322/IJSRP.8.10.2018.p8203.

Ilangumaran G, Smith DL (2017) Plant growth promoting rhizobacteria in amelioration of salinity stress: A systems biology perspective. *Front Plant Sci* 8:1768.

Jalil SU, Ansari MI (2018) Plant microbiome and its functional mechanism in response to environmental stress. *Int J Green Pharmacy* 12(1): S81–S92.

Jha P, Kumar A (2009) Characterization of novel plant growth promoting endophytic bacterium *Achromobacter xylosoxidans* from wheat plant. *Microb Ecol* 58(1):179–188.

Khan AL, Hussain J, Al-Harrasi A, Al-Rawahi A, Lee IJ (2015) Endophytic fungi: Resource for gibberellins and crop abiotic stress resistance. *Crit Rev Biotechnol* 35(1):62–74.

Khan TA, Mazid M, Quddusi S (2014) Role of organic and inorganic chemicals in plant-stress mitigation. In: Gaur RK, Sharma P (eds.), *Approaches to Plant Stress and Their Management*. Springer, New Delhi, pp. 39–52.

Kuklinsky-Sobral J, Araújo WL, Mendes R, Geraldi IO, Pizzirani-Kleiner AA, Azevedo JL (2004) Isolation and characterization of soybean-associated bacteria and their potential for plant growth promotion. *Environ Microbiol* 6(12):1244–1251.

Machado RMA, Serralheiro RP (2017) Soil salinity: Effect on vegetable crop growth: Management practices to prevent and mitigate soil salinization. *Horticulturae* 3(2):30. doi:10.3390/horticulturae3020030.

Mapelli F, Marasco R, Rolli E, Barbato M, Cherif H, Guesmi A, Ouzari I, Daffonchio D, Borin S (2013) Potential for plant growth promotion of rhizobacteria associated with Salicornia growing in Tunisian hypersaline soils. *BioMed Res Int.* doi:10.1155/2013/24807 8.

Mattos KA, Pádua VL, Romeiro A, Hallack LF, Neves BC, Ulisses TM, Barros CF, Todeschini AR, Previato JO, Mendonça-Previato L (2008) Endophytic colonization of rice (*Oryza sativa* L.) by the diazotrophic bacterium *Burkholderia kururiensis* and its ability to enhance plant growth. *An Acad Bras Cienc* 80(3):477–493.

Meritxell G, Jesús C, Sebastià O, Jordi M, Maarten WS, Carles A, Antoni L (2009) Salinity is reduced below the evaporation front during soil salinization: Estudios en la Zona no Saturada del Suelo. Vol IX, O. Silva et al. Barcelona, 18 a 20.

Moya JL, Primo-Millo E, Talon M (1999) Morphological factors determining salt tolerance in citrus seedling: The shoot-to-root ratio modulates passive root uptake of chloride ions and their accumulation in leaves. *Plant Cell Environ* 22:1425–1433.

Munns R (2002) Comparative physiology of salt and water stress. *Plant Cell Environ* 25:239–250.

OECD (2016) *Alternative Futures for Global Food and Agriculture*, OECD Publishing, Paris, France. doi:10.1787/9789264247826-en.

Pérez-Alfocea F, Albacete A, Ghanem ME, Dodd IC (2010) Hormonal regulation of source–sink relations to maintain crop productivity under salinity: A case study of root-to-shoot signaling in tomato. *Fun Plant Biol* 37:592–603.

Qin Y, Druzhinina IS, Pan X, Yuan Z (2016) Microbially mediated plant salt tolerance and microbiome-based solutions for saline agriculture. *Biotechnol Adv* 34:1245–1259.

Reiter B, Pfeifer U, Schwab H, Sessitsch A (2002) Response of endophytic bacterial communities in potato plants to infection with *Erwinia carotovora* subsp: *atroseptica*. *Appl Environ Microbiol* 68(5):2261–2268.

Rim SO, Lee JH, Choi WY, Hwang SK, SUH SJ, Lee IJ, Rhee IK, Kim JG (2005) *Fusarium proliferatum* KGL0401 as a new gibberellin-producing fungus. *J Microbiol Biotechnol* 15(4):809–814.

Rodriguez R, Redman R (2008) More than 400 million years of evolution and some plants still can't make it on their own: Plant stress tolerance via fungal symbiosis. *J Exp Bot* 59:1109–1114.

Roesch LFW, de Quadros PD, Camargo FA, Triplett EW (2007) Screening of diazotrophic bacteria *Azopirillum* spp. for nitrogen fixation and auxin production in multiple field sites in southern Brazil. *World J Microbiol Biotechnol* 23(10):1377–1383.

Ruppel S, Franken P, Witzel K (2013) Properties of the halophyte microbiome and their implications for plant salt tolerance. *Funct Plant Biol* 40:940–951.

Saini P, Gangwar M, Kalia A, Singh N, Narang D (2016) Isolation of endophytic actinomycetes from *Syzygium cumini* and their antimicrobial activity against human pathogens. *J Appl Nat Sci* 8(1):416–422.

Schmidhuber J, Tubiello FN (2007) Global food security under climate change. *Proc Natl Acad Sci USA*. 104(50):19703–19708.

Sharma A, Vaishnav A, Jamali H, Srivastava AK, Saxena AK, Srivastava AK (2016) Halophilic bacteria: Potential bioinoculant for sustainable agriculture and environment management under salt stress. In: Choudhary DK et al. (eds.), *Plant-microbe Interaction: An Approach to Sustainable Agriculture*. Springer, Singapore, pp. 297–325. doi:10.1007/978-981-10-2854-0_14.

Shen H, He X, Liu Y, Chen Y, Tang J, Guo T (2016) A complex inoculant of N2-fixing, P- and K-solubilizing bacteria from a purple soil improves the growth of Kiwifruit (*Actinidia chinensis*) plantlets. *Front Microbiol* 7:841. doi:10.3389/fmicb.2016.00841.

Shrivastava P, Kumar R (2014) Soil salinity: A serious environmental issue and plant growth promoting bacteria as one of the tools for its alleviation. *Saudi J Biol Sci* 22(2):123–131.

Singh D, Rajawat MVS, Kaushik R, Prasanna R, Saxena AK (2017) Beneficial role of endophytes in biofortification of Zn in wheat genotypes varying in nutrient use efficiency grown in soils sufficient and deficient in Zn. *Plant Soil*. doi:10.1007/s1110 4-017-3189-x.

Sorty AM, Meena KK, Choudhary K, Bitla UM, Minhas PS, Krishnani KK (2016) Effect of plant growth promoting bacteria associated with halophytic weed (*Psoralea corylifolia* L.) on germination and seedling growth of wheat under saline conditions. *Appl Biochem Biotechnol* 180:872–882.

Souza Rd, Ambrosini A, Passaglia LM (2015) Plant growth-promoting bacteria as inoculants in agricultural soils. *Genet Mol Biol* 38(4):401–419.

Sun OL, Gyung JC, Yong HC et al. (2008) Isolation and characterization of endophytic actinomycetes from Chinese cabbage roots as antagonists to *Plasmodiophora brassicae*. *J Microbiol Biotechnol* 18(11): 1741–1746. doi:10.4014/jmb.0800.108.

Sun Y, Cheng Z, Glick BR (2009) The presence of a 1-aminocyclopropane-1-carboxylate (ACC) deaminase deletion mutation alters the physiology of the endophytic plant growth-promoting bacterium *Burkholderia phytofirmans* PsJN. *FEMS Microbiol Lett* 296:131–136.

Taechowisan T, Chanaphat S, Ruensamran W, and Phutdhawong WS (2013) Antibacterial activity of Decursin from Streptomyces sp. GMT-8; an endophyte in *Zingiber officinale* Rosc. *J Appl Pharm Sci* 3(10):74–78. doi:10.7324/JAPS.2013.31012.

Tan HM, Cao LX, He ZF, Su GJ, Lin B, Zhou SN (2006) Isolation of endophytic actinomycetes from different cultivars of tomato and their activities against *Ralstonia solanacearum in vitro*. *World J Microbiol Biotechnol* 22(12):1275–1280.

Thornton PK, Ericksen PJ, Herrero M, Challinor AJ (2014) Climate variability and vulnerability to climate change: A review. *Glob Chang Biol* 20(11):3313–3328.

Ting AS, Meon S, Kadir J, Radu S, Singh G (2008) Endophytic microorganisms as potential growth promoters of banana. *BioControl* 53(3):541–553.

Tsavkelova EA, Cherdyntseva TA, Botina SG, Netrusov AI (2007) Bacteria associated with orchid roots and microbial production of auxin. *Microbiol Res* 162(1):69–76.

Tuteja N, Mahajan S (2007) Calcium signaling network in plants: An overview. *Plant Signal Behav* 2(2):79–85.

Vaishnav A, Jain S, Kasotia A, Kumari S, Gaur RK, Choudhary DK (2014) Molecular mechanism of benign microbe-elicited alleviation of biotic and abiotic stresses for plants. In: Gaur RK et al. (eds.), *Approaches to Plant Stress and Their Management*. Springer, New Delhi. doi:10.1007/978-81-322-1620-9_16.

Vaishnav A, Varma A, Tuteja N, Choudhary DK (2017) Characterization of bacterial volatiles and their impact on plant health under abiotic stress. In: Choudhary D, Sharma A, Agarwal P, Varma A, Tuteja N (eds.), *Volatiles and Food Security*. Springer, Singapore, pp. 15–24.

Verma VC, Gond SK, Kumar A, Mishra A, Kharwar RN, Gange AC (2009) Endophytic actinomycetes from *Azadirachta indica* A. Juss.: Isolation, diversity, and anti-microbial activity. *Microb Ecol* 57(4):749–756.

Waadt R, Hsu PK, Schroeder JI (2015) Abscisic acid and other plant hormones: Methods to visualize distribution and signaling. *Bioessays* 37(12):1338–1349.

Waheeda K, Shyam KV (2017) Formulation of novel surface sterilization method and culture media for the isolation of endophytic actinomycetes from medicinal plants and its antibacterial activity. *J Plant Pathol Microbiol* 8(399):2.

Waqas M, Khan AL, Hamayun M, Shahzad R, Kang SM, Kim JG, Lee IJ (2015) Endophytic fungi promote plant growth and mitigate the adverse effects of stem rot: An example of *Penicillium citrinum* and *Aspergillus terreus*. *J Plant Interact* 10(1):280–287.

Waqas M, Khan AL, Kamran M, Hamayun M, Kang SM, Kim YH, Lee IJ (2012) Endophytic fungi produce gibberellins and indoleacetic acid and promotes host-plant growth during stress. *Molecules* 17(9):10754–10773.

Wáskiewicz A, Gładysz O, Goliñski P (2016) Participation of phytohormones in adaptation to salt stress. In: Ahammed GJ, Yu J-Q (eds.), *Plant Hormones Under Challenging Environmental Factors*. Springer, Dordrecht, the Netherlands, pp. 75–115.

Weyens N, Beckers B, Schellingen K, Ceulemans R, Croes S, Janssen J, Haenen S, Witters N, Vangronsveld J (2013) Plant-associated bacteria and their role in the success or failure of metal phytoextraction projects: First observations of a field-related experiment. *Microb Biotechnol* 6(3):288–299.

Xiong L, Schumaker KS, Zhu JK (2002) Cell signaling during cold, drought, and salt stress. *Plant Cell* 14(Suppl):S165–S183.

Zhang YF, He LY, Chen ZJ, Wang QY, Qian M, Sheng XF (2011) Characterization of ACC deaminase-producing endophytic bacteria isolated from copper-tolerant plants and their potential in promoting the growth and copper accumulation of *Brassica napus*. *Chemosphere* 83(1):57–62.

Zhao K, Penttinen P, Guan T, Xiao J, Chen Q, Xu J, Lindström K, Zhang L, Zhang X, Strobel GA (2011) The diversity and anti-microbial activity of endophytic actinomycetes isolated from medicinal plants in Panxi plateau, China. *Curr Microbiol* 62(1):182–190.

13

Microbial Mitigation of Salinity and Water Stress in Legume Crops

Sanket Ray, Dhruti Amin, and Nafisa Patel

CONTENTS

13.1 Introduction

Legumes are broadly defined by their unique flower structure and podded fruits, and around 88% of the species is able to form nodules (de Faria et al. 1989). Environmental stress factors such as temperature, salinity, and water affect almost all parameters of the legume crop—starting from germination to maturity (Xu et al. 2014; Negrão, Schmöckel, & Tester et al. 2017; Nadeem et al. 2018, 2019). The legumes are well known for their significant role in increasing soil carbon sequestration, nitrogen fixation, and as high protein feed (Abberton 2010). The role of legumes in soil fertility is unanimously accepted already. The quest of the human population for animal-derived protein has led to a global warming problem that is being discussed on the major scientific forums nowadays. The dependency on animal-derived protein can be reduced by diverting the preferences toward legume crop-derived protein as a sustainable approach. The legume grain and forage legumes are estimated to be grown on around 180 million Ha, which accounts for 12%–15% of the Earth's arable land (Graham and Vance 2003). Legume is an important candidate in agroforestry and natural ecosystems. Industrial and medicinal properties of legumes impart their socioeconomic importance.

Biotic and abiotic stress cause negative effects in legumes and can significantly reduce crop productivity. Organisms such as fungi, bacteria, and viruses posed biotic stress on legumes. Various abiotic stresses such as excess water, water deficit, soil salinity, extreme temperature, and insufficient mineral nutrients in the soil and high- or low-light radiation damage legume crops (Bohnert & Sheveleva 1998; Bob et al. 2002). This chapter emphasizes the various microbial strategies on food legume plants to reduce the effect of salinity stress and water stress.

13.2 Salinity

Soil salinity and sodicity are a major abiotic constraint to legume crops, especially in areas with irrigated farming. About 60 million hectares or 20% of the total irrigated land area in the world is negatively affected by salt stress (Food and Agriculture Organization of the United Nations 2015). Moreover, to face the problem of feeding a persistent population, it is important to enhance legume productivity around 87% by 2050 (Kromdijk et al. 2016). Salt stress is due to accumulation of sodium chloride (NaCl) either from irrigation or natural accumulation (Flowers and Flowers, 2005).

Legumes are evidence of high sensitivity toward salt stress (Ashraf & O'Leary 1999). They are mainly affected during vegetative and reproductive phases by disturbance of the nutritional disparities, multi-level interaction of hormones, specific ion toxicity, and osmotic effects (He et al. 2015; Patil et al. 2016). Therefore, this chapter discusses the influence, mechanisms, and management strategies that may lead to the development of salt-tolerant cultivars under salinity as well as to fulfill the target of 2050 as discussed previously.

Salinity has negative effects on plants by affecting germination, growth, reproduction, and also the ability to maintain biological nitrogen fixation in grain legumes (Van et al. 2001). Due to salinity, there are also devastating impacts to physiological functions, nutritional balance, and hormonal regulation (Yadav et al. 1989; El Sayed 2011). Salinity reduces the ability of legume plants to take up water, causing immediate fall in growth and an array of metabolic changes identical to those caused by water stress (Munns 2002). It also disturbs composition of grains and grain yield (Lopez-Aguilar et al. 2003; Manchanda and Garg 2008). Salinity causes ionic imbalance that disrupts homeostasis of Na^+, Cl^-, K^+, and Ca^{+2} ions (Rodríguez-Navarro 2000; Hasegawa et al. 2000; Serrano et al. 1999; Niu et al. 1995).

13.2.1 Types Salinity

Salinity is the concentration of dissolved mineral salts present in the soils (soil solution) and waters. The dissolved mineral salts consist of the electrolytes of cations and anions. The major cations in saline soil solutions consist of Na^+, Ca^{2+}, Mg^{2+}, and K^+; the major anions are Cl^-, SO_4^{2-}, HCO^{3-}, CO_3^{2-}, and NO_3^-. Other constituents contributing to salinity in hypersaline soils and waters include B, Sr^{2+}, SiO_2, Mo, Ba^{2+}, and Al^{3+}. Water soluble salts accumulate in the soil solum (the upper part of the soil profile) or regolith (the layer or mantle of fragmental and unconsolidated rock material) to a level that impacts on agricultural crop, environmental health, and economic welfare (United States Salinity Laboratory Staff 1954; Rengasamy 2006). Rainfall contains very low amounts of salt, but over a period of time, salt deposited by rain can store in the landscape. Poor-quality irrigation water also contributes to salt accumulation in irrigated soils. Due to tsunami, salt deposition drastically increases in coastal soil as seawater intrusion onto land. With the combined effect of various climatic conditions, geographic landscape, and human activities this leads to a change in salt accumulation in the agricultural land (Rengasamy 2006). Salinity has been categorized as "primary" or "secondary" salinity by Ghassemi et al. (1995). Primary salinity is due to the accumulation of salts over long periods of time, from two natural progressions. One is the weathering of rocks comprising soluble salts of various types, mainly chlorides of sodium, calcium, and magnesium, and to a lesser extent, sulfates and carbonates. The second is the deposition of oceanic salt carried inland by wind and rain. The salt composition of this deposited salt is that of seawater, that is, mainly sodium chloride (Munns 2002). Naturally salt-affected areas occur widely in arid and semiarid areas (Rengasamy et al. 2003). They are often obvious as salt lakes but can also occur out of sight, in the subsoil, where they are not associated with groundwater or with rising water tables (Rengasamy 2002). This type of salinity is named "transient salinity" because it can move up and down the soil profile, and in and out of the root zone, depending on the season.

Secondary salinity results from human activities that change the hydrologic balance in the soil between water applied (irrigation or rainfall) and water used by crops (transpiration). The most common causes are (1) land clearing and the replacement of perennial vegetation with annual crops, and (2) irrigation schemes using salt-rich irrigation water or having insufficient drainage. Prior to human activities, in arid or semiarid climates, the natural vegetation would have used almost all the rainfall, with the deep roots of native vegetation ensuring that the water tables were well below the surface. Clearing and irrigation change this balance, so that rainfall on the one hand, and irrigation water on the other, provide more water than the crops can use. Water that "leaks" below the roots causes water tables to rise. This mobilizes salts previously stored in the subsoil and moves them up to the root zone. Plants use the water and leave the salt behind until the soil water becomes too salty for further water uptake by roots. The water table continues to rise, and when it comes close to the surface, water evaporates and leaves salts behind on the surface, forming a salt scald. The mobilized salt can also move laterally to water courses and increase their salinity (Ghassemi et al. 1995).

13.2.2 Influence of Salinity

Salinity negatively affects the soil physicochemical properties, agricultural production of legume, and ecological stability of the area. The damaging impacts of salinity include less agricultural production, which leads to low economic returns due to higher cost of cultivation, recovery, and agricultural management. Legume crops show a wide range of responses toward salt. Salt effects are the collective consequence of the complex interaction among different biochemical, physiological, and morphological processes (Singh and Chatrath 2001). Effects of salt stress reflect on change in morphological symptoms are indications of the harmful effects. The cell division and enlargement may directly or indirectly inhibit the plant's growing point due to salt stress. Salinity negatively affects shoot growth in growing tissues, not in mature photosynthetic tissues. According to Singh and Chatrah (2001), stunted growth of stems and leaves of the affected legumes was observed under salinity. Chloride tends to elongate palisade cells, which leads to leaves of legumes becoming succulent. Salt stress reduces dry matter content, increases root: shoot ratio, and diminishes leaf size; as a result, grain yield is reduced (Singh and Chatrath 2001). It affects many aspects of plant metabolism and, as a result, growth is reduced. Salinity decreases the rate of photosynthesis and respiration in legumes by its adverse effect on leaf expansion rate, leaf area, and leaf duration, as well as reduces water movement through the root with a decrease in hydraulic conductivity. Mineral uptake by roots is affected as a result of an imbalance in the availability of different ions. Proportions of Ca in the medium that are adequate under non-saline conditions become inadequate under saline conditions. Growth inhibition by salt arises due to the alteration of energy from growth to maintenance (Nieman and Maas 1978). Energy is utilized for the regulation of ion concentration in various parts and within the cell, for maintenance of membrane integrity, and the synthesis of organic solutes for osmoregulation or protection of macromolecules (Jain et al. 2001). Osmoregulation ensures that adequate turgor is maintained in the cell. In plant cells, accumulated organic compounds in the cytoplasm protect the conformation of macromolecules in the changing salinity condition (Borowitzak 1981; Wyn Jones and Pollard 1983). Since salt damage has a wide-ranging physiological spectrum affecting multiple metabolic processes, it is challenging to address the impact of individual processes on plant death or on the ultimate damage done to the plant.

13.3 Water Stress

The biggest consumer of water worldwide is the agriculture sector, covering up to 90% in developing countries. The total land cultivated in the world is 1474 million ha; 86% falls under the rainfed cultivation (Kumar 2005). Many times, these regions receive rainfall below the average level, consistently resulting in a condition referred to as drought. Thus, drought can be defined as an extended abnormal dry period. The global climate change includes drought and floods besides temperature, which account for the decrease in the yield of food crops. Among the risks in agriculture classified by the USDA, yield risk is the most common having a direct impact on agriculture, which mainly depends on the water supply and temperature (Motha and Menzie 2007). The other impact of global warming is the increased events of flooding, causing submergence of many ecosystems throughout the globe. Flooding can be natural or human-made, resulting in crop loss. An understanding of the major physiological and molecular mechanisms inducing plant tolerance to these calamities could result in developing strategies to reduce crop loss.

Legume plants are exposed to many kinds of environmental stress throughout the world. Among the entire stress conditions, water deficit is considered the most important factor reducing the crop productivity (Micheletto et al. 2007; Fang et al. 2010). Drought and its effect have been studied in grain legumes; the most important grain legumes are chickpea (*Cicerarietinum* L.), mashbean (*Vignamungo* L.), common bean (*Phaseolusvulgaris* L.), mung bean (*Vignaradiata* L.), and soyabean (*Glycine max* L.). Grain legume production is threatened by various environmental stresses (Vyas 2014), of which water deficit has been identified as an important factor limiting crop productivity worldwide. The mechanisms underlying the reduction in the yield are associated with leaf senescence (Lopez et al. 1997), oxidative damage to photosynthetic apparatus (Farooq et al. 2009), reduced carbon fixation and assimilate translocation (Mondal et al. 2011, Zlatev and Lidon 2012), inhibition of flowering and reproductive development (Samarah et al. 2009), sterility of pollen grains (Al-Ghzawi et al. 2009), reduced grain set and development (Croser et al. 2003, Liu et al. 2004, Vadez et al. 2012), and reduced sink capacity (Andersen et al. 2002).

Water stress is a major constraint limiting grain legume production, particularly in the arid and semiarid tropics. Different climate models have predicted changes in rainfall distribution and frequent drought spells for the future. Although drought impedes the productivity of grain legumes at all growth stages, its occurrence during reproductive and grain development stages (terminal drought) is more critical and usually results in significant loss in grain yield. However, the extent of yield loss depends on the duration and intensity of the stress. A reduction in the rate of net photosynthesis, and poor grain set and grain development are the principal reasons for terminal drought-induced loss in grain yield. Insight into the impact and resistance mechanism of terminal drought is required for effective crop improvement programs aiming to improve resistance to terminal drought in grain legumes.

13.3.1 Drought

Drought is categorized into major three types: viz. agricultural drought, meteorological drought, and hydrological drought. Less frequent rainfall or above-normal evaporation results in agriculture drought while reduced water supply or water levels are referred to as hydrological drought. Large-scale changes in the atmospheric patterns cause meteorological drought. Other factors such as high temperature, poor irrigation management, and erosion might also result in drought conditions.

13.3.1.1 Legume Response Under Drought Condition

Drought resistance mechanisms can be classified broadly into three categories (Taiz and Zeiger 2002) that include (i) Desiccation postponement, an ability to maintain tissue hydration or drought tolerance at high water potential; (ii) Desiccation tolerance, an ability to function while dehydrated or drought tolerance at low water potential; and (iii) Drought escape, which means avoidance of drought by completing its life cycles before the onset of dry periods to sustain reproduction. Dehydration tolerance or avoidance is achieved by developing large and deep root systems to acquire water from soil, closure of stomata, and impermeable leaf cuticle to avoid transpiration.

Drought resistance is also attributed to the development of some physiological traits such as osmotic adjustment, ABA content, chlorophyll content, proline, soluble sugars, and peroxidase or superoxide dismutase activity to remove toxic substances, which may cause drought tolerance (LUO 2010). The physiological traits affecting crop response include early vigor or faster phenological development (Cairns et al. 2009); completing the reproductive stage before the period of possible drought; seed size and early seedling vigor; early flowering; changes in the root architecture with an increase depth and density of roots; negative leaf water potential (LWP) causing cavitation in the xylem, leading to the loss of turgor and wilting of plants; leaf relative water content (RWC) by the balance between water supply and transpiration, as plant water status controls crop performance during drought; stomatal conductance and its reduction, which improves the water use efficiency and productivity. Plants exhibit stomatal closure to combat drought conditions; plants with high stomatal conductance transpire more and thus maintain a cooler canopy temperature (Lie et al. 2003).

Other modifications include anatomical changes to reduce the water loss such as sunken stomata, glaucousness (a waxy covering of the plant cuticle), and leaf pubescence density. Cell membrane stability and high phospholipids content cause an increase in membrane permeability, which leads to efflux of electrolytes. Reactive oxygen species (ROS) acts as positive signaling molecules to regulate response of plant growth to water stress. Osmotic adjustment, an active intracellular accumulation of organic solutes that maintains cellular turgor, preserves stomatal conductance and photosynthesis, delayed leaf senescence, and death. It also reduces flower abortion, improved root growth as well as increases water extraction from soil. Water use efficiency (WUE) is the ratio between the amount of dry matter produced per unit of water applied.

13.3.1.2 Legume Plants Adaptations

Three mechanisms were studied in grain legumes, i.e., soyabean (*Glycine max*), black gram (*Vigna mungo*), green gram (*V. radiata*), and cowpea (*V. unguiculata*) (Lawn 1982b). Dehydration avoidance was mainly responsible to combat with the water loss by stomatal control and paraheliotropic leaf movement. Reduction in the short-term growth was also seen due to lower rates of exchange of the gases as well as photosynthesis. Drought escape was another mechanism by developing plasticity in the form of rapid development and occasionally as delayed development. Depending on the stress periods, the plants adapted to either shortened or elongated crop duration.

Combating drought resistance in the legumes shows some interesting phenomenon, each depending on the seasonal pattern of water availability. There is as such no particular and specific adaptation to drought resistance in the grain legumes; rather, several alternative pathways are adapted to increase the productivity. soyabean adapts to the drought conditions by having least developmental plasticity and dehydration, depletes most of its soil water before pod filling, reducing the seed size and lowering the harvest index (HI) and water use efficiency (WUE). The soyabean strategy of "productivity maintenance" resulted in the most effective utilization of available water and the highest yields. Whereas, cowpea shows the highest developmental plasticity, dehydration, extreme avoidance response, and low HI and WUE. Although cowpea and soyabean had contrasting stress response strategies, the WUE was observed to be the same. In black gram, reduced use of water during the early season caused more available water during pod filling, with a higher HI and WUE. "Conservation" strategies of the Vigna species cause a lower production. Green gram exhibited strong dehydration avoidance and strong developmental plasticity in the form of more rapid phenological development under stress.

13.3.2 Flooding

Just as water scarcity affects the growth and yield of the crops, so too does waterlogging because excess water affects the morphology as well as the functionality of plants by affecting the root growth, root morphology, and metabolism. The establishment of a symbiotic relationship between the legumes and the rhizobia involves signaling between the plant and the rhizobia, infection of root, low oxygen microenvironment needed for the enzyme nitrogenase in the nodules, substrate for respiration, and energy generation. These processes are affected due to the anoxia and low soil redox affecting the growth of legumes. Also, the O_2 deficiency inhibits nutrient uptake by roots and can result in deficiencies in the shoots (Herzog et al. 2013). Water submersion reduces O_2 availability leading to barrier to gas diffusion and restriction in the aerobic respiration. Such environments also have low-light intensity—reducing photosynthesis—and a high amount of toxic compounds in soil and a low level of carbohydrates and ATP synthesis.

13.3.2.1 Legume Response Under Flood Condition

Plant adaptation to submergence has two underlying strategies, i.e., low oxygen quiescence syndrome (LOQS) and low oxygen escape syndrome (LOES) (Colmer and Voesenek 2009). In LOQS, there is reduction of underwater shoots growth, which conserves the substrate availability until the water level decreases. In LOES, there is a rapid underwater development of shoot organs, which reaches the water surface to establish a gas exchange. Plants adapt and acclimatize depending on their genotype, which includes altered petiole/internode elongation rates, cell ultrastructure modifications, development of lateral and adventitious roots, and formation of aerenchyma. Along with this, there are some metabolic adaptations where the deficiency of O_2 leads to a tight regulation in the ATP synthesis, which occurs by switching from aerobic to anaerobic respiration. The energy generation in complete anoxic conditions depends on the glycolytic flux, the availability of the polysachharides, and catabolizing capacity of the plants (Gibbs and Greenway 2003).

13.4 Salinity and Water Stress Management Using Microbial Mitigation

Nitrogen is the prime requirement for the legume growth. The plant growth-promoting rhizobacteria (PGPR) plays a greater role in handing the salinity stress in crop management. The use of microbial inoculants in legumes crops increases successful adaption and improves soil fertility and structure (Alori et al. 2017). The PGPR traits were observed to produce deaminase enzyme, promote nodulation and other physiological activities at the rhizosphere, which ultimately help the legumes in survival against salinity stress (Saleem et al. 2015). There are certain microbial inoculants available that produce different biological substances such as siderophore (Stajković-Srbinović et al. 2014), PO_4^{-2} solublizing enzymes (Kumari and Khanna 2016; Verma et al. 2013), Salicyclic acid (Ekinci et al. 2014), plant hormone Indole acetic acid (Gujral et al. 2013), deaminase enzyme (Saleem et al. 2015), and microbiocidal/biostatic enzyme (Moustaine et al. 2017). Rhizobacteria contribute legumes nutritionally by trapping and integrating nitrogen through nitrogen fixation. These organisms devise a way of penetrating the root tissue of legumes and establish themselves as endophytes within the plant, which enhances the plant's survival chances and adaptation in various stress conditions, mostly salinity and water stress (Latif Khan et al. 2016). Certain microbial inoculants contribute the intrinsic resistance of plants to pathogenic and environmental challenges. Specific PGPR bacteria are able to form biofilm and secretion of polysaccharides, which enhance stability of legumes under stress conditions (Kasim et al. 2016). Reports suggest that salt stress can be handled during germination in crops using *Pseudomonas* sp., which has the ability to synthesize exopolysaccharides (Sen and Chandrasekhar 2014). In salinity stress-containing regions, microorganisms are able to contribute significantly in biohydrometallurgy by the formulation of metal-rich solutions using microbial oxidation of sulphur containing ore. Further, siderophore-producing microorganisms are able to immobilize metals or cations and render them to bioavailable using polymeric compound.

13.4.1 Phytohormones

Phytohormones play an important role in legume defense mechanism, and plants react and adapt to stress by maintaining balance the level of phytohormones. Plant growth-promoting bacteria (PGPB) improve legume crop development by direct or indirect systems. In direct mechanism, microbes produce indole acetic acid, gibberellins, cytokinins, and ethylene. These plant hormones boost balanced growth of legumes as well as act as antagonists to phytopathogens. In indirect mechanism, the PGPB motivate plant resistance by producing chemicals that can regulate the hormone level.

13.4.2 Quorum Sensing

Quorum sensing is the cross talk between cells in bacteria by chemical means. This facilitates microorganisms to react quickly, increase supplement uptake, and change ecological condition. N-acyl-homoserine lactones in *Proteobacteria*, gamma-butyrolactones in *Streptomyces*, cis-11-methyl-2-dodecanoic acid in *Xanthomonas*, and oligopeptides in Gram-positive microorganisms are utilized as a part of cell-cell communication within the bacterial group (Danhorn and Fuqua 2007).

13.4.3 Root Colonization

Rhizobacteria are colonizing at root nodules in legumes in various phases of crop development, and they can establish mutual association among legume plant and microorganisms. Certain reports suggest the entry of Gram-positive microbes into the root through the primary root, horizontal roots, and root hair (Baldotto et al. 2011; Prieto et al. 2011). The successful colonized microorganisms enhance legume ability to defend against salinity and water stress conditions. Additional benefits of root colonization provide immediate response against plant pathogens and plant phenology (Wagner et al. 2014).

13.4.4 Microbial Inoculants in Legumes

Azospirillumbrasilense and *Rhizobium* spp. promote plant growth, shoot and root size in beans (Dardanelli et al. 2008). *Mesorhizobium* strains help legumes to sustain in salinity stress by providing growth (Wani et al. 2009). *Pseudomonas* spp., was reported for the production of ACC deaminase in drought stress conditions, which helped the crop to survive (Ali et al. 2014; Sergeeva et al. 2006). *Brevibacterium iodinum*, *Bacillus licheniformis*, and *Zhihengliuella alba* were able to produce excess ACC deaminase in salt stress conditions (Siddikee et al. 2011). In chickpea crop, *Pseudomonas putida* MTCC5279 (RA) was found to enhance the expression of stress response gene, maintain water content, osmolyte, membrane structure, and germination rate (Tiwari et al. 2016). *Halomonas maura* and *Ensifer meliloti* were reported for increasing the weight of shoot dry weight, yield, and water content of alfalfa plant in salt stress conditions.

13.5 Conclusion

Substantial effects of plant tolerance to abiotic stress will result in promoting yield and production of crops to feed ever-increasing human and livestock populations. This can be accomplished via the search, selection, and engineering of plant species capable of resisting salinity and drought stress. The practice of PGPR application will go a long way in assisting the plant to develop both intrinsic and extrinsic abilities to tolerate stress conditions and sustain crop yield. A stress management strategy of using microbes (PGPR) could aid greatly in ensuring nonstop and effective agricultural practices that will relieve the stress and boost the crop yield.

REFERENCES

Abberton, M. (2010). Enhancing the role of legumes: Potential and obstacles. *Integrated Crop Management*, *11*, 177–187.

Al-Ghzawi, A. A. M., Zaitoun, S., Gosheh, H., & Alqudah, A. (2009). Impacts of drought on pollination of *Trigonellamoabitica* (*Fabaceae*) via bee visitations. *Archives of Agronomy and Soil Science*, *55*(6), 683–692.

Ali, S. Z., Sandhya, V., & Rao, L. V. (2014). Isolation and characterization of drought-tolerant ACC deaminase and exopolysaccharide-producing fluorescent *Pseudomonas* sp. *Annals of Microbiology*, *64*(2), 493–502.

Alori, E. T., Dare, M. O., & Babalola, O. O. (2017). Microbial inoculants for soil quality and plant fitness, in *Sustainable Agriculture Review*, ed. E. Lichtfouse (Berlin: Springer), pp. 181–308. doi:10.1007/978-3-319-48006-0.

Andersen, M. N., Asch, F., Wu, Y., Jensen, C. R., Næsted, H., Mogensen, V. O., & Koch, K. E. (2002). Soluble invertase expression is an early target of drought stress during the critical, abortion-sensitive phase of young ovary development in maize. *Plant Physiology*, *130*(2), 591–604.

Ashraf, M., & O'Leary, J. W. (1999). Changes in soluble proteins in spring wheat stressed with sodium chloride. *Biologiaplantarum*, *42*(1), 113–117.

Baldotto, L. E. B., Olivares, F. L., & Bressan-Smith, R. (2011). Structural interaction between GFP-labeled diazotrophic endophytic bacterium *Herbaspirillum seropedicae* RAM10 and pineapple plantlets "Vitória" *Brazilian Journal of Microbiology*, *42*(1), 114–125.

Bob, B. B., Wilhelm, G., & Russell, L. J. (2000). *Biochemistry & Molecular Biology of Plants*. Rockville, MD: American Society of Plant Physiologists.

Bohnert, H. J., & Sheveleva, E. (1998). Plant stress adaptations—Making metabolism move. *Current Opinion in Plant Biology*, *1*(3), 267–274.

Borowitzak, L. J. (1981). Solute accumulation and regulation of cell water activity, in *The Physiology and Biochemistry of Drought Resistance in Plants*, eds. L. G. Paleg, D. Aspinall (Sidney: Academic Press), pp. 97–130.

Cairns, J. E., Audebert, A., Mullins, C. E., & Price, A. H. (2009). Mapping quantitative trait loci associated with root growth in upland rice (*Oryzasativa* L.) exposed to soil water-deficit in fields with contrasting soil properties. *Field Crops Research*, *114*(1), 108–118.

Colmer, T. D., & Voesenek, L. A. C. J. (2009). Flooding tolerance: Suites of plant traits in variable environments. *Functional Plant Biology*, *36*(8), 665–681.

Croser, J. S., Ahmad, F., Clarke, H. J., & Siddique, K. H. M. (2003). Utilisation of wild Cicer in chickpea improvement—Progress, constraints, and prospects. *Australian Journal of Agricultural Research*, *54*(5), 429–444.

Danhorn, T., & Fuqua, C. (2007). Biofilm formation by plant-associated bacteria. *Annual Review of Microbiology*, *61*, 401–422.

Dardanelli, M. S., de Cordoba, F. J. F., Espuny, M. R., Carvajal, M. A. R., Díaz, M. E. S., Serrano, A. M. G., ... Megías, M. (2008). Effect of *Azospirillum brasilense* co inoculated with Rhizobium on *Phaseolus vulgaris* flavonoids and Nod factor production under salt stress. *Soil Biology and Biochemistry*, *40*(11), 2713–2721.

De Faria, S. M., Lewis, G. P., Sprent, J. I., & Sutherland, J. M. (1989). Occurrence of nodulation in the Leguminosae. *New Phytologist*, *111*(4), 607–619.

Ekinci, M., Turan, M., Yildirim, E., Güneş, A., Kotan, R., & Dursun, A. (2014). Effect of plant growth promoting rhizobacteria on growth, nutrient, organic acid, amino acid and hormone content of cauliflower (*Brassica oleracea* L. var. *botrytis*) transplants. *Acta Scientiarum Polonorum-Hortorum Cultus*, *13*(6), 71–85.

Fang, X., Turner, N. C., Yan, G., Li, F. & Siddique, K. H. M. (2010). Flower numbers, pod production, pollen viability, and pistil function are reduced and flower and pod abortion increased in chickpea (*Cicerarietinum* L.) under terminal drought. *Journal of Experimental Botany*, *61*, 335–345.

FAO. FAO Statistical Pocketbook. Food and Agriculture Organization of the United Nations; 2015. http:/www.fao.org/documents/card/en/c/383d384a-28e6-47b3-a1a2-2496a9e017b2/. (Accessed 6 February 2016).

Farooq, M., Wahid, A., Kobayashi, N., Fujita, D., & Basra, S. M. A. (2009). Plant drought stress: Effects, mechanisms and management. *Agronomy for Sustainable Development*, *29*, 185–212.

Flowers, T. J., & Flowers, S. A. (2005). Why does salinity pose such a difficult problem for plant breeders? *Agricultural Water Management*, *78*(1–2), 15–24.

Ghassemi, F., Jakeman, A. J., & Nix, H. A. (1995). *Salinisation of Land and Water Resources: Human causes, Extent Management & Case Studies*. Sydney: University of New South Wales, p. 526.

Gibbs, J., & Greenway, H. (2003). Mechanisms of anoxia tolerance in plants. I. Growth, survival and anaerobic catabolism. *Functional Plant Biology*, *30*(1), 1–47.

Graham, P. H., & Vance, C. P. (2003). Legumes: Importance and constraints to greater use. *Plant Physiology*, *131*(3), 872–877.

Gujral, M. S., Agrawal, P., Khetmalas, M. B., & Pandey, R. (2013). Colonization and plant growth promotion of Sorghum seedlings by endorhizospheric *Serratia* sp. *Acta BiologicaIndica*, *2*(1), 343–352.

Hasegawa, P. M., Bressan, R. A., Zhu, J.-K., & Bohnert, H. J. (2000). Plant cellular and molecular responses to high salinity. *Annual Review of Plant Biology*, *51*(1), 463–499.

Herzog, C., Peter, M., Pritsch, K., Günthardt-Goerg, M. S., & Egli, S. (2013). Drought and air warming affects abundance and exoenzyme profiles of *Cenococcum geophilum* associated with *Quercusrobur*, *Q. petraea* and *Q. pubescens*. *Plant Biology*, *15*, 230–237.

He, Y., Fu, J., Yu, C., Wang, X., Jiang, Q., Hong, J., Lu, K., Xue, G., Yan, C., James, A., & Xu, L. (2015). Increasing cyclic electron flow is related to Na^+ sequestration into vacuoles for salt tolerance in soybean. *Journal of Experimental Botany*, *66*(21), 6877–6889.

Jain, M., Mathur, G., Koul, S. Sarin, & N. B. (2001). Ameliorative effects of proline on salt stress induced lipid peroxidation in cell lines of groundnut (*Arachis hypogea* L.). *Plant Cell Reports*, *20*, 463–468.

Kasim, W. A., Gaafar, R. M., Abou-Ali, R. M., Omar, M. N., & Hewait, H. M. (2016). Effect of biofilm forming plant growth promoting rhizobacteria on salinity tolerance in barley. *Annals of Agricultural Sciences*, *61*(2), 217–227.

Kromdijk, J., Głowacka, K., Leonelli, L., Gabilly, S. T., Iwai, M., Niyogi, K. K., & Long, S. P. (2016). Improving photosynthesis and crop productivity by accelerating recovery from photoprotection. *Science*, *354*(6314): 857–861.

Kumar, D. (2005). Breeding for drought resistance, in *Abiotic Stresses*, eds. P. Ashraf, J. Harris (New York: The Howarth Press), pp. 167–198.

Kumari, P., & Khanna, V. (2016). Biodiversity of Pseudomonas and Bacillus possessing both bioantagonistic and plant growth promoting traits in chickpea rhizosphere. *International Journal of Science and Nature*, *7*(1), 153–158.

Latif Khan, A., Ahmed Halo, B., Elyassi, A., Ali, S., Al-Hosni, K., Hussain, J., ... Lee, I. J. (2016). Indole acetic acid and ACC deaminase from endophytic bacteria improves the growth of *Solarium lycopersicum*. *Electronic Journal of Biotechnology*, *19*(3), 58–64.

Lawn, R. J. (1982b). Response of four grain legumes to water stress in south-eastern Queensland. II: Plant growth and soil water extraction patterns. *Australian Journal of Agricultural Research*, *33*(3), 497–509.

Lie, F., Anderson, M.N., & Jensen, C.P. (2003). Loss of pod set caused by drought stress is associated with water status and ABA content of reproductive structure in soybean. *Functional Plant Biology*, *30*, 271–280.

Liu, F., Jensen, C. R., & Andersen, M. N. (2004). Pod set related to photosynthetic rate and endogenous ABA in soyabeans subjected to different water regimes and exogenous ABA and BA at early reproductive stages. *Annals of Botany*, *94*(3), 405–411.

Lopez-Aguilar, R., Orduno-Cruz, A., Lucero-Arce, A., Murillo-Amador, B., & Troyo-Dieguez, E. (2003). Response to salinity of three grain legumes for potential cultivation in arid areas. *Soil Science and Plant Nutrition*, *49*(3), 329–336.

Lopez, F. B., Chauhan, Y. S., & Johansen, C. (1997). Effects of timing of drought stress on leaf area development and canopy light interception of short-duration pigeonpea. *Journal of Agronomy and Crop Science*, *178*, 1–7.

Luo, L. J. (2010). Breeding for water-saving and drought-resistance rice (WDR) in China. *Journal of Experimental Botany*, *61*(13), 3509–3517.

Manchanda, G., & Garg, N. (2008). Salinity and its effects on the functional biology of legumes. *Acta Physiologiae Plantarum*, *30*(5), 595–618.

Micheletto, S., Rodriguez-Uribe, L., Hernandez, R., Richins, R. D., Curry, J., & O'Connell, M. A. (2007). Comparative transcript profiling in roots of *Phaseolusacutifolius* and *P. vulgaris* under water deficit stress. *Plant Science*, *173*(5), 510–520.

Mondal, M. M. A., Fakir, M. S. A., Juraimi, A. S., Hakim, M. A., Islam, M. M., & Shamsuddoha, A. T. M. (2011). Effects of flowering behavior and pod maturity synchrony on yield of mungbean [*Vignaradiata* (L.) Wilczek]. *Australian Journal of Crop Science*, *5*(8), 945–953.

Motha, R. P., & Menzie, K. L. (2007). Meteorological information for evaluating agrometerological risk and uncertainty for agricultural marketing systems. *Bulletin of the World Meteorological Organization*, *56*(1), 30–33.

Moustaine, M., Elkahkahi, R., Benbouazza, A., Benkirane, R., & Achbani, E. H. (2017). Effect of plant growth promoting rhizobacterial (PGPR) inoculation on growth in tomato (*Solanumlycopersicum* L.) and characterization for direct PGP abilities in Morocco. *International Journal of Environment, Agriculture and Biotechnology*, *2*(2), 590–595.

Munns, R. (2002). Comparative physiology of salt and water stress. *Plant, Cell & Environment* *25*(2), 239–250.

Nadeem, M., Li, J., Wang, M., Shah, L., Lu, S., Wang, X., & Ma, C. (2018). Unraveling field crops sensitivity to heat stress: Mechanisms, approaches, and future prospects. *Agronomy*, *8*, 128.

Nadeem, M., Li, J., Yahya, M., Wang, M., Ali, A., Cheng, A., Wang, X., & Ma, C. (2019). Grain legumes and fear of salt stress: Focus on mechanisms and management strategies. *International Journal of Molecular Sciences*, *20*, 799.

Negrão, S., Schmöckel, S.M., & Tester, M. (2017). Evaluating physiological responses of plants to salinity stress. *Annals of Botany*, *119*, 1–11.

Niu, X., Bressan, R. A., Hasegawa, P. M., & Pardo, J. M. (1995). Ion homeostasis in NaCl stress environments. *Plant Physiology, 109*(3), 735.

Nieman, R. H. & Maas, E. V. (1978). 6th Intl. Biophys. Cong. Abst. pp. 121 (cited in Salinity Tolerances in Plants Strategies for Crop Improvement (R.C. Staples & G.H. Toenniesen, Eds.). 1984. John Wiley & Sons, New York.

Patil, G., Do, T., Vuong, T. D., Valliyodan, B., Lee, J. D., Chaudhary, J., Shannon J., & Nguyen, H. T. (2016). Genomic-assisted haplotype analysis and the development of high-throughput SNP markers for salinity tolerance in soybean. *Scientific Reports, 6*(1), 1–13.

Prieto, P., Schilirò, E., Maldonado-González, M. M., Valderrama, R., Barroso-Albarracín, J. B., & Mercado-Blanco, J. (2011). Root hairs play a key role in the endophytic colonization of olive roots by *Pseudomonas* spp. with biocontrol activity. *Microbial Ecology, 62*(2), 435–445.

Rengasamy, P. (2002). Transient salinity and subsoil constraints to dryland farming in Australian sodic soils: An overview. *Australian Journal of Experimental Agriculture, 42*(3), 351–361.

Rengasamy, P., Chittleborough, D., & Helyar, K. (2003). Root-zone constraints and plant-based solutions for dryland salinity. *Plant and Soil, 257*(2), 249–260.

Rodríguez-Navarro, A. (2000). Potassium transport in fungi and plants. *Biochimicaet BiophysicaActa, Reviews on Biomembranes, 1469*(1), 1–30.

Saleem, A. R., Bangash, N., Mahmood, T., Khalid, A., Centritto, M., & Siddique, M. T. (2015). Rhizobacteria capable of producing ACC deaminase promote growth of velvet bean (*Mucuna pruriens*) under watermstress condition. *International Journal of Agriculture and Biology, 17*, 663–667.

Samarah, N. H., Haddad, N., & Alqudah, A. M. (2009). Yield potential evaluation in chickpea genotypes under late terminal drought in relation to the length of reproductive stage. *Italian Journal of Agronomy, 3*, 111–117.

Sen, S., Chandrasekhar, C. N. (2014). Effect of PGPR on growth promotion of rice (*Oryzasativa* L.) under salt stress. *Asian Journal of Plant Science and Research, 4*, 62–67.

Sergeeva, E., Shah, S., & Glick, B. R. (2006). Growth of transgenic canola (*Brassica napus* cv. Westar) expressing a bacterial 1-aminocyclopropane-1-carboxylate (ACC) deaminase gene on high concentrations of salt. *World Journal of Microbiology and Biotechnology, 22*(3), 277–282.

Serrano, R., Mulet, J. M., Rios, G., Marquez, J. A., De Larrinoa, I. F., Leube, M. P., Mendizabal, I. Mendizabal, I., Pascual-Ahuir, A., Proft, M. et al. (1999). A glimpse of the mechanisms of ion homeostasis during salt stress. *Journal of Experimental Botany, 50*, 1023–1036.

Siddikee, M. A., Glick, B. R., Chauhan, P. S., JongYim, W., & Sa, T. (2011). Enhancement of growth and salt tolerance of red pepper seedlings (*Capsicum annuum* L.) by regulating stress ethylene synthesis with halotolerant bacteria containing 1-aminocyclopropane-1-carboxylic acid deaminase activity. *Plant Physiology and Biochemistry, 49*(4), 427–434.

Singh, K., & Chatrath R. (2001). Salinity tolerance. Chapter 8, in *Application of Physiology in Wheat Breeding*, eds. M. P. Reynolds, J. I. Ortiz-Monasterio, A. McNab (Mexico: CIMMYT Crop Improvement Division), pp. 101–110.

Stajković-Srbinović, O. L. I. V. E. R. A., Delić, D. U. Š. I. C. A., Kuzmanović, D. J. O. R. D. J. E., Protić, N. A. D. A., Rasulić, N. A. T. A. Š. A., & Knežević-Vukčević, J. E. L. E. N. A. (2014). Growth and nutrient uptake in oat and barley plants as affected by rhizobacteria. *Romanian Biotechnological Letters, 19*, 9429–9436.

Taiz, L., & Zeiger, E. (2002). *Plant Physiology*. Sunderland, MA: Sinauer Associates.

Tiwari, S., Lata, C., Chauhan, P. S., & Nautiyal, C. S. (2016). *Pseudomonas putida* attunes morphophysiological, biochemical and molecular responses in *Cicerarietinum* L. during drought stress and recovery. *Plant Physiology and Biochemistry, 99*, 108–117.

United States Salinity Laboratory Staff (1954). Diagnosis and Improvement of Saline and Alkali Soils. U.S. Department of Agriculture Handbook, 60. U.S. Government Printing Office, Washington, DC, p. 160.

Vadez, V., Berger, J. D., Warkentin, T., Asseng, S., Ratnakumar, P., Rao, K. P. C., ... Sharma, H. C. (2012). Adaptation of grain legumes to climate change: A review. *Agronomy for Sustainable Development, 32*(1), 31–44.

Van Hoorn, J. W., Katerji, N., Hamdy, A., & Mastrorilli, M. (2001). Effect of salinity on yield and nitrogen uptake of four grain legumes and on biological nitrogen contribution from the soil. *Agricultural Water Management*, 51(2), 87–98.

Verma, J. P., Yadav, J., Tiwari, K. N., & Kumar, A. (2013). Effect of indigenous Mesorhizobium spp. and plant growth promoting rhizobacteria on yields and nutrients uptake of chickpea (*Cicerarietinum* L.) under sustainable agriculture. *Ecological Engineering*, *51*, 282–286.

Vyas, S. P. (2014). Impact and strategies for yield improvement of arid legumes under drought. *The International Journal of Applied Life Science Engineering*, *1*, 12–19.

Wagner, M. R., Lundberg, D. S., Coleman-Derr, D., Tringe, S. G., Dangl, J. L., & Mitchell-Olds, T. (2014). Natural soil microbes alter flowering phenology and the intensity of selection on flowering time in a wild *Arabidopsis* relative. *Ecology Letters*, *17*(6), 717–726.

Wani, P. A., Zaidi, A., & Khan, M. S. (2009). Chromium reducing and plant growth promoting potential of *Mesorhizobium* species under chromium stress. *Bioremediation Journal*, *13*(3), 121–129.

Wyn Jones, R. G., & Pollard, A. (1983). Proteins, enzymes and inorganic ions. In: Lauchli A, Bieleski RL (eds) *Encyclopedia of Plant Physiology* (Inorganic Plant Nutrition), vol. 15 B (New Series). Springer, Berlin, pp. 528–562.

Xu, J., Yuan, Y., Xu, Y., Zhang, G., Guo, X., Wu, F., Wang, Q., Rong, T., Pan, G., Cao, M. et al. (2014). Identification of candidate genes for drought tolerance by whole-genome resequencing in maize. *BMC Plant Biology*, *14*, 83.

Yadav, H. D., Yadav, O. P., Dhankar, O. P., Oswal, M. C. (1989). Effect of chloride salinity and boron on germination, growth, and mineral composition of chickpea (*Cicer arietinum* L.). *Annals Arid Zone*, 28, 63–67.

Zlatev, Z., Lidon, F. C. (2012). An overview on drought induced changes in plant growth, water relations and photosynthesis. *Emirates Journal of Food and Agriculture*, 24: 57–72.

14

Plant Growth-Promoting Rhizobacteria (PGPR) for the Mitigation of Salt Stress and Growth Promotion of Chickpea

Kusum Sharma and Senthilkumar Murugesan

CONTENTS

14.1 Introduction

Chickpea (*Cicer arietinum* L.) is an important food legume with high adaptability to climatic conditions (Roy et al., 2010). At a global level after dry beans, chickpea is ranked as the second important legume (FAOSTAT, 2016). Chickpea seed is a protein-rich supplement and plays an important role in sustentation of soil fertility through nitrogen fixation mechanism, particularly in the arid and low rainfall areas (Jukanti et al., 2012; Atnaf et al., 2015). Earlier research reported that the growth of the legume crop was particularly restricted by abiotic factors such as salinity (Egamberdieva et al., 2017). According to a report by the Food and Agriculture Organization, it was estimated that approximately more than 6% of land area has been affected by salinity (FAO, 2008). A large part of lands with saline soil are not under cultivation, but a considerable proportion (20% of total) constitutes the cultivation land while 32 million ha under dryland agriculture have been affected by salinity (Munns and Tester, 2008). Soil salinization worldwide is a very serious threat to agriculture especially in arid and semiarid regions, and approximately one-third of irrigated arable land is affected by the salinity. As compared to dry lands, irrigated lands are more prone to salinity because of the salt deposition through irrigated water that occurs year after year. A vast tract of irrigated lands has degenerated by secondary salinization and is no longer economical for cultivation. During evapotranspiration, plants are often supplied with more water than they require, which causes secondary salinization. The proportion of salinized land area might increase, owing to climate change conditions conducive for salt accumulation (Othman et al., 2006). Hypertonic conditions and ion toxicity (occurs over several hours to days and weeks) are the result of toxic ions

(Na^+ and Cl^-) accumulating in the cells caused osmotic stress as an immediate impact of salinity (occurs within minutes). Perturbed water balance and ion homeostasis affected physiological conditions such as hormonal status, transpiration, photosynthesis, translocation of nutrients, and other metabolic processes in many plants that limits the plant growth (Munns, 2002; Latrach et al., 2014; Kim et al., 2017). The consequences of the excess of soluble salts in the soil lead to plant demise (Rout and Shaw, 2001) due to the osmotic stress, which results in specific ion toxicity and ionic imbalances (Munns and James, 2003). A high salt concentration especially NaCl in the soil can hinder germination, vegetative growth, yield, and quality of crops (Li et al., 2014). Soil salinization is considered as one of the most serious threats for the agriculture because it limits optimal production (Munns and Tester, 2008). Many legume crops such as chickpea are salt sensitive and are seriously affected by soil salinization, mainly by the chloride salts (Manchanda and Sharma, 1990). Chickpea is mainly affected at early stages during germination (Khan et al., 2013). It was also reported by many researchers that chickpea is sensitive to the salt and affected by the salinity in its growth stages such as germination (Khalid et al., 2001), vegetative growth (Khan et al., 2015), and especially during reproductive processes (Turner et al., 2013; Khan et al., 2017). Predominantly chickpea is adversely affected by the saline soil due to the Na^+ toxicity (i.e., metabolism inhibition, cellular damage) in leaf tissues (Khan et al., 2016). Plants under salt stress conditions were reported with reduced photosynthetic rate with limited availability of photosynthates that further diminished both vegetative and reproductive phases of the plants (Khan et al., 2017).

14.2 Causes of Soil Salinization

Soil salinization is caused by natural (physical or chemical weathering or transport) or human activities (use of salt rich irrigation) that increase the dissolved salts, predominantly sodium chloride in the soil. Primary salinity occurs due to the natural process leading high accumulation of salt in soil as well as groundwater over extended periods of time, which results in the formation of salt lakes, salt marshes, marine sediments, and salt scalds in the landscape. Weathering of rocks and minerals is the source of salts for primary salinity, which releases soluble salts. These salts are washed downstream by precipitation; wind-borne salts are deposited inland; and influx of seawater is followed by subsequent retreat (Pitman and Lauchli, 2002; Rengasamy, 2002). The main source of secondary salinity is the cultivation operations such as clearing of lands, inadequate drainage, etc. Salt deposition occurs in the topsoil while the water is evaporated very easily; this results in salinity of dry land and may eventually form a salt scald. The effect of salinity will be more harmful if the groundwater table is high in arid and coastal areas where only salt-tolerant plants (halophytes) can grow (Doering and Sandoval, 1981; Rengasamy, 2002). Salinity of the soil depends on the soil type; it will be low in sandy soil and high for the clay due to the binding of Na^+ to the negatively charged clay minerals. Salinity depends on the electric conductivity (ECe) of soil, and according to a report by the USDA Salinity Laboratory, soil with ECe 4 dS m^{-1} is defined as saline soil. On average, irrigated lands are twice as productive as rainfed land and produce approximately one-third of global food (Munns and Tester, 2008); hence, the salinity due to the irrigation has been recognized as a serious problem because salinity and water are inseparably linked together.

14.3 Effects of Soil Salinity on Plant Growth and Physiology

Soil salinity is a serious threat that affects almost all the stages of the plant growth but in early stages showed its major consequences. Crops such as cereals, forages, or the horticultural are relatively susceptible to salinity that should be either the soluble salt in irrigated water or present in rhizospheric soil solution. Most of the crop species were affected by ECe less than 4 dS m^{-1} and thus, saline soil inhibits the yield of crops. Seed germination and early seedling growth are the most salt-sensitive stages of the plants because plant tissues at these stages are in direct contact with soil and affected (Rahman et al., 2000; Jamil et al., 2006). Many studies have demonstrated that salinity inhibits the early stages of various crops such as wheat (Hasana and Miyake, 2017), chickpea (Plá and Cobos-Porras, 2015), faba bean (Rabie and Almadini, 2005), rice (Xu et al., 2011), maize (Khodarahmpour et al., 2012), and soybean

(Essa, 2002). A significant reduction in germination percentage as well as length of seedling root and shoot of cabbage, sugar beet, paniculate amaranth, and pak choi was observed (Jamil et al., 2006). High salt content reduced the shoot length up to 50% and root length up to 70% of bean seedlings grown in a gnotobiotic sand system (Egamberdieva, 2011). Similar observation was reported in groundnut (Mensah et al., 2006) and chickpea (Mensah et al., 2006). Demir and Arif (2003) have reported that the root growth of safflower was inhibited more than shoot growth due to salinity effect. A significant reduction in shoot and root growth as well as in content of essential oil in *Ammolei majus*, *Hyoscyamus niger*, and *Matricaria chamomile* was reported by Ashraf and Harris (2004) and Razmjoo et al. (2008). Salinity inhibits many plant physiological processes including photosynthesis, stomatal conductance, chlorophyll content, and mineral uptake of basil (*Ocimum basilicum*) (Golpayegani and Tilebeni, 2011). Inhibition of root growth due to salinity is considered a result of salt-induced reduction in water availability and unbalanced nutrient uptake by the seedling (Dolatabadian et al., 2011). In addition, osmotic pressure due to higher salinity results in reduction of germination percentage (Atak et al., 2006; Neamatollahi et al., 2009). Reduction in accessibility and uptake of important elements such as N, P, K, and Mg by the plant was due to the abundance of Na^+ and Cl^- ions of salt (Heidari and Jamshid, 2010). Synthesis of phytohormone and plant maturation was also inhibited by the salinity stress (Xiong et al., 2002). Adverse effects of soil salinity such as reduction in nodulation, total nitrogen content, and yield of legumes had been reported by Singleton and Bohlool (1984).

During the development of soil salinity, most of the major plant mechanisms such as photosynthesis, protein synthesis, metabolism of energy and lipid are negatively affected (Parida and Das, 2005). Partial closure of stomata and the osmotic stress are the main cause for the inhibition of photosynthesis (Meloni et al., 2003; Reddy et al., 2004). It was also reported that due to the nutrient imbalance, the crops suffer from membrane destabilization (Hasegawa et al., 2000). Further, the plants that suffered due to osmotic stress have experienced reduction in cell growth and development, reduction in leaf area and chlorophyll, high defoliation, and senescence (Shannon and Grieve, 1999). Uptake of excess amounts of Na^+ or the consumption of limited amounts of Ca^{2+} and K^- can cause the imbalance of nutrient in leaves because a low amount of Na^+ and the high amount of K^+ or Ca^{2+} are involved for the optimum function of plants that may cause death of plants in severe cases (Rajput et al., 2016; Acosta et al., 2017). Plant enzymes are affected by the excess amount of sodium ions and more importantly from the chloride ion, which causes the swelling of cells and tissues and various physiological changes—this has a consequence of less energy production in plants. Accumulation of Cl^- ion by the plants can interrupt the mechanism of photosynthesis by inhibiting the nitrate reductase activity (Xu et al., 2000). Under excessive Na^+ and Cl^- at rhizosphere, competitive interactions occur between the nutrient ions concentration for binding sites and transport proteins in root cells (Tester and Davenport, 2003). Once the capacity of the plant to store excess salt is exhausted, gradual accumulation of salts occurs in the intracellular space that results in dehydration and death of cells (White and Broadley, 2001). Under salt stress, osmotic pressure will increase and influence the water and nutrient uptake. With increasing salinity by NaCl, uptake of almost all the micronutrient and macronutrient is decreased in the roots and shoots. The plants having perturbed nutrients relations are most susceptible to invasion by the biotic and abiotic stresses. Such physiological dysfunctions cause reduction in size and self-life of edible parts such as fruits and further resulted in less nutritional and economical value (Romic et al., 2008). Much secondary stress such as oxidative stress is generated by the primary salinity effect, which causes the accumulation of reactive oxygen species that have adverse effect on the bio-membranes, proteins, nucleic acids, and enzymes (Porcel et al., 2012).

14.4 Effect of Plant Growth-Promoting Rhizobacteria (PGPR) on Plant Growth

Bacteria associated with plants can be classified on the basis of behavior and effect into beneficial, deleterious, and neutral groups (Dobbelaere et al., 2003). Free-living beneficial soil bacteria are termed as plant growth-promoting rhizobacteria (PGPR; Kloepper et al., 1989). Only 1%–2% bacteria are identified as PGPR that are helpful in enhancing the plant growth promotion (Antoun and Kloepper, 2001). Bacteria associated with rhizosphere, rhizoplane, and endosphere can play a very important role in improving the plant's resistance to salinity stress (Etesami and Beattie, 2017, 2018; Etesami and Maheshwari, 2018). The use of beneficial microorganism to enhance plant growth by facilitating them to absorb greater amounts of nutrient and water is an important practice in sustainable agricultural systems, particularly in saline soils (Basu et al., 2018). Earlier researches demonstrate the efficiency of PGPR to enhance plant growth promotion to improve the crop production under saline stress conditions (Bashan et al., 2014; Egamberdieva and Lugtenberg, 2014; Kang et al., 2014; Islam et al., 2016; Parray et al., 2016; Egamberdieva et al., 2017). Various genera have been identified as the PGPR in which *Bacillus* and *Pseudomonas* spp. are predominant (Podile and Kishore, 2006). Commercially, PGPR and their interaction are exploited (Podile and Kishore, 2006) and play a beneficial role for plant growth promotion and sustainable agriculture. Application of PGPR inoculation has been reported in many crops such as chickpea, maize, wheat, oat, barley, peas, canola, soy, potatoes, tomatoes, lentils, radicchio, and cucumber (Gray and Smith, 2005). According to many research reports, it is widely proved that these beneficial soil microbes—by the signaling of phytohormone and several other mechanisms—can mitigate salinity and enhance stress tolerance in plants (Dodd and Perez-Alfocea, 2012). Crops inoculated with potential PGPR are gaining importance in the agricultural area because these microbes have the ability to enhance plant growth promotion even under the saline conditions; they can mitigate salinity and improve salt tolerance, which results in restoring yield (Lugtenberg et al., 2013). The bacteria isolated from extreme regions such as salty area, desert, and ocean are found to have more potential to induce salt tolerance in the plant in comparison to normal soil. For example, the *Pseudomonas fluorescens* isolated from the rhizosphere of the date palm was reported to increase root growth in maize (*Zea mays*) under salt stress conditions (Zerrouk et al., 2016). Halophillic bacterium *Serratia* sp. Sl-12, isolated from the salt lake, has been found to mitigate salt stress and increase biomass of shoots in wheat plants (Singh and Jha, 2016). The PGPR-inoculated seed of many crops such as tomato, pepper, canola, bean, and lettuce have been found to enhance plant growth by promoting increased root and shoot growth, dry weight, fruit and seed yield under salt stress conditions by improving salt stress (Glick et al., 1997; Mayak et al., 2004; Yildirim and Taylor, 2005; Barassi et al., 2006; Egamberdieva et al., 2013). It was reported by Creus et al. (2004) that the PGPR show enhanced osmotic adjustment by altering the plant–water relationships.

14.5 Mitigation of Salt Stress in Crops by Plant Growth-Promoting Rhizobacteria

Salinity has a huge negative impact on the growth and yield of several crop plants; in addition, it also affects fertility of the soil. In semiarid environments, aggregation of the soil stability is an important property for controlling the growth of plants. Hence, improvement of soil structural stability for the proper crop production is one of the most important factors. Contribution of microbial populations associated with plants either as endophytes or free-living forms has great importance (Jastrow and Miller, 1991) for rendering saline soils suitable for agriculture. Occurrence and functional diversity of beneficial microbes that have importance in the agricultural field to enhance the plant growth promotion and soil remediation were researched and reviewed (Zahran, 1999; Paul and Park, 2013; Etesami and Beattie, 2018; Shrivastava and Kumar, 2015). From the stressed environment such as the desert, saline region, or hill slopes, etc., several beneficial microbes such as *Rhizobium*, *Bradyrhizobium*, *Azotobacter*, and *Azospirillum* have been reported to enhance the plant growth and are assumed to be involved in the natural reclamation process of the soil (Selvakumar et al., 2009; Upadhyay and Singh, 2015). Apart from the salt tolerance mechanism of their own, these beneficial microbes can also impart in the tolerance of plants under salt stress conditions. Under saline stress conditions, interaction of the beneficial microbes with several crops can reduce the poor growth by mitigating salinity and hence improve the plants' sustainability under adverse conditions (Dimkpa et al., 2009). The beneficial effects of microbial inoculation on plant physiology and plant growth development under salt stress conditions have been published in many studies. PGPR or the beneficial microbes can have direct and indirect mechanisms for the plant growth promotion. Some of the beneficial microbes exert direct stimulation on plant growth by the mechanism of nitrogen fixation, phytohormone production, iron sequestered by bacterial strain, and phosphate solubilization (Hayat et al., 2010; Rodríguez and Fraga, 1999). Some of the microbes can enhance growth promotion by indirect mechanism such as protecting the plants against the phytopathogens causing plant diseases. Beneficial microbes can protect crops against soilborne diseases that are mostly caused by pathogenic fungi (Lugtenberg and Kamilova, 2009) to withstand the plants under adverse conditions such as with abiotic and biotic stresses. Soilborne microbe pseudomonads have received significant attention because of their catabolic versatility, great root-colonizing ability, and their ability to produce a wide range of enzymes and metabolites (Vessey, 2003).

14.6 PGPR-Mediated Plant Root Proliferation and Plant Vigor

Beneficial microbes found in the rhizospheric region are helpful in plant growth promotion through various kinds of mechanisms. By using software GS Root® (PP Systems, Amesbury, Massachusetts), Paul and Sarma (2006) demonstrated the enhanced root biomass of black pepper vines. Under salt stress soil, root proliferation—mediated by the rhizobacteria—has been well proven (Diby et al., 2005a). Hence, to alleviate the negative effect of salinity in the plants, the co-inoculation of beneficial PGPR with seeds will be a fruitful strategy. In various plant species, inoculation of beneficial microbes leads to enhanced lateral roots and root hairs, which results in improved tolerance of plants under abiotic stress. It was reported that the potential of root colonization of salt-tolerant *Pseudomonas* strain was not hampered even under salinity stress (Paul and Nair, 2008). Enhanced root growth promotion can result in a larger surface area of leaves that promotes the nutrient uptake and water acquisition (Diby et al., 2005b; Paul and Sarma, 2006), which was expected to mitigate negative salt stress effects on the plants. Under saline stress conditions, it was reported that PGPR-inoculated plants of lettuce were found to be more hydrated than the PGPR non-inoculated control (Kohler et al., 2009). An extent induction of hydration in plants by the PGPR strains might be used as attributable of water use efficiency (Saravanakumar and Samiyappan, 2007). It was observed that when the roots of tomato plants were treated with the suspension of beneficial bacterial strains, an extent of growth suspension was reduced due to the stress of salinity, and the plants treated with bacterial strains were found to have more fresh and increased dry weight untreated plants Mayak et al. (2004). The same result was concluded by Fu et al. (2010) in eggplant and was reported that during the increases, salinity growth of the plants was reduced; but when the same plant was treated with *Pseudomonas* sp. DW1, an extent growth and more dry weight of the plants were observed than the untreated plants, which shows the role of beneficial microbes to enhance plant growth promotion by mitigating the salinity. Beneficial microbes PGPR can improve seed germination under salt stress conditions; the same result was also concluded by Barassi et al. (2006) in lettuce seeds inoculated with *Azospirillum* spp. in salt stress conditions. Biopriming is an integrated process of biological (seeds inoculated with beneficial microorganisms) and physiological aspects (seed hydration). Under salt stress conditions, application of biopriming with beneficial microbes in redish crop was reported to have increased seed germination (Kaymak et al., 2009). Maize crop treated with *Azotobacter* strains found to have increased chlorophyll content admits the positive effect of PGPR on the growth and the development of the plants (Rojas-Tapias et al., 2012). The similar result of enhanced total chlorophyll contents (a, b, and carotenoids) was also observed in rhizobacterial strain-treated maize and canola crops (Glick et al., 1998; Nadeem et al., 2007). When the inoculation of *A. brasilense* and *Pantoea dispersa* was done in pepper plants under salt stress, higher dry matter accumulation was observed. This could be due to the higher stomatal conductance and photosynthesis than uninoculated plants, and this occurs without affecting the concentration of chlorophyll (del Amor and Cuadra-Crespo, 2012).

14.7 Production of Phytohormones by PGPR

Some of the beneficial microbes have the ability to produce phytohormones, which helps in plant growth promotion. Phytohormone production such as indole acetic acid, gibberellins, and other growth regulators by PGPR is assumed to improve the root growth, shoot growth, and root tips number can enhance the nutrient uptake by the plants believed to improved plant growth and health under salt stress conditions (Egamberdieva and Kucharova, 2009). IAA production is a common trait of PGPR, and these microbes are assumed to counteract salt stress in plants. It has been observed that repeated inoculation of *Azospirillum* in plants has morphological change in roots and the attribution of plant growth production of plant-growth-promoting substances such as auxins, cytokinins, and gibberellins (Spaepen et al., 2008). It was demonstrated by Sadeghi et al. (2012) that inoculation of *Streptomyces* in wheat can increase plant growth and is also capable of producing IAA in the presence of salt stress. During drying it was observed that soil microbial colonizing in the rhizosphere of lettuce plants stimulates shoot biomass and enhances plant growth promotion through various means, and it helps in mitigation of salt stress. Plant and beneficial microbial interaction in several crops under salt stress conditions can reduce the poor growth and health of the plant and helps the plant to sustain under adverse conditions. Phytohormone cytokine (Paul and Lade) producing PGPR *B. subtilis* (Arkhipova et al., 2007) have been implicit considerable signaling of cytokinin root to shoot. IAA found within the plant known as the growth- promoting hormone and the abscisic acid is a well-known stress-inducing phytohormone that works as a growth inhibitor of the plant. Inoculation of bacterial strain *P. putida Rs-198* in seeds showed increased production of IAA while inhibiting the abscisic acid production in the same plant, which resulted in enhanced growth parameter in cotton seedlings. PGPR having phytohormone production appears to be a relatively common trait, so researchers should pay more attention to searching for and applying selected microbes in agriculture, which can enhance growth and ameliorate salt stress.

14.8 ACC Production by PGPR

Beneficial microbes found in rhizosphere under salt stress conditions act as sink for ACC (Saleem et al., 2007) in plants. Increased levels of ACC results in enhanced ethylene concentration that lastly increases plant damage (Botella et al., 2000). To overcome the problem of salinity, common chemical inhibitors of ethylene synthesis, such as cobalt ions and amino-ethoxyvinyle glycine, are often used. The synthetic chemicals used in agriculture are not very costly but have negative impacts on the environment (Dodd and Cann, 2009). In the presence of bacterial strain-producing 1-aminocyclopropane-1-carboxylate deaminase, the same compound is sequestered and degraded by the bacterial strains and used for the nitrogen and energy (Mayak et al., 2004), facilitating growth promotion in plants under salt stress condition (Nadeem et al., 2014; Siddikee et al., 2010). In addition, bacteria can reduce the deleterious effect of ethylene, ameliorate plant stress, and promote plant growth promotion (Glick et al., 2007). The beneficial role of 1-aminocyclopropane-1-carboxylate deaminase- producing rhizobacteria for enhancing salt tolerance in tomato, rice, and various other crops has been reported; these microbes can improve the above plant growth under salinity stress (Mayak et al. 2004; Bal et al., 2013; Etesami and Beattie, 2018). According to research reported by Siddikee et al. (2010), microbes isolated from the coastal soil have the potential to enhance growth promotion even under salt stress conditions through their 1-aminocyclopropane-1-carboxylate deaminase activity. Halo-tolerant species of microbes vary from different genera such as *Bacillus, Brevibacterium, Planococcus, Zhihengliuella, Halomonas, Exiguobacterium, Oceanimonas, Corynebacterium, Arthrobacter*, and *Micrococcus* (Siddikee et al., 2010). The enzyme 1-aminocyclopropane-1-carboxylate deaminase production was found in a wide range of rhizobacterial strains as well as in many genera such as *Achromobacter, Acidovorax, Alcaligenes, Enterobacter, Klebsiella, Methylobacterium, Pseudomonas, Rhizobium*, and *Variovorax* (Esquivel-Cote et al., 2010).

14.9 Accumulation of Osmolytes

Alleviating the stress to enhance plant growth and yield of salt water homeostasis and proper functioning of organs of the photosynthesis process are essential. Overproduction of different compatible organic solutes such as proline and glycine betaine in the plant is one of the most common stress responses of the plant (Serraj and Sinclair, 2002). Accumulation of proline in the plants is a sensitive physiological signal for salts and other stresses. Under saline stress conditions, increased accumulation of proline has been reported in soybean plants; the inoculation of beneficial microbes in plants can mitigate the salinity stress and enhance the plant growth promotion (Han and Lee, 2005). In wheat plants, increased accumulation of proline was reported upon root colonization with PGPRs (Zarea et al., 2012). In plants, accumulation of proline by *Azospirillum* as an osmoprotectant was also reported (Bashan, 1999; Casanovas et al., 2003). In higher plants, it is reported that proline protects against salt/osmotic stresses, by adjusting osmotic pressure as well as stabilizing many functional units (Makela et al., 2000). By stabilizing the plant structure such as membranes and proteins, scavenging free radicals and buffering cellular redox potential proline can also help in alleviating salt stress under salinity (Ashraf and Foolad, 2007; Kohler et al., 2009).

Under salinity stress, an important defense strategy is the increased total soluble sugar (TSS) content of plants, which enables them to cope with salinity stress. It was reported by Upadhyay et al. (2012) that wheat plants treated with PGPR showed increased levels of proline and total soluble sugar, which significantly contributed to their osmotolerance. Similarly, rhizobia that have trehalose metabolism also seem to have improving growth of plant and their yield, and the adaptation to sustain under abiotic stress of leguminous plants (Suárez et al., 2008).

14.10 Anti-oxidative Enzymes

In plants during the salt shock, there is the formation of reactive oxygen species, which can damage the lipids, protein, and nucleic acids. It was reported that under salinity stress, limiting of photosynthesis due to the enhanced reduction in photosynthetic electron chain can produce reactive oxygen species (Johnson et al., 2005; Hichem and Mounir, 2009). Antioxidants of the plants are reported to have great resistance against the oxidative damage (Spychalla and Desborough, 1990). In salinity, antioxidant enzymes such as catalase, guaicol peroxidase, and superoxide dismutase increase; hence, the correlation of these enzymes' levels and salt tolerance exist (Mittova et al., 2002, 2003). Treatment of plants with beneficial microbes have shown a significant increase in several plant defense enzymes such as superoxide dismutase, peroxidase, catalase, polyphenol oxidase, phenylalanine ammonia-lyase, lipoxygenase, and phenolics (Nautiyal et al., 2008; Liang et al., 2011; Chakraborty et al., 2013), which should be helpful for the good health of the plants. These antioxidant enzymes induced by the PGPR are believed to control the salinity and increase salt tolerance in the plants, the same process also by eliminating hydrogen peroxide from salt-stressed roots (Kim et al., 2005).

14.11 Production of Extracellular Polymeric Substance (EPS)

It has been reported that extracellular polymeric substance-producing PGPR can significantly enhance soil macropores' volume and rhizosphere soil aggregation; hence, EPS-inoculating plants can enhance water and fertilizer availability, which conclude the plants for better management of salt shock. On the aggregation of root-adhering soils, the influence of extracellular polymeric substance-producing PGPR has been reported by (Alami et al., 2000; Upadhyay et al., 2011). It was reported that extracellular polymeric substance-producing by bacteria possess water-holding and cementing properties; this may play a vital role in formation and stabilization of soil aggregates as well as in nutrient regulation and water flow across plant roots by forming a layer of biofilm.

14.12 Enhancement of Plant Nutrient Uptake

In plants, mineral nutrition has a great importance to survive and to adaptation under adverse environmental conditions, impairment of which exacerbates the adverse effects of abiotic stresses. Salinity-induced nutritional disorders can adversely affect the crop performance related to growth and development. Nitrogen element in plants accounts for approximately 80% of minerals absorbed by the plants; hence, inadequate nitrogen in plants often causes growth-limiting nutritional stress (Marschner, 1995). Previous studies indicate that the N uptake/accumulation in plants can be reduced by the salinity stress (Feigin and

Tweedie, 1985). Phosphate accumulation in plants can also adversely affect by the salinity, resulting in P-deficiency symptoms. It was reported that phosphate uptake/accumulation in crops grown in soils is primarily reduced by the salinity by reducing phosphate availability (Sharpley et al., 1992). Maintenance of K^+ for the plants under salinity is needed for their survival, and it was reported in various crops about the sodium-induced K^+ deficiency (Botella et al., 1997). PGPR play a vital role for the circulation of nutrients in plants in many ways that have an excellent result in reducing the use of chemical fertilizers in agricultural crops. Along with the fixation of nitrogen, PGPR can directly enhance the plant growth by solubilizing inorganic phosphate, improving nutrient uptake, and mineralizing organic phosphate (Dobbelaere et al., 2003; Ogut and Kandemir, 2010). Phosphate solubilization by PGPR in rhizosphere is the most common way to enhance the nutrient availability to the host plant (Rashid et al., 2004). In black paper vines, enhanced nutrient mobilization in the rhizosphere was reported by Diby et al. (2005b), while black paper vines treated with beneficial microbes have root proliferation and enhanced plant growth, which may be due to the significant uptake of nitrogen (N) and phosphorus (P). Rhizobacterial strains with efficient phosphorus-solubilizing activity even under high salt stress (60 g L^{-1} NaCl) conditions have been reported (Son et al., 2006; Upadhyay et al., 2011). Inoculation of *A. Brasilense* on seedlings can result in reducing damaging effects of NaCl on wheat seedlings (Creus et al., 1997). Better seed germination and vegetative growth were reported in lettuce seed inoculated by *Azospirillum* spp. than in non-inoculated seed under salinity (Barassi et al., 2006). Under salt stress conditions, *Phaseolus–Rhizobium* interaction during the nodulation was affected, while in salt stress plant, secondary inoculation with *Azospirillum* can result in an extended exudation of plant flavonoids compared to *Rhizobium* alone (Dardanelli et al., 2008). This co-inoculation of plants with different bacterial strains contributed to relieving the abiotic stress.

14.13 Conclusion

The cost associated with reclaiming soil salinity is potentially enormous, and salinity may negatively impact agriculture, biodiversity, and the environment. Across the globe, as the saline area of agricultural fields increases every year, it is of much public concern. Since salinity has a detrimental impact on the plant growth development and production, it is necessary to search for a traditional, eco-friendly tool to reduce the salinity. Worldwide, PGPR are getting more attention in agriculture because of their ecologically compatible and environmentally friendly nature even under saline soil agriculture. Application of PGPR to improve plant growth even under adverse conditions reveals their positive impact on agriculture. Development and utilization of PGPR are an important way for the organic farming, and now it is the need of the day to practice this widely for sustaining agricultural productivity.

REFERENCES

Acosta-Motos, J., M. Ortuño, A. Bernal-Vicente, P. Diaz-Vivancos, M. Sanchez-Blanco, and J. Hernandez. "Plant responses to salt stress: Adaptive mechanisms." *Agronomy* 7.1 (2017): 18.

Alami, Y., et al. "Rhizosphere soil aggregation and plant growth promotion of sunflowers by an exopolysaccharide-producing *Rhizobium* sp: Strain isolated from sunflower roots." *Applied and Environmental Microbiology* 66.8 (2000): 3393–3398.

Antoun, H. and J. Kloepper. 2001. "Plant growth promoting rhizobacteria (PGPR)." In: *Encyclopedia of Genetics*, S. Brenner, and J. Miller (eds.), Academic Press, pp. 1477–1480.

Arkhipova, T. N., et al. "Cytokinin producing bacteria enhance plant growth in drying soil." *Plant and Soil* 292.1–2 (2007): 305–315.

Ashraf, M. F. M. R., and M. Foolad. "Roles of glycine betaine and proline in improving plant abiotic stress resistance." *Environmental and Experimental Botany* 59.2 (2007): 206–216.

Ashraf, M. P. J. C., and P. J. C. Harris. "Potential biochemical indicators of salinity tolerance in plants." *Plant Science* 166.1 (2004): 3–16.

Atak, M., et al. "Effects of NaCl on the germination, seedling growth and water uptake of triticale." *Turkish Journal of Agriculture and Forestry* 30.1 (2006): 39–47.

Atnaf, M., K. Tesfaye, and D. Kifle. "The Importance of legumes in the Ethiopian farming system and overall economy: An overview." *American Journal of Experimental Agriculture* 7.6 (2015): 347–358.

Bal, H. B., et al. "Isolation of ACC deaminase producing PGPR from rice rhizosphere and evaluating their plant growth promoting activity under salt stress." *Plant and Soil* 366.1–2 (2013): 93–105.

Barassi, C. A., et al. "Seed inoculation with *Azospirillum* mitigates NaCl effects on lettuce." *Scientia Horticulturae* 109.1 (2006): 8–14.

Bashan, Y. "Interactions of *Azospirillum* spp. in soils: A review." *Biology and Fertility of Soils* 29.3 (1999): 246–256.

Bashan, Y., L. de-Bashan, S.R. Prabhu, and J.-P. Hernandez. "Advances in plant growth-promoting bacterial inoculant technology: Formulations and practical perspectives (1998–2013)." *Plant Soil* 378 (2014): 1–33. doi:10.1007/s11104-013-1956-x.

Basu, S., R. C. Rabara, and S. Negi. "AMF: The future prospect for sustainable agriculture." *Physiological and Molecular Plant Pathology* 102 (2018): 36–45.

Botella, M. A., et al. "Salinity induced potassium deficiency in maize plants." *Journal of Plant Physiology* 150.1–2 (1997): 200–205.

Botella, M. Á., et al. "Polyamine, ethylene and other physico-chemical parameters in tomato (*Lycopersicon esculentum*) fruits as affected by salinity." *Physiologia Plantarum* 109.4 (2000): 428–434.

Casanovas, E. M., C. A. Barassi, and R. J. Sueldo. "*Azospirillum* inoculation mitigates water stress effects in maize seedlings." *Cereal Research Communications*, 31.3 (2002): 343–350.

Casanovas, E. M., et al. "*Azospirillum*-inoculated maize plant responses to irrigation restraints imposed during flowering." *Cereal Research Communications* 31 (2003): 395–402.

Chakraborty, U., et al. "Water stress amelioration and plant growth promotion in wheat plants by osmotic stress tolerant bacteria." *World Journal of Microbiology and Biotechnology* 29.5 (2013): 789–803.

Creus, C. M., R. J. Sueldo, and C. A. Barassi. "Shoot growth and water status in *Azospirillum*-inoculated wheat seedlings grown under osmotic and salt stresses." *Plant Physiology and Biochemistry (France)* 35 (1997): 939–944.

Creus, C. M., R. J. Sueldo, and C. A. Barassi. "Water relations and yield in *Azospirillum*-inoculated wheat exposed to drought in the field." *Canadian Journal of Botany* 82.2 (2004): 273–281.

Dardanelli, M. S., et al. "Effect of *Azospirillum brasilense* coinoculated with *Rhizobium* on *Phaseolus vulgaris* flavonoids and Nod factor production under salt stress." *Soil Biology and Biochemistry* 40.11 (2008): 2713–2721.

Del Amor, F. M., and P. Cuadra-Crespo. "Plant growth-promoting bacteria as a tool to improve salinity tolerance in sweet pepper." *Functional Plant Biology* 39.1 (2012): 82–90.

Demir, M. and I. Arif. "Effects of different soil salinity levels on germination and seedling growth of safflower (*Carthamus tinctorius* L.)." *Turkish Journal of Agriculture* 27.4 (2003): 221–227.

Diby, P., et al. "Pseudomonas fluorescens mediated vigour in black pepper (*Piper nigrum* L.) under green house cultivation." *Annals of Microbiology* 55.3 (2005a): 171–174.

Dimkpa, C., T. Weinand, and F. Asch. "Plant–rhizobacteria interactions alleviate abiotic stress conditions." *Plant, Cell & Environment* 32.12 (2009): 1682–1694.

Dobbelaere, S., J. Vanderleyden, and Y. Okon. "Plant growth-promoting effects of diazotrophs in the rhizosphere." *Critical Reviews in Plant Sciences* 22.2 (2003): 107–149.

Dodd, D., and I. K.O. Cann. "Enzymatic deconstruction of xylan for biofuel production." *GCB Bioenergy* 1.1 (2009): 2–17.

Dodd, I. C., and F. Pérez-Alfocea. "Microbial amelioration of crop salinity stress." *Journal of Experimental Botany* 63.9 (2012): 3415–3428.

Doering, E. J., and F. M. Sandoval. "Chemistry of seep drainage in southwestern North Dakota." *Soil Science* 132.2 (1981): 142–149.

Dolatabadian, A., S. Ali Mohammad Modarres Sanavy, and F. Ghanati. "Effect of salinity on growth, xylem structure and anatomical characteristics of soybean." *Notulae Scientia Biologicae* 3.1 (2011): 41–45.

Egamberdieva, D., and B. Lugtenberg. 2014. "Use of plant growth-promoting rhizobacteria to alleviate salinity stress in plants." In: *Use of Microbes for the Alleviation of Soil Stresses*, M. Miransari (ed.), Springer, New York, pp. 73–96.

Egamberdieva, D., S. Wirth, D. Jabborova, L. A. Räsänen, and H. Liao. "Coordination between *Bradyrhizobium* and *Pseudomonas* alleviates salt stress in soybean through altering root system architecture." *Journal of Plant Interactions* 12.1 (2017) 100–107. doi: 10.1080/17429145.2017.1294212.

Egamberdieva, D., et al. "Alleviation of salt stress of symbiotic *Galega officinalis* L.(goat's rue) by co-inoculation of *Rhizobium* with root-colonizing *Pseudomonas*." *Plant and Soil* 369.1–2 (2013): 453–465.

Egamberdieva, D., and Z. Kucharova. "Selection for root colonising bacteria stimulating wheat growth in saline soils." *Biology and Fertility of Soils* 45.6 (2009): 563–571.

Esquivel-Cote, R., et al. "1-aminocyclopropane-1-carboxylic acid deaminase improves early growth of tomato seedlings under nitrogen deficiency." *Plant Soil* 337 (2010): 65–75.

Essa, T. A. "Effect of salinity stress on growth and nutrient composition of three soybean (*Glycine max L.* Merrill) cultivars." *Journal of Agronomy and Crop Science* 188.2 (2002): 86–93.

Etesami, H., and G. A. Beattie. 2017. "Plant-microbe interactions in adaptation of agricultural crops to abiotic stress conditions." In: *Probiotics and Plant Health*, K. Vivek, K. Manoj, S. Shivesh, and P. Ram (eds.), Springer, Singapore, pp. 163–200.

Etesami, H., and G. A. Beattie. "Mining halophytes for plant growth-promoting halotolerant bacteria to enhance the salinity tolerance of non-halophytic crops." *Front Microbiology* 9 (2018):148. 1–20. doi:10.3389/fmicb.2018.00148.

Etesami, H., and D. K. Maheshwari. "Use of plant growth promoting rhizobacteria (PGPRs) with multiple plant growth promoting traits in stress agriculture: Action mechanisms and future prospects." *Ecotoxicology and Environmental Safety* 156 (2018):225–246. doi:10.1016/j.ecoenv.2018.03.013.

FAOSTAT, FAO. http://www.fao.org/faostat/en/-data/QC. Food and Agriculture Organization of the United Nations (FAO) (2016).

Feigin, P. D., and R. L. Tweedie. "Random coefficient autoregressive processes: A Markov chain analysis of stationarity and finiteness of moments." *Journal of Time Series Analysis* 6.1 (1985): 1–14.

Food and Agriculture Organisation of the United Nations, "*FAO Land and Plant Nutrition Management Service*," 2008. http://www.fao.org/ag/agl/agll/spush/ World reference base for soil resources (1998). FAO, Rome, Italy.

Fu, Q., et al. "Ameliorative effects of inoculation with the plant growth-promoting rhizobacterium *Pseudomonas* sp. DW1 on growth of eggplant (*Solanum melongena* L.) seedlings under salt stress." *Agricultural Water Management* 97.12 (2010): 1994–2000.

Glick, B. R., D. M. Penrose, and J. Li. "A model for the lowering of plant ethylene concentrations by plant growth-promoting bacteria." *Journal of Theoretical Biology* 190.1 (1998): 63–68.

Glick, B. R., et al. 2007. "Promotion of plant growth by ACC deaminase-producing soil bacteria." In: *New Perspectives and Approaches in Plant Growth-Promoting Rhizobacteria Research*. Springer, Dordrecht, the Netherlands, pp. 329–339.

Glick, B. R., C. Liu, S. Ghosh, and E. B. Dumbrof. "Early development of canola seedlings in the presence of the plant growth-promoting rhizobacterium *Pseudomonas putida* GR12-2." *Soil Biology and Biochemistry* 29 (1997): 1233–1239.

Golpayegani, A., and H. G. Tilebeni. "Effect of biological fertilizers on biochemical and physiological parameters of basil (*Ociumum basilicm* L.) medicine plant." *American-Eurasian Journal of Agricultural and Environmental Science*11 (2011): 411–416.

Gray, E. J., and D. L. Smith. "Intracellular and extracellular PGPR: Commonalities and distinctions in the plant–bacterium signaling processes." *Soil Biology and Biochemistry* 37.3 (2005): 395–412.

Han, H. S., and K. D. Lee. "Plant growth promoting rhizobacteria effect on antioxidant status, photosynthesis, mineral uptake and growth of lettuce under soil salinity." *Research Journal of Agricultural and Biological Sciences* 1.3 (2005): 210–215.

Hasana, R., and H. Miyake. "Salinity stress alters nutrient uptake and causes the damage of root and leaf anatomy in maize." *KNE Life Sciences* 3.4 (2017): 219–225.

Hasegawa, P. M., et al. "Plant cellular and molecular responses to high salinity." *Annual Review of Plant Biology* 51.1 (2000): 463–499.

Hayat, R., et al. "Soil beneficial bacteria and their role in plant growth promotion: A review." *Annals of Microbiology* 60.4 (2010): 579–598.

Heidari, M., and P. Jamshid. "Interaction between salinity and potassium on grain yield, carbohydrate content and nutrient uptake in pearl millet." *ARPN Journal of Agricultural and Biological Science* 5.6 (2010): 39–46.

Hichem, H., and D. Mounir. "Differential responses of two maize (*Zea mays* L.) varieties to salt stress: Changes on polyphenols composition of foliage and oxidative damages." *Industrial Crops and Products* 30.1 (2009): 144–151.

Islam, F., Yasmeen, T., Arif, M. S., Ali, S., Ali, B., Hameed, S., and Zhou, W. "Plant growth promoting bacteria confer salt tolerance in *Vigna radiata* by upregulating antioxidant defense and biological soil fertility." *Plant Growth Regulation* 80 2016: 23–36.

Jamil, M., et al. "Effect of salt (NaCl) stress on germination and early seedling growth of four vegetables species." *Journal of Central European Agriculture* 7.2 (2006): 273–282.

Jastrow, J. D., and R. M. Miller. "Methods for assessing the effects of biota on soil structure." *Agriculture, Ecosystems & Environment* 34.1–4 (1991): 279–303.

Johnson, G. N. "Cyclic electron transport in C3 plants: Fact or artefact?." *Journal of Experimental Botany* 56.411 (2005): 407–416.

Jukanti, A. K., Gaur, P. M., Gowda, C. L. L., and Hibbar, R. N. "Nutritional quality and health benefits of chickpea (*Cicer arietinum* L.): A review." *British Journal of Nutrition*,108 (2012): S11–S26. doi:10.1017/S0007114512000797.

Kang, S.-M., et al. "Gibberellin-secreting rhizobacterium, *Pseudomonas putida* H-2-3 modulates the hormonal and stress physiology of soybean to improve the plant growth under saline and drought conditions." *Plant Physiology and Biochemistry* 84 (2014): 115–124.

Kaymak, H. Ç., et al. "The effects of bio-priming with PGPR on germination of radish (*Raphanus sativus* L.) seeds under saline conditions." *Turkish Journal of Agriculture and Forestry* 33.2 (2009): 173–179.

Khalid, A., M. Arshad, Z. A. Zahir, and M. Khalid. "Relative efficiency of rhizobacteria for auxin biosynthesis." *Journal of Biological Sciences* 1.8 (2001): 750–754.

Khan, A. L., et al. "Endophytic fungi from Frankincense tree improves host growth and produces extracellular enzymes and indole acetic acid." *PloS One* 11.6 (2016): e0158207.

Khan, A. L., et al. "Plant growth-promoting endophyte *Sphingomonas* sp. LK11 alleviates salinity stress in *Solanum pimpinellifolium*." *Environmental and Experimental Botany* 133 (2017): 58–69.

Khan, A. M. E. E. R., and M. U. N. A. W. E. R. Aziz. "Influence of foliar application of potassium on wheat (*Triticum aestivum* L.) under saline conditions." *Science, Technology and Development* 32.4 (2013): 285–289.

Khan, M. I. R., et al. "Salicylic acid-induced abiotic stress tolerance and underlying mechanisms in plants." *Frontiers in Plant Science* 6 (2015): 462.

Khodarahmpour, Z., M. Ifar, and M. Motamedi. "Effects of NaCl salinity on maize (*Zea mays* L.) at germination and early seedling stage." *African Journal of Biotechnology* 11.2 (2012): 298–304.

Kim, M.J., et al. "Overexpression of AtSZF2 from *Arabidopsis* showed enhanced tolerance to salt stress in soybean." *Plant Breeding and Biotechnology* 5 (2017): 1–15. doi:10.9787/PBB.2017.5.1.001.

Kloepper, J.W., Lifshitz, R., and Zablotowicz, R.M. "Free-living bacterial inocula for enhancing crop productivity." *Trends Biotechnology*, 7 (1989): 39–43.

Kohler, J., et al. "Induction of antioxidant enzymes is involved in the greater effectiveness of a PGPR versus AM fungi with respect to increasing the tolerance of lettuce to severe salt stress." *Environmental and Experimental Botany* 65.2–3 (2009): 245–252.

Latrach, L., M. Farissi, M. Mouradi, B. Makoudi, A. Bouizgaren, and C. Ghoulam. "Growth and nodulation of alfalfa-rhizobia symbiosis under salinity: Electrolyte leakage, stomatal conductance, and chlorophyll fluorescence." *Turkish Journal of Agricultural and Forestry* 38 (2014): 320–326.

Li, J., Pu, L., Han, M., Zhu, M., Zhang, R., and Xiang, Y. "Soil salinization research in China: Advances and prospects." *Journal of Geographical Sciences*, 24.5 (2014): 943–960.

Liang, C., et al. "Directed evolution of a thermophilic endoglucanase (Cel5A) into highly active Cel5A variants with an expanded temperature profile." *Journal of Biotechnology* 154.1 (2011): 46–53.

Lugtenberg, B. J. J., et al. "Plant growth promotion by microbes." *Molecular Microbial Ecology of the Rhizosphere* 2 (2013): 561–573.

Lugtenberg, B., and F. Kamilova. "Plant-growth-promoting rhizobacteria." *Annual Review of Microbiology* 63 (2009): 541–556.

Makela, J., et al. "Communication device with two modes of operation." U.S. Patent No. 6,047,196. April 4, 2000.

Manchanda, H. R., and S. K. Sharma. "Influence of different chloride: Sulphate ratios on yield of chickpea (*Cicer arietinum*) at comparable salinity levels." *Indian Journal of Agricultural Sciences* 60.8 (1990): 553–555.

Marschner, H. 1995. *Mineral Nutrition of Higher Plants*. 2nd edn. Academic Press, New York.

Mayak, S., T. Tirosh, and B. R. Glick. "Plant growth-promoting bacteria confer resistance in tomato plants to salt stress." *Plant Physiology and Biochemistry* 42.6 (2004): 565–572.

Meloni, D. A., et al. "Photosynthesis and activity of superoxide dismutase, peroxidase and glutathione reductase in cotton under salt stress." *Environmental and Experimental Botany* 49.1 (2003): 69–76.

Mensah, J. K., et al. "Effects of salinity on germination, growth and yield of five groundnut genotypes." *African Journal of Biotechnology* 5.20 (2006): 1973–1979.

Mittova, V., M. Guy, M. Tal, and M. Volokita. "Response of the cultivated tomato and its wild salt-tolerant relative *Lycopersicon pennellii* to salt dependent oxidative stress: Increased activities of antioxidant enzymes in root plastids." Free *Radical Research* 36 (2002): 195–202.

Mittova, V., et al. "Up-regulation of the leaf mitochondrial and peroxisomal antioxidative systems in response to salt-induced oxidative stress in the wild salt-tolerant tomato species *Lycopersicon pennellii*." *Plant, Cell & Environment* 26.6 (2003): 845–856.

Munns, R., and M. Tester. "Mechanisms of salinity tolerance." *Annual Review of Plant Biology* 59 (2008): 651–681.

Munns, R., and R. A. James. "Screening methods for salinity tolerance: A case study with tetraploid wheat." *Plant and Soil* 253.1 (2003): 201–218.

Munns, R. "Comparative physiology of salt and water stress." *Plant, Cell & Environment* 25.2 (2002): 239–250.

Nadeem, S. M., et al. "The role of mycorrhizae and plant growth promoting rhizobacteria (PGPR) in improving crop productivity under stressful environments." *Biotechnology Advances* 32.2 (2014): 429–448.

Nadeem, S. M. et al. "Preliminary investigations on inducing salt tolerance in maize through inoculation with rhizobacteria containing ACC deaminase activity." *Canadian Journal of Microbiology* 53.10 (2007): 1141–1149.

Nautiyal, K. M., et al. "Brain mast cells link the immune system to anxiety-like behavior." *Proceedings of the National Academy of Sciences* 105.46 (2008): 18053–18057.

Neamatollahi, E., et al. "Does hydro and osmo-priming improve fennel (*Foeniculum vulgare*) seeds germination and seedlings growth?" *Notulae Botanicae Horti Agrobotanici Cluj-Napoca* 37.2 (2009): 190–194.

Ogut, M., F. Er, and N. Kandemir. "Phosphate solubilization potentials of soil *Acinetobacter* strains." *Biology and Fertility of Soils* 46.7 (2010): 707–715.

Othman, Y., et al. "Variation in germination and ion uptake in barley genotypes under salinity conditions." *World Journal of Agricultural Sciences* 2.1 (2006): 11–15.

Parida, A. K., and A. B. Das. "Salt tolerance and salinity effects on plants: A review." *Ecotoxicology and Environmental Safety* 60.3 (2005): 324–349.

Parray, J. A., S. Jan, A. N. Kamili, R. A. Qadri, D. Egamberdieva, and P. Ahmad. Current perspectives on plant growth promoting rhizobacteria. *Plant Growth Regulation* 35 (2016): 877–902.

Paul, D., and H. Lade. "Plant-growth-promoting rhizobacteria to improve crop growth in saline soils: A review." *Agronomy for Sustainable Development* 34.4 (2014): 737–752.

Paul, D., and K. Park. "Identification of volatiles produced by *Cladosporium cladosporioides* CL-1, a fungal biocontrol agent that promotes plant growth." *Sensors* 13.10 (2013): 13969–13977.

Paul, D., and Y. R. Sarma. "Plant growth promoting rhizhobacteria (PGPR)-mediated root proliferation in black pepper (*Piper nigrum* L.) as evidenced through GS Root software." *Archives of Phytopathology and Plant Protection* 39.4 (2006): 311–314.

Paul, D., and S. Nair. "Stress adaptations in a plant growth promoting rhizobacterium (PGPR) with increasing salinity in the coastal agricultural soils." *Journal of Basic Microbiology* 48.5 (2008): 378–384.

Pitman, M. G., and A. Läuchli. 2002. "Global impact of salinity and agricultural ecosystems." In: *Salinity: Environment-Plants-Molecules*, A. Lauchli and U. Luttge (eds.). Springer, Dordrecht, the Netherlands, pp. 3–20.

Plá, C. L., and L. Cobos-Porras. 2015. "Salinity: Physiological impacts on legume nitrogen fixation." In: *Legume Nitrogen Fixation in a Changing Environment*, Springer, Cham, pp. 35–65.

Plá, C. L., and L. Cobos-Porras. 2015. "Salinity: Physiological impacts on legume nitrogen fixation." *Legume Nitrogen Fixation in a Changing Environment*. Springer, Cham, Switzerland S. Sulieman and LP. Tran (Eds.), pp. 35–65.

Podile, A. R., and G. K. Kishore. 2006. "Plant-associated bacteria." In: *Plant Growth Promoting Rhizobacteria*, S.S. Gnanamanickam (ed.), Springer, Amsterdam, the Netherlands, pp. 195–230.

Porcel, R., R. Aroca, and J. M. Ruiz-Lozano. "Salinity stress alleviation using arbuscular mycorrhizal fungi: A review." *Agronomy for Sustainable Development* 32.1 (2012): 181–200.

Rabie, G. H., and A. M. Almadini. "Role of bioinoculants in development of salt-tolerance of *Vicia faba* plants under salinity stress." *African Journal of Biotechnology* 4.2 (2005): 210–222.

Rahman, S., et al. "Salinity-induced ultrastructural alterations in leaf cells of rice (*Oryza sativa* L.)." *Plant Production Science* 3.4 (2000): 422–429.

Rajput, A., et al. "Impact of different crop geometries and depths of planting on growth and yield of rice in system of rice intensification." *Bioscience Biotechnology Research Communications* 9.3 (2016): 512–516.

Rashid, A., et al. "On-farm seed priming reduces yield losses of mungbean (*Vigna radiata*) associated with mungbean yellow mosaic virus in the North West Frontier Province of Pakistan." *Crop Protection* 23.11 (2004): 1119–1124.

Razmjoo, K. H. O. R. S. H. I. D., P. A. R. I. S. A. Heydarizadeh, and M. R. Sabzalian. "Effect of salinity and drought stresses on growth parameters and essential oil content of *Matricaria chamomile*." *International Journal of Agricultural and Biology* 10.4 (2008): 451–454.

Reddy, A. R., K. V. Chaitanya, and M. Vivekanandan. "Drought-induced responses of photosynthesis and antioxidant metabolism in higher plants." *Journal of Plant Physiology*, 161.11 (2004): 1189–1202.

Rengasamy, P. "Transient salinity and subsoil constraints to dryland farming in Australian sodic soils: An overview." *Australian Journal of Experimental Agriculture* 42.3 (2002): 351–361.

Rodríguez, H., and R. Fraga. "Phosphate solubilizing bacteria and their role in plant growth promotion." *Biotechnology Advances* 17.4–5 (1999): 319–339.

Rojas-Tapias, D., et al. "Effect of inoculation with plant growth-promoting bacteria (PGPB) on amelioration of saline stress in maize (*Zea mays*)." *Applied Soil Ecology* 61 (2012): 264–272.

Romic, D., et al. "Salinity and irrigation method affect crop yield and soil quality in watermelon (*Citrullus lanatus* L.) growing." *Irrigation and Drainage: The Journal of the International Commission on Irrigation and Drainage* 57.4 (2008): 463–469.

Rout, N. P., and B. P. Shaw. "Salt tolerance in aquatic macrophytes: Ionic relation and interaction." *Biologia Plantarum* 44.1 (2001): 95–99.

Roy, F., J. I. Boye, and B. K. Simpson. "Bioactive proteins and peptides in pulse crops: Pea, chickpea and lentil." *Food Research International* 43.2 (2010): 432–442.

Kim, S. Y., J. H. Lim, M. R. Park, Y. J. Kim, T. I. Park, Y. W. Seo, K. G. Choi, and S. J. Yun. "Enhanced antioxidant enzymes are associated with reduced hydrogen peroxide in barley roots under saline stress." *Journal of Biochemistry and Molecular Biology* 38 (2005): 218–224.

Sadeghi, A., et al. "Plant growth promoting activity of an auxin and siderophore producing isolate of Streptomyces under saline soil conditions." *World Journal of Microbiology and Biotechnology* 28.4 (2012): 1503–1509.

Saleem, M., et al. "Perspective of plant growth promoting rhizobacteria (PGPR) containing ACC deaminase in stress agriculture." *Journal of Industrial Microbiology & Biotechnology* 34.10 (2007): 635–648.

Saravanakumar, D., and R. Samiyappan. "ACC deaminase from *Pseudomonas fluorescens* mediated saline resistance in groundnut (*Arachis hypogea*) plants." *Journal of Applied Microbiology* 102.5 (2007): 1283–1292.

Selvakumar, S., D. Sivasankaran, and V. K. Singh. "Enantioselective Henry reaction catalyzed by C 2-symmetric chiral diamine–copper (II) complex." *Organic & Biomolecular Chemistry* 7.15 (2009): 3156–3162.

Serraj, R., and T. R. Sinclair. "Osmolyte accumulation: Can it really help increase crop yield under drought conditions?" *Plant, Cell & Environment* 25.2 (2002): 333–341.

Shannon, M. C. and C. M. Grieve. "Tolerance of vegetable crops to salinity." *Scientia Horticulturae* 78 (1999): 5–38. doi:10.1016/S0304-4238(98)00189-7

Sharpley, A. N., et al. "The transport of bioavailable phosphorus in agricultural runoff." *Journal of Environmental Quality* 21.1 (1992): 30–35.

Shrivastava, P., and R. Kumar. "Soil salinity: A serious environmental issue and plant growth promoting bacteria as one of the tools for its alleviation." *Saudi Journal of Biology Science* 22 (2015): 123–131.

Siddikee, M. D. A., et al. "Isolation, characterization, and use for plant growth promotion under salt stress, of ACC deaminase-producing halotolerant bacteria derived from coastal soil." *Journal of Microbiology and Biotechnology* 20.11 (2010): 1577–1584.

Singh, R. P., and P. N. Jha. "Alleviation of salinity-induced damage on wheat plant by an ACC deaminase-producing halophilic bacterium *Serratia* sp. SL-12 isolated from a salt lake." *Symbiosis* 69.2 (2016): 101–111.

Singleton, P. W., and B. Ben Bohlool. "Effect of salinity on nodule formation by soybean." *Plant Physiology* 74.1 (1984): 72–76.

Son, H. J., G. T. Park, M. S. Cha, and M. S. Heo. "Solubilization of insoluble inorganic phosphates by a novel salt-and pH-tolerant *Pantoea agglomerans* R-42 isolated from soybean rhizosphere." *Bioresource Technology*, 97.2 (2006), 204–210.

Spaepen, S., et al. "Effects of *Azospirillum brasilense* indole-3-acetic acid production on inoculated wheat plants." *Plant and Soil* 312.1–2 (2008): 15–23.

Spychalla, J. P., and S. L. Desborough. "Superoxide dismutase, catalase, and α-tocopherol content of stored potato tubers." *Plant Physiology* 94.3 (1990): 1214–1218.

Suárez, R., et al. "Improvement of drought tolerance and grain yield in common bean by overexpressing trehalose-6-phosphate synthase in rhizobia." *Molecular Plant-Microbe Interactions* 21.7 (2008): 958–966.

Tester, M., and R. Davenport. "Na^+ tolerance and Na^+ transport in higher plants." *Annals of Botany* 91.5 (2003): 503–527.

Turner, N. C., et al. "Salinity tolerance and ion accumulation in chickpea (*Cicer arietinum* L.) subjected to salt stress." *Plant and Soil* 365.1–2 (2013): 347–361.

Turner, T. R., E. K. James, and P. S. Poole. "The plant microbiome." *Genome Biology* 14.6 (2013): 209.

Upadhyay, S. K., and D. P. Singh. "Effect of salt-tolerant plant growth-promoting rhizobacteria on wheat plants and soil health in a saline environment." *Plant Biology* 17.1 (2015): 288–293.

Upadhyay, S. K., et al. "Impact of PGPR inoculation on growth and antioxidant status of wheat under saline conditions." *Plant Biology* 14.4 (2012): 605–611.

Upadhyay. S. K., J. S. Singh, A. K. Saxena, and D. P. Singh. "Impact of PGPR inoculation on growth and antioxidant status of wheat under saline conditions." *Plant Biology* 14 (2011): 605–611.

Vessey, J. K. "Plant growth promoting rhizobacteria as biofertilizers." *Plant and Soil* 255.2 (2003): 571–586.

White, P. J., and M. R. Broadley. "Chloride in soils and its uptake and movement within the plant: A review." *Annals of Botany* 88.6 (2001): 967–988.

Xiong, L., K. S. Schumaker, and J.-K. Zhu. "Cell signaling during cold, drought, and salt stress." *The Plant Cell* 14.1 (2002): S165–S183.

Xu, S., et al. "Hydrogen-rich water alleviates salt stress in rice during seed germination." *Plant and Soil* 370.1–2 (2013): 47–57.

Yildrim, E., M. F. Donmez, and M. Turan. "Use of bioinoculants in ameliorative effects on radish plants under salinity stress." *Journal of Plant Nutrition* 31.12 (2008): 2059–2074.

Yildirim, E., and A. G. Taylor. "Effect of biological treatments on growth of bean plants under salt stress." *Science* 123.1 (2005) *Journal of plant nutrition*, 31(12), 2059–2074.

Zahran, H. H. "Rhizobium-legume symbiosis and nitrogen fixation under severe conditions and in an arid climate." *Microbiology and Molecular Biology Reviews* 63.4 (1999): 968–989.

Zarea, M. J., et al. "Effect of *Piriformospora indica* and *Azospirillum* strains from saline or non-saline soil on mitigation of the effects of NaCl." *Soil Biology and Biochemistry* 45 (2012): 139–146.

Zerrouk, I. Z., et al. "A *Pseudomonas* strain isolated from date-palm rhizospheres improves root growth and promotes root formation in maize exposed to salt and aluminum stress." *Journal of Plant Physiology* 191 (2016): 111–119.

15

Contribution of Arbuscular Mycorrhizal Symbiosis to Salinity Tolerance in Leguminous Plants

Hassan Etesami and Saeid Shafiei

CONTENTS

15.1 Introduction

Over the last few decades, food security has received worldwide attention due to the ever–increasing population and, as a result, increased demand for food. One of the major challenges facing modern societies is to respond to this demand and ensure food security through sustainable agricultural development. It is hardly possible to expand agriculture by increasing the areas under cultivation due to the lack of arable land, which are affected by salinity (Farooq et al., 2017). Salinity is one of the most important abiotic stresses in agriculture and one of the major limiting factors in plant growth and crop production in the world (Munns and Tester, 2008). This abiotic stress affects all stages of plant development and causes hyperosmotic stress, which inhibits the plant's ability to take up water (physiological drought); homeostasis disruption; nutritional imbalances; oxidative stress; and ionic toxicity in plants (Figure 15.1) (Deinlein et al., 2014; Farooq et al., 2015). Salinity has also become one of the most serious global problems for providing water and land to meet the world's food needs (Etesami and Maheshwari, 2018). Saline soils exist in practically all climatic regions, from the moist tropics to the polar regions. However, the salinization process is more prevalent in arid and semiarid regions and is increasing steadily in these regions (Porcel et al., 2012). More than 6% of the world's total land area (more than 800 million hectares of land throughout the world) are affected by salinity stress (Etesami and Maheshwari, 2018; FAO, 2005). Salinity problems are likely to become more severe in the future as we face climate change at a global level (Khan and Basha, 2015).

Legumes, which represent the second largest and most diverse family of dicotyledonous flowering plants (Doyle and Luckow, 2003), are important plants both ecologically and agriculturally in many parts of the world (occupation of 12%–15% of arable land worldwide) (Mishra et al., 2014) and are very important in human and animal nutrition, due to their high grain protein contents (17%–40%), complex carbohydrates, essential amino acids, vitamins, fiber and minerals, and growth in a wide range of global climates (Farooq et al., 2017). These plants form important symbiotic associations with rhizobial bacteria in specialized organs called nodules and are recognized as the most important system for biological nitrogen fixation (BNF) (Bruning and Rozema, 2013). In agriculture, these plants are widely used as green manure, and crop rotations that include these plants are considered an effective strategy for improvement of soil fertility and crop yields by increasing soil organic matter content and nitrogen (Zahran, 1999). The BNF derived from the rhizobia–legumes symbioses can benefit not only the host legume crop, but it may also have positive effects for succeeding non-legume crops (Lupwayi et al., 2004). Besides, rhizobia

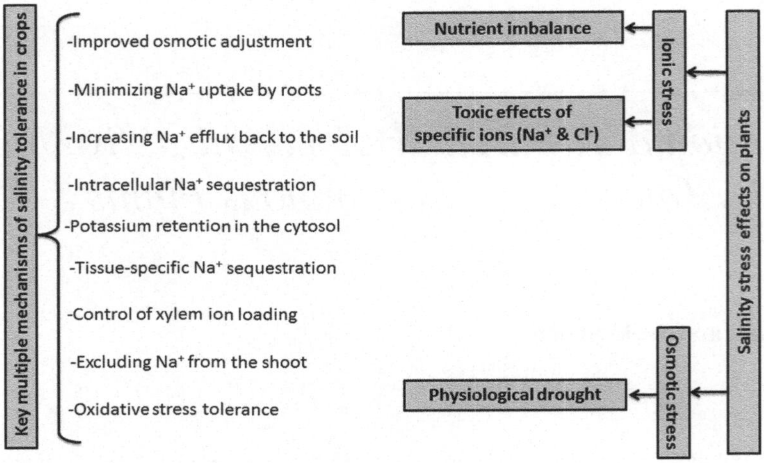

FIGURE 15.1 Main effects of salinity on plants and key mechanisms used by the plants for enhancing salinity tolerance.

associated with legume plants may also function as non-symbiotic plant growth–promoting bacteria as in the case of economically influential non-legume crops such as wheat or rice, which are the best examined examples that benefit from rhizobial bacteria as endophytes (Biswas et al., 2000; Chaintreuil et al., 2000).

Like many agricultural crops (glycophytes), the legumes are also sensitive to salinity (Maas and Hoffman, 1977) and are expected to be significantly affected (Khan and Basha, 2015). The majority of legumes are sensitive to salinity (80% of biomass production compared to control at ~3 dS/m) or moderately sensitive to salinity (80% of biomass production at ~6 dS/m) (Khan and Basha, 2015), and only a few agronomical legumes can grow in salt–affected soils (Ashraf and McNeilly, 2004) by some tolerance mechanisms (Figure 15.2) such as hormonal regulation, osmoregulation and osmotic adjustment, ion homeostasis, hormonal regulations, and activation of the antioxidant defense system (Farooq et al., 2017). Many studies indicate that N_2-fixing legume plants are more sensitive to salinity than those relying only on the mineral forms of nutrients (Lauter et al., 1981; Yousef and Sprent, 1983).

Effects of salinity on various leguminous crops have been well reviewed by Khan and Basha (2015) and Farooq et al. (2017). In general, salinity stress disturbs photosynthesis and hormonal regulation and brings about nutritional imbalance, specific ion toxicity (hyperionic–stress), and osmotic (hyperosmotic–stress) effects in legumes, diminishing grain yield and quality of these plants up to 12%–100% (Farooq et al., 2017). It has been well known that the symbiotic association between legumes and rhizobial bacteria is also susceptible to salinity stress than either partner alone (Bouhmouch et al., 2005; Egamberdieva et al., 2013; Farooq et al., 2017; Slattery et al., 2001). In other words, salinity stress affects nitrogen fixation in legumes. Salinity can adversely affect the emergence and the growth of root and root hairs and potential infection sites, the number of rhizobial cells, the establishment of rhizobia in the rhizosphere/root surfaces, various phases of the initiation of symbiosis (infection thread initiation), colonization and infection of root hairs, nodule formation, root nodule development, nodule activity, photosynthate transport toward nodules and the size of nodules, biosynthesis of nitrogenase, nitrogenase activity, supply of respiratory substrates to the bacteroids, shrinkage of root hairs (root hair curling), photosynthate supply to the nodules, nodule respiration, nodular conductance of O_2 (diffusion of O_2 in the nodule), enzymes associated with sucrose degradation, and synthesis of green pigments from leghaemoglobin (Bekki et al., 1987; Bouhmouch et al., 2005; Bruning and Rozema, 2013; Cordovilla et al., 1995, 1999; Delgado et al., 1993, 1994; Egamberdieva et al., 2013; Frechilla et al., 2001; González et al., 2001; Khan and Basha, 2015; López et al., 2008; Maas and Hoffman, 1977; Marcar et al., 1991; Rabie and Almadini, 2005; Rekha et al., 2007; Sarath et al., 1986; Serraj et al., 1994, 2001; Singleton and Bohlool, 1984; Subbarao et al., 1990; Tejera et al., 2004; Tu, 1981; Valdenegro et al., 2001; Van Hoorn et al., 2001; Zahran and

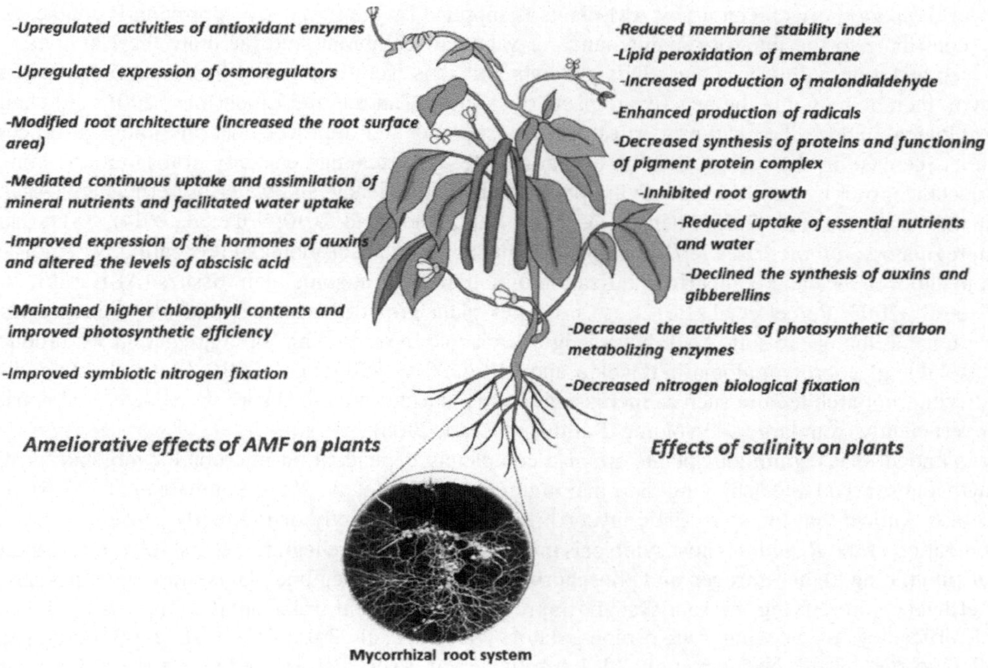

FIGURE 15.2 Effect of salinity on plant and ameliorative roles of arbuscular mycorrhizal fungi (AMF) on salinity–stressed legume plant.

Sprent, 1986). Thus, improved growth and yield of leguminous crops and developed salt–tolerant symbioses to feed the growing population and to enable cultivation of leguminous crops in saline soils are a demanding need today, owing to the augmenting threat to crop productivity of global warming and changing climatic patterns (diminished area of arable land), in addition to the anticipated human population explosion in the near future.

Agriculture under saline conditions has already caused major challenges in many countries for a variety of reasons. Some strategies such as the production of salt–tolerant legume plants through genetic engineering (the selection and breeding of improved cultivars) (Bliss, 1993; Etesami and Beattie, 2018), leaching of salt from root zone, use of soil amendments, shifting the crop calendars, and drainage and improved irrigation practices have been recommended to cope with this problem (salinity stress). Although successful, they are not economical (expensive and/or very time consuming) (Munns, 2002; Shrivastava and Kumar, 2015). In contrast, one of the important, environmentally friendly, and effective strategies for using saline soils' potential—mitigating the adverse effects of salinity on plant growth and, as a result, increasing productivity in salt–affected land—is application of bio–fertilizers such as salinity–resistant microorganisms (Etesami and Beattie, 2017, 2018; Etesami and Maheshwari, 2018; Noori et al., 2018). Several reports have shown that arbuscular mycorrhizal fungi (AMF), as a pervasive member of soil microbiota with significant roles in natural and disturbed ecosystems (Navarro et al., 2014; Tang et al., 2009), can help host plants to cope with different environmental stresses (Abeer et al., 2014; Al-Karaki et al., 2001; Feng et al., 2002; Giri et al., 2003; Miransari, 2010; Porcel et al., 2012; Smith and Read, 2010). More than 80% of vascular plant families (terrestrial plants including most agricultural, horticultural, and hardwood crop species) in natural ecosystems seem to be symbiotic with mycorrhizal fungi (Ruiz-Lozano et al., 2012; Smith and Read, 2010). These ecologically compatible symbiosis mycorrhizal fungi can have profound effects on host tolerance and plant adaptability under different environmental stresses including salinity stress. AMF can improve salt tolerance in AM plants by (i) providing macronutrients (N, P, K, and Mg) and micronutrients (Cu, Fe, Mn, and Zn) in poor soils (Farooq et al., 2017) especially phosphorus (P) nutrition. According to Marschner and Dell (1994),

80% of the phosphorus taken up by AM plants is supplied by the fungi. By improving P uptake, AMF may contribute to the integrity maintenance of vacuolar membrane and facilitate the Na^+ ions compartmentalization within vacuoles. This prevents Na^+ ions from interfering in metabolic pathways of growth, thereby reducing the negative impacts of salinity (Cantrell and Linderman, 2001); (ii) changes physiological and biochemical properties of the host plants and improves host physiological processes such as increase in the carbon dioxide exchange rate, gas exchange capacity, transpiration, stomatal conductance, root hydraulic conductivity, photosynthetic efficiency, and water use efficiency (Ahanger et al., 2014a; Aroca et al., 2013; Sheng et al., 2008; Smith and Read, 2010; Wu et al., 2014); (iii) produces osmoregulators; (iv) increases K^+/Na^+ ratios by enhancing K^+ absorption and preventing Na^+ translocation to shoot tissues and compartmentalization of sodium within some plant tissues (Al-Karaki, 2006; Giri et al., 2003; Porcel et al., 2012); (v) produces plant growth hormones; and (vi) improves rhizospheric condition of the soil. AMF have a significant role in improving soil aggregation via producing an insoluble glycoprotein glomalin (Gadkar and Rillig, 2006; Rillig et al., 2010; Singh, 2012); and (vii) modifying root architecture such as increasing the root surface area to absorb more water and nutrients more efficiently from large soil volume (Smith and Read, 2008).

It is known that leguminous plants are also completely dependent on microbial symbionts for more growth and survival and achieving their maximum growth (Jia et al., 2004; Soumare et al., 2015). It has also been known that the synergistic mycorrhizal fungi can directly or indirectly cause infection and colonization of the *Rizobium*–host symbiosis processes and improve legume plants' tolerance to salinity by accumulating higher nitrogen and phosphorus in the shoots of legume plants, improving phosphorus use efficiency, modifying the root system, improving nutrient and water uptake, improving photosynthetic efficiency, and accumulating osmoregulators (Ashrafi et al., 2014; Biró et al., 2000; Farooq et al., 2017; Guo et al., 2010; Nadeem et al., 2014; Nautiyal et al., 2010; Patreze and Cordeiro, 2004; Requena et al., 2001; Soumare et al., 2015; Tajini et al., 2012; Tavasolee et al., 2011).

Considering the severity of the salinization as an environmental problem and a serious threat for legume production worldwide, this chapter provides an overview of the contribution of arbuscular mycorrhizal symbiosis to salinity tolerance in leguminous plants. The main conclusions and future trends for research in this area will then be presented.

15.2 Arbuscular Mycorrhizal Fungi (AMF)

Mycorrhiza, the symbiotic association generated by fungi with plant roots, is one of the most common and long–standing symbiotic associations in the plant kingdom, which has a global expansion and is found in a wide range of ecological conditions such as aquatic environments, hot and cold deserts, and even saline areas (Smith and Read, 2010). AMF, the fungi that colonize the host plant's root tissues intracellularly and form intraradical and extraradical structures (Porcel et al., 2012), form a momentous integral component of the natural ecological systems (Giri et al., 2003). These fungi belong to the fungal phylum Glomeromycota (SchÜßLer et al., 2001). The application of AMF in augmenting crop yield started a long time ago, and there are many published studies about how these beneficial AMF can augment plant growth, development, nutrient uptake, and yield by various mechanisms (Abeer et al., 2015; Latef and Miransari, 2014; Miransari, 2010; Nadeem et al., 2014; Porcel et al., 2012; Porras-Soriano et al., 2009; Smith and Read, 2010; Talaat and Shawky, 2014). AMF increase uptake of nutrients (especially phosphorous, which is extremely immobile element in soils) with low availability for plants from the soil by the extraradical fungal hyphae. Fungal mycelium outstretches the plant's effective absorption surfaces beyond the nutrient-depleted zone that develops around the root caused by direct root uptake processes (Cardoso and Kuyper, 2006; Medina et al., 2007; Nadeem et al., 2014; Yasmeen et al., 2019). These fungi improve quality of soil structure (the formation of water-stable soil aggregates) by producing glomalin, a glycoprotein produced by the external hyphae. Glomalin contributes to the initiation and stabilization of soil aggregates, owing to its glue–like hydrophobic nature (Miller and Jastrow, 2000; Nadeem et al., 2014; Requena et al., 2001; Wright and Upadhyaya, 1998). Synergistic interactions of nitrogen-fixing bacteria (NFB) and phosphate–solubilizing microorganisms (PSM) with mycorrhizal fungi have been demonstrated (Nadeem et al., 2014). These fungi can also increase the activity of these microorganisms. Increased activity of these beneficial microorganisms, in turn, leads to the improvement of the bioavailability of the major plant nutrients N and P for mycorrhizal plants (Barea et al., 2002).

It is known that some plants are unable to endure habitat–imposed environmental stresses in the absence of mycorrhizal fungi (Porcel et al., 2012; Whipps, 2004). These fungi could confer AM plant tolerance to abiotic stresses such as drought (Miransari, 2010; Ruiz-Lozano, 2003), heat (Compant et al., 2010), heavy metal toxicity (Biró et al., 2000; Burd et al., 2000; He and Yang, 2007; Ismaiel et al., 2014; Latef, 2011, 2018; Redon et al., 2009), nutrient deficiency (Wang et al., 2011) and salinity (Evelin et al., 2009; Latef and Miransari, 2014; Miransari, 2010; Porcel et al., 2012), and biotic stresses (Linderman, 2000; Pozo and Azcón-Aguilar, 2007). In addition to non-leguminous plants, the AMF have great potential to improve growth and nodulation—and hence nitrogen fixation in leguminous plants under non-saline and saline conditions (Farooq et al., 2017; Nadeem et al., 2014).

15.3 Effect of AMF on Growth of Salinity–Stressed Leguminous Plants

Although AMF are known to exist in saline environments (Giri et al., 2003), there are some reports that salinity can also negatively affect the colonization capacity, spore germination, fungal development; reduce the number of arbuscules; and impact the growth of fungal hyphae of the AMF (Abeer et al., 2015; Alqarawi et al., 2014; Giri et al., 2007; Hajiboland, 2013; Hashem et al., 2016a; Juniper and Abbott, 2006; Latef and Chaoxing, 2011, 2014; Nisha et al., 2014; Sheng et al., 2008). However, in some studies, salt–tolerant AMF species that could maintain their colonization capacity and symbiosis efficiency under saline conditions have also been reported (Aliasgharzadeh et al., 2001; Yamato et al., 2008). These fungi could improve the performance of salinity–stressed plants (Abd_Allah et al., 2015; Alqarawi et al., 2014; Aroca et al., 2013; Chen et al., 2017; Giri et al., 2007; Hajiboland, 2013; Hajiboland et al., 2010; Latef and He, 2014; Latef and Chaoxing, 2011; Nadeem et al., 2014; Porcel et al., 2012, 2014; Zuccarini and Okurowska, 2008) and increase the fitness of the host plants by various mechanisms such as (i) increased uptake and acquisition of nutrients (Asghari et al., 2005; Cantrell and Linderman, 2001; Evelin et al., 2012) (Kim et al., 2010; Smith and Read, 2008) such as P (Garg and Manchanda, 2008; Kapoor et al., 2013; Sharifi et al., 2007) and N (Garg and Manchanda, 2008) by diminishing the toxic effects of Na ions and thereby maintaining the chlorophyll content of the plant, enhancing production of enzymes controlling the primary nitrogen fixation in the extra–radical mycelia and upregulating arginine catabolism–controlling enzymes in the intraradical mycelia, and assimilating nitrate in the extra radical mycelia via the activity of nitrate reductase located in the arbucular–containing cells (Evelin et al., 2009, 2012; Kapoor et al., 2013); K (Evelin et al., 2012; Latef and Chaoxing, 2011; Zuccarini and Okurowska, 2008) by the regulation of genes encoding membrane transport proteins involved in K^+/Na^+ homeostasis (Chen et al., 2017), Ca (Evelin et al., 2009; Rabie et al., 2005; Sharifi et al., 2007); Mg (Giri et al., 2003; Giri and Mukerji, 2004; Latef and Chaoxing, 2011), Cu (Giri et al., 2007), and Zn (Sharifi et al., 2007); (ii) accumulated osmoregulator compounds (Evelin et al., 2013; Hamdia and Shaddad, 2010) such as inorganic ions such as K^+ and compatible organic solutes like proline (Jindal et al., 1993; Sharifi et al., 2007) glycine betaine (Al-Garni, 2006; Sheng et al., 2011), polyols, trehalose, and quaternary ammonium compounds (i.e., pipecolatebetaine, prolinebetaine, alaninebetaine, glycinebetaine, and hydroxyprolinebetaine), and soluble sugars (Al-Garni, 2006; Feng et al., 2002; Garg and Manchanda, 2009; Khan et al., 2014; Latef and Chaoxing, 2014; Moghaieb et al., 2004; Sharifi et al., 2007; Sheng et al., 2011; Talaat and Shawky, 2014). These compatible solutes conserve subcellular structures, attenuate oxidative damage caused by free radicals, augment plant tolerance against salt stress–induced water loss from the cells, buffer cellular redox potential under stress, and maintain the membrane integrity, and enzymes and proteins stabilization under salt stress (Hoseini, 2010; Latef and Chaoxing, 2014; Masood et al., 2013); (iii) produced plant growth hormones (Abeer et al., 2015; Hashem et al., 2016b; Navarro et al., 2014; Waqas et al., 2012). It has been reported that mycorrhization can alter the abscisic acid (ABA) levels, a plant growth and development–regulating phytohormone that is involvement in the response of the plant to abiotic stress including salinity stress, in the host plant (Estrada-Luna and Davies Jr, 2003; Evelin et al., 2009; Jahromi et al., 2008); (iv) increased photosynthetic rate

(Aroca et al., 2013; Evelin et al., 2009; Hajiboland et al., 2010; Kaya et al., 2009; Latef and Chaoxing, 2011, 2014; Zuccarini, 2007) via maintaining higher chlorophyll contents (i.e., by supporting uptake of Mg from soil, an important component of chlorophyll molecule) (Sheng et al., 2008); (v) improved water use efficiency (Hajiboland et al., 2010) by inducing an increase in root hydraulic conductivity of the host plants, changing expression of aquaporin genes (Aroca et al., 2007; Jahromi et al., 2008; Ouziad et al., 2006; Ruiz-Lozano et al., 2012), altering morphology of root system, improving stomatal conductance, which increases the demand for transpiration, and accumulating solutes, which improve plant osmotic adjustment (Aroca et al., 2007; Kapoor et al., 2008; Kothari et al., 1990; Latef and Chaoxing, 2014; Marulanda et al., 2003; Sánchez-Blanco et al., 2004; Sheng et al., 2008); (vi) enhanced the activities of antioxidant enzymes (i.e., superoxide dismutases, catalase, peroxidases, ascorbate peroxidase, and glutathione reductase) (Abeer et al., 2014; Ahmad et al., 2015; Alguacil et al., 2003; He et al., 2007; Latef and Chaoxing, 2011; Manchanda and Garg, 2011a; Wu et al., 2014). These enzymes minimize the occurrence of oxidative damage caused by salinity-induced reactive oxygen species (ROS) (i.e., free radicals such as superoxide anion and hydroxyl radical, and nonradical molecules such as single oxygen and hydrogen peroxide) (Abeer et al., 2015; Evelin and Kapoor, 2014; Hajiboland et al., 2010; Latef and Chaoxing, 2011, 2014). Increased activity of enzymatic antioxidants of AM plants may be attributed toward improved uptake of important nutrients in AM plants that help to enhance synthesis of biomolecules that serve as cofactors of enzymes (Yasmeen et al., 2019); (vii) induced the expression of some plant genes related to salt stress responses (Porcel et al., 2012) (inducing systemic tolerance to stress) (Ruiz-Lozano et al., 2012); and (viii) improved rhizospheric and soil conditions. It is known that soil salinity can also alter soil structure, resulting in reduced soil porosity, aeration, and water conductance. AMF can improve the quality of soil structure by producing glomalin (Asghari et al., 2005).

The role of AMF for improving growth and yield of legume plants under saline conditions has received attention (Farooq et al., 2017). There are many studies available in the literature, which signify the potential of AMF for augmenting growth and development of legume plants under salinity–stress conditions (Table 15.1). In addition to enhanced growth and yield of leguminous plants, AMF also improved legume–rhizobia symbiotic performance and enhanced nodulation and nitrogen fixation in legumes, such as faba bean (*Vicia faba*) (Jia et al., 2004), lucerne (*Medicago sativa*) (Ardakani et al., 2009), common bean (*Phaseolus vulgaris*) (Tajini et al., 2012), mung bean (*Vigna radiata*) (Rabie et al., 2005; Singh et al., 2011), pigeon pea (*Cajanus cajan* L.) (Bhattacharjee and Sharma, 2012), and lentil (*Lens culinaris*) (Xavier and Germida, 2002), in saline environments by various mechanisms such as improving nutrient uptake, the nodulation process, water uptake, photosynthesis, etc. (Abd_Allah et al., 2015; Abeer et al., 2014, 2015; Ashrafi et al., 2014; Chen et al., 2017; Evelin et al., 2009; Farooq et al., 2017; Garg, 2008; Garg and Chandel, 2011; Garg and Manchanda, 2008; Hashem et al., 2016a, 2016b, 2019; Nadeem et al., 2014; Sharifi et al., 2007).

It is well known that soil salinity can significantly reduce phosphorus uptake by plants due to forming calcium phosphate precipitation (Bano and Fatima, 2009). Phosphorus plays a significant role in the nitrogen fixation process as an essential nutrient. Several authors have shown that the increase in the availability of phosphorus is essential for the establishment of nodulation and efficiency of nitrogen fixation (Sakamoto et al., 2013; Tajini et al., 2012). For example, plants deficient in P show decreased nodule–number and biomass (Vadez et al., 1996). In addition, N_2 fixation by rhizobial bacteria in soils with high phosphate-fixing capacity has been shown to be weak due to a limited supply of phosphorus and other micronutrients. Since AMF are generally associated with legumes in such soils, they can augment nutrient uptake by the plant (Antunes et al., 2006; Javaid, 2010). For example, improved P nutrition in AM plants enhanced nodulation and nitrogen fixation in legumes (Garg and Manchanda, 2008; Kapoor et al., 2013). By increasing the surface area for absorption, solubilizing P through the release of organic acids and phosphatase enzymes, and uptake through mycorrhizal hyphae (Bolan, 1991), AMF may improve P uptake, which in turn would result in more energy available for nitrogen fixation by rhizobia (Fitter and Garbaye, 1994; Mortimer et al., 2008) and finally increase the tolerance of legumes plants to salinity stress (Biró et al., 2000; Patreze and Cordeiro, 2004). AMF also augment P uptake in plants via their extensive hyphal network, which explores more soil volume and quickly transfers P to the cortical cells, thus diminishing the impact of P depletion in the rhizosphere (Garg and Pandey, 2015; Smith and Smith, 2011).

TABLE 15.1

Ameliorative Role of Arbuscular Mycorrhizal Fungi (AMF) against Salinity Stress in Leguminous Plants

AMF	Legume Plant	Salinity Level	Effect/Action Mechanism of AMF on Salinity Stressed Legume	References
Glomus mosseae	Alfalfa (*Medicago sativa* L.)	0, 60, 120 and 180 mM NaCl	Increased mycorrhizal infection, nodulation, root and shoot P contents, the value of K/Na ratio, the shoot proline content and the root Ca content and decreased leaf Na content of AM inoculated alfalfa plants compared to control plants	Ashrafi et al. (2014)
Rhizophagus irregularis	Black locust (*Robinia pseudoacacia* L.)	0, 100, and 200 mM NaCl	Increased photosynthetic capacity, water status, and K^+/Na^+ homeostasis, reduced the root–to–shoot allocation of Na^+, upregulated aquaporin genes in leaves and roots that facilitate water uptake by roots, water transport to special tissues and CO_2 diffusion across the plasma membrane, and facilitated Na^+ secretion from root cells, Na^+ unloading from the xylem, and K^+ translocation to the shoots	Chen et al. (2017)
Claroideoglomus etunicatum, *Rh. intraradices*, and *Funneliformis mosseae*	*Acacia gerrardii* Benth.	0 and 250 Mm NaCl	Increased shoot and root dry weight, nodule number, and leghemoglobin content and enhanced the N, P, K, Mg, and Ca contents and phosphatase activities in salt–stressed *A. gerrardii* tissues and reduced Na and Cl concentration	Hashem et al. (2016a)
F. mosseae (syn. *Glomus mosseae*), *Rh. intraradices* (syn. *Glomus intraradices*), and *Claroideoglomus etunicatum* (syn. *Glomus etunicatum*)	Common bean (*Phaselous vulgaris* L.)	0, 0.15, and 0.25 M NaCl	Alleviated the adverse impact of salinity on the plant growth, anabolic physiological attributes, and nutrient uptake by reducing the oxidative damage of salt through strengthening and modulation of the systemic acquired resistance	Abd_Allah and Alqarawi (2015)
F. mosseae, *Rh. Intraradices*, and *C. etunicatum*	Cowpea (*Vigna unguiculata* L.)	0 and 200 mM NaCl	Ameliorated the negative impact of salinity on the growth parameters, enhanced antioxidant enzyme activity, strengthened the plant's defense system, increased proline content, and increased uptake of mineral elements	Abeer et al. (2015)
F. mosseae, *Rh. Intraradices*, and *C. etunicatum*	Faba bean (*Vicia faba* L.)	0, 50, and 100 mM NaCl	Mitigates the effect of NaCl stress, improved the growth and yield, enhancement in nodule activity and pigment content, increased phytoconstituents, minimized the effect of salinity on lipid peroxidation, restored the potassium and calcium contents and maintained their ratio, and alleviated the negative effect of sodium chloride	Abeer et al. (2014)
G. clarum	Faba bean (*Vicia faba* L.)	0 and 6.0 dSm^{-1}	Increased tolerance of salinity, mycorrhizal dependency, phosphorus level, phosphatases enzymes, nodule number, nitrogen level, protein content and nitrogenase enzymes, and enhanced K^+/Na^+, Mg^+/Na^+, and Ca^+/Na^+ ratios	Rabie and Almadini (2005)
G. mosseae	*Cajanus cajan* (Pigeon pea)	0, 4, 6, and 8 dSm^{-1}	Improved nodulation, leghemoglobin content, and nitrogenase activity and enhanced activities of superoxide dismutase (SOD), catalase (CAT), ascorbate peroxidase (APX), peroxidase (POD), and glutathione reductase (GR)	Garg and Manchanda (2008)
G. mosseae	Chickpea (*Cicer arietinum* L.)	0, 4, 6, and 8 dS m^{-1}	Enhanced growth, biomass accumulation, nodule numbers, dry weights, and nitrogenase activity	Garg and Chandel (2011)

(*Continued*)

TABLE 15.1 (*Continued*)
Ameliorative Role of Arbuscular Mycorrhizal Fungi (AMF) against Salinity Stress in Leguminous Plants

AMF	Legume Plant	Salinity Level	Effect/Action Mechanism of AMF on Salinity Stressed Legume	References
G.mosseae, G. intraradices, and *G. etunicatum*	Cowpea [*Vigna unguiculata* (L.) Walp.]	0 and 200 Mm NaCl	Enhanced activity of antioxidant enzymes such as SOD, CAT, APX, POD, and (GR), increased proline content, increased uptake of mineral elements, and improved status of osmoregulation	Abeer et al. (2015)
F. mosseae, Rh. Intraradices, and *C. etunicatum*	Soybean (*Glycine max* L. Merrill)	0 and 200 Mm NaCl	Improved the symbiotic performance of soybean by improving nodule formation, leghemoglobin content, nitrogenase activity and auxin synthesis, protected soybean from salt–induced membrane damage, reduced the production of hydrogen peroxide, reduced the production of TBARS, and reduced lipid peroxidation	Hashem et al. (2019)
G. etunicatum	Soybean (*Glycine max* L. Merrill)	0, 50, 100, 150, and 200 Mm NaCl	Enhanced fresh and dry weight, root proline, P, K, and Zn; and diminished shoot proline and Na concentrations	Sharifi et al. (2007)
G. mosseae, G. intraradices, and *G. etunicatum*	Lupine (*Lupinus termis* Forsik)	0 and 250 Mm NaCl	Increased the activities of antioxidant enzymes of SOD, CAT, APX, and POD; increased accumulation of osmoregulators (proline, glycine betaine, and sugars); reduced lipid peroxidation; and increased membrane stability	Hashem et al. (2016b)

Salinity can also inhibit the absorption of Ca^{2+} by plants. This cation is involvement in cell division, elongation, and membrane function and structure (Bigot et al., 1983; Torimitsu et al., 1985). Ca ion also serves as a momentous cellular messenger for downstream signaling (Ahanger et al., 2014b). Deficiency of this nutrient reduces the emergence and the growth of roots and root hairs and decreases potential infection sites (Zahran and Sprent, 1986). In addition, due to its involvement in cell walls, salinity–induced Ca deficiency causes breaks in certain zones of the walls of cells from nodules of plants grown under salt stress (Bolaños et al., 2003). It has been reported that AMF can increase Ca^{2+} uptake, which in turn increases node–gene induction and expression, and concurrently affects the attachment of rhizobia to root hairs and nodule development (Smith et al., 1992).

It has been known that salinity, by decreasing CO_2 availability due to the limitations of diffusion through the stomata and mesophyll (Chaves et al., 2009) and hindering the photosynthetic electron flow through photosystems II and I (PSII and PSI) (Allakhverdiev and Murata, 2008; Lu and Vonshak, 2002), negatively affect photosynthesis, the fundamental process of primary metabolism (Farooq et al., 2017). It has been proposed that the inhibition of the photosynthesis in salinity–stressed plants led to a restriction of photosynthate transport toward nodules (a decrease in proportion of photosynthates available to the rhizobium nodules) and a reduction of the size of these organs (Bekki et al., 1987). Previous studies have demonstrated that mycorrhizal symbiosis can augment photosynthesis in salinity–stressed plants (Hajiboland et al., 2010; Porcel et al., 2015; Sheng et al., 2008) and make a greater portion of photosynthates accessible to the rhizobium nodules (Mortimer et al., 2008).

Other mechanisms by which AMF made improvements to the undesirable effects of salinity on the nodulation process and, consequently, augmented nodule numbers and N fixation in legumes include: improved Mg content, which affects chlorophyll (Ahanger et al., 2014b); improved K uptake, which results in an augmented K^+/Na^+ ratio, an important aspect for the maintenance of physiological cellular functioning (Abd_Allah et al., 2015; Ahanger et al., 2015; Garg and Bhandari, 2016); augmented N uptake in legumes by establishing a plant ammonium transporter (Farooq et al., 2017; Garg and Manchanda, 2008); augmented leghemoglobin content and nitrogenase activity (Manchanda and Garg, 2011b); augmented osmotic adjustment (i.e., increase in proline content) (Garg and Bhandari, 2016; Jindal et al., 1993; Rabie et al., 2005; Sharifi et al., 2007); and produced trehalose in the extraradical mycelium, which acts as the main storage carbohydrate and also as an abiotic–stress protectant in plants inoculated with AMF and nitrogen-fixing bacteria (Ocón et al., 2007).

Overall, based on the above studies and available findings, it can be concluded that AMF can protect leguminous plants and other plants against the detrimental effects of salinity stress and also improve symbiotic nitrogen fixation in these plants for an agricultural development that would be friendly to the environment and consumers.

15.4 Conclusions

Leguminous plants as food crops play a substantial nutritional function in the diet of millions of people living in developing countries, and legumes are sometimes referred to as the poor man's meat. Many agriculturally important legumes are salt sensitive. Symbionts such as AMF and rhizobial bacteria are commonly found to be more salinity tolerant than their host plants. It is well known that AMF can improve tolerance of their host plants to salinity stress by various mechanisms such as production of different solutes, plant hormones, antioxidant products, the adjusted rate of K^+/Na^+, extensive network of the mycorrhizal plant roots, and enhanced nutrient water and uptake. Among all the processes, mineral nutrient improvement appears to be the common mechanism under salinity stress by AMF. Additionally, the legume–rhizobium system can also benefit from the presence of the AMF because these fungi make improvements not only to phosphorus deficiency but also any other macronutrient and micronutrient deficiencies that might be restricting to rhizobial bacteria.

Despite available findings, developing more detailed fundamental knowledge about the interactions between leguminous plants and specific AMF isolates adapted to high concentrations of salinity would facilitate a better understanding of diminishing salinity stress or abiotic stresses and possibly would permit better predictions regarding the plant response. We anticipated that there would be a growing interest to study the mechanisms of AMF–mediated salinity stress tolerance in leguminous plants in the near future. Taken together, well–designed, large–scale, and long–term field trials are required to evaluate the effect of AMF application in alleviating abiotic and biotic stresses in leguminous plants and should also be investigated.

REFERENCES

Abd_Allah, E. F., Hashem, A., Alqarawi, A. A., Bahkali, A. H., and Alwhibi, M. S. (2015). Enhancing growth performance and systemic acquired resistance of medicinal plant *Sesbania sesban* (L.) Merr using arbuscular mycorrhizal fungi under salt stress. *Saudi Journal of Biological Sciences* **22**, 274–283.

Abeer, H., Abd_Allah, E. F., Alqarawi, A. A., and Egamberdieva, D. (2015). Induction of salt stress tolerance in cowpea [*Vigna unguiculata* (L.) Walp.] by arbuscular mycorrhizal fungi. *Legume Research: An International Journal* **38**, 579–588.

Abeer, H., Abd_Allah, E. F., Alqarawi, A. A., El-Didamony, G., Alwhibi, M., Egamberdieva, D., and Ahmad, P. (2014). Alleviation of adverse impact of salinity on faba bean (*Vicia faba* L.) by arbuscular mycorrhizal fungi. *Pakistan Journal of Botany* **46**, 2003–2013.

Ahanger, M. A., Agarwal, R. M., Tomar, N. S., and Shrivastava, M. (2015). Potassium induces positive changes in nitrogen metabolism and antioxidant system of oat (*Avena sativa* L cultivar Kent). *Journal of Plant Interactions* **10**, 211–223.

Ahanger, M. A., Hashem, A., Abd-Allah, E. F., and Ahmad, P. (2014a). Arbuscular mycorrhiza in crop improvement under environmental stress. In *Emerging Technologies and Management of Crop Stress Tolerance*, pp. 69–95. Elsevier, San Diego, CA.

Ahanger, M. A., Tyagi, S. R., Wani, M. R., and Ahmad, P. (2014b). Drought tolerance: Role of organic osmolytes, growth regulators, and mineral nutrients. In *Physiological Mechanisms and Adaptation Strategies in Plants Under Changing Environment: Volume 1* (P. Ahmad and M. R. Wani, Eds.), pp. 25–55. Springer, New York.

Ahmad, P., Hashem, A., Abd-Allah, E. F., Alqarawi, A. A., John, R., Egamberdieva, D., and Gucel, S. (2015). Role of *Trichoderma harzianum* in mitigating NaCl stress in Indian mustard (*Brassica juncea* L) through antioxidative defense system. *Frontiers in Plant Science* **6**, 868.

Al-Garni, S. M. S. (2006). Increasing NaCl-salt tolerance of a halophytic plant *Phragmites australis* by mycorrhizal symbiosis. *American-Eurasian Journal of Agricultural and Environmental Science* **1**, 119–126.

Al-Karaki, G. N. (2006). Nursery inoculation of tomato with arbuscular mycorrhizal fungi and subsequent performance under irrigation with saline water. *Scientia Horticulturae* **109**, 1–7.

Al-Karaki, G. N., Hammad, R., and Rusan, M. (2001). Response of two tomato cultivars differing in salt tolerance to inoculation with mycorrhizal fungi under salt stress. *Mycorrhiza* **11**, 43–47.

Alguacil, M., Hernandez, J., Caravaca, F., Portillo, B., and Roldan, A. (2003). Antioxidant enzyme activities in shoots from three mycorrhizal shrub species afforested in a degraded semi-arid soil. *Physiologia Plantarum* **118**, 562–570.

Aliasgharzadeh, N., Rastin, S. N., Towfighi, H., and Alizadeh, A. (2001). Occurrence of arbuscular mycorrhizal fungi in saline soils of the Tabriz Plain of Iran in relation to some physical and chemical properties of soil. *Mycorrhiza* **11**, 119–122.

Allakhverdiev, S. I., and Murata, N. (2008). Salt stress inhibits photosystems II and I in cyanobacteria. *Photosynthesis Research* **98**, 529–539.

Alqarawi, A. A., Abd Allah, E. F., and Hashem, A. (2014). Alleviation of salt-induced adverse impact via mycorrhizal fungi in *Ephedra aphylla* Forssk. *Journal of Plant Interactions* **9**, 802–810.

Antunes, P. M., De Varennes, A., Rajcan, I., and Goss, M. J. (2006). Accumulation of specific flavonoids in soybean (*Glycine max* (L.) Merr.) as a function of the early tripartite symbiosis with arbuscular mycorrhizal fungi and *Bradyrhizobium japonicum* (Kirchner) Jordan. *Soil Biology and Biochemistry* **38**, 1234–1242.

Ardakani, M. R., Pietsch, G., Moghaddam, A., Raza, A., and Friedel, J. K. (2009). Response of root properties to tripartite symbiosis between lucerne (*Medicago sativa* L.), rhizobia and mycorrhiza under dry organic farming conditions. *American Journal of Agricultural and Biological Sciences* **4**, 266–277.

Aroca, R., Porcel, R., and Ruiz-Lozano, J. M. (2007). How does arbuscular mycorrhizal symbiosis regulate root hydraulic properties and plasma membrane aquaporins in *Phaseolus vulgaris* under drought, cold or salinity stresses? *New Phytologist* **173**, 808–816.

Aroca, R., Ruiz-Lozano, J. M., Zamarreño, Á. M., Paz, J. A., García-Mina, J. M., Pozo, M. J., and López-Ráez, J. A. (2013). Arbuscular mycorrhizal symbiosis influences strigolactone production under salinity and alleviates salt stress in lettuce plants. *Journal of Plant Physiology* **170**, 47–55.

Asghari, H. R., Marschner, P., Smith, S. E., and Smith, F. A. (2005). Growth response of *Atriplex nummularia* to inoculation with arbuscular mycorrhizal fungi at different salinity levels. *Plant and Soil* **273**, 245–256.

Ashraf, M., and McNeilly, T. (2004). Salinity tolerance in *Brassica* oilseeds. *Critical Reviews in Plant Sciences* **23**, 157–174.

Ashrafi, E., Zahedi, M., and Razmjoo, J. (2014). Co-inoculations of arbuscular mycorrhizal fungi and rhizobia under salinity in alfalfa. *Soil Science and Plant Nutrition* **60**, 619–629.

Bano, A., and Fatima, M. (2009). Salt tolerance in *Zea mays* (L). following inoculation with *Rhizobium* and *Pseudomonas*. *Biology and Fertility of Soils* **45**, 405–413.

Barea, J. M., Toro, M., Orozco, M. O., Campos, E., and Azcón, R. (2002). The application of isotopic (32 P and 15 N) dilution techniques to evaluate the interactive effect of phosphate-solubilizing rhizobacteria, mycorrhizal fungi and *Rhizobium* to improve the agronomic efficiency of rock phosphate for legume crops. *Nutrient Cycling in Agroecosystems* **63**, 35–42.

Bekki, A., Trinchant, J.-C., and Rigaud, J. (1987). Nitrogen fixation (C_2H_2 reduction) by Medicago nodules and bacteroids under sodium chloride stress. *Physiologia Plantarum* **71**, 61–67.

Bhattacharjee, S., and Sharma, G. D. (2012). Effect of dual inoculation of arbuscular mycorrhiza and rhizobium on the chlorophyll, nitrogen and phosphorus contents of pigeon pea (*Cajanus cajan* L.). *Advances in Microbiology* **2**, 561.

Biró, B., Köves-Péchy, K., Vörös, I., Takács, T., Eggenberger, P., and Strasser, R. J. (2000). Interrelations between *Azospirillum* and *Rhizobium* nitrogen-fixers and arbuscular mycorrhizal fungi in the rhizosphere of alfalfa in sterile, AMF-free or normal soil conditions. *Applied Soil Ecology* **15**, 159–168.

Biswas, J. C., Ladha, J. K., Dazzo, F. B., Yanni, Y. G., and Rolfe, B. G. (2000). Rhizobial inoculation influences seedling vigor and yield of rice. *Agronomy Journal* **92**, 880–886.

Bliss, F. A. (1993). Breeding common bean for improved biological nitrogen fixation. In *Enhancement of Biological Nitrogen Fixation of Common Bean in Latin America: Results from an FAO/IAEA Co-ordinated Research Programme, 1986–1991* (F. A. Bliss and G. Hardarson, Eds.), pp. 71–79. Springer, Dordrecht, the Netherlands.

Bolan, N. S. (1991). A critical review on the role of mycorrhizal fungi in the uptake of phosphorus by plants. *Plant and Soil* **134**, 189–207.

Bolaños, L., El-Hamdaoui, A., and Bonilla, I. (2003). Recovery of development and functionality of nodules and plant growth in salt-stressed *Pisum sativum*––*Rhizobium leguminosarum* symbiosis by boron and calcium. *Journal of Plant Physiology* **160**, 1493–1497.

Bouhmouch, I., Souad-Mouhsine, B., Brhada, F., and Aurag, J. (2005). Influence of host cultivars and *Rhizobium* species on the growth and symbiotic performance of *Phaseolus vulgaris* under salt stress. *Journal of Plant Physiology* **162**, 1103–1113.

Bruning, B., and Rozema, J. (2013). Symbiotic nitrogen fixation in legumes: Perspectives for saline agriculture. *Environmental and Experimental Botany* **92**, 134–143.

Burd, G. I., Dixon, D. G., and Glick, B. R. (2000). Plant growth-promoting bacteria that decrease heavy metal toxicity in plants. *Canadian Journal of Microbiology* **46**, 237–245.

Cantrell, I. C., and Linderman, R. G. (2001). Preinoculation of lettuce and onion with VA mycorrhizal fungi reduces deleterious effects of soil salinity. *Plant and Soil* **233**, 269–281.

Cardoso, I. M., and Kuyper, T. W. (2006). Mycorrhizas and tropical soil fertility. *Agriculture, Ecosystems & Environment* **116**, 72–84.

Chaintreuil, C., Giraud, E., Prin, Y., Lorquin, J., Bâ, A., Gillis, M., de Lajudie, P., and Dreyfus, B. (2000). Photosynthetic bradyrhizobia are natural endophytes of the African wild rice *Oryza breviligulata*. *Applied and Environmental Microbiology* **66**, 5437–5447.

Chaves, M. M., Flexas, J., and Pinheiro, C. (2009). Photosynthesis under drought and salt stress: Regulation mechanisms from whole plant to cell. *Annals of Botany* **103**, 551–560.

Chen, J., Zhang, H., Zhang, X., and Tang, M. (2017). Arbuscular mycorrhizal symbiosis alleviates salt stress in black locust through improved photosynthesis, water status, and K^+/Na^+ homeostasis. *Frontiers in Plant Science* **8**, 1739.

Compant, S., Van Der Heijden, M. G. A., and Sessitsch, A. (2010). Climate change effects on beneficial plant–microorganism interactions. *FEMS Microbiology Ecology* **73**, 197–214.

Cordovilla, M. D. P., Ligero, F., and Lluch, C. (1999). Effect of salinity on growth, nodulation and nitrogen assimilation in nodules of faba bean (*Vicia faba* L.). *Applied Soil Ecology* **11**, 1–7.

Cordovilla, M. P., Ocana, A., Ligero, F., and Lluch, C. (1995). Salinity effects on growth analysis and nutrient composition in four grain legumes-rhizobium symbiosis. *Journal of Plant Nutrition* **18**, 1595–1609.

Deinlein, U., Stephan, A. B., Horie, T., Luo, W., Xu, G., and Schroeder, J. I. (2014). Plant salt-tolerance mechanisms. *Trends in Plant Science* **19**, 371–379.

Delgado, M. J., Garrido, J. M., Ligero, F., and Lluch, C. (1993). Nitrogen fixation and carbon metabolism by nodules and bacteroids of pea plants under sodium chloride stress. *Physiologia Plantarum* **89**, 824–829.

Delgado, M. J., Ligero, F., and Lluch, C. (1994). Effects of salt stress on growth and nitrogen fixation by pea, faba-bean, common bean and soybean plants. *Soil Biology and Biochemistry* **26**, 371–376.

Doyle, J. J., and Luckow, M. A. (2003). The rest of the iceberg. Legume diversity and evolution in a phylogenetic context. *Plant Physiology* **131**, 900–910.

Egamberdieva, D., Jabborova, D., and Wirth, S. (2013). Alleviation of salt stress in legumes by co-inoculation with *Pseudomonas* and *Rhizobium*. In *Plant Microbe Symbiosis: Fundamentals and Advances*, pp. 291–303. Springer, Berlin, Germany.

Estrada-Luna, A. A., and Davies Jr, F. T. (2003). Arbuscular mycorrhizal fungi influence water relations, gas exchange, abscisic acid and growth of micropropagated chile ancho pepper (*Capsicum annuum*) plantlets during acclimatization and post-acclimatization. *Journal of Plant Physiology* **160**, 1073–1083.

Etesami, H., and Beattie, G. A. (2017). Plant-microbe interactions in adaptation of agricultural crops to abiotic stress conditions. In *Probiotics and Plant Health*, pp. 163–200. Springer, Singapore.

Etesami, H., and Beattie, G. A. (2018). Mining halophytes for plant growth-promoting halotolerant bacteria to enhance the salinity tolerance of non-halophytic crops. *Frontiers in Microbiology* **9**, 1–20.

Etesami, H., and Maheshwari, D. K. (2018). Use of plant growth promoting rhizobacteria (PGPRs) with multiple plant growth promoting traits in stress agriculture: Action mechanisms and future prospects. *Ecotoxicology and Environmental Safety* **156**, 225–246.

Evelin, H., Giri, B., and Kapoor, R. (2012). Contribution of *Glomus intraradices* inoculation to nutrient acquisition and mitigation of ionic imbalance in NaCl-stressed *Trigonella foenum-graecum*. *Mycorrhiza* **22**, 203–217.

Evelin, H., Giri, B., and Kapoor, R. (2013). Ultrastructural evidence for AMF mediated salt stress mitigation in *Trigonella foenum-graecum*. *Mycorrhiza* **23**, 71–86.

Evelin, H., and Kapoor, R. (2014). Arbuscular mycorrhizal symbiosis modulates antioxidant response in salt-stressed *Trigonella foenum-graecum* plants. *Mycorrhiza* **24**, 197–208.

Evelin, H., Kapoor, R., and Giri, B. (2009). Arbuscular mycorrhizal fungi in alleviation of salt stress: A review. *Annals of Botany* **104**, 1263–1280.

FAO (2005). The State of Food and Agriculture http://www.fao.org.

Farooq, M., Gogoi, N., Hussain, M., Barthakur, S., Paul, S., Bharadwaj, N., Migdadi, H. M., Alghamdi, S. S., and Siddique, K. H. M. (2017). Effects, tolerance mechanisms and management of salt stress in grain legumes. *Plant Physiology and Biochemistry* **118**, 199–217.

Farooq, M., Hussain, M., Wakeel, A., and Siddique, K. H. (2015). Salt stress in maize: Effects, resistance mechanisms, and management. A review. *Agronomy for Sustainable Development* **35**, 461–481.

Feng, G., Zhang, F., Li, X., Tian, C., Tang, C., and Rengel, Z. (2002). Improved tolerance of maize plants to salt stress by arbuscular mycorrhiza is related to higher accumulation of soluble sugars in roots. *Mycorrhiza* **12**, 185–190.

Fitter, A. H., and Garbaye, J. (1994). Interactions between mycorrhizal fungi and other soil organisms. *Plant and Soil* **159**, 123–132.

Frechilla, S., Lasa, B., Ibarretxe, L., Lamsfus, C., and Aparicio-Tejo, P. (2001). Pea responses to saline stress is affected by the source of nitrogen nutrition (ammonium or nitrate). *Plant Growth Regulation* **35**, 171–179.

Gadkar, V., and Rillig, M. C. (2006). The arbuscular mycorrhizal fungal protein glomalin is a putative homolog of heat shock protein 60. *FEMS Microbiology Letters* **263**, 93–101.

Garg, N. (2008). Effect of Arbuscular Mycorrhizal Inoculation on Salt-Induced Nodule Senescence in *Cajanus cajan* (Pigeonpea). *Journal of Plant Growth Regulation* **27**, 115.

Garg, N., and Bhandari, P. (2016). Silicon nutrition and mycorrhizal inoculations improve growth, nutrient status, K^+/Na^+ ratio and yield of *Cicer arietinum* L. genotypes under salinity stress. *Plant Growth Regulation* **78**, 371–387.

Garg, N., and Chandel, S. (2011). Effect of mycorrhizal inoculation on growth, nitrogen fixation, and nutrient uptake in *Cicer arietinum* (L.) under salt stress. *Turkish Journal of Agriculture and Forestry* **35**, 205–214.

Garg, N., and Manchanda, G. (2008). Effect of arbuscular mycorrhizal inoculation on salt-induced nodule senescence in *Cajanus cajan* (pigeonpea). *Journal of Plant Growth Regulation* **27**, 115.

Garg, N., and Manchanda, G. (2009). Role of arbuscular mycorrhizae in the alleviation of ionic, osmotic and oxidative stresses induced by salinity in *Cajanus cajan* (L.) Millsp. (pigeonpea). *Journal of Agronomy and Crop Science* **195**, 110–123.

Garg, N., and Pandey, R. (2015). Effectiveness of native and exotic arbuscular mycorrhizal fungi on nutrient uptake and ion homeostasis in salt-stressed *Cajanus cajan* L. (Millsp.) genotypes. *Mycorrhiza* **25**, 165–180.

Giri, B., Kapoor, R., and Mukerji, K. G. (2003). Influence of arbuscular mycorrhizal fungi and salinity on growth, biomass, and mineral nutrition of *Acacia auriculiformis*. *Biology and Fertility of Soils* **38**, 170–175.

Giri, B., Kapoor, R., and Mukerji, K. G. (2007). Improved tolerance of *Acacia nilotica* to salt stress by arbuscular mycorrhiza, *Glomus fasciculatum* may be partly related to elevated K/Na ratios in root and shoot tissues. *Microbial Ecology* **54**, 753–760.

Giri, B., and Mukerji, K. G. (2004). Mycorrhizal inoculant alleviates salt stress in *Sesbania aegyptiaca* and *Sesbania grandiflora* under field conditions: Evidence for reduced sodium and improved magnesium uptake. *Mycorrhiza* **14**, 307–312.

González, E. M., Gálvez, L., Royuela, M., Aparicio-Tejo, P. M., and Arrese-Igor, C. (2001). Insights into the regulation of nitrogen fixation in pea nodules: Lessons from drought, abscisic acid and increased photoassimilate availability. *Agronomie* **21**, 607–613.

Guo, Y., Ni, Y., and Huang, J. (2010). Effects of rhizobium, arbuscular mycorrhiza and lime on nodulation, growth and nutrient uptake of lucerne in acid purplish soil in China. *Tropical Grasslands* **44**, 109–114.

Hajiboland, R. (2013). Role of arbuscular mycorrhiza in amelioration of salinity. In *Salt Stress in Plants*, pp. 301–354. Springer, New York.

Hajiboland, R., Aliasgharzadeh, N., Laiegh, S. F., and Poschenrieder, C. (2010). Colonization with arbuscular mycorrhizal fungi improves salinity tolerance of tomato (*Solanum lycopersicum* L.) plants. *Plant and Soil* **331**, 313–327.

Hamdia, M. A., and Shaddad, M. A. K. (2010). Salt tolerance of crop plants. *Journal of Stress Physiology & Biochemistry* **6**, 64–90.

Hashem, A., Abd_Allah, E. F., Alqarawi, A. A., Al-Huqail, A. A., Wirth, S., and Egamberdieva, D. (2016a). The interaction between arbuscular mycorrhizal fungi and endophytic bacteria enhances plant growth of *Acacia gerrardii* under salt stress. *Frontiers in Microbiology* **7**, 1089.

Hashem, A., Abd_Allah, E. F., Alqarawi, A. A., Wirth, S., and Egamberdieva, D. (2016b). Arbuscular mycorrhizal fungi alleviate salt stress in lupine (*Lupinus termis* Forsik) through modulation of antioxidant defense systems and physiological traits. *Legume Research: An International Journal* **39**, 198–207.

Hashem, A., Abd_Allah, E. F., Alqarawi, A. A., Wirth, S., and Egamberdieva, D. (2019). Comparing symbiotic performance and physiological responses of two soybean cultivars to arbuscular mycorrhizal fungi under salt stress. *Saudi Journal of Biological Sciences* **26**, 38–48.

He, Z., and Yang, X. (2007). Role of soil rhizobacteria in phytoremediation of heavy metal contaminated soils. *Journal of Zhejiang University Science B* **8**, 192–207.

He, Z., He, C., Zhang, Z., Zou, Z., and Wang, H. (2007). Changes of antioxidative enzymes and cell membrane osmosis in tomato colonized by arbuscular mycorrhizae under NaCl stress. *Colloids and Surfaces B: Biointerfaces* **59**, 128–133.

Hoseini, S. M. (2010). Studying effect of salinity stress on germination, proline and carbohydrate content in thyme (*Thymus vulgaris* L.) seedlings. *International Journal of Agriculture and Crop Sciences (IJACS)* **2**, 34–38.

Ismaiel, A. A., Hegazy, H. S., and Azb, M. A. (2014). Physiological response of '*Vicia faba*' L. to inoculation with 'Rhizobium' and arbuscular mycorrhizal fungi: Comparative study for irrigation with Nile water and wastewater. *Australian Journal of Crop Science* **8**, 781.

Jahromi, F., Aroca, R., Porcel, R., and Ruiz-Lozano, J. M. (2008). Influence of salinity on the in vitro development of *Glomus intraradices* and on the in vivo physiological and molecular responses of mycorrhizal lettuce plants. *Microbial Ecology* **55**, 45–53.

Javaid, A. (2010). Role of arbuscular mycorrhizal fungi in nitrogen fixation in legumes. In *Microbes for Legume Improvement*, pp. 409–426. Springer, Wien.

Jia, Y., Gray, V. M., and Straker, C. J. (2004). The influence of *Rhizobium* and arbuscular mycorrhizal fungi on nitrogen and phosphorus accumulation by *Vicia faba*. *Annals of Botany* **94**, 251–258.

Jindal, V., Atwal, A., Sekhon, B. S., and Singh, R. (1993). Effect of vesicular-arbuscular mycorrhizae on metabolism of mung plants under NaCl salinity. *Plant Physiology and Biochemistry (France)* **31**, 475–481.

Juniper, S., and Abbott, L. K. (2006). Soil salinity delays germination and limits growth of hyphae from propagules of arbuscular mycorrhizal fungi. *Mycorrhiza* **16**, 371–379.

Kapoor, R., Evelin, H., Mathur, P., and Giri, B. (2013). Arbuscular mycorrhiza: Approaches for abiotic stress tolerance in crop plants for sustainable agriculture. In *Plant Acclimation to Environmental Stress*, pp. 359–401. Springer, New York.

Kapoor, R., Sharma, D., and Bhatnagar, A. K. (2008). Arbuscular mycorrhizae in micropropagation systems and their potential applications. *Scientia Horticulturae* **116**, 227–239.

Kaya, C., Ashraf, M., Sonmez, O., Aydemir, S., Tuna, A. L., and Cullu, M. A. (2009). The influence of arbuscular mycorrhizal colonisation on key growth parameters and fruit yield of pepper plants grown at high salinity. *Scientia Horticulturae* **121**, 1–6.

Khan, M. I. R., Asgher, M., and Khan, N. A. (2014). Alleviation of salt-induced photosynthesis and growth inhibition by salicylic acid involves glycinebetaine and ethylene in mungbean (*Vigna radiata* L.). *Plant Physiology and Biochemistry* **80**, 67–74.

Khan, P. S. S. V., and Basha, P. O. (2015). Salt stress and leguminous crops: Present status and prospects. *Legumes under Environmental Stress: Yield, Improvement and Adaptations*, pp. 21–51. John Wiley & Sons, Ltd.

Kim, K., Yim, W., Trivedi, P., Madhaiyan, M., Boruah, H. P. D., Islam, M. R., Lee, G., and Sa, T. (2010). Synergistic effects of inoculating arbuscular mycorrhizal fungi and *Methylobacterium oryzae* strains on growth and nutrient uptake of red pepper (*Capsicum annuum* L.). *Plant and Soil* **327**, 429–440.

Kothari, S. K., Marschner, H., and George, E. (1990). Effect of VA mycorrhizal fungi and rhizosphere microorganisms on root and shoot morphology, growth and water relations in maize. *New Phytologist* **116**, 303–311.

Latef, A., and He, C. X. (2014). Does inoculation with *Glomus mosseae* improve salt tolerance in pepper plants? *Journal of Plant Growth Regulation* **33**, 644–653.

Latef, A. A. (2018). Growth and some physiological activities of pepper (*Capsicum annuum* L.) in response to cadmium stress and mycorrhizal symbiosis.

Latef, A. A. H. A. (2011). RETRACTED ARTICLE: Influence of arbuscular mycorrhizal fungi and copper on growth, accumulation of osmolyte, mineral nutrition and antioxidant enzyme activity of pepper (*Capsicum annuum* L.). *Mycorrhiza* **21**, 495–503.

Latef, A. A. H. A., and Chaoxing, H. (2011). Effect of arbuscular mycorrhizal fungi on growth, mineral nutrition, antioxidant enzymes activity and fruit yield of tomato grown under salinity stress. *Scientia Horticulturae* **127**, 228–233.

Latef, A. A. H. A., and Chaoxing, H. (2014). Does inoculation with *Glomus mosseae* improve salt tolerance in pepper plants? *Journal of Plant Growth Regulation* **33**, 644–653.

Latef, A. A. H. A., and Miransari, M. (2014). The role of arbuscular mycorrhizal fungi in alleviation of salt stress. In *Use of Microbes for the Alleviation of Soil Stresses*, pp. 23–38. Springer, New York.

Lauter, D. J., Munns, D. N., and Clarkin, K. L. (1981). Salt response of chickpea as influenced by N Supply 1. *Agronomy Journal* **73**, 961–966.

Linderman, R. G. (2000). Effects of mycorrhizas on plant tolerance to diseases. In *Arbuscular Mycorrhizas: Physiology and Function*, pp. 345–365. Springer, Dordrecht, the Netherlands.

López, M., Herrera-Cervera, J. A., Iribarne, C., Tejera, N. A., and Lluch, C. (2008). Growth and nitrogen fixation in *Lotus japonicus* and *Medicago truncatula* under NaCl stress: Nodule carbon metabolism. *Journal of Plant Physiology* **165**, 641–650.

Lu, C., and Vonshak, A. (2002). Effects of salinity stress on photosystem II function in cyanobacterial *Spirulina platensis* cells. *Physiol Plant* **114**, 405–413.

Lupwayi, N. Z., Clayton, G. W., Hanson, K. G., Rice, W. A., and Biederbeck, V. O. (2004). Endophytic rhizobia in barley, wheat and canola roots. *Canadian Journal of Plant Science* **84**, 37–45.

Maas, E. V., and Hoffman, G. J. (1977). Crop salt tolerance–current assessment. *Journal of the Irrigation and Drainage Division* **103**, 115–134.

Manchanda, G., and Garg, N. (2011a). Alleviation of salt-induced ionic, osmotic and oxidative stresses in *Cajanus cajan* nodules by AM inoculation. *Plant Biosystems* **145**, 88–97.

Manchanda, G., and Garg, N. (2011b). Plant biosystems: An international journal dealing with all aspects of plant biology. *Official Journal of the Societa Botanica Italiana*, 88–97.

Marcar, N. E., Dart, P., and Sweeney, C. (1991). Effect of root-zone salinity on growth and chemical composition of *Acacia ampliceps* BR Maslin, *A. auriculiformis* A. Cunn. ex Benth. and *A. mangium* Willd at two nitrogen levels. *New Phytologist* **119**, 567–573.

Marschner, H., and Dell, B. (1994). Nutrient uptake in mycorrhizal symbiosis. *Plant and Soil* **159**, 89–102.

Marulanda, A., Azcon, R., and Ruiz-Lozano, J. M. (2003). Contribution of six arbuscular mycorrhizal fungal isolates to water uptake by *Lactuca sativa* plants under drought stress. *Physiologia Plantarum* **119**, 526–533.

Masood, A., Iqbal, N., Asgher, M., Khan, M. I. R., Fatma, M., and Khan, N. A. (2013). Variation in carbohydrate accumulation in two cultivars of mustard and its association with salt tolerance. *Journal of Functional and Environmental Botany* **3**, 94–102.

Medina, A., Jakobsen, I., Vassilev, N., Azcón, R., and Larsen, J. (2007). Fermentation of sugar beet waste by Aspergillus niger facilitates growth and P uptake of external mycelium of mixed populations of arbuscular mycorrhizal fungi. *Soil Biology and Biochemistry* **39**, 485–492.

Miller, R. M., and Jastrow, D. J. (2000). Mycorrhizal fungi influence soil structure. Chapter 1. *Arbuscular Mycorrhizas: Physiology and Function* (Y. Kapulnik and D. D. Douds. Jr., Eds), pp. 3–18. Springer Nature Switzerland AG.

Miransari, M. (2010). Contribution of arbuscular mycorrhizal symbiosis to plant growth under different types of soil stress. *Plant Biology* **12**, 563–569.

Mishra, S., Panda, S. K., and Sahoo, L. (2014). Transgenic Asiatic grain legumes for salt tolerance and functional genomics. *Reviews in Agricultural Science* **2**, 21–36.

Moghaieb, R. E. A., Saneoka, H., and Fujita, K. (2004). Effect of salinity on osmotic adjustment, glycinebetaine accumulation and the betaine aldehyde dehydrogenase gene expression in two halophytic plants, *Salicornia europaea* and *Suaeda maritima*. *Plant Science* **166**, 1345–1349.

Mortimer, P. E., Pérez-Fernández, M. A., and Valentine, A. J. (2008). The role of arbuscular mycorrhizal colonization in the carbon and nutrient economy of the tripartite symbiosis with nodulated *Phaseolus vulgaris*. *Soil Biology and Biochemistry* **40**, 1019–1027.

Munns, R. (2002). Comparative physiology of salt and water stress. *Plant, Cell & Environment* **25**, 239–250.

Munns, R., and Tester, M. (2008). Mechanisms of salinity tolerance. *Annual Review of Plant Biology* **59**, 651–681.

Nadeem, S. M., Ahmad, M., Zahir, Z. A., Javaid, A., and Ashraf, M. (2014). The role of mycorrhizae and plant growth promoting rhizobacteria (PGPR) in improving crop productivity under stressful environments. *Biotechnology Advances* **32**, 429–448.

Nautiyal, C. S., Chauhan, P. S., DasGupta, S. M., Seem, K., Varma, A., and Staddon, W. J. (2010). Tripartite interactions among *Paenibacillus lentimorbus* NRRL B-30488, *Piriformospora indica* DSM 11827, and *Cicer arietinum* L. *World Journal of Microbiology and Biotechnology* **26**, 1393–1399.

Navarro, J. M., Pérez-Tornero, O., and Morte, A. (2014). Alleviation of salt stress in citrus seedlings inoculated with arbuscular mycorrhizal fungi depends on the rootstock salt tolerance. *Journal of Plant Physiology* **171**, 76–85.

Nisha, K., Kuldeep, Y., and Ashok, A. (2014). Application of AM Fungi with *Bradyrhizobium japonicum* in improving growth, nutrient uptake and yield of *Vigna radiata* L. under saline soil. *Journal of Stress Physiology & Biochemistry* **10**, 134–152.

Noori, F., Etesami, H., Najafi Zarini, H., Khoshkholgh-Sima, N. A., Hosseini Salekdeh, G., and Alishahi, F. (2018). Mining alfalfa (*Medicago sativa* L.) nodules for salinity tolerant non-rhizobial bacteria to improve growth of alfalfa under salinity stress. *Ecotoxicology and Environmental Safety* **162**, 129–138.

Ocón, A., Hampp, R., and Requena, N. (2007). Trehalose turnover during abiotic stress in arbuscular mycorrhizal fungi. *New Phytologist* **174**, 879–891.

Ouziad, F., Wilde, P., Schmelzer, E., Hildebrandt, U., and Bothe, H. (2006). Analysis of expression of aquaporins and Na^+/H^+ transporters in tomato colonized by arbuscular mycorrhizal fungi and affected by salt stress. *Environmental and Experimental Botany* **57**, 177–186.

Patreze, C. M., and Cordeiro, L. (2004). Nitrogen-fixing and vesicular–arbuscular mycorrhizal symbioses in some tropical legume trees of tribe Mimoseae. *Forest Ecology and Management* **196**, 275–285.

Porcel, R., Aroca, R., and Ruiz-Lozano, J. M. (2012). Salinity stress alleviation using arbuscular mycorrhizal fungi. A review. *Agronomy for Sustainable Development* **32**, 181–200.

Porcel, R., Redondo-Gómez, S., Mateos-Naranjo, E., Aroca, R., Garcia, R., and Ruiz-Lozano, J. M. (2015). Arbuscular mycorrhizal symbiosis ameliorates the optimum quantum yield of photosystem II and reduces non-photochemical quenching in rice plants subjected to salt stress. *Journal of Plant Physiology* **185**, 75–83.

Porcel, R., Zamarreño, Á. M., García-Mina, J. M., and Aroca, R. (2014). Involvement of plant endogenous ABA in *Bacillus megaterium* PGPR activity in tomato plants. *BMC Plant Biology* **14**, 36.

Porras-Soriano, A., Soriano-Martín, M. L., Porras-Piedra, A., and Azcón, R. (2009). Arbuscular mycorrhizal fungi increased growth, nutrient uptake and tolerance to salinity in olive trees under nursery conditions. *Journal of Plant Physiology* **166**, 1350–1359.

Pozo, M. J., and Azcón-Aguilar, C. (2007). Unraveling mycorrhiza-induced resistance. *Current Opinion in Plant Biology* **10**, 393–398.

Rabie, G. H., Aboul-Nasr, M. B., and Al-Humiany, A. (2005). Increased salinity tolerance of cowpea plants by dual inoculation of an arbuscular mycorrhizal fungus *Glomus clarum* and a nitrogen-fixer *Azospirillum brasilense*. *Mycobiology* **33**, 51–60.

Rabie, G. H., and Almadini, A. M. (2005). Role of bioinoculants in development of salt-tolerance of *Vicia faba* plants under salinity stress. *African Journal of Biotechnology* **4**, 210–222.

Redon, P.-O., Béguiristain, T., and Leyval, C. (2009). Differential effects of AM fungal isolates on *Medicago truncatula* growth and metal uptake in a multimetallic (Cd, Zn, Pb) contaminated agricultural soil. *Mycorrhiza* **19**, 187–195.

Rekha, P. D., Lai, W.-A., Arun, A. B., and Young, C.-C. (2007). Effect of free and encapsulated *Pseudomonas putida* CC-FR2-4 and *Bacillus subtilis* CC-pg104 on plant growth under gnotobiotic conditions. *Bioresource Technology* **98**, 447–451.

Requena, N., Perez-Solis, E., Azcón-Aguilar, C., Jeffries, P., and Barea, J.-M. (2001). Management of indigenous plant-microbe symbioses aids restoration of desertified ecosystems. *Applied and Environmental Microbiology* **67**, 495–498.

Rillig, M. C., Mardatin, N. F., Leifheit, E. F., and Antunes, P. M. (2010). Mycelium of arbuscular mycorrhizal fungi increases soil water repellency and is sufficient to maintain water-stable soil aggregates. *Soil Biology and Biochemistry* **42**, 1189–1191.

Ruiz-Lozano, J. M. (2003). Arbuscular mycorrhizal symbiosis and alleviation of osmotic stress. New perspectives for molecular studies. *Mycorrhiza* **13**, 309–317.

Ruiz-Lozano, J. M., Porcel, R., Azcon, C., and Aroca, R. (2012). Regulation by arbuscular mycorrhizae of the integrated physiological response to salinity in plants: New challenges in physiological and molecular studies. *Journal of Experimental Botany* **63**, 4033–4044.

Sakamoto, K., Ogiwara, N., and Kaji, T. (2013). Involvement of autoregulation in the interaction between rhizobial nodulation and AM fungal colonization in soybean roots. *Biology and Fertility of Soils* **49**, 1141–1152.

Sánchez-Blanco, M. J., Ferrández, T., Morales, M. A., Morte, A., and Alarcón, J. J. (2004). Variations in water status, gas exchange, and growth in *Rosmarinus officinalis* plants infected with *Glomus deserticola* under drought conditions. *Journal of Plant Physiology* **161**, 675–682.

Sarath, G., Pfeiffer, N. E., Sodhi, C. S., and Wagner, F. W. (1986). Bacteroids are stable during dark-induced senescence of soybean root nodules. *Plant Physiology* **82**, 346–350.

SchÜßLer, A., Schwarzott, D., and Walker, C. (2001). A new fungal phylum, the Glomeromycota: Phylogeny and evolution. *Mycological Research* **105**, 1413–1421.

Serraj, R., Roy, G., and Drevon, J. J. (1994). Salt stress induces a decrease in the oxygen uptake of soybean nodules and in their permeability to oxygen diffusion. *Physiologia Plantarum* **91**, 161–168.

Serraj, R., Vasquez-Diaz, H., Hernandez, G., and Drevon, J.-J. (2001). Genotypic difference in the short-term response of nitrogenase activity (C_2H_2 reduction) to salinity and oxygen in the common bean. *Agronomie* **21**, 645–651.

Sharifi, M., Ghorbanli, M., and Ebrahimzadeh, H. (2007). Improved growth of salinity-stressed soybean after inoculation with salt pre-treated mycorrhizal fungi. *Journal of Plant Physiology* **164**, 1144–1151.

Sheng, M., Tang, M., Chen, H., Yang, B., Zhang, F., and Huang, Y. (2008). Influence of arbuscular mycorrhizae on photosynthesis and water status of maize plants under salt stress. *Mycorrhiza* **18**, 287–296.

Sheng, M., Tang, M., Zhang, F., and Huang, Y. (2011). Influence of arbuscular mycorrhiza on organic solutes in maize leaves under salt stress. *Mycorrhiza* **21**, 423–430.

Shrivastava, P., and Kumar, R. (2015). Soil salinity: A serious environmental issue and plant growth promoting bacteria as one of the tools for its alleviation. *Saudi Journal of Biological Sciences* **22**, 123–131.

Singh, L. P., Gill, S. S., and Tuteja, N. (2011). Unraveling the role of fungal symbionts in plant abiotic stress tolerance. *Plant Signaling & Behavior* **6**, 175–191.

Singh, P. K. (2012). Role of glomalin related soil protein produced by arbuscular mycorrhizal fungi: A review. *Agricultural Science Research Journal* **2**, 119–125.

Singleton, P. W., and Bohlool, B. B. (1984). Effect of salinity on nodule formation by soybean. *Plant Physiology* **74**, 72–76.

Slattery, J. F., Coventry, D. R., and Slattery, W. J. (2001). Rhizobial ecology as affected by the soil environment. *Australian Journal of Experimental Agriculture* **41**, 289–298.

Smith, S. E., and Read, D. J. (2008). Mineral nutrition, toxic element accumulation and water relations of arbuscular mycorrhizal plants. *Mycorrhizal Symbiosis* **3**, 145–18.

Smith, S. E., and Read, D. J. (2010). *Mycorrhizal Symbiosis*. Academic Press, Amsterdam, the Netherlands.

Smith, S. E., Robson, A. D., and Abbott, L. K. (1992). The involvement of mycorrhizas in assessment of genetically dependent efficiency of nutrient uptake and use. *Plant and Soil* **146**, 169–179.

Smith, S. E., and Smith, F. A. (2011). Roles of arbuscular mycorrhizas in plant nutrition and growth: New paradigms from cellular to ecosystem scales. *Annual Review of Plant Biology* **62**, 227–250.

Soumare, A., Diop, T., Manga, A., and Ndoye, I. (2015). Role of arbuscular mycorrhizal fungi and nitrogen fixing bacteria on legume growth under various environmental stresses. *International Journal of Biosciences* **7**, 31–46.

Subbarao, G. V., Johansen, C., Jana, M. K., and Rao, J. K. (1990). Comparative salinity tolerance of symbiotically dependent and nitrogen-fed pigeonpea (*Cajanus cajan*) and its wild relative *Atylosia platycarpa. Biology and Fertility of Soils* **10**, 11–16.

Tajini, F., Trabelsi, M., and Drevon, J.-J. (2012). Combined inoculation with *Glomus intraradices* and *Rhizobium tropici* CIAT899 increases phosphorus use efficiency for symbiotic nitrogen fixation in common bean (*Phaseolus vulgaris* L.). *Saudi Journal of Biological Sciences* **19**, 157–163.

Talaat, N. B., and Shawky, B. T. (2014). Protective effects of arbuscular mycorrhizal fungi on wheat (*Triticum aestivum* L.) plants exposed to salinity. *Environmental and Experimental Botany* **98**, 20–31.

Tang, M., Chen, H., Huang, J. C., and Tian, Z. Q. (2009). AM fungi effects on the growth and physiology of *Zea mays* seedlings under diesel stress. *Soil Biology and Biochemistry* **41**, 936–940.

Tavasolee, A., Aliasgharzad, N., SalehiJouzani, G., Mardi, M., and Asgharzadeh, A. (2011). Interactive effects of Arbuscular mycorrhizal fungi and rhizobial strains on chickpea growth and nutrient content in plant. *African Journal of Biotechnology* **10**, 7585–7591.

Tejera, N. A., Campos, R., Sanjuan, J., and Lluch, C. (2004). Nitrogenase and antioxidant enzyme activities in *Phaseolus vulgaris* nodules formed by *Rhizobium tropici* isogenic strains with varying tolerance to salt stress. *Journal of Plant Physiology* **161**, 329–338.

Torimitsu, K., Hayashi, M., Ohta, E., and Sakata, M. (1985). Effect of K^+ and H^+ stress and role of Ca^{2+} in the regulation of intracellular K^+ concentration in mung bean roots. *Physiologia Plantarum* **63**, 247–252.

Tu, J. C. (1981). Effect of salinity on rhizobium-root-hair interaction, nodulation and growth of soybean. *Canadian Journal of Plant Science* **61**, 231–239.

Vadez, V., Rodier, F., Payre, H., and Drevon, J. J. (1996). Nodule permeability to O_2 and nitrogenase-linked respiration in bean genotypes varying in the tolerance of N_2 fixation to P deficiency. *Plant Physiology and Biochemistry (France)* 34, 871–878.

Valdenegro, M., Barea, J. M., and Azcón, R. (2001). Influence of arbuscular-mycorrhizal fungi, *Rhizobium meliloti* strains and PGPR inoculation on the growth of *Medicago arborea* used as model legume for re-vegetation and biological reactivation in a semi-arid mediterranean area. *Plant Growth Regulation* **34**, 233–240.

Van Hoorn, J. W., Katerji, N., Hamdy, A., and Mastrorilli, M. (2001). Effect of salinity on yield and nitrogen uptake of four grain legumes and on biological nitrogen contribution from the soil. *Agricultural Water Management* **51**, 87–98.

Wang, X., Pan, Q., Chen, F., Yan, X., and Liao, H. (2011). Effects of co-inoculation with arbuscular mycorrhizal fungi and rhizobia on soybean growth as related to root architecture and availability of N and P. *Mycorrhiza* **21**, 173–81.

Waqas, M., Khan, A. L., Kamran, M., Hamayun, M., Kang, S.-M., Kim, Y.-H., and Lee, I.-J. (2012). Endophytic fungi produce gibberellins and indoleacetic acid and promotes host-plant growth during stress. *Molecules* **17**, 10754–10773.

Whipps, J. M. (2004). Prospects and limitations for mycorrhizas in biocontrol of root pathogens. *Canadian Journal of Botany* **82**, 1198–1227.

Wright, S. F., and Upadhyaya, A. (1998). A survey of soils for aggregate stability and glomalin, a glycoprotein produced by hyphae of arbuscular mycorrhizal fungi. *Plant and Soil* **198**, 97–107.

Wu, Q.-S., Zou, Y.-N., and Abd-Allah, E. F. (2014). Mycorrhizal association and ROS in plants. In *Oxidative Damage to Plants*, pp. 453–475. Elsevier, Saint Louis, MO.

Xavier, L. J. C., and Germida, J. J. (2002). Response of lentil under controlled conditions to co-inoculation with arbuscular mycorrhizal fungi and rhizobia varying in efficacy. *Soil Biology and Biochemistry* **34**, 181–188.

Yamato, M., Ikeda, S., and Iwase, K. (2008). Community of arbuscular mycorrhizal fungi in a coastal vegetation on Okinawa island and effect of the isolated fungi on growth of sorghum under salt-treated conditions. *Mycorrhiza* **18**, 241–249.

Yasmeen, T., Tariq, M., Iqbal, S., Arif, M. S., Riaz, M., Shahzad, S. M., Ali, S., Noman, M., and Li, T. (2019). Ameliorative capability of plant growth promoting rhizobacteria (PGPR) and arbuscular mycorrhizal fungi (AMF) against salt stress in plant. In *Plant Abiotic Stress Tolerance*, pp. 409–448. Springer, Cham, Switzerland.

Yousef, A. N., and Sprent, J. I. (1983). Effects of NaCl on growth, nitrogen incorporation and chemical composition of inoculated and NH_4NO_3 fertilized *Vicia faba* (L.) plants. *Journal of Experimental Botany* **34**, 941–950.

Zahran, H. H. (1999). Rhizobium-legume symbiosis and nitrogen fixation under severe conditions and in an arid climate. *Microbiology and Molecular Biology Reviews* **63**, 968–989.

Zahran, H. H., and Sprent, J. I. (1986). Effects of sodium chloride and polyethylene glycol on root-hair infection and nodulation of *Vicia faba* L. plants by *Rhizobium leguminosarum*. *Planta* **167**, 303–309.

Zuccarini, P. (2007). Mycorrhizal infection ameliorates chlorophyll content and nutrient uptake of lettuce exposed to saline irrigation. *Plant Soil and Environment* **53**, 283.

Zuccarini, P., and Okurowska, P. (2008). Effects of mycorrhizal colonization and fertilization on growth and photosynthesis of sweet basil under salt stress. *Journal of Plant Nutrition* **31**, 497–513.

16

Microbial-Mediated Abiotic Stress Tolerance in Soybean Plants

Abhishek Bharti, Hemant S. Maheshwari, Mahaveer P. Sharma, and Anil Prakash

CONTENTS

16.1 Introduction

The abiotic and biotic stresses are the potential hindrance for attaining the sustainability in crop production to feed an ever-growing population (Lesk et al., 2016). Abiotic stresses include drought, low/high temperature, salinity and acidity, light intensity, water submergence, low oxygen, and nutrient starvation, which severely limit the plant growth (Bailey-Serres et al., 2008). Global climate change again escalates the chances of these stresses occurring, leading to a potential threat to food security (Anjum et al., 2011). Among all known abiotic stresses, drought and salinity stress are a major threat in soybean-growing areas across the world. The soybean crop is the most drought-sensitive crop (Valliyodan et al., 2017) and may decrease the soybean yield up to 40% (Lamaoui et al., 2018). But these losses may increase up to 80% depending upon the stage of the crop and severity of drought (Dias et al., 2012). Whereas, salinity level higher than 5 dS m^{-1} severely damages crop yield in soybean (Miransari et al., 2013). Flooding stress also contributes to major abiotic stress in soybean, which impairs plant growth and production (Khatoon et al., 2012). Heavy metals negatively affect crop productivity and quality of the produce (Shahid et al., 2015).

The soybean [*Glycine max* (L.) Merrill] is a leading oilseed legume crop in the world as well as in India (Tidke et al., 2015). The soybean contributes >60% oilseed production and >70% total protein consumption in the world. Soybean world production is projected at about 341.76 million metric tons of in 2019/20, which is 16.88 million metric tons lower than the previous year crop season (Global Market Analysis, 2020). The soybean seed is highly nutritious and named a golden bean due to its high oil (20%) and protein contents (40%) and acts as the cheapest source of protein (Toorchi et al., 2009). Globally, major soybean-producing countries are the United States, Brazil, Argentina, China, India, Japan, and Vietnam (Tran & Mochida, 2010). In India, it is grown predominantly as a rainfed crop in central and peninsular parts of India in vertisols soils (Agarwal et al., 2013). Its area and production have increased considerably after introduction in the late 1960s, but the productivity remained at 1.06 t/ha^{-1}, whereas the world average is over 2.0 t/ha. The low productivity is mainly due to various biotic, abiotic, and socioeconomic causes (Bhatia et al., 2014).

The real challenges before the plant-soil scientists are global food security and sustainability in production without damaging the natural resources and ecosystem functioning at low cost (Altieri et al., 2004). The harmful impact of non-judicious use of chemical fertilizers can be experienced on altered soil biodiversity and the presence of a contaminant in the food chain (González-Rodríguez et al., 2011). Now, considering the present scenario of climatic aberrations and abiotic stresses in soybean, the exploitation of a microbial entity for counteracting abiotic stress challenge is a need of an hour. This chapter highlights the effect of abiotic stresses on soybean plants. Key bacterial and fungal-mediated mechanisms are involved in conferring tolerance against abiotic stress and its exploitation in soybean. What are the gaps in the existing knowledge and how can we direct our future research programs against abiotic stress without damaging the natural resources? In a nutshell, this chapter will help readers by providing a comprehensive understanding of microbial interventions to confer tolerance to abiotic stresses in soybean.

16.2 Abiotic Stresses Limiting Soybean Production

Multiple environmental stresses such as drought, salinity, high temperature, chilling, waterlogging, metal toxicity, and ultraviolet radiation affect soybean production adversely. The detailed review of work is provided under the following subheads.

16.2.1 Drought and Heat Stress

The drought stress and heat stress are linked together, and the cumulative effect is more detrimental to plants (Dreesen et al., 2012). Environmental prediction models indicate the increasing severity and frequency of the drought in coming years due to global climate change (Lesk et al., 2016). The water requirements for soybean crop are about 450–700 mm (Dogan et al., 2007). The effect of drought stress response of plants is largely determined by the magnitude of stress and developmental stage of plant growth and plant species; even genotype within the species differs for tolerance (Demirevska et al., 2009). When the evapotranspiration rate exceeds moisture retained in the soil, the plants undergo drought stress (Anjum et al., 2011). The drought affects the absorption of water and nutrients from the soil, which results in reduced root growth and uptake of immobile elements such as phosphorus. If sufficient moisture is not present in the soil, then it will affect the germination process and seedlings will not emerge out—and thereby poor crop stands in the field (Farooq et al., 2009). Drought stress in soybean decreases the leaf water potential, photosynthetic rate, stomatal conductance rate, and transpiration rate (Ohashi et al., 2006). The growth parameters such as root and shoot length, fresh and dry weight, and vegetative growth are found to be decreased in crop under drought stress (Manikavelu et al., 2006). The chlorophyll content and relative water content in leaf were decreased under water stress (Munne-Bosch and Alegre, 2000). The accumulation of ABA hormone occurs in the root, which translocated via the xylem to leaf and thereby closed the stomata to prevent transpiration losses (Vishwakarma et al., 2017). During drought stress, stomatal conductance decreases; consequently, less water is transpired from the leaves, resulting in heat stress. Such a combined effect is more pronounced in tropical and subtropical countries (De Boeck et al., 2015). The leaves formed under drought stress are small and have a high stomatal density (Tisne et al., 2010). Under drought stress, the biological nitrogen fixation by *Bradyrhizobium japonicum* is retarded and consequently, there are decreases in 30%–40% nitrogen content in leaves and pods (Streeter, 2003). When drought coincides with reproductive stages, it has been reported that flowering, pod initiation, and pod filling that have medium-sized seeds and reduction in germination percentage up to 80% (Samarah et al., 2009). Drought results in decrease in the number of pods per plants and test weight of the seed (Pedersen and Lauer, 2004). Drought combined with heat stress significantly reduced soybean production by reducing seed size and number, test weight, and seed quality (Sehgal et al., 2018). Drought may also affect root- associated microbiome both directly by modulating moisture availability, as well as indirectly by altering soil chemistry and plant phenotypes (Naylo and Coleman-Derr, 2018). The key mechanisms of PGPR involved in drought tolerance are by the production of exopolysaccharides (EPS), phytohormones such as IAA and ACC-deaminase, accumulation of osmolytes, antioxidant enzymes synthesis, modulation of drought-responsive genes, and altering the root system architecture (Ngumbi and Kloepper, 2016; Vurukonda et al., 2016). Moreover, the high trehalose accumulated in plants triggers reactive oxygen species (ROS) and/or helps in the protection of macromolecular structures against the destabilizing effect of anhydrobiotic conditions (Bharti et al., 2016). Reduction in plant ethylene level during stress by bacterial ACC-deaminase-producing bacteria help in drought stress alleviation (Arshad et al., 2008). Many PGPR or endophytes experimentally tested in soybean and their mechanisms have been compiled in Table 16.1.

TABLE 16.1

Examples of Bacterial Formulations Tested to Alleviate Abiotic Stress in Soybean

S/N.	Microorganisms	Stress	Mechanisms	References
1	*Bacillus subtilis* and *Bacillus licheniformis*	Drought	Photosynthetic parameters and radiation use efficiency improved.	Mondani et al. (2019)
2	*Bradyrhizobium japonicum* and AM Fungi	Drought	Root system improved and photosynthetic efficiency increased.	Takács et al. (2018)
3	*Bacillus amyloliquefaciens* H-2-5	Salinity	Enhanced phosphate solubilization, GAs, salicylic acid, jasmonic acid, and proline production	Kim et al. (2017)
4	*Sphingomonas* spp. LK11	Drought	Phytohormones and synthesizes trehalose Gene expression of (GmDREBa and GmDREB2) and MYB (myeloblastosis) transcription factor (GmMYBJ1) were modulated.	Asaf et al. (2017)
5	Endophytic bacterium *Pseudomonas putida PIR3C* and *Roultella terrigena* PCM8	Salinity	Improved physiological parameters and bacterium enhanced ACC-deaminase activity and lowered ethylene levels.	Simarmata et al. (2018)

(Continued)

TABLE 16.1 (*Continued*)

Examples of Bacterial Formulations Tested to Alleviate Abiotic Stress in Soybean

S/N.	Microorganisms	Stress	Mechanisms	References
6	*Bacillus aryabhattai* SRB02	Oxidative and nitrosative stress	ABA-induced stomatal closure, increased levels of IAA, Jasmonic acid and antioxidant enzymes	Park et al. (2017)
7	*Bradyrhizobium japonicum* BDYD1 and the endophyte *Stenotrophomonas rhizophila* ep-17	Salt stress	Enhanced plant biomass and nutrient uptake	Egamberdieva et al. (2016)
8	*Bradyrhizobium japonicum* strains	Drought	Phosphorus solubilization, siderophore production	Marinkovi et al. (2013)
9	*Pseudomonas* sp. strain AK-1, and *Bacillus* spp. strain SJ-5	Salinity	IAA, EPS, and ACC-deaminase production and increased accumulation of proline and other photosynthetic and biochemical parameters	Kumari et al. (2015)
10	*Pseudomonas* sp. strain AK-1	Salinity	EPS production, proline accumulation, chlorophyll formation, modulation of K± transporter (HKT1) gene, and upregulation in shoot and downregulation in root	Kasotia et al. (2016)
11	*Bacillus subtilis* UFGS1, *Bacillus thuringiensis* UFGS2, and *Bacillus cereus* UFGRB2 and UFGRB3	Drought	Increased photosynthetic and physiological performance Differential expression of drought stress genes such as Gmp5cs, Gmgols, Gmdreb1a, and Gmereb	Martins et al. (2018)
12	Endophytes Fungus *Paecilomyces formosus* LHL 10 and Bacteria *Sphingomonas* sp. LK11	Aluminum and Zinc Stresses	Physiological and photosynthetic parameters improved, antioxidant enzymes increased macronutrient uptake, modulate gene expression levels increased gibberellins levels.	Bilal et al., (2018)
13	Endophyte *Pseudomonas putida* PIR3C and *Roultella terrigena* PCM8	Salinity	ACC-deaminase activity and improved physiological parameters	Simarmata et al. (2018)
14	*Pseudomonas simiae*	Salinity	VOCs, 4-nitoguaiacol and quinolone help in germination of seed under salinity.	Vaishnav et al. (2016)
15	*Azotobacter* and *Azospirillum*	Drought	Antioxidant enzymes such as CAT, SOD, and glutathione peroxidase activity and proline accumulation increased.	Zakikhani et al. (2012)

16.2.2 Salinity

Soybean is classified under moderately salt-tolerant crop and sensitive to salinity stress depending upon the stage of the crop. The soybean seedling stage is more sensitive than the seed germination stage (Hosseini et al., 2002). Physiological and morphological parameters such as plant length, biomass, leaf size, number of internode and branches, and test weight of soybean seeds are adversely affected (Miransari and Smith, 2009). The protein and oil content of soybean seed is decreased under salinity stress (Wan et al., 2002). Salinity results in accumulation of excess Na^+/ Cl^- ion in cytoplasm and causes osmotic imbalances and thus, lesser water absorption into the root (Carillo et al., 2011). These interfere with many plant physiological processes such as transpiration, photosynthesis, and nutrients transportation inside the plants (Munns, 2002). The decrease in chlorophyll content, leaf surface area, stomatal conductance, and reduced PS-II activity occur (Jamil et al., 2007). Osmotic stress results in closure of the stomata due to abscisic acid (ABA) synthesis, which results in less transpiration in the plants (Fricke et al., 2006). Salinity stress results in the accumulation of reactive oxygen species (ROS), which causes membrane damage and lipid peroxidation (Islam et al., 2016). Higher salinity restricts plant growth, nitrogen fixation rate, and rhizobial population in the soil (Kim et al., 2017). Salinity interferes with the signal exchange between soybean and *Bradyrhizobium japonicum* during symbiotic nitrogen fixation process (Miransari., 2015 b).

PGPR or endophytic microorganism alleviates salinity stress by osmolytes accumulation, producing ROS-scavenging enzymes, phytohormone Indole-3-acetic acid (IAA) and cytokinins production, and enhancing the efficiency of micronutrients and macronutrients absorption, thereby managing osmotic or ionic stress caused by salinity in the soil (Ilangumaran et al., 2017; Vaishnav et al., 2018). Salt-tolerant PGPR upon root colonization enhanced root-shoot communication and maintained water and nutrients regulation, modified root system architecture (RSA), delayed leaf senescence and carbohydrate metabolism. PGPR increased the activity of K^+ transporter and H^+ exchangers for salt exclusion (Ilangumaran and Smith, 2017).

PGPR *Bacillus firmus* SW5 help in the alleviation of salinity stress in soybean by modification of root system architecture and upregulating antioxidant and other stress-related genes. This bacterium produces phytohormone IAA and 1-aminocyclopropane-1-carboxylic acid (ACC) deaminase, antioxidant enzymes, and volatile organic compounds (El-Esawi et al., 2018). Similarly, PGPR, *Pseudomonas* sp. strain VS1 produces IAA hormone, which forms more lateral roots and helps soybean in better seed emergence under salinity stress (Kasotia et al., 2012). ACC deaminase-producing endophytic bacterial reduces the ethylene levels and promotes plant health under salinity stress (Vaishnav et al., 2018). Bacterium *Pseudomonas simiae* AU-M4 (MTCC-12057) enhances salt tolerance in soybean and mung bean. The key mechanism used by this bacterium was the accumulation of total soluble sugar, proline, salicylic acid, and by lowering ethylene and ABA levels in the inoculated plants. Moreover, DREB2, P5CS, and HKT1 genes were found to be upregulated upon inoculation (Vaishnav et al., 2018). Co-inoculation of *Bradyrhizobium japonicum* USDA110 and *Pseudomonas putida* TSAU1 in soybean modify root system architecture for better nitrogen and phosphorus absorption and help in higher nodule formation (Egamberdieva et al., 2013).

16.2.3 Acidity

Soybean crop production was severely hampered below pH 5.8 (Pedersen, 2004). Whereas, the nodulation process is affected below pH 4.7 (Mubarik and Sunatmo, 2014). Nodule formation is decreased by more than 90% and nodule dry weight by more than 50% in acidic soils (Alva et al., 1990). The acid soils cause nutrients imbalances and have the toxic concentration of iron, aluminum, and deficiency of calcium, magnesium, and phosphorus in the soil (Zheng et al., 2010). Nitrogen-fixing rhizobia growth, survivability, abundance, and competitiveness are influenced by low pH (Vargas and Graham, 1988). Acid tolerant *Bradyrhizobium* spp. isolated from soybean plant were evaluated for symbiotic efficiency under low pH for N2 fixation. The isolates showing maximum tolerance to acidity were better in fixing nitrogen biologically (Appunu and Dhar, 2006). The calcium is essential for rhizobial attachment, infection. and infection thread formation to root hairs (Caetano-Anolles et al., 1989). The combination of liming and mycorrhizal application, *Glomus intraradices* application was found to enhance phosphorus uptake and plant growth in tropical acid soils (Nurlaeny et al., 1996).

16.2.4 Low Temperature

The optimum temperature for soybean growth, nodulation, and symbiotic nitrogen fixation is 20°C–30°C (Dashti et al., 1997). Low temperature or temperature below freezing is sensitive to soybean plant growth (Meyer et al., 2001). The physiological process such as photosynthesis and nitrogen fixation are severely affected (Krüger et al., 2014). Cold stress during the flowering stage leads to a lesser number of pod formation and poor seed setting (van Heerden et al., 2003). Under cold stress, the formation of ROS occurs, which causes oxidative stress and severely affects the metabolic activity of plants (Jajic et al., 2015). Cold-tolerant PGPR are *Pantoea dispersa, Pseudomonas fragi, and Pseudomonas lurida, Serratia marcescens* (Selvakumar et al., 2012). Three *Bacillus* strains, *B. subtilis* NEB4, *B. subtilis* NEB5, and *B. thuringiensis* NEB17 isolated from a nodule of soybean were co-inoculated with *Bradyrhizobium japonicum* enhanced nodule number, nodule weight, plant biomass, and grain yield in soybean under low root-zone temperatures. Out of three bacterial cultures, only *B. thuringiensis* NEB17 showed consistent results (Bai et al., 2003). Co-inoculation of isoflavonoid genistein and *Bradyrhizobium japonicum* negated the effect of suboptimum root temperature (Miransari and Smith, 2008). Similarly, the combination of genistein, *Bradyrhizobium*, or PGPR has a positive influence on soybean plant growth and symbiotic nitrogen fixation; photosynthetic rate under suboptimum root temperature even below 15°C was also reported (Dashti et al., 2016).

16.2.5 Water-Submergence

The soybean plant underwater submergence creates anaerobic or anoxia condition in the plants, which hinders root growth and affects the survival of plants (Nakayama et al., 2004). Across the globe, the losses occurred in soybean due to flooding estimated to 25% (Mustafa and Komatsu, 2014). This impairs photosynthesis process, nitrogen fixation, pod formation, and grain filling, which ultimately resulted in lower yields (Hasanuzzaman et al., 2016). Flooding stress leads to starch accumulations in the leaves, which affected photosynthesis rate (Mutava et al., 2015). The important critical crop growth stages such as seed germination, vegetative (V2-V4), and reproductive phases are adversely affected due to flooding stress (VanToai et al., 1994). Water submergence effect on soybean plant biomass and pod formation stage is more damaging than the blooming stage (Ara et al., 2015). Under low oxygen, ACC synthase enzyme is produced, which forms ethylene in the plant (Tewari and Arora, 2013). Underwater submergence, ACC-deaminase positive bacterial isolates lower the amount of plant ethylene by the breakdown of ACC (Grichko and Glick, 2001). Soybean crop inoculated with *Bradyrhizobium japonicum* enhances nitrogen fixation under flooding stress (Beutler et al., 2014). Microaerophilic PGPR, *Azospirillum* can also promote plant growth under flooding stress (Sahoo et al., 2014). Some species of arbuscular mycorrhizal fungi (AMF) can survive under flooding and colonize soybean root (Isobe et al., 2011). The use of AMF for flooding tolerance should be investigated in soybean.

16.2.6 Heavy Metal/Toxic Metal

Heavy metals such as cadmium, chromium, copper, lead, and mercury at higher concentrations cause chlorosis and oxidative stress in soybean (Mao et al., 2018) The decrease in overall plant size, destruction of chlorophyll, impairment of enzyme activity, and decrease in uptake of nutrients from soil were reported under heavy metal stress. Heavy metals affect the symbiotic nitrogen fixation process in the soybean plant (Xiong et al., 2010), and particularly cadmium has a negative impact on soybean plant growth (Chmielowska-Bąk et al., 2018). Soil microbial community in rhizosphere plays a significant role in bioremediation of heavy metal-contaminated soils (Lasat, 2002). Microorganisms tolerate heavy metals by effluxing metal ions, extracellular and metal sequestration, and transforming or reducing to less toxic form (Ianeva et al., 2009).

PGPR affects mobility and availability of toxic metal by altering soil pH and secreting chelating compound and siderophores (Kuiper et al., 2004). AMF tolerate heavy metal stress by absorbing metals, providing nutrients from soils, altering soil conditions, and interacting with beneficial microbes such as *Bradyrhizobium* spp. and many PGPR (Miransari., 2015 a). The mycorrhizal fungi help in alleviating stress by reduced uptake of cadmium and increase uptake of phosphorus (Cui et al., 2019). AMF-inoculated soybean plants with a lower uptake of manganese and higher uptake of phosphorus in root were observed (Nogueira et al., 2007). Co-inoculation of PGPR *Bradyrhizobium japonicum* E109 and *Azospirillum brasilense* Az39 were examined as a synergistic effect in tolerating arsenic toxicity in soybean plants, and inoculation improved the plant growth; nitrogen content and lesser arsenic translocation to the aerial part were recorded (Armendariz et al., 2019). A compatible phytohormone-producing consortium of endophytic fungal and bacterial symbionts *Paecilomyces formosus* LHL10 and *Sphingomonas* sp. LK11 were also tested in soybean against aluminum and zinc stress. The inoculation improved plant physiological parameters, reduced oxidative stress via synthesis of antioxidative enzymes and gibberellic acid, and lowered ABA level and jasmonate levels. Moreover, downregulation of metal ATPase genes and upregulation of ariadne such as ubiquitin ligase gene GmARI1 occurred under stress and were also observed (Bilal et al., 2018). Phytohormone IAA and GA3-producing fungal isolate *Galactomyces geotrichum* WLL1 were mixed with pyrolyzed biochar derived from pine hardwoods and were treated in soybean, which ameliorates heavy metal stress by inducing systemic resistance (Waqas et al., 2014).

16.2.7 Nutritional Deficiency

Plant nutrients' (micronutrients and macronutrients) availability is higher at neutral soil pH than higher pH. Phosphorus gets fixed in iron or aluminum phosphate in acidic soils, and calcium and magnesium phosphate in alkaline soils (Ch'ng et al., 2014). Thus, phosphorus availability is higher at 6.5–7.5 pH. Similarly, in acidic soils, the biological nitrogen fixation process is retarded. Molybdenum is an important element in formation of the nitrogen-fixing nitrogenase enzyme. However, its availability is affected by acidic soils. However, reclamation of acidic soils by lime or gypsum is not cost effective for reclaiming subsoil acidity. Alternatively, microbial means of nutrient management and soil reclamation has gained attention in the past. The beneficial PGPR, endophytic bacteria and fungi, nitrogen fixer *Bradyrhizobium* spp. and AMF play a role in the mobilization of micronutrients and macronutrients for better plant growth in soybean (Sharma et al., 2012, 2016). They provide nitrogen via symbiotic *Bradyrhizobium japonicum* and associative or free-living PGPR in the soybean. Two *Bacillus aryabhattai* strains MDSR7 and MDSR14 were also tested in soybean for their increased zinc solubilization, mobilization, and plant growth (Ramesh et al., 2014).

16.3 Microbe-Mediated Alleviation of Abiotic Stresses and Their Mechanism

16.3.1 Role of PGPR and Bacterial Endophytes for Abiotic Stress Tolerance and Their Mechanisms

The plant growth-promoting rhizobacteria (PGPR) and bacterial endophytes help in the plant growth promotion by various direct and indirect mechanisms (Vurukonda et al., 2016) (Figure 16.1). They confer tolerance to abiotic and biotic stresses by various direct mechanisms through the production of phytohormones, nutrient acquisition via phosphorus solubilization and siderophore production, biologically nitrogen fixation. Whereas, indirect mechanisms are via inhibition of pathogen by competition for nutrients, production of HCN or by antibiotic production, and induction of systemic acquired resistance (SAR) or induced systemic resistance (ISR) (Etesami and Maheshwari, 2018). Induced Systemic Tolerance (IST) is the term being used for beneficial microbe-mediated physical and chemical changes in the plants against abiotic stresses (Vurukonda et al., 2016). The importance of microorganisms to alleviate abiotic

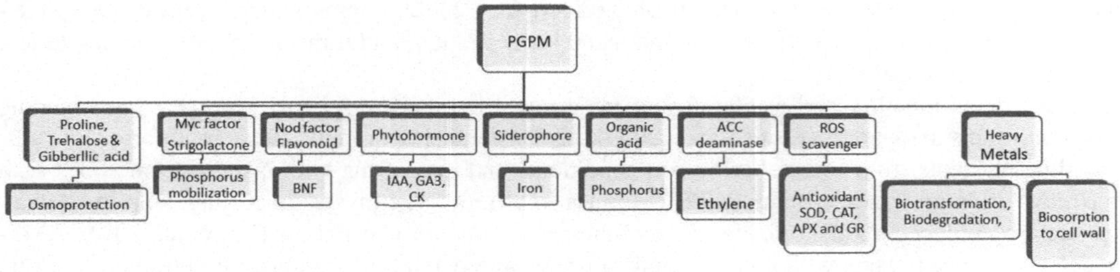

FIGURE 16.1 Microbes mediated abiotic stress tolerance mechanism.

stresses in plants continues to be a very important area of research (Souza et al., 2015). The detailed investigation of PGPR and endophytic bacteria for abiotic stress tolerance can be summarized as follows: (i) modifications in root system architecture (RSA) via microbially produced IAA and ACC-deaminase phytohormone, and nitric acid (Long et al., 2008); (ii) osmolyte accumulations for osmotic adjustments such as glycine betaine (Yuwono et al., 2005); (iii) maintaining higher K^+/Na^+ ratios under salt stress by *Azospirillum* (Hamdia et al., 2004); (iv) induction of induced systemic resistance (ISR) in the host (Walters and Fountaine, 2009); (v) reducing membrane potential of cell membrane by altering saturation pattern of phospholipid and thereby, changes in proton efflux occur due to change in elasticity, e.g., *Azospirillum* (Pereyra et al., 2006), (vi) increased availability of nutrients by modulating rhizospheric pH via organic acid secretion (Gadd et al., 2004); (vii) bacterial exopolysachharides against drought, salinity, and metal toxicity (Ashraf et al., 2004); (viii) biological nitrogen fixation and siderophore production for iron nutrition (Vurukonda et al., 2016), (ix) heavy metals detoxification by intracellular sequestration, metal chelator siderophore, and sorption in cellular components (Dimpkpa et al., 2009a); and (x) bacterial migration from rhizoplane to the rhizosphere and binding heavy metals in to unavailable forms (Pishchik et al., 2002). Multifarious plant growth-promoting soybean endophyte *Pseudomonas aeruginosa* (LSE-2) is an IAA producer, phosphorus solubilizer, ACC-deaminase, and biofilm former that can be used with nitrogen-fixing *Bradyrhizobium* sp. (LSBR-3) for nutrient acquisition and better plant and soil health (Kumawat et al., 2019). The PGPR bacterium isolated from soybean seed, *Bacillus amyloliquefaciens* strain LL2012 was co-inoculated with natural symbiont *Bradyrhizobium japonicum* and was found to enhance nodulation and crop yield due to their higher compatibility *of B. japonicum* in soybean (Masciarelli et al., 2014). Similarly, endophytic bacteria *Bacillus megaterium* LNL6 and *Methylobacterium oryzae* CBMB20 isolated from nodules of *Lesperdeza* spp. and rice leaf, respectively, were highly compatible when co-inoculated with *Bradyrhizobium japonicum* MN110 in soybean plants. The results showed higher leghemoglobin content, ARA, and total plant nitrogen content (Subramanian et al., 2015). We can say that the interactive effect of PGPR or any endophyte with *Bradyrhizobium japonicum* improves overall symbiotic nitrogen fixation synergistically. We discussed the key mechanism utilized by PGPR or endophytic bacteria in the soybean plant.

16.3.2 Phytohormonal Activity

The PGPR produces many phytohormones such as auxins, gibberellins, cytokinin, and ACC-deaminase. The key hormone IAA is produced by many PGPR such as *Azospirillum*, which helps in the modification of root system architecture by forming lateral roots (Dimpka et al., 2009b). The *Bacillus subtilis* produces cytokinin under drought stress by maintaining high water content (Liu et al., 2013). The bacterium *Pseudomonas Putida* H-2-3 strain secretes gibberellic acid under drought and salinity stress in the soybean plant. The inoculation helps in the improvement of the plant fresh and dry weight of shoots and leaf chlorophyll content. The modulation of the hormone-like ABA, salicylic acid, and jasmonic acid were also observed. Moreover, this bacterium produces antioxidant enzymes that scavenge toxic reactive oxygen species (ROS) compounds (Kang et al., 2014). Similarly, ACC-deaminase enzyme produced by many bacteria, which cleaves ACC, an ethylene precursor into α-ketobutyrate and ammonia and thereby, lowers the ethylene level in the plants (Glick, 2014). Reduced ethylene levels help in better root elongation and higher physical colonization of beneficial bacteria in the roots (Hardoim et al., 2008). The ACC-deaminase positive bacteria alleviate metal stress by lengthening the plant-root (Gontia-Mishra et al., 2014). Ali et al. (2014) reported that, drought tolerance imparted by *Pseudomonas* spp. through the production of EPS, ACC-deaminase, and other PGP traits. The *Azosprillum* bacterium releases a volatile compound, e.g., nitric acid, which acts as a signal for more IAA production in the plants (Molina-Favero et al., 2008). The combination of *Azospirillum brasilense* with two PGPR namely, *Rhizobium japonicum* and *Azotobacter chroococcum*, was tested in soybean under water stress and lower accumulation of ABA hormone and polyamines, viz. putrescine and spermidine, were observed (Zahedi and Abbasi, 2015).

16.3.3 Exopolysaccharide Production

The bacterial exopolysaccharide (EPS) production helps in the alleviation of drought, salinity stress, and heavy metal stress. EPS is the important trait required for the survival of bacteria under drought stress, microbial aggregation, and beneficial plant-microbe interaction (Naseem et al., 2018). Bacterial EPS has also played a role in biodetoxification of heavy metal-contaminated sites through biosorption mechanism (Gupta and Diwan, 2017). It has been reported that the bacteria such as *Pseudomonas, Bacillus*, and *Rhizobia* produce EPS, which maintains higher soil moisture and helps in macroaggregates formation in soil and enhanced plant biomass. For example, the EPS-forming *Azospirillum* improves soil structure under drought stress by forming soil aggregation and rhizo-sheath (Sandhya et al., 2009). When EPS-forming *Proteus penneri* (Pp1), *Pseudomonas aeruginosa* (Pa2), and *Alcaligenes faecalis* (AF3) were used in maize crop, they improved morphological, physiological, and biochemical parameters of plants. These EPSs also bind with sodium cation and improve root and shoot biomass in wheat and thereby, impart salinity tolerance (Ashraf et al., 2004). Under simulated drought conditions, the mixed inoculation of catalase and EPS-producing *Rhizobium leguminosarum* (LR-30), *Mesorhizobium ciceri* (CR-30 and CR-39), and *Rhizobium phaseoli* (MR-2) showed higher growth, biomass, and drought tolerance index of the wheat seedlings due to enhanced production of IAA, which increases root length of the seedling (Hussain et al., 2014).

16.3.4 P Solubilization and Siderophore Production

The phosphate solubilizing bacteria (PSB) helps in the solubilizing and mineralizing fixed form of phosphorus. In acidic soils, phosphorus gets fixed with iron and aluminum ions. Whereas, under the alkaline condition with calcium and magnesium ions (Jones et al., 1991). The PSB excrete large amounts of low-molecular weight organic acids such as citric, oxalic, malic, and fumaric acids. These enzymes help in the chelating elements and thereby release available forms of phosphate for the plants (Khan et al., 2007). Phosphorus solubilizing activity under drought has been reported in *Pseudomonas* spp. (Sandhya et al., 2010) and *Azospirillum* spp. (Arzanesh et al., 2011). Phosphorus solubilizer *Bacillus cereus* GS6 amended with phosphate-enriched compost (PEC) were found to improve nodulation and nitrogen-fixing efficiency in the soybean plant on arid soils. The combined inoculated plants have higher N, P, and K content in the nodule, shoot, and seed. Moreover, soil analysis showed higher soil organic content, soil dehydrogenase activity, phosphomonoesterase activity, and higher availability of N, P, and K elements in the soil (Arif et al., 2017). Phosphate solubilizing bacteria *Pantoea agglomerans* R-42 isolated from soybean rhizosphere was found to have tolerance to high temperature (5°C–45°C), salt concentration (1%–5%), and pH value (3–11) (Son et al., 2006). Phosphate solubilizing bacteria belonging to *Burkholderia* spp., *Enterobacter* spp., and *Bradyrhizobium* spp. were tested in soybean under greenhouse, and it was found that *Burkholderia* spp. PER2F has higher plant length and N/P ratio (Fernández et al., 2007). Similarly, iron-chelating compound, a siderophore producing PGPR, provides direct benefit to plants by iron nutrition and indirectly by phytopathogen inhibition. Siderophore producing *Bradyrhizobium japonicum* NCIM 2746 strain produces two types of siderophore—hydroxamate and catecholate—which increases germination percentage, chlorophyll content, root length, number of pods, oil, and protein content in soybean (Khandelwal et al., 2002).

16.3.5 Modulation of Antioxidant Enzymes

PGPR application helps in improving drought stress tolerance by modulating the levels of anti-oxidant enzymes favorably (Gusain et al., 2015). Under abiotic stresses, ROS such as hydrogen peroxide (H_2O_2), superoxide anions (O_2 (•–), hydroxyl radical (OH•), and singlet oxygen ($^1O_2$) formation occur inside the plant. This ROS causes oxidative damages and lipid peroxidation of cellular components, which ultimately leads to cell death or apoptosis (El-Beltagi & Mohamed, 2013). The plants have various enzymatic and nonenzymatic enzymes, which play a pivotal role in various abiotic stresses for scavenging toxic ROS (Das and Roychoudhury, 2014). The enzymatic enzymes include superoxide dismutase (SOD), ascorbate peroxidase (APX), catalase (CAT), and glutathione peroxidase (GR), etc. Whereas, nonenzymatic includes carotenoids, tocopherols, and flavonoids (Gill and Tuteja, 2010). The PGPR modulate the expression of ROS scavenging enzymes under various abiotic stresses (Jha and Subramanian, 2014). The fungal and bacterial endophytes were inoculated in chickpea plant, and it was observed that the expression of antioxidant genes, SOD, and dehydrin were modulated for scavenging ROS (Kumari et al., 2018).

16.3.6 Compatible Solute Production

The plant accumulates nontoxic compatible solute, also termed as osmoprotectants, for an osmotic adjustment (Farooq et al., 2009). This helps in stabilization of protein and cell membrane from being denaturation under abiotic stress without interfering with normal cellular function. These compounds are glycine-betaine; amino acids such as proline, ectoine; and sugar molecule such as trehalose (Hussain Wani et al., 2013). The glycine-betaine has a role under salt, drought, cold, and temperature stress (Chen and Murata, 2008). Inoculation of PGPB produces higher amounts of proline amino acid in chickpea under drought conditions (Kumari, 2018). The osmolyte such as proline, sugars, and free amino acids were increased under drought stress upon inoculation (Sandhya et al., 2010). The ABA mediates the synthesis of compatible solute such as proline, which gets accumulated under stress response (Kishor et al., 2005). The trehalose accumulation under salinity stress by *Bradyrhizobium japonicum* bacterium helps in maintaining symbiotic nitrogen fixation (Sugawara et al., 2010). Soybean rhizobia alleviate salinity by accumulating osmolites such as glutamate, ectoine, betaine, and sugars (Gouffi and Blanco, 2000).

16.3.7 Volatile Compounds

The volatile compounds are also shown to impart drought tolerance. The volatile compound, 2R, 3R-butanediol produced by *Pseudomonas chlororaphis* O6 has a role in drought stress tolerance by decreasing stomatal apertures and activating salicylic acid-dependent mechanisms in *Arabidopsis thaliana* plants (Cho et al., 2008). The PGPR bacterium *Pseudomonas simiae* strain AU enhanced seedling growth in soybean by means of volatile compounds, which elicits IST (Vaishnav et al., 2015). The volatiles produced by *B. subtilis* GB03 trigger jasmonic acid and salicylic acid pathways, which scavenges toxic ROS by antioxidant enzymes (Kwon et al., 2010). Similarly, the bacterial role in sodium homeostasis in arabidopsis was also reported (Zhang et al., 2008). It is known that bacterial volatile compounds have a greater role in abiotic stress tolerance, but work on this aspect is less exploited.

16.4 Role of Fungal Endophytes and Their Mechanism for Abiotic Stress Tolerance in Soybean

Fungal endophytes are fungi, which colonize and live inside the plant without producing any observable symptoms (Chandra et al., 2012). Endophytic fungi ameliorate the adverse effect of abiotic stresses such as salt, drought, and heat stress (Khan et al., 2011). Endophytic fungi promote plant growth via improving physiological parameters, modulating phytohormones, altering root exudation pattern, and modifying soil parameters. An endophytic fungus *Piriformospora indica* helps in alleviating abiotic stress in soybean by providing nutrients, and enhanced rhizobial nodulation was confirmed via transcriptomics analysis (Bajaj et al., 2018). This fungus helps in the uptake of nitrogen, phosphorus, and various micronutrients (Yadav et al., 2010). The modulation of phytohormones such as jasmonate, ethylene, salicylic acid, and ABA levels in the plant in response to *Piriformospora indica* for mitigating various abiotic stresses has also been reported (Gill et al., 2016). The drought tolerance potential of endophyte *Piriformospora indica* has been tested in soybean genotypes on the basis of chlorophyll content as a fluorescence characteristic (Rathod et al., 2011). Endophytic fungus *Porostereum spadiceum* AGH786 imparts salinity tolerance in soybean by modulating endogenous gibberellic acid and iso flavones (daidzen and genistein) level (Hamayun et al., 2017). Soybean endophyte *Fusarium verticillioides* RK01 alleviates salinity stress in soybean plants by producing gibberellic acid and scavenging ROS via various antioxidant enzymes such as SOD, CAT, and peroxidase. Endophyte-treated plants showed higher plant biomass and protein content (Radhakrishnan et al., 2013). Phosphate solubilizing fungal endophyte *Fusarium verticillioides* RK01 impart salinity tolerance via increasing salicylic acid content and lowering ABA level in the soybean plants (Radhakrishnan et al., 2015). Endophytic fungus *Penicillium minioluteum* LHL09 inoculated to soybean plant provide salinity tolerance by modulating plant phytohormone levels and flavonoids (Khan et al., 2011). An endophyte isolated from *Chenopodium album* root, *Aspergillus flavus* CHS1 strain showed improved plant growth promotion of soybean plant under salinity. This fungus lowers the ABA and jasmonate level and increases the levels of antioxidant enzymes SOD, CAT, peroxidases, and polyphenol oxidase inside the soybean plant (Asaf et al., 2018). Gibberellins producing endophytic fungus *Penicillium funiculosum* LHL06 were found to be tolerant to higher levels of copper when applied to soybean plant. Under copper stress, fungal-treated soybean plant had higher plant and root biomass, which helped plants in the absorption of carbon, hydrogen, and nitrogen elements. Moreover, the treated plant had higher proline and glutamate levels to scavenge toxic ROS (Khan et al., 2013).

Arbuscular mycorrhizal fungi (AMF) help in alleviating drought-stressed soybean plants by reducing oxidative stress, accumulation of soluble sugar, and proline content (Grümberg et al., 2015). AMF-mediated drought stress tolerance in the plants has been studied by many researchers (Saia et al., 2014). AMF-inoculated soybean plants have higher root-shoot biomass, leaf water potential, and accumulation of proline under drought stress (Porcel and Ruiz-Lozano, 2004). Co-inoculation of AMF and *Bradyrhizobium japonicum* in improving soybean root development, nutrient acquisition, and nodule biomass has been reviewed (Sharma et al., 2016; Meena et al., 2018). AMF *Rhizophagus intraradices* inoculation in soybean plant reduces the arsenic toxicity level by reducing uptake in contaminated soil (Spagnoletti and Lavado, 2015). Herbicide paraquat causes oxidative stress in soybean plant and was regulated by inoculation of AMF, *Glomus mosseae*, and *Glomus intraradices* (Bressano et al., 2010).

16.5 Future Prospects

Understanding underground chemical communication between plant-microbe is pivotal for enhancing the efficiency of PGPR or PGPF or endophytes under abiotic stress. The microbial ecology with respect to growth, multiplication, dispersal, and survivability should be considered for exploiting microorganisms under any abiotic stresses. That is why microbes should be preferably isolated from a stressed soil for bio-inoculants due to a greater possibility of survivability or adaptability. All the dimensions influencing colonization and population dynamics should be checked in stressed soil prior to release in the fields before obtaining the desired result. To date, more emphasis on bacterial and fungal microorganisms has been focused on abiotic stress management. Other soil microbial microorganisms need to be explored as an alternative means of abiotic stress alleviation. Microorganisms should be isolated from plant grows better in the extremely stressed environment for getting abiotic stress tolerance. Multiomics approaches such as genomics, proteomics, metabolomics, and metagenomics should be included for understanding plant-microbe interactions, metabolic pathways, and gene regulations under abiotic stresses.

16.6 Conclusion

This chapter systematically discussed the impacts of various abiotic stresses in soybean growth and production, and how these abiotic stresses are mitigated by various bacterial and fungal bioinoculant using direct and indirect mechanisms in the plants. The performance of PGPR, PGPF, and endophyte can be best exploited if we understand rhizosphere ecology, signalomics, interactomics, colonization behavior, and consortia approaches for abiotic stress alleviation. The holistic approach may help in desired rhizocompetence or endocompetence for enhancing tolerance to stress.

REFERENCES

Agarwal, D. K., Billore, S. D., Sharma, A. N., Dupare, B. U., & Srivastava, S. K. (2013). Soybean: Introduction, improvement, and utilization in India—problems and prospects. *Agricultural Research*, *2*(4), 293–300.

Ali, S. Z., Sandhya, V., & Rao, L. V. (2014). Isolation and characterization of drought-tolerant ACC deaminase and exopolysaccharide-producing fluorescent *Pseudomonas* sp. *Annals of Microbiology*, *64*(2), 493–502.

Altieri, M. A. (2004). Linking ecologists and traditional farmers in the search for sustainable agriculture. *Frontiers in Ecology and the Environment*, *2*(1), 35–42.

Alva, A. K., Asher, C. J., & Edwards, D. G. (1990). Effect of solution pH, external calcium concentration, and aluminium activity on nodulation and early growth of cowpea. *Australian Journal of Agricultural Research*, *41*(2), 359–365.

Anjum, S. A., Xie, X. Y., Wang, L. C., Saleem, M. F., Man, C., & Lei, W. (2011). Morphological, physiological and biochemical responses of plants to drought stress. *African Journal of Agricultural Research*, *6*(9), 2026–2032.

Appunu, C., & Dhar, B. (2006). Symbiotic effectiveness of acid-tolerant *Bradyrhizobium* strains with soybean in low pH soil. *African Journal of Biotechnology*, *5*(10), 842–845.

Ara, R., Mannan, M. A., Khaliq, Q. A., & Miah, M. U. (2015). Waterlogging Tolerance of Soybean. *Bangladesh Agronomy Journal*, *18*(2), 105–109.

Arif, M. S., Muhammad, R. I. A. Z., Shahzad, S. M., Yasmeen, T., Shafaqat, A. L. I., & Akhtar, M. J. (2017). Phosphorus-mobilizing rhizobacterial strain *Bacillus cereus* GS6 improves symbiotic efficiency of soybean on an Aridisol amended with phosphorus-enriched compost. *Pedosphere*, *27*(6), 1049–1061.

Armendariz, A. L., Talano, M. A., Nicotra, M. F. O., Escudero, L., Breser, M. L., Porporatto, C., & Agostini, E. (2019). Impact of double inoculation with *Bradyrhizobium japonicum* E109 and *Azospirillum brasilense* Az39 on soybean plants grown under arsenic stress. *Plant Physiology and Biochemistry*, *138*, 26–35.

Arshad, M., Shaharoona, B., & Mahmood, T. (2008). Inoculation with *Pseudomonas* spp. containing ACC-deaminase partially eliminates the effects of drought stress on growth, yield, and ripening of pea (*Pisum sativum* L.). *Pedosphere*, *18*(5), 611–620.

Arzanesh, M. H., Alikhani, H. A., Khavazi, K., Rahimian, H. A., Miransari, M. (2011) Wheat (*Triticum aestivum* L.) growth enhancement by *Azospirilum* sp. under drought stress. *World Journal of Microbiology and Biotechnology*, 27:197–205.

Asaf, S., Hamayun, M., Khan, A. L., Waqas, M., Khan, M. A., Jan, R., ... Hussain, A. (2018). Salt tolerance of *Glycine max*. L induced by endophytic fungus *Aspergillus flavus* CSH1, via regulating its endogenous hormones and antioxidative system. *Plant Physiology and Biochemistry*, *128*, 13–23.

Asaf, S., Khan, A. L., Khan, M. A., Imran, Q. M., Yun, B. W., & Lee, I. J. (2017). Osmoprotective functions conferred to soybean plants via inoculation with *Sphingomonas* sp. LK11 and exogenous trehalose. *Microbiological Research*, *205*, 135–145.

Ashraf, M., Hasnain, S., Berge, O., & Mahmood, T. (2004). Inoculating wheat seedlings with exopolysaccharide-producing bacteria restricts sodium uptake and stimulates plant growth under salt stress. *Biology and Fertility of Soils*, *40*(3), 157–162.

Bai, Y., Zhou, X., & Smith, D. L. (2003). Enhanced soybean plant growth resulting from coinoculation of Bacillus strains with *Bradyrhizobium japonicum*. *Crop Science*, *43*(5), 1774–1781.

Bailey-Serres, J., & Voesenek, L. A. C. J. (2008). Flooding stress: Acclimations and genetic diversity. *Annual Review of Plant Biology*, *59*, 313–339.

Bajaj, R., Huang, Y., Gebrechristos, S., Mikolajczyk, B., Brown, H., Prasad, R., ... Bushley, K. E. (2018). Transcriptional responses of soybean roots to colonization with the root endophytic fungus *Piriformospora indica* reveals altered phenylpropanoid and secondary metabolism. *Scientific Reports*, *8*(1), 10227.

Beutler, A. N., Giacomeli, R., Alberto, C. M., Silva, V. N., da Silva Neto, G. F., Machado, G. A., & Santos, A. T. L. (2014). Soil hydric excess and soybean yield and development in Brazil. *Australian Journal of Crop Science*, *8*(10), 1461.

Bharti, A., Garg, S., Satpute, G., Maheshwari, H. S., Anwar, K., Pareek, A., Prakash, A., & Sharma, M.P. (2016). *B. daqingense* novel high N2–fixing rhizobia recovered from high-trehalose accumulating soybean line. In Proceedings of 57th AMI Annual conference of association of microbiologists of India and International Symposium on Microbes and Biosphere What's New What's Next held in Guhawati University, Guhawati from Nov 24–27, 2016, p. 517.

Bhatia, V. S., Jumrani, K., & Pandey, G. P. (2014). Developing drought tolerance in soybean using physiological approaches. *Soybean Research*, *12*, 1–19.

Bilal, S., Shahzad, R., Khan, A. L., Kang, S. M., Imran, Q. M., Al-Harrasi, A., ... Lee, I. J. (2018). Endophytic microbial consortia of phytohormones-producing fungus *Paecilomyces formosus* LHL10 and bacteria *Sphingomoñas* sp. LK11 to *Glycine max* L. regulates physio-hormonal changes to attenuate aluminum and zinc stresses. *Frontiers in Plant Science*, *9*, 1273.

Bressano, M., Curetti, M., Giachero, L., Gil, S. V., Cabello, M., March, G., ... Luna, C. M. (2010). Mycorrhizal fungi symbiosis as a strategy against oxidative stress in soybean plants. *Journal of Plant Physiology*, *167*(18), 1622–1626.

Caetano-Anollés, G., Lagares, A., & Favelukes, G. (1989). Adsorption of *Rhizobium meliloti* to alfalfa roots: Dependence on divalent cations and pH. *Plant and Soil*, *117*(1), 67–74.

Carillo, P., Annunziata, M. G., Pontecorvo, G., Fuggi, A., & Woodrow, P. (2011). Salinity stress and salt tolerance. In *Abiotic stress in Plants-Mechanisms and Adaptations*. Arun Shanker, B Venkateswarlu. (Eds.), Rijeka, Croatia: IntechOpen.

Ch'ng, H. Y., Ahmed, O. H., & Majid, N. M. A. (2014). Improving phosphorus availability in an acid soil using organic amendments produced from agroindustrial wastes. *The Scientific World Journal* 6.

Chandra, S. (2012). Endophytic fungi: Novel sources of anticancer lead molecules. *Applied Microbiology and Biotechnology*, *95*(1), 47–59.

Chen, T. H., & Murata, N. (2008). Glycinebetaine: An effective protectant against abiotic stress in plants. *Trends in Plant Science*, *13*(9), 499–505.

Chmielowska-Bąk, J., Izbiańska, K., Ekner-Grzyb, A., Bayar, M., & Deckert, J. (2018). Cadmium stress leads to rapid increase in RNA oxidative modifications in soybean seedlings. *Frontiers in Plant Science*, *8*, 2219.

Cho, S. M., Kang, B. R., Han, S. H., Anderson, A. J., Park, J. Y., Lee, Y. H., ... Kim, Y. C. (2008). 2R, 3R-butanediol, a bacterial volatile produced by *Pseudomonas chlororaphis* O6, is involved in induction of systemic tolerance to drought in *Arabidopsis thaliana*. *Molecular Plant-Microbe Interactions*, *21*(8), 1067–1075.

Cui, G., Ai, S., Chen, K., & Wang, X. (2019). Arbuscular mycorrhiza augments cadmium tolerance in soybean by altering accumulation and partitioning of nutrient elements, and related gene expression. *Ecotoxicology and Environmental Safety*, *171*, 231–239.

Das, K., & Roychoudhury, A. (2014). Reactive oxygen species (ROS) and response of antioxidants as ROS-scavengers during environmental stress in plants. *Frontiers in Environmental Science*, *2*, 53.

Dashti, N. H., Cherian, V. M., & Smith, D. L. (2016). Soybean production and suboptimal root zone temperatures. In *Abiotic and Biotic Stresses in Soybean Production*, Mohammad Miransari (Ed.), pp. 217–240. London, UK, Academic Press.

Dashti, N., Zhang, F., Hynes, R., & Smith, D. L. (1997). Application of plant growth-promoting rhizobacteria to soybean (*Glycine max* [L.] Merr.) increases protein and dry matter yield under short-season conditions. *Plant and Soil*, *188*(1), 33–41.

Demirevska, K., Zasheva, D., Dimitrov, R., Simova-Stoilova, L., Stamenova, M., & Feller, U. (2009). Drought stress effects on Rubisco in wheat: Changes in the Rubisco large subunit. *Acta Physiologiae Plantarum*, *31*(6), 1129.

De Boeck, E., Jacxsens, L., Bollaerts, M., & Vlerick, P. (2015). Food safety climate in food processing organizations: Development and validation of a self-assessment tool. *Trends in Food Science and Technology*, 46, 242–251.

Dias, M. A. S., Dias, J., Carvalho, L. M. V., Freitas, E. D., & Dias, P. L. S. (2013). Changes in extreme daily rainfall for São Paulo, Brazil. *Climatic Change*, 116, 705–722. doi:10.1007/s10584-012-0504-7.

Dimkpa, C. O., Merten, D., Svatoš, A., Büchel, G., & Kothe, E. (2009a). Siderophores mediate reduced and increased uptake of cadmium by *Streptomyces tendae* F4 and sunflower (*Helianthus annuus*), respectively. *Journal of Applied Microbiology, 107*(5), 1687–1696.

Dimkpa, C., Weinand, T., & Asch, F. (2009b). Plant–rhizobacteria interactions alleviate abiotic stress conditions. *Plant, Cell & Environment, 32*(12), 1682–1694.

Dogan, E., Kirnak, H., & Copur, O. (2007). Deficit irrigations during soybean reproductive stages and CROPGRO-soybean simulations under semi-arid climatic conditions. *Field Crops Research, 103*(2), 154–159.

Dreesen, F. E., De Boeck, H. J., Janssens, I. A., & Nijs, I. (2012). Summer heat and drought extremes trigger unexpected changes in productivity of a temperate annual/biannual plant community. *Environmental and Experimental Botany, 79*, 21–30.

Egamberdieva, D., Berg, G., Lindström, K., & Räsänen, L. A. (2013). Alleviation of salt stress of symbiotic *Galega officinalis* L. (goat's rue) by co-inoculation of Rhizobium with root-colonizing *Pseudomonas. Plant and Soil, 369*(1–2), 453–465.

Egamberdieva, D., Jabborova, D., & Berg, G. (2016). Synergistic interactions between *Bradyrhizobium japonicum* and the endophyte *Stenotrophomonas rhizophila* and their effects on growth, and nodulation of soybean under salt stress. *Plant and Soil, 405*(1–2), 35–45.

El-Beltagi, H. S., & Mohamed, H. I. (2013). Reactive oxygen species, lipid peroxidation and antioxidative defense mechanism. *Notulae Botanicae Horti Agrobotanici Cluj-Napoca, 41*(1), 44–57.

El-Esawi, M. A., Alaraidh, I. A., Alsahli, A. A., Alamri, S. A., Ali, H. M., & Alayafi, A. A. (2018). *Bacillus firmus* (SW5) augments salt tolerance in soybean (*Glycine max* L.) by modulating root system architecture, antioxidant defense systems and stress-responsive genes expression. *Plant Physiology and Biochemistry, 132*, 375–384.

Etesami, H., & Maheshwari, D. K. (2018). Use of plant growth promoting rhizobacteria (PGPRs) with multiple plant growth promoting traits in stress agriculture: Action mechanisms and future prospects. *Ecotoxicology and Environmental Safety, 156*, 225–246.

Farooq, M., Wahid, A., Kobayashi, N., Fujita, D., & Basra, S. M. A. (2009). Plant drought stress: Effects, mechanisms and management. In *Sustainable Agriculture*, Eric Lichtfouse (Ed.), pp. 153–188, Springer, Dordrecht, the Netherlands.

Fernández, L. A., Zalba, P., Gómez, M. A., & Sagardoy, M. A. (2007). Phosphate-solubilization activity of bacterial strains in soil and their effect on soybean growth under greenhouse conditions. *Biology and Fertility of Soils, 43*(6), 805–809.

Fricke, W., Akhiyarova, G., Wei, W., Alexandersson, E., Miller, A. et al. (2006). The short-term growth response to salt of the developing barley leaf. *Journal of Experimental Botany*, 57, 1079–1095.

Gadd, G. M. (2004). Microbial influence on metal mobility and application for bioremediation. *Geoderma*, 122, 109–119.

Gill, S. S., & Tuteja, N. (2010). Reactive oxygen species and antioxidant machinery in abiotic stress tolerance in crop plants. *Plant Physiology and Biochemistry, 48*(12), 909–930.

Gill, S. S., Gill, R., Trivedi, D. K., Anjum, N. A., Sharma, K. K., Ansari, M. W., ... Varma, A. (2016). *Piriformospora indica*: Potential and significance in plant stress tolerance. *Frontiers in Microbiology, 7*, 332.

Global Market Analysis, FAS, USDA Foreign Agricultural Service/USDA, March 2020. p. 32. https://apps.fas.usda.gov/psdonline/circulars/production.pdf.

Glick, B. R. (2014). Bacteria with ACC deaminase can promote plant growth and help to feed the world. *Microbiological Research, 169*(1), 30–39.

Gontia-Mishra, I., Sasidharan, S., & Tiwari, S. (2014). Recent developments in use of 1–aminocyclopropane-1–carboxylate (ACC) deaminase for conferring tolerance to biotic and abiotic stress. *Biotechnology Letters, 36*(5), 889–898.

González-Rodríguez, R. M., Rial-Otero, R., Cancho-Grande, B., Gonzalez-Barreiro, C., & Simal-Gándara, J. (2011). A review on the fate of pesticides during the processes within the food-production chain. *Critical Reviews in Food Science and Nutrition, 51*(2), 99–114.

Gouffi, K., & Blanco, C. (2000). Is the accumulation of osmoprotectant the unique mechanism involved in bacterial osmoprotection? *International Journal of Food Microbiology, 55*(1–3), 171–174.

Grichko, V. P., & Glick, B. R. (2001). Amelioration of flooding stress by ACC deaminase-containing plant growth-promoting bacteria. *Plant Physiology and Biochemistry, 39*(1), 11–17.

Grümberg, B. C., Urcelay, C., Shroeder, M. A., Vargas-Gil, S., & Luna, C. M. (2015). The role of inoculum identity in drought stress mitigation by arbuscular mycorrhizal fungi in soybean. *Biology and Fertility of Soils*, *51*(1), 1–10.

Gupta, P., & Diwan, B. (2017). Bacterial exopolysaccharide mediated heavy metal removal: A review on biosynthesis, mechanism and remediation strategies. *Biotechnology Reports*, *13*, 58–71.

Gusain, Y. S., Singh, U. S., & Sharma, A. K. (2015). Bacterial mediated amelioration of drought stress in drought tolerant and susceptible cultivars of rice (*Oryza sativa* L.). *African Journal of Biotechnology*, *14*(9), 764–773.

Hamayun, M., Hussain, A., Khan, S. A., Kim, H. Y., Khan, A. L., Waqas, M., ... Lee, I. J. (2017). Gibberellins producing endophytic fungus *Porostereum spadiceum* AGH786 rescues growth of salt affected soybean. *Frontiers in Microbiology*, *8*, 686.

Hamdia, M. A. E. S., Shaddad, M. A. K., & Doaa, M. M. (2004). Mechanisms of salt tolerance and interactive effects of *Azospirillum brasilense* inoculation on maize cultivars grown under salt stress conditions. *Plant Growth Regulation*, *44*(2), 165–174.

Hardoim, P. R., van Overbeek, L. S., & van Elsas, J. D. (2008). Properties of bacterial endophytes and their proposed role in plant growth. *Trends in Microbiology*, *16*(10), 463–471.

Hasanuzzaman, M., Nahar, K., Rahman, A., Mahmud, J. A., Hossain, M. S., & Fujita, M. (2016). Soybean production and environmental stresses. In *Environmental Stresses in Soybean Production*, Mohammad Miransari (Ed.), pp. 61–102, Amsterdam, the Netherlands, Academic Press.

Hosseini, M. K., Powell, A. A., & Bingham, I. J. (2002). Comparison of the seed germination and early seedling growth of soybean in saline conditions. *Seed Science Research*, *12*(3), 165–172.

Hussain Wani, S., Brajendra Singh, N., Haribhushan, A., & Iqbal Mir, J. (2013). Compatible solute engineering in plants for abiotic stress tolerance-role of glycine betaine. *Current Genomics*, *14*(3), 157–165.

Hussain, M. B., Zahir, Z. A., Asghar, H. N., & Asgher, M. (2014). Can catalase and exopolysaccharides producing rhizobia ameliorate drought stress in wheat? *International Journal of Agriculture and Biology*, *16*(1), 3–13.

Ianeva, O. D. (2009). Mechanisms of bacteria resistance to heavy metals. *Mikrobiolohichnyi zhurnal (Kiev, Ukraine: 1993)*, *71*(6), 54–65.

Ilangumaran, G., & Smith, D. L. (2017). Plant growth promoting rhizobacteria in amelioration of salinity stress: A systems biology perspective. *Frontiers in Plant Science*, *8*, 1768.

Islam, F., Yasmeen, T., Arif, M. S., Ali, S., Ali, B., Hameed, S., & Zhou, W. (2016). Plant growth promoting bacteria confer salt tolerance in *Vigna radiata* by up-regulating antioxidant defense and biological soil fertility. *Plant Growth Regulation*, *80*(1), 23–36.

Isobe, K., Maruyama, K., Nagai, S., Higo, M., Maekawa, T., Mizonobe, G., ... Ishii, R. (2011). Arbuscular mycorrhizal fungal community structure in soybean roots: Comparison between Kanagawa and Hokkaido, Japan. *Advances in Microbiology*, *1*(01), 13.

Jajic, I., Sarna, T., & Strzalka, K. (2015). Senescence, stress, and reactive oxygen species. *Plants*, *4*(3), 393–411.

Jamil, M., Lee, K. J., Kim, J. M., Kim, H. S., & Rha, E. S. (2007). Salinity reduced growth PS2 photochemistry and chlorophyll content in radish. *Scientia Agricola*, *64*(2), 111–118.

Jha, Y., & Subramanian, R. B. (2014). PGPR regulate caspase-like activity, programmed cell death, and antioxidant enzyme activity in paddy under salinity. *Physiology and Molecular Biology of Plants*, *20*(2), 201–207.

Jones, D. A., Smith, B. F. L., Wilson, M. J., & Goodman, B. A. (1991). Solubilizator fungi of phosphate in rise soil. *Mycological Research*, *95*, 1090–1093.

Kang, S. M., Radhakrishnan, R., Khan, A. L., Kim, M. J., Park, J. M., Kim, B. R., ... Lee, I. J. (2014). Gibberellin secreting rhizobacterium, Pseudomonas putida H-2–3 modulates the hormonal and stress physiology of soybean to improve the plant growth under saline and drought conditions. *Plant Physiology and Biochemistry*, *84*, 115–124.

Kasotia, A., Jain, S., Vaishnav, A., Kumari, S., Gaur, R. K., & Choudhary, D. K. (2012). Soybean growth promotion by *Pseudomonas* sp. strain VS-1 under salt stress. *Pakistan Journal of Biological Sciences*, *15*(14), 698– 701.

Kasotia, A., Varma, A., Tuteja, N., & Choudhary, D. K. (2016). Amelioration of soybean plant from saline-induced condition by exopolysaccharide producing Pseudomonas-mediated expression of high affinity K± transporter (HKT1) gene. *Current Science*, *111*, 1961–1967.

Khan, M.S., Zaidi, A., Wani, P.A. (2007). Role of phosphate-solubilizing microorganisms in sustainable agriculture—a review. *Agronomy for Sustainable Development 27*, 29–43.

Khan, A. L., & Lee, I. J. (2013). Endophytic *Penicillium funiculosum* LHL06 secretes gibberellin that reprograms *Glycine max* L. growth during copper stress. *BMC Plant Biology*, *13*(1), 8.

Khan, A. L., Hamayun, M., Ahmad, N., Hussain, J., Kang, S. M., Kim, Y. H., ... Hwang, Y. H. (2011). Salinity stress resistance offered by endophytic fungal interaction between *Penicillium minioluteum* LHL09 and *Glycine max*. *Journal of Microbiology and Biotechnology*, *21*, 893–902.

Khan, A. L., Hamayun, M., Kim, Y. H., Kang, S. M., Lee, J. H., & Lee, I. J. (2011). Gibberellins producing endophytic *Aspergillus fumigatus* sp. LH02 influenced endogenous phytohormonal levels, isoflavonoids production and plant growth in salinity stress. *Process Biochemistry*, *46*(2), 440–447.

Khandelwal, S. R., Manwar, A. V., Chaudhari, B. L., & Chincholkar, S. B. (2002). Siderophoregenic *bradyrhizobia* boost yield of soybean. *Applied Biochemistry and Biotechnology*, *102*(1–6), 155–168.

Khatoon, A., Rehman, S., Hiraga, S., Makino, T., & Komatsu, S. (2012). Organ-specific proteomics analysis for identification of response mechanism in soybean seedlings under flooding stress. *Journal of Proteomics*, *75*(18), 5706–5723.

Kim, M. J., Radhakrishnan, R., Kang, S. M., You, Y. H., Jeong, E. J., Kim, J. G., & Lee, I. J. (2017). Plant growth promoting effect of *Bacillus amyloliquefaciens* H-2–5 on crop plants and influence on physiological changes in soybean under soil salinity. *Physiology and Molecular Biology of Plants*, *23*(3), 571–580.

Kishor, P. K., Sangam, S., Amrutha, R. N., Laxmi, P. S., Naidu, K. R., Rao, K. R. S. S., ... Sreenivasulu, N. (2005). Regulation of proline biosynthesis, degradation, uptake and transport in higher plants: Its implications in plant growth and abiotic stress tolerance. *Current Science*, *88*(3), 424–438.

Krüger, G. H. J., De Villiers, M. F., Strauss, A. J., De Beer, M., Van Heerden, P. D. R., Maldonado, R., & Strasser, R. J. (2014). Inhibition of photosystem II activities in soybean (*Glycine max*) genotypes differing in chilling sensitivity. *South African Journal of Botany*, *95*, 85–96.

Kuiper, I., Lagendijk, E. L., Bloemberg, G. V., & Lugtenberg, B. J. (2004). Rhizoremediation: A beneficial plant-microbe interaction. *Molecular Plant-Microbe Interactions*, *17*(1), 6–15.

Kumari, S., Vaishnav, A., Jain, S., Varma, A., & Choudhary, D. K. (2015). Bacterial-mediated induction of systemic tolerance to salinity with expression of stress alleviating enzymes in soybean (*Glycine max* L. Merrill). *Journal of Plant Growth Regulation*, *34*(3), 558–573.

Kumari, V., Germida, J., &Vujanovic, V. (2018). Legume endosymbionts: Drought stress tolerance in second-generation chickpea (*Cicer arietinum*) seeds. *Journal of Agronomy and Crop Science*, *204*(6), 529–540.

Kumawat, K. C., Sharma, P., Sirari, A., Singh, I., Gill, B. S., Singh, U., & Saharan, K. (2019). Synergism of *Pseudomonas aeruginosa* (LSE-2) nodule endophyte with *Bradyrhizobium* sp. (LSBR-3) for improving plant growth, nutrient acquisition and soil health in soybean. *World Journal of Microbiology and Biotechnology*, *35*(3), 47.

Kwon, Y. S., Ryu, C. M., Lee, S., Park, H. B., Han, K. S., Lee, J. H., ... Bae, D. W. (2010). Proteome analysis of *Arabidopsis* seedlings exposed to bacterial volatiles. *Planta*, *232*(6), 1355–1370.

Lamaoui, M., Jemo, M., Datla, R., & Bekkaoui, F. (2018). Heat and drought stresses in crops and approaches for their mitigation. *Frontiers in Chemistry*, *6*, 26.

Lasat, H. A. (2002). Phytoextraction of toxic metals: A review of biological mechanisms. *Journal of Environmental Quality*, 31, 109–120.

Lesk, C., Rowhani, P., & Ramankutty, N. (2016). Influence of extreme weather disasters on global crop production. *Nature*, *529*(7584), 84.

Liu, F., Xing, S., Ma, H., Du, Z., & Ma, B. (2013). Cytokinin-producing, plant growth-promoting rhizobacteria that confer resistance to drought stress in *Platycladus orientalis* container seedlings. *Applied Microbiology and Biotechnology*, *97*(20), 9155–9164.

Long, H. H., Schmidt, D. D., & Baldwin, I. T. (2008). Native bacterial endophytes promote host growth in a species-specific manner: Phytohormone manipulations do not result in common growth responses. *PLoS One*, *3*(7), e2702.

Manikavelu, A., Nadarajan, N., Ganesh, S. K., Gnanamalar, R. P., & Babu, R. C. (2006). Drought tolerance in rice: Morphological and molecular genetic consideration. *Plant Growth Regul*, *50*, 121–138.

Mao, F., Nan, G., Cao, M., Gao, Y., Guo, L., Meng, X., & Yang, G. (2018). The metal distribution and the change of physiological and biochemical process in soybean and mung bean plants under heavy metal stress. *International Journal of Phytoremediation*, *20*(11), 1113–1120.

Marinkovic, J., Djordjevic, V., Baleševic Tubic, S., Bjelic, D., Vucelic Radovic, B., & Josic, D. (2013). Osmotic stress tolerance, PGP traits and RAPD analysis of *Bradyrhizobium japonicum* strains. *Genetika*, *45*(1), 75–86.

Martins, S. J., Rocha, G. A., de Melo, H. C., de Castro Georg, R., Ulhôa, C. J., de Campos Dianese, É., ... Vaz, K. S. (2018). Plant-associated bacteria mitigate drought stress in soybean. *Environmental Science and Pollution Research*, *25*(14), 13676–13686.

Masciarelli, O., Llanes, A., & Luna, V. (2014). A new PGPR co-inoculated with *Bradyrhizobium japonicum* enhances soybean nodulation. *Microbiological Research*, *169*(7–8), 609–615.

Meena, R. S., Vijayakumar, V., Yadav, G. S., & Mitran, T. (2018). Response and interaction of *Bradyrhizobium japonicum* and arbuscular mycorrhizal fungi in the soybean rhizosphere. *Plant Growth Regulation*, *84*(2), 207–223.

Meyer, D. W., & Badaruddin, M. (2001). Frost tolerance of ten seedling legume species at four growth stages. *Crop Science*, *41*(6), 1838–1842.

Miransari, M. (2015a). Phytoremediation using microbial communities: I. In *Phytoremediation*, Springer, Cham, Switzerland, pp. 177–182.

Miransari, M. (Ed.). (2015b). *Abiotic and Biotic Stresses in Soybean Production: Soybean Production* (Vol. 1). Cambridge, MA, Academic Press, Mohammad Miransari (Ed).

Miransari, M., & Smith, D. (2008). Using signal molecule genistein to alleviate the stress of suboptimal root zone temperature on soybean-*Bradyrhizobium* symbiosis under different soil textures. *Journal of Plant Interactions*, *3*(4), 287–295.

Miransari, M., & Smith, D. L. (2009). Alleviating salt stress on soybean (*Glycine max* (L.) Merr.)–*Bradyrhizobium japonicum* symbiosis, using signal molecule genistein. *European Journal of Soil Biology*, *45*(2), 146–152.

Miransari, M., Riahi, H., Eftekhar, F., Minaie, A., & Smith, D. L. (2013). Improving Soybean (*Glycine max* L.) N 2 Fixation under Stress. *Journal of Plant Growth Regulation*, *32*(4), 909–921.

Molina-Favero, C., Creus, C. M., Simontacchi, M., Puntarulo, S., & Lamattina, L. (2008). Aerobic nitric oxide production by *Azospirillum brasilense* Sp245 and its influence on root architecture in tomato. *Molecular Plant-Microbe Interactions*, *21*(7), 1001–1009.

Mondani, F., Khani, K., Honarmand, S. J., & Saeidi, M. (2019). Evaluating effects of plant growth-promoting rhizobacteria on the radiation use efficiency and yield of soybean (*Glycine max*) under water deficit stress condition. *Agricultural Water Management*, *213*, 707–713.

Mubarik, N. R., & Sunatmo, T. I. (2014). Symbiotic of nitrogen fixation between acid aluminium tolerant *Brayrhizobium japonicum* and soybean. *Advances in Biology and Ecology of Nitrogen Fixation*, Rijeka, Croatia, InTech, pp. 259–274.

Munné-Bosch, S., & Alegre, L. (2000). Changes in carotenoids, tocopherols and diterpenes during drought and recovery, and the biological significance of chlorophyll loss in Rosmarinus officinalis plants. *Planta*, *210*, 925–931.

Munns, R. (2002). Comparative physiology of salt and water stress. *Plant, Cell & Environment*, *25*(2), 239–250.

Mustafa, G., & Komatsu, S. (2014). Quantitative proteomics reveals the effect of protein glycosylation in soybean root under flooding stress. *Frontiers in Plant Science*, *5*, 627.

Mutava, R. N., Prince, S. J. K., Syed, N. H., Song, L., Valliyodan, B., Chen, W., & Nguyen, H. T. (2015). Understanding abiotic stress tolerance mechanisms in soybean: A comparative evaluation of soybean response to drought and flooding stress. *Plant Physiology and Biochemistry*, *86*, 109–120.

Nakayama, N., Hashimoto, S., Shimada, S., Takahashi, M., Kim, Y. H., Oya, T., & Arihara, J. (2004). The effect of flooding stress at the germination stage on the growth of soybean [*Glycine max*] in relation to initial seed moisture content. *Japanese Journal of Crop Science (Japan)*, 73, 323–329.

Naseem, H., Ahsan, M., Shahid, M. A., & Khan, N. (2018). Exopolysaccharides producing rhizobacteria and their role in plant growth and drought tolerance. *Journal of Basic Microbiology*, *58*(12), 1009–1022.

Ngumbi, E., & Kloepper, J. (2016). Bacterial-mediated drought tolerance: Current and future prospects. *Applied Soil Ecology*, *105*, 109–125.

Nogueira, M. A., Nehls, U., Hampp, R., Poralla, K., & Cardoso, E. J. B. N. (2007). Mycorrhiza and soil bacteria influence extractable iron and manganese in soil and uptake by soybean. *Plant and Soil*, *298*(1–2), 273–284.

Nurlaeny, N., Marschner, H., & George, E. (1996). Effects of liming and mycorrhizal colonization on soil phosphate depletion and phosphate uptake by maize (*Zea mays* L.) and soybean (*Glycine max* L.) grown in two tropical acid soils. *Plant and Soil*, *181*(2), 275–285.

Ohashi, Y., Nakayama, N., Saneoka, H., & Fujita, K. (2006). Effects of drought stress on photosynthetic gas exchange, chlorophyll fluorescence and stem diameter of soybean plants. *Biologia Plantarum*, *50*(1), 138–141.

Park, Y. G., Mun, B. G., Kang, S. M., Hussain, A., Shahzad, R., Seo, C. W., ... Lee, I. J. (2017). Bacillus aryabhattai SRB02 tolerates oxidative and nitrosative stress and promotes the growth of soybean by modulating the production of phytohormones. *PloS One*, *12*(3), e0173203.

Pedersen, P. (2004). Soil pH and Plant Population Effects on Soybean Yield. Iowa State Research Farm Progress Reports. 1326. http://lib.dr.iastate.edu/farms_reports/1326.

Pedersen, P., & Lauer, J. G. (2004). Response of soybean yield components to management system and planting date. *Agronomy Journal*, *96*(5), 1372–1381.

Pereyra, M. A., Zalazar, C. A., & Barassi, C. A. (2006). Root phospholipids in *Azospirillum*-inoculated wheat seedlings exposed to water stress. *Plant Physiology and Biochemistry*, *44*(11–12), 873–879.

Pishchik, V. N., Vorobyev, N. I., Chernyaeva, I. I., Timofeeva, S. V., Kozhemyakov, A. P., Alexeev, Y. V., & Lukin, S. M. (2002). Experimental and mathematical simulation of plant growth promoting rhizobacteria and plant interaction under cadmium stress. *Plant and Soil*, *243*(2), 173–186.

Porcel, R., & Ruiz-Lozano, J. M. (2004). Arbuscular mycorrhizal influence on leaf water potential, solute accumulation, and oxidative stress in soybean plants subjected to drought stress. *Journal of Experimental Botany*, *55*(403), 1743–1750.

Radhakrishnan, R., Khan, A. L., & Lee, I. J. (2013). Endophytic fungal pre-treatments of seeds alleviates salinity stress effects in soybean plants. *Journal of Microbiology*, *51*(6), 850–857.

Radhakrishnan, R., Khan, A. L., Kang, S. M., & Lee, I. J. (2015). A comparative study of phosphate solubilization and the host plant growth promotion ability of *Fusarium verticillioides* RK01 and *Humicola* sp. KNU01 under salt stress. *Annals of Microbiology*, *65*(1), 585–593.

Ramesh, A., Sharma, S. K., Sharma, M. P., Yadav, N., & Joshi, O. P. (2014). Inoculation of zinc solubilizing *Bacillus aryabhattai* strains for improved growth, mobilization and biofortification of zinc in soybean and wheat cultivated in Vertisols of central India. *Applied Soil Ecology*, *73*, 87–96.

Rathod, D. P., Brestic, M., & Shao, H. B. (2011). Chlorophyll a fluorescence determines the drought resistance capabilities in two varieties of mycorrhized and non-mycorrhized *Glycine max* Linn.

Sahoo, R. K., Ansari, M. W., Pradhan, M., Dangar, T. K., Mohanty, S., & Tuteja, N. (2014). Phenotypic and molecular characterization of native *Azospirillum* strains from rice fields to improve crop productivity. *Protoplasma*, *251*(4), 943–953.

Saia, S., Benítez, E., García-Garrido, J. M., Settanni, L., Amato, G., & Giambalvo, D. (2014). The effect of arbuscular mycorrhizal fungi on total plant nitrogen uptake and nitrogen recovery from soil organic material. *The Journal of Agricultural Science*, *152*(3), 370–378.

Samarah, N. H., Mullen, R. E., Goggi, S., & Gaul, A. (2009). Effect of drying treatment and temperature on soybean seed quality during maturation. *Seed Science and Technology*, *37*(2), 469–473.

Sandhya, V., Ali Sk, Z., Grover, M., Reddy, G., Venkateswarlu, B. (2009). Alleviation of drought stress effects in sunflower seedlings by exopolysaccharides producing *Pseudomonas putida* strain P45. *Biology and Fertility of Soils*, *46*:17–26.

Sandhya, V. S. K. Z., Ali, S. Z., Grover, M., Reddy, G., & Venkateswarlu, B. (2010). Effect of plant growth promoting *Pseudomonas* spp. on compatible solutes, antioxidant status and plant growth of maize under drought stress. *Plant Growth Regulation*, *62*(1), 21–30.

Sehgal, A., Sita, K., Siddique, K. H., Kumar, R., Bhogireddy, S., Varshney, R. K., ... Nayyar, H. (2018). Drought or/and heat-stress effects on seed filling in food crops: Impacts on functional biochemistry, seed yields, and nutritional quality. *Frontiers in Plant Science*, 9, 1–19.

Selvakumar, G., Panneerselvam, P., & Ganeshamurthy, A. N. (2012). Bacterial mediated alleviation of abiotic stress in crops. In *Bacteria in Agrobiology: Stress Management*, Springer, Berlin, Heidelberg, pp. 205–224.

Shahid, M., Khalid, S., Abbas, G., Shahid, N., Nadeem, M., Sabir, M., ... Dumat, C. (2015). Heavy metal stress and crop productivity. In Crop *P*roduction and *G*lobal *E*nvironmental *I*ssues, D. K Maheshwari (Ed.), Springer, Cham, Switzerland, pp. 1–25.

Sharma, M.P., Jaisighani, K., Sharma, S.K. & Bhatia, V.S. (2012). Effect of native soybean rhizobia and AM fungi in the improvement of nodulation, growth, soil enzymes and physiological status of soybean under microcosm conditions. *Agricultural Research* 1(4):346–351. doi:10.1007/s40003-012-0038-2.

Sharma, M.P., Singh, S., Sharma, S.K., Ramesh, A., Bhatia, V.S. (2016). Co-inoculation of resident AM fungi and soybean rhizobia enhanced nodulation, yield, soil biological parameters and saved fertilizer inputs in vertisols under microcosm and field conditions. *Soybean Research 14*(2): 39–53.

Simarmata, R., Ngadiman, N., Rohman, S., & Simanjuntak, P. (2018). Amelioration of salt tolerance in soybean (*Glycine max*. L) by plant-growth promoting endophytic bacteria produce 1–Aminocyclopropane-1–Carboxylase Deaminase. *Annales Bogorienses, 22*(2), 81–93.

Son, H. J., Park, G. T., Cha, M. S., & Heo, M. S. (2006). Solubilization of insoluble inorganic phosphates by a novel salt and pH-tolerant *Pantoea agglomerans* R-42 isolated from soybean rhizosphere. *Bioresource Technology, 97*(2), 204–210.

Souza, R. D., Ambrosini, A., and Passaglia, L. M. P. (2015). Plant growth-promoting bacteria as inoculants in agricultural soils. *Genetics and Molecular Biology, 38*, 401–419. doi:10.1590/S1415-475738420150053.

Spagnoletti, F., & Lavado, R. (2015). The arbuscular mycorrhiza *Rhizophagus intraradices* reduces the negative effects of arsenic on soybean plants. *Agronomy, 5*(2), 188–199.

Streeter, J. G. (2003). Effects of drought on nitrogen fixation in soybean root nodules. *Plant, Cell & Environment, 26*(8), 1199–1204.

Subramanian, P., Kim, K., Krishnamoorthy, R., Sundaram, S., & Sa, T. (2015). Endophytic bacteria improve nodule function and plant nitrogen in soybean on co-inoculation with *Bradyrhizobium japonicum* MN110. *Plant Growth Regulation, 76*(3), 327–332.

Sugawara, M., Cytryn, E. J., & Sadowsky, M. J. (2010). Functional role of *Bradyrhizobium japonicum* trehalose biosynthesis and metabolism genes during physiological stress and nodulation. *Applied and Environmental Microbiology, 76*(4), 1071–1081.

Takács, T., Cseresnyés, I., Kovács, R., Parádi, I., Kelemen, B., Szili-Kovács, T., & Füzy, A. (2018). Symbiotic effectivity of dual and tripartite associations on soybean (*Glycine max* L. Merr.) cultivars inoculated with *Bradyrhizobium japonicum* and AM fungi. *Frontiers in Plant Science, 9*, 1631.

Tewari, S., & Arora, N. K. (2013). Plant growth promoting rhizobacteria for ameliorating abiotic stresses triggered due to climatic variability. *Climate Change and Environmental Sustainability, 1*(2), 95–103.

Tidke, S. A., Ramakrishna, D., Kiran, S., Kosturkova, G., & Ravishankar, G. A. (2015). Nutraceutical Potential of Soybean. *Asian Journal of Clinical Nutrition, 7*, 22–32.

Tisne, S., Schmalenbach, I., Reymond, M., Dauzat, M., Pervent, M., Vile, D., & Granier, C. (2010). Keep on growing under drought: Genetic and developmental bases of the response of rosette area using a recombinant inbred line population. *Plant, Cell & Environment, 33*(11), 1875–1887.

Toorchi, M., Yukawa, K., Nouri, M. Z., & Komatsu, S. (2009). Proteomics approach for identifying osmotic-stress-related proteins in soybean roots. *Peptides, 30*(12), 2108–2117.

Tran, L. S. P., & Mochida, K. (2010). Functional genomics of soybean for improvement of productivity in adverse conditions. *Functional & Integrative Genomics, 10*(4), 447–462.

Vaishnav, A., Kumari, S., Jain, S., Varma, A., & Choudhary, D. K. (2015). Putative bacterial volatile-mediated growth in soybean (*Glycine max* L. Merrill) and expression of induced proteins under salt stress. *Journal of Applied Microbiology, 119*(2), 539–551.

Vaishnav, A., Kumari, S., Jain, S., Varma, A., Tuteja, N., & Choudhary, D. K. (2016). PGPR-mediated expression of salt tolerance gene in soybean through volatiles under sodium nitroprusside. *Journal of Basic Microbiology, 56*(11), 1274–1288.

Vaishnav, A., Shukla, A. K., Sharma, A., Kumar, R., & Choudhary, D. K. (2018). Endophytic bacteria in plant salt stress tolerance: Current and future prospects. *Journal of Plant Growth Regulation*, 38, 650–668.

Valliyodan, B., Lee, S. H., & Nguyen, H. T. (2017). Sequencing, assembly, and annotation of the soybean genome. In *The Soybean Genome*, Henry T Nguyen, Madan Kumar Bhattacharyya (Eds.), Springer, Cham, Switzerland, pp. 73–82.

Van Heerden, P. D., Tsimilli-Michael, M., Krüger, G. H., & Strasser, R. J. (2003). Dark chilling effects on soybean genotypes during vegetative development: Parallel studies of CO_2 assimilation, chlorophyll a fluorescence kinetics O-J-I-P and nitrogen fixation. *Physiologia Plantarum, 117*(4), 476–491.

VanToai, T. T., Beuerlein, A. F., Schmitthenner, S. K., & St Martin, S. K. (1994). Genetic variability for flooding tolerance in soybeans. *Crop Science, 34*(4), 1112–1115.

Vargas, A. A., & Graham, P. H. (1988). *Phaseolus vulgaris* cultivar and *Rhizobium* strain variation in acid-pH tolerance and nodulation under acid conditions. *Field Crops Research*, *19*(2), 91–101.

Vishwakarma, K., Upadhyay, N., Kumar, N., Yadav, G., Singh, J., Mishra, R. K., ... Sharma, S. (2017). Abscisic acid signaling and abiotic stress tolerance in plants: A review on current knowledge and future prospects. *Frontiers in Plant Science*, *8*, 161.

Vurukonda, S. S. K. P., Vardharajula, S., Shrivastava, M., & SkZ, A. (2016). Enhancement of drought stress tolerance in crops by plant growth promoting rhizobacteria. *Microbiological Research*, *184*, 13–24.

Walters, D. R., & Fountaine, J. M. (2009). Practical application of induced resistance to plant diseases: An appraisal of effectiveness under field conditions. *The Journal of Agricultural Science*, *147*(5), 523–535.

Wan, C., Shao, G., Chen, Y., & Yan, S. (2002). Relationship between salt tolerance and chemical quality of soybean under salt stress. *Chinese Journal of Oil Sciences*, *24*(2), 67–72.

Waqas, M., Khan, A. L., Kang, S. M., Kim, Y. H., & Lee, I. J. (2014). Phytohormone-producing fungal endophytes and hardwood-derived biochar interact to ameliorate heavy metal stress in soybeans. *Biolo*gy and Fertility of Soils*, *50*(7), 1155–1167.

Xiong, J., Fu, G., Tao, L., & Zhu, C. (2010). Roles of nitric oxide in alleviating heavy metal toxicity in plants. *Archives of Biochemistry and Biophysics*, *497*(1–2), 13–20.

Yadav, V., Kumar, M., Deep, D. K., Kumar, H., Sharma, R., Tripathi, T., ... Johri, A. K. (2010). A phosphate transporter from the root endophytic fungus *Piriformospora indica* plays a role in phosphate transport to the host plant. *Journal of Biological Chemistry*, *285*(34), 26532–26544.

Yuwono, T., Handayani, D., & Soedarsono, J. (2005). The role of osmotolerant rhizobacteria in rice growth under different drought conditions. *Australian Journal of Agricultural Research*, *56*(7), 715–721.

Zahedi, H., & Abbasi, S. (2015). Effect of plant growth promoting rhizobacteria (PGPR) and water stress on phytohormones and polyamines of soybean. *Indian Journal of Agricultural Research*, *49*(5), 427–431.

Zakikhani, H., Ardakani, M. R., Rejali, F., Gholamhoseini, M., Joghan, A. K., & Dolatabadian, A. (2012). Influence of diazotrophic bacteria on antioxidant enzymes and some biochemical characteristics of soybean subjected to water stress. *Journal of Integrative Agriculture*, *11*(11), 1828–1835.

Zhang, H., Kim, M. S., Sun, Y., Dowd, S. E., Shi, H., & Paré, P. W. (2008). Soil bacteria confer plant salt tolerance by tissue-specific regulation of the sodium transporter HKT1. *Molecular Plant-Microbe Interactions*, *21*(6), 737–744.

Zheng, S. J. (2010). Crop production on acidic soils: Overcoming aluminium toxicity and phosphorus deficiency. *Annals of Botany*, *106*(1), 183–184.

Section V

Crop Responses to Microbe-Mediated Biotic Stress Management

17

Bacterial Biological Control Agents for Soilborne Diseases Management in Pulses: Present Status and Future Prospects

Pranamita Kunda, Abhishek Mukherjee, and Paltu Kumar Dhal

CONTENTS

17.1 Introduction

17.1.1 Pulses

The word "pulse" is derived from the Latin word "puls" meaning pottage, i.e., porridge made from boiling seeds (Singh et al. 2018). Pulses are nutrient dense edible legumes that are an important source of proteins, minerals, and vitamins and play a significant role in our lives in providing nutrition.

Although pulses contain the same amount of calories as that of cereals, their protein content is equivalent to that of meat and poultry. Hence, pulses are also known as poor man's meat and rich person's vegetables (Singh et al. 2015). India is the world's leading producer as well as consumer of pulses. However, like other crops, to continue meeting the increasing internal demand, production of pulses also has to be increased. Major pulse-producing states in India are Madhya Pradesh, Maharashtra, Rajasthan, Uttar Pradesh, Karnataka, Andhra Pradesh, Chattisgarh, Bihar, and Tamil Nadu (Bhat et al. 2012; Sen Gupta 1974). The major pulse crops grown in India are *Cicer arietinum* (Gram), *Cicer arietinum* (Chickpea), *Lens culinaris* (Lentils), *Cajanus cajan* (Pigeon pea), *Vigna mungo* (Urad), *Vigna radiata* (Mung bean), *Phaseolus vulgaris* (Red kidney beans), *Lathyrus sativus* (White peas), etc. (Sen Gupta 1974). Several biotic and abiotic factors limit pulse production. Conditions such as unfavorable weather and soil conditions, improper agronomic techniques, poor quality of seed material, lack of high-yielding varieties, and pests and disease incidences can affect pulse production. In particular, the incidence of diseases further hampers the production. Fungi, bacteria, virus, and nematodes are the important constraints in the production of food materials as well as of pulses. Globally, on average, the crop suffers 26% yield losses due to plant diseases; in some areas or fields, yield loss is 100% due to crop failure (Khan et al. 2011). Pulse production is also hampered to a great extent because of these factors. From India, 10%–15% yield loss of pulses is reported because of diseases (Rana et al. 2016). Although in general, legumes are pest-free crops, if proper crop rotation is maintained, some major pests and diseases devastate their production (Singh et al. 2015). Among different diseases, soilborne diseases are a major cause of concern in pulse production and hence have to be addressed properly.

17.1.2 Soilborne Diseases

Every year crop production is limited due to several soilborne pathogens that attack different varieties of crops, and pulses are also severely affected. Soilborne diseases are caused by pathogens that reside on the soil surface as well as in the soil matrix. Soil serves as a reservoir for the pathogen inoculums, and these pathogens can survive in the soil even in the absence of their host. The identification of soilborne diseases is also difficult. Mostly the symptoms involve underground parts of root tissues and go unnoticed until severely affected plants show aboveground symptoms such as stunting, wilting, and death (Divya Rani and Sudini 2013). Symptoms of typical soilborne pathogen diseases are lesions, rot, and wilting—ultimately leading to death (Katan 2017). Management of soilborne diseases is difficult, as it affects the plant at various stages (Kala et al. 2016). Several biotic and abiotic factors including agricultural practices influence soilborne disease development. Unlike foliar diseases, which occur in natural habitats, soilborne diseases are extremely rare in natural habitats. Instead, soilborne diseases are a natural phenomenon that is aggravated by agricultural practices such as cultivation of susceptible crop varieties in the field. These diseases are monocyclic, and their occurrence is predominantly determined by the soil environment. Soilborne pathogens produce inoculums or resting structures that can remain inactive in the soil for a long period of time. Under suitable conditions or by the release of root exudates of a susceptible host, the inoculums germinate and infect the plant. For the successful establishment of the pathogen inside plants, apart from compatible pathogen and host, the environmental factors also have to be compatible. Although it is difficult to manage soilborne diseases, existing strategies to control these pathogens include cultural, chemical, physical, and biological control besides development of host plant resistance (Katan 2017). While use of a broad spectrum of chemical pesticides is effective and convenient in managing soilborne diseases, their excessive use can result in severe damage to the environment by affecting a wide range of beneficial flora, fauna, and microbes (Kala et al. 2016). Moreover, the use of chemical pesticides can also damage human health. In recent years, increasing awareness among farmers about the harmful effects of chemical pesticides have helped generate a trend in limiting the use of harmful chemicals in agroecosystems (Kumar et al. 2010). Therefore, the use of biological control agents and biopesticides are gaining importance since they serve as the perfect alternative for harmful chemical pesticides (Compant et al. 2005; Cawoy et al. 2011).

17.1.3 Biological Control Agents

Biological control agents or biocontrol agents (BCA) can be defined as natural enemies that use parasitism, predation, or other mechanisms to control pests (Hokkanen and Pimentel 1984). Some examples of BCA are bacteria, fungus, virus, insects, etc. Using bacteria as BCA is an effective way in disease management since they are eco-friendly, have no side effects, are easy to use, narrow spectrum, sustainable, rational, and are safe to use as they do not harm the environment like chemical pesticides or insecticides (Kumar et al. 2010; Cawoy et al. 2011). Moreover, BCA can also help us to get rid of our overdependence on chemicals (Bahroun et al. 2016). Therefore, management of diseases by using BCA is gaining importance worldwide (Bhavani et al.; Adhilakshmi et al. 2014). In this aspect, bacteria can serve as a potential candidate. Studies have been performed throughout the world where rhizospheric bacteria, i.e., bacteria that remain in close association with the plant in the rhizosphere and endophytic bacteria—i.e., bacteria that reside in the plant—have been used successfully in managing diseases caused by various pests and pathogens (Bahroun et al. 2018). Species of *Rhizobium*, *Bacillus*, *Pseudomonas*, *Burkholderia*, and *Xanthomonas* are known antagonists of pests and pathogens (Kumar et al. 2010; Satya et al. 2011). Both endophytic as well as rhizospheric bacteria not only can promote plant growth and development but can also prevent the incidence of disease inside plants. Bacteria can promote plant growth either directly or indirectly (Bahroun et al. 2018). Direct influence of plant growth is imparted by nitrogen fixation, mobilization of nutrients such as phosphorus and iron, as well as by production of phytohormones (Bahroun et al. 2018; Hardoim et al. 2008; Akhtar and Siddiqui 2008). Simultaneously, bacteria can also protect plants from the attack of various pests and pathogens and thus stimulate plant growth indirectly (Aravind et al. 2010). These bacteria are capable of inducing systemic resistance in plants and play an important role in priming plant's defense-related genes during pathogen attacks (Saravanakumar et al. 2007). Generally, to control infection by a single pathogen, a single biological control agent is used. However, using a consortium of two or more bacterium has often shown better results in plant colonization and disease suppression rather than usage of a single bacterium (Akhtar and Siddiqui 2008). This happens because when taxonomically different biocontrol agents are used, they can colonize the root more aggressively because of their different requirement of temperature, pH, and moisture conditions. Therefore, use of specific beneficial bacteria as bioformulations or biofertilizers will be of great use to farmers. These bacteria can effectively control a few major diseases of pulses as discussed henceforth. There lies a shortcoming since studies of bacteria as biocontrol agents in managing pulse diseases are mainly restricted to wilt and rot diseases. Except for these two diseases, not much work has been done on other major diseases of pulses; thus, control of other soilborne diseases by bacteria still remains an unexplored field.

17.2 Major Diseases of Pulses in India

Pulses are attacked by a number of soilborne pathogens, most of them being fungus. A few root gall nematodes such as *Meloidogyne* sp. also attack pulses. The list is mentioned in Table 17.1. Some major soilborne diseases of pulses are as follows.

17.2.1 Wilt

The most common disease of pulses caused by the soilborne fungus, *Fusarium oxysporum*, which affects nearly all types of pulse crops including gram, pigeon pea, chick pea, lentils, red kidney beans as well as black-eyed pea, results in substantial yield reduction. This disease is predominant in the states of Punjab, Madhya Pradesh, Uttar Pradesh, Bihar, Madras, and parts of Mysore, Maharashtra, and West Bengal. The aboveground symptoms noticed are wilting, yellowing, and premature dropping of leaves; within 7–10 days after external symptoms appear, the plant dies. The incidence of the disease is increased by high temperature during sowing and flowering. Use of resistant varieties is the best preventive measure.

TABLE 17.1
Major Diseases of Various Pulses Caused by Soilborne Pathogens

Sl. No.	Name of the Disease	Causal Agent(s)	Pulses Affected (Common Name)	Pulses Affected (Scientific Name)
1	Wilt	*Fusarium oxysporum*	Gram	*Cicer arietinum*
			Pigeon pea	*Cajanus cajan*
			Red kidney beans	*Phaseolus vulgaris*
			Lentils	*Lens culinaris*
			Black-eyed pea	*Vigna unguiculata*
2	Rot	*Sclerotina sclerotiorum*	Gram	*Cicer arietinum*
		Rhizoctonia solani		
		Macrophomina phaseolina	Green beans	*Vigna radiata*
		Macrophomina phaseolina	Black matpe	*Vigna mungo*
		Fusarium solani	Red kidney beans	*Phaseolus vulgaris*
		Macrophomina phaseolina	Black eyed peas	*Vigna unguilata*
3	Blight	*Phytophthora drechsleri*	Pigeon pea	*Cajanus cajan*
		Alternaria alternata	Pigeon pea	*Cajanus cajan*
		Sclerotium rolfsii	Green beans	*Vigna radiata*
		Rhizoctonia solani	Black-eyed pea	*Vigna unguiculata*
		Sclerotium rolfsii	Black-eyed pea	*Vigna unguiculata*
4	Powdery mildew	*Leveillula taurica*	Pigeon pea	*Cajanus cajan*
		Erisyphe polygoni	Black matpe	*Vigna mungo*
		Erisyphe polygoni	Green beans	*Vigna radiata*
		Erisyphe polygoni	Red kidney beans	*Phaseolus vulgaris*
		Erisyphe polygoni	Black-eyed pea	*Vigna unguiculata*
5	Root gall and root knot	*Meloidogyne incognita*	Green gram	*Vigna radiata*
		Meloidogyne incognita	Chickpea	*Cicer arietinum*
		Meloidogyne javanica	Mung bean	*Vigna radiata*
		Meloidogyne incognita	Black gram	*Vigna mungo*

17.2.1.1 Biocontrol Measure

There are many reports available where bacteria have acted as BCA to prevent the occurrence of wilt disease in different pulses. Some examples are:

1. Chickpea wilt caused by *Fusarium oxysporum* f.sp. *ciceri* can be controlled by the endophytic bacteria *Pseudomonas fluorescens*. Seed treatment was found to be the best method of application resulting in the most effective result. The efficiency in suppressing the disease increases with the combined application of the endophyte with organic amendments such as mustard cake (Kala et al. 2016).
2. Rhizospheric *Pseudomonas fluorescens* has also been effective in controlling chickpea wilt. Talc-based formulations of this endophyte proved to be effective. *P. fluorescens* is also a plant growth-promoting bacterium (Vidhyasekaran and Muthamilan 1995).
3. In greengram (*Vigna radiata*), diseases such as root gall caused by *Meloidogyne incognita* and wilt caused by *Fusarium oxysporum* were controlled by the rhizobacteria *Pseudomonas fluorescens*. The bacteria could suppress nematode infection, as it had the capability to alter root exudates that in turn changed nematode behavior and suppress nematode multiplication (Akhtar et al. 2005).
4. *Fusariumudum* is the most important soilborne pathogen causing wilt in pigeon pea. Use of endophytic *Sinorhizobium fredii* KCC5 isolated from the root nodules of *Cajanus cajan* and

Pseudomonas fluorescens LPK2 isolated from tomato rhizosphere were able to suppress wilt in pigeon pea (Kumar et al. 2010).

5. Wilt of lentils caused by *Fusarium oxysporum* f.sp. *lentis*, can be reduced by seed treatment with *Pseudomonas fluorescens* in combination with a fungus, *Trichoderma harzianum* (Garkoti et al. 2014).

17.2.2 Root Rot

Rotting of root is caused predominantly by the soilborne fungus *Macrophomina phaseolina*. This disease attacks the root of plants and caused poor growth, wilted leaves, early leaf drop, branch dieback, and eventually death. Although the main cause of root rot is poorly drained soil, eventually the weakened roots can be attacked by soilborne fungi. The spores of these fungi reside in soil in dormant conditions and once waterlogged, the spores germinate and attack the weak roots of the plant, causing severe disease. The disease affects the following pulses: gram, green beans, red kidney beans, black matpe, and black-eyed peas.

17.2.2.1 Biocontrol Measure

The different biological control measures reported in controlling root rot of pulses are as follows:

1. Root rot of mung bean is caused by the fungus *Macrophomina phaseolina*. Different strains of *Streptomyces* sp. isolated from groundnut rhizosphere were able to inhibit the mycelial growth of the pathogen. The strains that were able to reduce the pathogen growth also promoted growth of the plants by increasing germination percentage, root, and shoot length. Using powdered formulations of *Streptomyces* sp. in seed treatment or in soil caused reduction of the disease, but a combination of both the methods increased the efficiency of disease control even in field trials (Adhilakshmi et al. 2014).
2. *Pseudomonas alcaligenes*, *Bacillus pumilus* along with *Glomus intraradices* has been found to be effective in controlling root knot nematode *Meloidogyne incognita* and root-rot fungus *Macrophomina phaseolina* in chickpea. The assessment was done based on shoot dry mass, pod number, nodulation, shoot content of chlorophyll, nitrogen, potassium, and phosphorus on plants after inoculation with the pathogen over untreated plants. It was observed that *P. alcaligenes* was the most effective in promoting plant growth after infection by the pathogen than the other two. Combination of all three caused a greater increase in all the parameters as compared to any of the two mixtures. Bacteria belonging to pseudomonads were able to suppress the growth of both parasitic as well as nonparasitic pathogens, as they were capable of producing certain biologically active substances such as antibiotics (Akhtar and Siddiqui 2008).
3. Effects of *Pseudomonas putida, P. alcaligenes, and Pseudomonas* isolate (Ps28) were evaluated on the hatching and penetration of *Meloidogyne incognita* in chickpea roots and on the fungus *Macrophomina phaseolina*. Other traits such as siderophore production, Indole acetic acid production, and hydrogen cyanide production were also studied. *P. putida* was the most efficient among them and not only colonized roots better than the rest but was also the best to minimize infection by both the pathogens. Efficiency of the bacterial strains are as follows: *P. putida, P. alcaligenes*, and *Pseudomonas* isolate Ps28. *P. putida* has the ability to act as a biocontrol agent controlling root rot disease complex in chickpea. It also promoted plant growth along with inhibiting the growth of the fungal pathogen and reducing the incidence of gall by preventing nematode infection as well as gall formation (Akhtar et al. 2005).
4. *Rahnella aquatilis* B16C, *Pseudomonas yamanorum* B12 and *Pseudomonas fluorescens* B8P were analyzed for their potential to control root rot of faba bean caused by the fungus *Fusarium solani*. Endophytic bacteria were isolated from faba bean (*Vicia faba*) nodules. Among the three isolates, *Rahnella aquatilis* B16C showed the best anatagonist potential against *Fusarium solani* and could be used as a potential biocontrol agent in the field (Bahroun et al. 2018).

5. The soilborne fungus *Rhizoctonia bataticola* causes dry root rot in chickpea and is a major limiting factor for chickpea production in India. The endophyte of chickpea *Pseudomonas aeruginosa* has been found to be effective in controlling the growth of this pathogen (Bhavani et al.).
6. The fungus *Rhizoctonia bataticola* that causes dry root rot in chickpea can be effectively managed by actinomycetes isolates. The isolates were identified as *Streptomyces tsusimaensis, Streptomyces caviscabies, Streptomyces setonii,* and *Streptomyces africanus.* The isolates were able to inhibit the growth of the fungus *Macrophomina phaseolina* and *Rhizoctonia bataticola* in vitro. Moreover, all these isolates were also able to reduce the incidence of disease under greenhouse conditions (Wei et al. 1991).
7. Root rot of common bean (*Phaseolus vulgaris*) is caused by the fungus, *Fusarium solani.* The study investigates the role of mycorrhiza *Glomus mosseae* and rhizospheric bacteria *Rhizobium leguminosarum* against the fungus. It was found that *Glomus* can effectively reduce the colonization of the pathogen, and its activity is more pronounced in presence of the bacteria. Although both the fungus and bacteria alone are capable of reducing pathogenicity, they serve as a better biological control agent when applied in combination (Dar et al. 1997).
8. Charcoal rot of mung bean caused by *Macrophomina phaseolina* can be reduced by coating the seeds with biocontrol agents such as *Streptomyces* sp. and *Rhizobium meliloti.* Both the bacteria were successful in preventing the disease in 30-day-old mung bean plants and reduced incidence of disease in 60-day-old plants (Hussain et al. 1990).
9. Root rot in mung bean caused by *Macrophomina phaseolina* can be reduced by treating with powdered bioformulations of *Pseudomonas fluorescens* Pf1. This bioformulation along with chitin was most effective in reducing the incidence of root rot (Saravanakumar et al. 2007).
10. Root rot root-knot disease in mung bean is caused by a group of pathogens *Macrophomina phaseolina, Fusarium solani,* and *Rhizoctonia solani.* Rhizospheric bacterial strains such as *Pseudomonas fluorescens* and *Bacillus subtilis* can inhibit growth of fungal mycelium. When used as a seed dressing or as soil drench, it can reduce infection of roots significantly by also preventing growth of nematodes. Thus, improves plant growth and yield in mung bean (Ali Siddiqui et al. 2001).
11. Collar rot of chickpea caused by the fungus, *Sclerotium rolfsii,* can be inhibited by the bacteria *Bacillus* and *Pseudomonas,* isolated from composts and macrofauna. The isolates of *Pseudomonas* sp. were also compatible with chickpea *Rhizobium* sp. When applied together by seed-priming or seed-coating methods, they reduced the disease incidence in plants as well as promoted plant growth (Hameeda et al. 2010).
12. Root rot disease complex in chickpea caused by *Macrophomina phaseolina* and *Meloidogyne incognita* can be controlled with the use of phosphate solubilizing bacteria, *Pseudomonas aeruginosa.* These bacteria can convert insoluble form of phosphates into soluble form and thus promote growth and yield of plants (Siddiqui and Akhtar 2007).

17.2.3 Blight

This disease can be caused by different fungus such as *Phytophthora drechsleri* causing Phytophthora blight, *Alternaria alternata* causing Alternaria blight, *Sclerotium rolfsii* causing Sclerotium blight, *Macrophomina phaseolina* causing Macrophomina blight, and *Ascochyta* sp. causing Ascochyta blight. This disease occurs in epidemic form primarily in the states of Punjab, Bihar, and Jammu. The aerial parts of plants are attacked by the fungus that causes the appearance of circular spots on the leaves and pods and elongated spots on petioles (Sen Gupta 1974).

17.2.3.1 Biocontrol Measure

There is no report of bacterial biological control of blight in pulses.

17.2.4 Powdery Mildew

The disease is primarily caused by the fungus *Erisyphe polygoni*. At first, symptoms such as small, white, circular powdery spots appear on the upper surface of leaves; at advanced stages, the whole leaf gets covered with talcum powder. Then leaves become smaller and chlorotic and finally drop off (Sen Gupta 1974).

17.2.4.1 Biocontrol Measure

There is no report of biological control of powdery mildew in pulses by bacteria.

17.2.5 Root Gall

Galls in root of the plants are caused by a sedentary endoparasitic nematode known as *Meloidogyne incognita* (Akhtar et al. 2005). The nematode attacks a wide range of crops and vegetables. Root knot caused by nematodes generally is associated with root rot or wilt caused by fungus. The list of different biological control agents reducing disease incidence in various pulses is mentioned in Table 17.2.

TABLE 17.2

Different Bacterial Biological Control Agents Which Have Been Reported to Control Various Disease in Pulses

Sl. No.	Disease	Pulses Affected	Bacterial Biocontrol Agent	Mode of Action	References
1	Fusarium wilt	*Cicer arietinum*	*Pseudomonas fluorescens*	Direct antagonism	Kala et al. (2016)
2	Fusarium wilt	*Cicer arietinum*	*Pseudomonas fluorescens*	Colonization	Vidhyasekaran & Muthamilan (1995)
3	Wilt and root gall	*Vigna radiata*	*Pseudomonas fluorescens*	Antibiosis, lytic enzymes, and siderophore production	Akhtar et al. (2005)
4	Fusarium wilt	*Cajanus cajan*	*Sinorhizobium fredii* and *P. fluorescens*	Phytostimulation and chitinolysis	Kumar et al. (2010)
5	Root rot	*Vigna radiata*	*Streptomyces* sp.	Induced systemic resistance and phytostimulation	Adhilakshmi et al. (2014)
6	Root rot and root gall	*Cicer arietinum*	*Pseudomonas alcaligenes* and *Bacillus pumilus*	Antibiosis and phytostimulation	Siddiqui & Akhtar (2007)
7	Root rot and root gall	*Cicer arietinum*	*Pseudomonas putida* and *Pseudomonas alcaligenes*	Siderophore and HCN production, phytostimulation, colonization	Akhtar & Siddique (2008)
8	Root rot	*Vicia faba*	*Rahnella aquatilis*, *Pseudomonas yamanorum* and *Pseudomonas fluorescens*	Antibiosis and siderophore production	Bahroun et al. (2018)
9	Dry root rot	*Cicer arietinum*	*Pseudomonas aeruginosa*	Induced systemic resistance	Bhavani et al. (2015)
10	Root rot	*Phaseolus vulgaris*	*Rhizobium leguminosarum*	Rhizobitoxine production and phytostimulation	Dar et al. (1997)
11	Charcoal rot	*Vigna radiata*	*Streptomyces* sp. and *Rhizobium meliloti*	Colonization	Hussain et al. (1990)
12	Root rot	*Vigna mungo*	*Pseudomonas fluorescens*	Induced systemic resistance and chitinolysis	Saravanakumar et al. (2007)
13	Root rot and root knot	*Vigna radiata*	*Pseudomonas fluorescens* and *Bacillus subtilis*	Production of antibiotics and siderophore	Siddiqui et al. (2001)
14	Collar rot	*Cicer arietinum*	*Bacillus* and *Pseudomonas*	Siderophore production and seed priming	Hameeda et al. (2010)
15	Root rot	*Cicer arietinum*	*Pseudomonas aeruginosa*	Phytostimulation	Siddiqui et al. (2001)
16	Root knot	*Vigna mungo*	*Bradyrhizobium* sp.	Phytostimulation	Bhat et al. (2012)
17	Wilt	*Lens culinaris*	*Pseudomonas fluorescens*	Not explored	Garkoti et al. (2014)

17.2.5.1 Biocontrol Measure

The bacterial biocontrol agent used in preventing root knot has also been used in preventing root rot or wilt diseases in pulses.

1. *Pseudomonas alcaligenes*, *Bacillus pumilus*, and *Glomus intraradices* have been found to be effective in controlling root knot nematode, *Meloidogyne incognita* and root-rot fungus, *Macrophomina phaseolina* in chickpea. It was observed that *P. alcaligenes* was the most effective in promoting plant growth after infection by the pathogen than the other two (Akhtar and Siddiqui 2008).
2. Effects of *Pseudomonas putida*, *P. alcaligenes*, and *Pseudomonas* isolate (Ps28) were evaluated on the hatching and penetration of *Meloidogyne incognita* in chickpea roots and on the fungus *Macrophomina phaseolina*. *P. putida* was the most efficient among them and not only colonized roots better than the rest but was also the best to minimize infection by both the pathogens (Akhtar et al. 2005).
3. Root-rot disease complex in chickpea caused by *Macrophomina phaseolina* and *Meloidogyne incognita* can be controlled with the use of phosphate solubilizing bacteria, *Pseudomonas aeruginosa*. These bacteria can convert insoluble form of phosphates into soluble form and thus promote growth and yield of plants (Siddiqui and Akhtar 2007).
4. In greengram (*Vigna radiata*) diseases such as root gall caused by *Meloidogyne incognita* and wilt caused by *Fusarium oxysporum* were able to be controlled by the rhizobacteria *Pseudomonas fluorescens*. The bacteria could suppress the nematode infection, as they had the capability to alter root exudates, which in turn changed nematode behavior and suppressed nematode multiplication (Akhtar et al. 2005).
5. In black gram (*Vigna mungo*), the interaction of root knot nematode, *Meloidogyne incognita*, with the rhizospheric bacteria, *Bradyrhizobium* sp. and the fungus *Paecilomyces lilacinus*, was studied. It was shown that inoculation of plants with the bacteria was efficient in improving growth of the plants mainly by fixing nitrogen (Bhat et al. 2012).

17.3 Mechanism of Action

Various workers have proposed different mechanisms of action used by the biological control agents against antagonistic pathogens. Bacteria adopt a variety of strategies to combat against fungus. The schematic representation of the mechanism of action adopted by bacteria in mitigating biotic stress is shown in Figure 17.1. The strategies followed by biocontrol agents are as following :-

17.3.1 Production of Antibiotics

Several reports suggest that the pseudomonad group of bacteria can act against pathogenic fungus, as they are capable of producing secondary metabolites such as antibiotics (Akhtar and Siddiqui 2008). Fluorescent pseudomonads are also reported to produce 2,4-diacetylphloroglucinol, an antifungal metabolite that can protect plants from invasion by fungal pathogens (Raaijmakers and Weller 1998). The bacteria, *Rahnella aquatilis*, along with *Pseudomonas* also had the potential to produce PRN (chlorinated phenylpyrrole antibiotics), which can suppress the growth of soilborne fungi. There are reports suggesting that endophytic bacteria possess a number of genes responsible for the synthesis of antibiotic compounds such as Pyrrolnitrin (PRN), Phenazine (PHZ), Phenazine-1-carboxylic acid (PCA), Pyoluteorin (PLt), 2,4-Diacetylphloroglucinol (DAPG), and Hydrogen cyanide (HCN). These molecules have the potential to either inhibit growth of the pathogens or kill them directly (Lugtenberg and Kamilova 2009). Antifungal activity of bacteria is mainly dependent on the production of a range of bioactive metabolites that have different effects on various pathogens (Duffy et al. 2003).

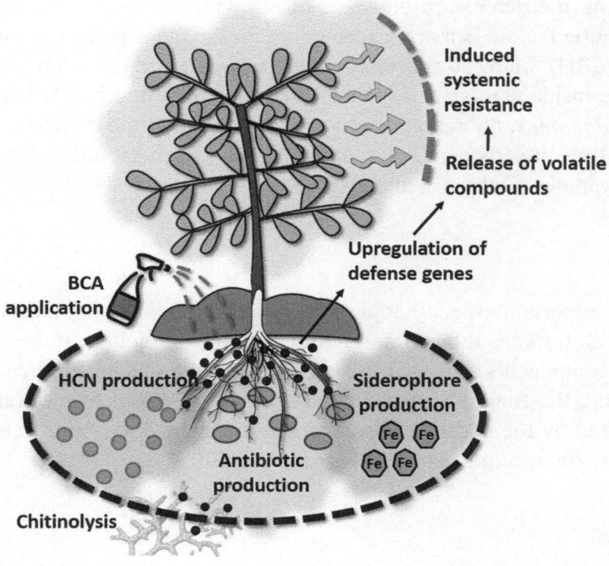

FIGURE 17.1 Schematic representation of the mechanism of action followed by bacteria in managing soilborne pathogens. In the soil, direct antagonism is achieved by synthesis and secretion of antibiotics, siderophores, and HCN. Chitinolysis also plays an important role in managing soilborne fungus. BCA also upregulates defense-related genes in the plant, leading to synthesis of volatile compounds and activating induced systemic resistance in plants thereby preventing pathogen attack indirectly.

17.3.2 Production of Siderophores

Iron is an essential element required for normal plant growth and development. Under iron-limiting conditions, bacteria synthesize siderophores that are chelating agents that sequester iron under low availability. Fungus can also produce siderophores, but those have low affinity for iron than bacterial siderophores (Compant et al. 2005). Fluorescent *Pseudomonas* produce siderophores and deprive the pathogen of available iron thereby causing antagonism (Voisard et al. 1989). The bacteria *Rahnella aquatilis* is also reported to produce siderophores. *Actinomycetes* (e.g., *Streptomyces* sp.) are also capable of producing siderophores and inhibit the growth of phytopathogens (Tokala et al. 2002). *Actinomycetes* that reside in the rhizosphere compete with other rhizospheric pathogenic fungi for iron, and by producing siderophores they can chelate iron better than the pathogen, and this may be a possible mode of inhibition of pathogen by these groups of bacteria (Gopalakrishnan et al. 2011).

17.3.3 Production of Hydrogen Cyanide (HCN)

Apart from antibiotics and siderophores, pseudomonad strains can also produce hydrogen cyanide by which they can directly inhibit the growth of other fungus. HCN is a broad spectrum antimicrobial compound that can control root infection (Akhtar and Siddiqui 2008). HCN, like the antibiotic PRN, can suppress growth of soilborne fungi (Bahroun et al. 2018). *Actinomycetes* are also reported to produce HCN, which further helps in disease suppression (Wei et al. 1991).

17.3.4 Induced Systemic Resistance

Streptomyces sp. are also able to reduce infection by inducing the expression of peroxidase and polyphenol oxidase in plants thereby activating the defense pathways in plants. *Actinomycetes*, when acting as biological control agents, are known to induce biochemical defense responses in the host. The levels of accumulation of plant defense responses are directly proportional with their ability to resist the attack of pathogens. Triggering the defense response in plants with the help of its own defense system by inducing biotic and abiotic factors is a novel technology in the management of plant defense responses (Gopalakrishnan et al. 2011). Many strains of *Pseudomonas* can also protect plants indirectly by inducing systemic resistance inside plants against various pests and pathogens (Akhtar and Siddiqui 2008). Another strain of *Pseudomonas*, fluorescent *Pseudomonas*, are reported to induce systemic resistance in plants besides direct killing of pathogens by producing secondary metabolites such as phenazines, acetyl phloroglucinols, and cyanides (Defago et al. 1990).

17.3.5 Chitinolysis

Chitinolysis is another important aspect that plays an important role in biological control, and increased control is achieved by chitin amendment of soil with chitinolytic biological control agents. Chitin and β-1,3-glucan are chief components of fungal cell wall. Bacteria producing chitinase and β-1,3-glucanase are capable of degrading the fungal hyphal wall. Studies have shown that extracellular chitinase and laminarinase synthesized by the bacteria *Pseudomonas strutzeri* digest and lyse the mycelium of the pathogenic fungus, *Fusarium solani* (Saravanakumar et al. 2007).

17.3.6 Rhizobitoxine

Rhizobium sp. is known to produce a chemical rhizobitoxine that prevents pathogenic fungi from entering plant tissues. Nodulation with *Rhizobium* thus helps in managing diseases (Dar et al. 1997).

17.3.7 Phytostimulation

There are some bacteria that are capable of phytostimulation, i.e., promoting plant growth by modulating a plant's hormonal balance apart from protecting the plant from the attack of pathogens. These microbes have dual properties and indirectly they help in increasing productivity of the plant by promoting its growth in addition to functioning as disease control mechanisms. *Pseudomonas putida* is also reported to produce high levels of auxin, Indole acetic acid (IAA), which promotes plant growth in addition to protecting plants from pathogenic fungus (Akhtar and Siddiqui 2008).

17.3.8 Colonization

Moreover, fluorescent *Pseudomonas* strains are also fierce colonizers of the plant root and outcompete the pathogens and prevent them from colonizing (Neilands and Leong 1986). Competition for ecological niche can be utilized in preventing attack of a broad range of pathogens (Compant et al. 2005).

17.4 Limitations

When applied directly to the field, biocontrol agents often fail to compete with the existing microflora and establish themselves in the soil and thus fail to control disease. Inefficiency of the system that fails to maintain the viability of the propagules during storage and marketing is partly responsible for the failure of biocontrol agents in soil. BCAs can provide effective control if their growth can be maintained in the formulations and the propagules maintain adequate viability in the formulations. For the massive propagations of microorganisms, a supportive food base is required in the formulations that would provide nutrition for the sustainability and development of propagules. Different techniques have been studied and used to deliver BCAs into soil such as in the form of pellets, granules, or liquid formulations. Fine granules or powdered forms are a more effective delivery system than the liquid formulations since they are cheap and also easy to handle. Although different solid material combinations were tested in the formation of powdered formulations, none of them provided higher CFU load or long shelf life of the formulation. In this regard, an effective formulation has been prescribed by a group of scientists who have used molasses as the carbon source for the growth of microbes along with sawdust to provide a porous and light environment for the multiplication of the microbes (Khan et al. 2011).

17.5 Future Prospects

The losses caused due to yield reduction need to be prevented to maintain food security and check economic losses. Therefore, sustainable agricultural management must be practiced to provide better food and fiber (Verma et al. 2013). To a great extent, it is in the hands of the farmers to maintain crop yield and production. Therefore, any procedure designed to help the resource-poor farmers to combat pests and pathogens needs to be handy and less complicated since this will facilitate the use of the method by farmers. Use of biofertilizers is an effective way in preventing attack of pathogens or in reducing the incidence of disease. But to make the effect long lasting, correct formulations of biofertilizers should be used. Studying the different mechanisms of actions used by the microbes in preventing disease will be helpful in formulating the correct combinations of microbes that can effectively act against the pathogen. In this context, it is notable that using indigenous microbes is advantageous, as they can establish themselves in the native soil easily and thus facilitate plant-microbe interactions (Verma et al. 2013). There are various aspects by which bacteria can be used as biofertilizers. One easy and effective way is to use them as bioformulation along with a carbon source that will provide easy multiplication of the microbes in soil thereby causing their enrichment (Khan et al. 2011). Another effective way is to do seed priming (Vidhyasekaran and Muthamilan 1995) or seed treatment with bacterial inoculum (Kala et al. 2016). Seed priming is the technique where seeds are soaked in water and then air dried before sowing (Hameeda et al. 2010). This is a low-cost technology that can be adopted by the farmers since it is cost-effective and can also mitigate suboptimal stress level or adverse physical conditions (Hameeda et al. 2010). In the future, strategies have to be adopted where native microbes having plant growth-promoting abilities are enriched and then used to protect the plants as well as to promote their growth. Since microbes residing in soil form a community and interact with each other as well as plants, it becomes of utmost importance to have a complete knowledge of the organisms inhabiting a particular environment. In this regard, metagenomics plays a key role in exploring the wide diversity of microbes inhabiting a particular environment (Kunda et al. 2018). Therefore, metagenome analysis of the soil can provide us with important information regarding the microbial community, which would help to design a better strategy in stress mitigation. Knowing the microbial community followed by enriching the beneficial microbes in the native soil can be an effective way in managing biotic stress. It is also worth mentioning that very little work has been done to screen biological control agents against pathogens affecting growth of pulses in India. Most of the studies have found *Bacillus* sp. and *Pseudomonas* sp. as the effective biological control agents (Bahroun et al. 2018). But nonsymbiotic endophytes may also remain in unique association with the plant and may provide more potent effect in growth promotion of hosts (Compant et al. 2005). There is a lot of scope for researchers to work in this emerging trend so that a clear knowledge can be obtained about the different biocontrol agents that can be used effectively to mitigate biotic stress in plants.

REFERENCES

Adhilakshmi, M., Paranidharan, V., Balachandar, D., Ganesamurthy, K. & Velazhahan, R. Suppression of root rot of mung bean (*Vigna radiata* L.) by *Streptomyces* sp. is associated with induction of peroxidase and polyphenol oxidase. *Archives of Phytopathology and Plant Protection* 47, 571–583 (2014).

Akhtar, H., Anita, S. & Kumar, S. P. Studies on the management of root-knot nematode, *Meloidogyne incognita*-wilt fungus, *Fusarium oxysporum* disease complex of green gram, *Vigna radiata* cv ML-1108. *Journal of Zhejiang University Science B* 6, 736–742 (2005).

Akhtar, M. S. & Siddiqui, Z. A. *Glomus intraradices*, *Pseudomonas alcaligenes*, and *Bacillus pumilus*: Effective agents for the control of root-rot disease complex of chickpea (*Cicer arietinum* L.). *Journal of General Plant Pathology* 74, 53–60 (2008).

Ali Siddiqui, I., Ehetshamul-Haque, S. & Shahid Shaukat, S. Use of rhizobacteria in the control of root rot–root knot disease complex of mungbean. *Journal of Phytopathology* 149, 337–346 (2001).

Aravind, R., Eapen, S., Kumar, A., Dinu, A. & Ramana, K. Screening of endophytic bacteria and evaluation of selected isolates for suppression of burrowing nematode (*Radopholus similis* Thorne) using three varieties of black pepper (*Piper nigrum* L.). *Crop Protection* 29, 318–324 (2010).

Bahroun, A., Jousset, A., Mhamdi, R., Mrabet, M. & Mhadhbi, H. Anti-fungal activity of bacterial endophytes associated with legumes against *Fusarium solani*: Assessment of fungi soil suppressiveness and plant protection induction. *Applied Soil Ecology* 124, 131–140 (2018).

Bhat, M. Y., Wani, A. H. & Fazal, M. Effect of *Paecilomyces lilacinus* and plant growth promoting rhizobacteria on *Meloidogyne incognita* inoculated black gram, *Vigna mungo* plants. *Journal of Biopesticides* 5, 36 (2012).

Bhavani, D. G., Kumari, A. M. & Kumar, A. K. Chickpea endophytic bacteria inhibiting dry root rot fungus *Rhizoctonia bataticola* International Journal of Scientific & Engineering Research, 6, 83–86 (2015).

Cawoy, H., Bettiol, W., Fickers, P. & Ongena, M. in *Pesticides in the Modern World-Pesticides Use and Management* (InTech, 2011).

Compant, S., Duffy, B., Nowak, J., Clément, C. & Barka, E. A. Use of plant growth-promoting bacteria for biocontrol of plant diseases: Principles, mechanisms of action, and future prospects. *Applied and Environmental Microbiology* 71, 4951–4959 (2005).

Dar, G. H., Zargar, M. & Beigh, G. Biocontrol of *Fusarium* root rot in the common bean (*Phaseolus vulgaris* L.) by using symbiotic *Glomus mosseae* and *Rhizobium leguminosarum*. *Microbial Ecology* 34, 74–80 (1997).

Defago, G., Haas, D., Bollag, J. & Stotzky, G. Pseudomonads as antagonists of soilborne plant pathogens: Modes of action and genetic analysis. *Soil Biochemistry* 6, 249–291 (1990).

Divya Rani, V. & Sudini, H. Management of soilborne diseases in crop plants: An overview. *International Journal of Plant, Animal and Environmental Sciences* 3, 156–164 (2013).

Duffy, B., Schouten, A. & Raaijmakers, J. M. Pathogen self-defense: Mechanisms to counteract microbial antagonism. *Annual Review of Phytopathology* 41, 501–538 (2003).

Garkoti, A., Kumar, V. & Tripathi, H. Control of wilt disease of lentil through bio control agents and organic amendments in Tarai region of Uttarakhand, India. *Journal of Environmental Biology* 35, 1067 (2014).

Gopalakrishnan, S. et al. Evaluation of actinomycete isolates obtained from herbal vermicompost for the biological control of Fusarium wilt of chickpea. *Crop Protection* 30, 1070–1078 (2011).

Hameeda, B., Harini, G., Rupela, O., Rao, J. K. & Reddy, G. Biological control of chickpea collar rot by co-inoculation of antagonistic bacteria and compatible Rhizobia. *Indian Journal of Microbiology* 50, 419–424 (2010).

Hardoim, P. R., van Overbeek, L. S. & van Elsas, J. D. Properties of bacterial endophytes and their proposed role in plant growth. *Trends in Microbiology* 16, 463–471 (2008).

Hokkanen, H. & Pimentel, D. New approach for selecting biological control agents. *The Canadian Entomologist* 116, 1109–1121 (1984).

Hussain, S., Ghaffar, A. & Aslam, M. Biological control of *Macrophomina phaseolina* charcoal rot of sunflower and mung bean. *Journal of Phytopathology* 130, 157–160 (1990).

Kala, C., Gangopadhyay, S. & Godara, S. Eco-friendly management of wilt caused by *Fusarium oxysporum* f.sp. *ciceri* in chickpea. *Legume Research* 39, 129–134 (2016).

Katan, J. Diseases caused by soilborne pathogens: Biology, management and challenges. *Journal of Plant Pathology* 99, 305–315 (2017).

Khan, M. R., Majid, S., Mohidin, F. A. & Khan, N. A new bioprocess to produce low cost powder formulations of biocontrol bacteria and fungi to control fusarial wilt and root-knot nematode of pulses. *Biological Control* 59, 130–140 (2011).

Kumar, H., Bajpai, V. K., Dubey, R., Maheshwari, D. & Kang, S. C. Wilt disease management and enhancement of growth and yield of *Cajanus cajan* (L) var. Manak by bacterial combinations amended with chemical fertilizer. *Crop Protection* 29, 591–598 (2010).

Kunda, P., Dhal, P. K. & Mukherjee, A. Endophytic bacterial community of rice (*Oryza sativa* L.) from coastal saline zone of West Bengal: 16S rRNA gene based metagenomics approach. *Meta Gene* 18, 79–86 (2018).

Lugtenberg, B. & Kamilova, F. Plant-growth-promoting rhizobacteria. *Annual Review of Microbiology* 63, 541–556 (2009).

Neilands, J. & Leong, S. A. Siderophores in relation to plant growth and disease. *Annual Review of Plant Physiology* 37, 187–208 (1986).

Raaijmakers, J. M. & Weller, D. M. Natural plant protection by 2,4-diacetylphloroglucinol-producing *Pseudomonas* spp. in take-all decline soils. *Molecular Plant-Microbe Interactions* 11, 144–152 (1998).

Rana, D., Dass, A., Rajanna, G. & Kaur, R. Biotic and abiotic stress management in pulses. *Indian Journal of Agronomy* 61, 238–248 (2016).

Saravanakumar, D. et al. Rhizobacterial bioformulation for the effective management of *Macrophomina* root rot in mungbean. *Archives of Phytopathology and Plant Protection* 40, 323–337 (2007).

Satya, V., Vijayasamundeeswari, A., Paranidharan, V. & Velazhahan, R. *Burkholderia* sp. strain TNAU-1 for biological control of root rot in mung bean (*Vigna radiata* L.) caused by *Macrophomina phaseolina*. *Journal of Plant Protection Research* 51, 273–278 (2011).

Sen Gupta, P. K. Diseases of major pulse crops in India. *PANS Pest Articles & News Summaries* 20, 409–415 (1974).

Siddiqui, Z. & Akhtar, M. Biocontrol of a chickpea root-rot disease complex with phosphate-solubilizing microorganisms. *Journal of Plant Pathology* 67–77 (2007).

Singh, A. K., Singh, S., Prakash, V., Kumar, S. & Dwivedi, S. Pulses production in India: Present status, bottleneck and way forward. *Journal of AgriSearch* 2, 75–83 (2015).

Singh, R., Singh, M. K., Singh, A. K. & Singh, C. Pulses production in India: Issues and elucidations. *The Pharma Innovation Journal* 7, 10–13 (2018).

Tokala, R. K. et al. Novel plant-microbe rhizosphere interaction involving *Streptomyces lydicus* WYEC108 and the pea plant (*Pisum sativum*). *Applied and Environmental Microbiology* 68, 2161–2171 (2002).

Verma, J. P., Yadav, J., Tiwari, K. N. & Kumar, A. Effect of indigenous *Mesorhizobium* spp. and plant growth promoting rhizobacteria on yields and nutrients uptake of chickpea (*Cicer arietinum* L.) under sustainable agriculture. *Ecological Engineering* 51, 282–286 (2013).

Vidhyasekaran, P. & Muthamilan, M. Development of formulations of *Pseudomonas fluorescens* for control of chickpea wilt. *Plant Disease (USA) 79, 782–786* (1995).

Voisard, C., Keel, C., Haas, D. & Dèfago, G. Cyanide production by *Pseudomonas fluorescens* helps suppress black root rot of tobacco under gnotobiotic conditions. *The EMBO Journal* 8, 351–358 (1989).

Wei, G., Kloepper, J. W. & Tuzun, S. Induction of systemic resistance of cucumber to *Colletotrichum orbiculare* by select strains of plant growth-promoting rhizobacteria. *Phytopathology* 81, 1508–1512 (1991).

Weller, D. M. Biological control of soilborne plant pathogens in the rhizosphere with bacteria. *Annual Review of Phytopathology* 26, 379–407 (1988).

18

Bacillus thuringiensis *Cry Gene Mediated Biological Pest Management in Grain Legumes*

Abhishek Sharma, Dhruti Amin, Rajesh Arora, and Ajay Kumar Mathur

CONTENTS

18.1 Introduction

The legume family constitutes of around 400 genera and 10,000 species. Many of these are grown worldwide to obtain mineral, micronutrients, and protein-rich seeds and grains (Bohra et al. 2015; Mishra et al. 2018). Grain legumes play a major role in constituting the food safety and nutritional security program around the globe, and especially in underdeveloped and developing countries, as *grain legumes* provide a high-quality protein, complementary diet to vegetarian populations. Therefore, *grain legumes* stand as the second most frugally valued and nutritionally imperative grain crop as per the International Legume Database and Information Service (ILDIS) 2006–2013. Legume crops are cultivated as either major crops throughout the year or as rotational crops. In both cases, legume crops not only deliver high pecuniary benefits for being rich in proteins and minerals but also provide socioeconomic, ecological, and environmental benefits by reducing soil erosion and degradation.

The 2012 report of the Consultative Group on International Agricultural Research (CGIAR) presented the global data of arable land used for legume cultivation worldwide. In the given data, Myanmar (36%), Niger (30.6%), Kenya (27%), Mozambique (22%), and Burkina (20%) are the major countries with maximum arable lands cultivating legume crops. The arable land under legume crop production as recorded in 2013 was about 80.75 million hectares producing 73 million tons with a typical production of 904 kg/hectare (Parankusam et al. 2018). The average legume crop production was about 63.9 million tons in the last decade (2003–2012), and Asia contributes the most with (45.64%) of total production (http://faostat3.fao.org/home.E). Legumes are grown in over 171 countries worldwide, from which beans account for 37.5% of global legume production, followed by chickpea (16.77%), cowpea (15.05%), peas (8.50%), pigeon pea (7.70%), lentils (5.38%), and other (11.1%). India ranks first for cultivating 35% of cultivated area and 25% of global production of legume crops as a major and rotational crop (Kumar et al. 2011; Parankusam et al. 2018). Being a developing country, India is suffering from several serious issues ranging from health to socioeconomic growth of the population. The malnutrition problem in India is very high and has made the country rank 97th out of 118 countries in the Global Hunger Index (GHI), with a serious hunger situation (GHI Report 2017). This alarming situation has raised an emergency, necessitating measures to deal with the nutritional security of the population worldwide, in which grain legumes are providing an extremely valued contribution for ensuring global food and nutritional security.

Legume crops not only improve the socioeconomic development of growing countries but also play a crucial role and ensure safety and balancing of environmental conditions. These crops have a major contribution to global, biological nitrogen cycle completion. The legume crops ensure the fixing of atmospheric nitrogen molecules into the soil with the help of rhizobia, residing with the crop in the soil in symbiotic association. This symbiotic association mediated nitrogen fixation (SAMNF) makes legume crops competent to grow in almost all kind of soils with very few requirements of fertilizers. Therefore, the legume cultivation as regular and rotational crops not only cuts down the fertilizer requirements for self-crop but also for the other crops (Goyal et al. 2018).

18.2 Pest Problems in Grain Legumes

Pest is the major threat and hurdle in the way of a global food and nutrition security mission worldwide. There are almost 200 insect species are used to attack grain legumes; however, almost 12 are considered as a major threat (Keneni et al. 2011). Pests infect and affect the grain legumes in two ways, directly and indirectly. The direct damage they cause is a morphological loss done by chewing the plant parts such as leaves, stems, fruits, and flowers or by sucking the flower sap, resulting in severe foliar damage. Insect pests generally attack and feed over pods, seeds, and flowers also, and therefore drastically affect the crop yield in the field as well as in storage. Major pests that infect grain legumes are Aphids such as *Aphis craccivora*, legume pod borer such as *Maruca vitrata*, and gram/spotted pod borer *Helicoverpa armigera*, whitefly *Bemisia tabaci* Genn., leafhopper *Empoasca* spp., caterpillar *Spodoptera* spp., cutworm *Agrotis ipsilon*, pulse beetle *Callosobruchus chinensis* and *C. maculatus*, thrips *Caliothrips indicus* Bag. and *Megaleurothrips* distalis, etc. (Reddy 2009). The indirect damage is caused by sucking pests that introduce the viral disease by working as virus transmitting vectors. Whitefly *Bemisia tabaci* Genn. also acts as a yellow mosaic virus (YMV) transmitting vector that carries YMV, which causes 30%–70% damage in crop yield. These pests root extensive impairment with an estimated average loss of at least 20 lakh ton grain legumes worth US$85 million. Combined together, there is an approximately 20% annual loss of grain legumes (Dhaliwal et al. 2010; Mishra et al. 2018).

In India, gram/spotted legume pod borer *Helicoverpa armigera* is the most harmful pest, causing destruction of pigeon pea and chickpea. The legume pod borer *Maruca vitrata* is the second major causative agent that reduces the pigeon pea yield up to 84%, worth US$30 million. Similarly, pod fly *Melanagromyza abtusa* causes more than 40% pigeonpea crop destruction (Margam et al. 2011; Singh et al. 2013). This loss is a massive hurdle in the socioeconomic development of grain legume-producing countries.

18.3 Pest Management in Grain Legumes

The grain legumes being the source of high proteins, micronutrients, and minerals are not only preferred by humankind to have as a staple food with good nutritional values but by other living organisms such as birds and insects. However, their unwanted interference of these organisms in field-grown crops creates massive destruction of food material cultivated for humankind. Pests play a major role in the destruction. Therefore, it is almost indispensable to apply eco-friendly pest control approaches to get rid of crop damage caused by pests. For this, an integrated pest management system strategy is the most suitable one. In agriculture, the integrated pest management system works on biological, chemical, and cultural systems. The biological means involves use of natural enemies of the pest as predators or parasites to pest. The chemical means applies the use of insecticides and pesticides to eliminate the pests, whereas cultural control involves exclusion, repulsion, and physical removal, etc. Among these, chemical-based pest management is more effective so far; however, being very costly the use of these hazardous pesticides has shown decisive paraphernalia on the surrounding environment, and have a drastic upshot over soil fertility and microbial inhabitants residing within. Continuous use of these harmful chemicals not only leads to the evolution of pesticides and insecticides resistant pest varieties but also potentially show carcinogenic, mutagenic, and teratogenic effects on human and animal health. Therefore, to overcome these hazardous effects of pesticides, biotechnology-based integrated pest management with the aim to develop genetically engineered pest-resistant plant varieties is preferably used. With the increased hazardous effects of chemical pesticides, the development of resistant varieties has forced us to search for alternative solutions that are safe for environment, soil, and human populations. For this, the eco-friendly biological control agent (BCA)-based formulations also known as biopesticides are used worldwide. These include the use of *Agrobacterium radiobactor, Pseudomonas* spp., *Trichoderma* spp., *Fusarium* spp., *Aspergillus niger, Coniothyrium* spp. *Beauveria bassiana, Metarhizium* spp. *Nuclear Polyhedrosis Virus (NPVs,)* and most importantly *Bt* in BCA-based integrated plant pest management (Keswani et al. 2015; Purkayastha et al. 2016; Mishra et al. 2018). Globally, 175 biopesticides and active ingredients forming almost 700 products are registered and sold globally, whereas in India only 15 biopesticides are registered under the Indian Insecticides Act, 1968, and sold (Tables 18.1 and 18.2). This chapter attempts to provide an update on the biotechnology-based interventions for grain legume pest management in the background of *Cry* (*Bt*) gene and its structural and functional diversity.

18.3.1 *Bacillus thuringiensis*-Mediated Pest Management in Grain Legumes

Bacillus thuringiensis, also acknowledged as *Bt*, is a rod-shaped gram-positive bacterium with the ability to secrete an array of insecticidal proteins that makes it a prevalent bio control agent in agriculture. *Bt israelensis* was the first *Bt* species found to be toxic against dipteran larvae of harmful species such as black flies, chironoid midges, and mosquitoes (Chattopadhyay and Banerjee 2018). *Bt* has an array of chemical arsenal providing the entomopathogenic activity to it. Therefore, it is in high demand as a biological control agent in an agricultural integrated pest management system (Deist et al. 2014). *Bt* has the ability to grow well and colonize in the gut of insects, where it synthesizes a range of toxins based on its physiological conditions during growth and colonization. These physiological states are *vegetative growth phase* and *vegetative stationary phase* (spore-forming stage). While in *vegetative growth phase, Bt* exudates **S**ecreted **i**nsecticidal **p**roteins (*Sip*) and **V**egetative **i**nsecticidal **p**roteins (*Vip*); the *vegetative stationary phase* exudates parasporal crystalline (structurally bipyramidal crystals that contain one or more similar proteins of about 135 kDa and are toxic to lepidopterous larvae) δ-endotoxins such as

TABLE 18.1

Biopesticides Registered in India under Indian Insecticides Act 1968

S.No.	Biopesticide	Targeted Pest
1	*Bacillus thuringiensis* var. *israelensis* (Bti)	Diamondback Moths
2	*Bacillus thuringiensis* var. *kurstaki* (Btk)	Diamondback Moths
3	*Bacillus thuringiensis* var. *galleriae* (Btg)	*Helocoverpa armigera*
4	*Bacillus sphaericus*	Diamondback Moths
5	*Bacillus firmus*	Diamondback Moths
6	*Beauveria bassiana*	Mango hoppers, Mealybugs, Coffee Pod borer
7	*Cymbopogon* spp.	Insect
8	*Heterorhabditis bacteriphora*	Borers
9	Neem-based biopesticides	Insect whitefly
10	*Trichoderma viride*	Root rots and wilts
11	*Tricoderma harzianum*	Root rots and wilts
12	*Trichogramma parasitoid*	Sugarcane borers
13	*Pseudomonas fluorescens*	Bacterial and fungal pathogen
14	Nuclear Polyhedrosis Virus of *Helicoverpa armigera*	*Helicoverpa armigera*
15	Nuclear Polyhedrosis Virus of *Spodoptera litura*	*Spodoptera litura*

TABLE 18.2
Commercial Formulations of Biopesticides Registered in India

S.No.	Biopesticides	Registered Formulations
1	*Beauvaria bassiana*	1.00% WP; 1.15% WP; 1.15% SC
2	*Bacillus thuringiensis* var. *israelensis* serotype H14	5.00% AS; 12.00% AS
3	*Bacillus thuringiensis* var. *kurstaki* serotype H3a, H3b (HD-1)	0.50% WP; 5.00% WP; 2.50% AS
4	*Bacillus thuringiensis* var. *kurstaki* serotype H2a, 3b (NRD-12)	3.50% AS & WG
5	*Bacillus thuringiensis* var. *galleriae*	1.30% FC
6	*Bacillus sphaericus* serotype H5a H5b	1.30% FC
7	*Bacillus sphaericus* serotype B101	1.30% FC
8	*Tricoderma viride*	5.00% WP
9	*Tricoderma harzianum*	5.00 % WP
10	*Tricoderma harzianum* (IIPRTh-1)	2.00% WP
11	*Tricoderma harzianum* + *Tricoderma viride* + *Tricoderma virens*	5.00% WP
12	*Metarhizium anisopliae* var. *acridum*	1.15% WP
13	HaNPV (*Helicoverpa armigera* Nuclear Polyhedrosis Virus)	0.43% AS; 2.00% AS
14	Microbial Consortia of *Bacillus thuringiensis* and *Tricoderma harzianum*	2.00% WP
15	*Verticilium lecanii*	1.15 % WP

Abbreviations: AS, Aerosol Dispersible; FC, Flowable Concentrate; SC, Suspension Concentrate; WP, Wettable Powder; WG, Water Dispersible Granules.

Crystalline toxins (*Cry*) and **Cyt**olytic toxins (*Cyt*) (de Maagd et al. 2003; Deist et al. 2014; Land and Miljand 2014; Palma et al. 2014; Palma 2015; Brasseur et al. 2015). All these toxins play a strategic role of effective insecticidal toxins; however, the role of *Cry* toxin was broadly used and applied in the biological control agent-mediated integrated pest management. *Cry* proteins are further classified into four major domains, viz. (i) *Ext_Mtx2-like toxin*, (ii) *Bin-like toxin*, and (iii) *Cry toxin* and (iv) unclassified *Cry toxin* (Chattopadhyay and Banerjee 2018). With our best knowledge it has to be clear that *Bt* insecticidal are limited to *Bt*, and not all *Bt* strains are capable of secreting *Bt* toxins with the insecticidal property.

Natural strain, conjugates strain, genetically engineered strain *Bt*, and its gene for different diversity of *Cry* proteins are used by expressing them in recombinant *E. coli* cells. The World Data Center for Microorganisms (WDCM) has shown the record of over 84 *Bt* strains, and many of them are with pesticidal activities (www.wfec.info/). The success of *Bt* has risen to a level where it is used in as much as 90% of the world's total microbial pest control agent (MCPA) (Chattopadhyay and Banerjee 2018). In grain legumes, the use of *Bt* gene/toxin is tested vastly against a variety of pests; however, this use is restricted to very few crops such as soybean, chickpea, pigeon pea, and cowpea, etc. (Stewart et al. 1996; Walker et al. 2000; Ramu et al. 2012; Ghosh et al. 2017; Das et al. 2016, 2017; Chattopadhyay and Banerjee 2018). Also, the use of cry proteins is mostly practiced so far (Figure 18.1). However, the major problem that remains is the high discerning and slender activity spectrum of *Bt Cry* gene and has made its utilization species and crop specific (Table 18.3) (Naranjo 2014).

18.3.1.1 Bt *Chickpea*

Chickpea (*Cicer arietinum*) is one of the major legumes that stands third in the global food legume production (Das et al. 2017). India controls a very major proportion of global chickpea production and contributes 70%–75% toward global production. Chickpea seeds are the major source of dietary proteins (20%–25%) and also help to improve the environment and soil fertility via microflora-associated

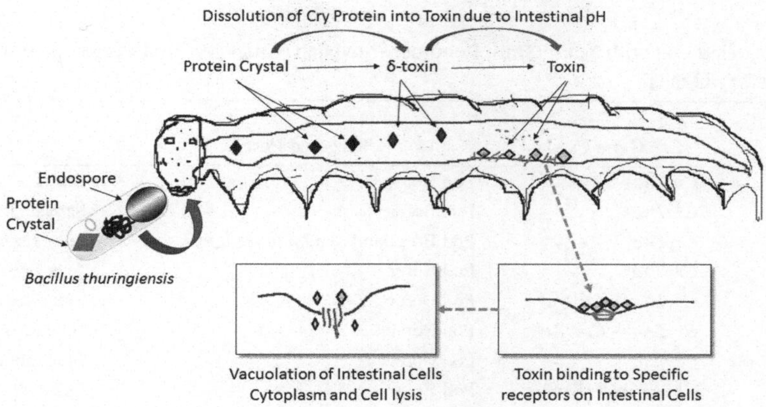

FIGURE 18.1 Physiological mechanism of action of *Cry* gene toxin.

TABLE 18.3

Bacillus thuringiensis (Bt) and Other (*Bt*-based) Genetically Engineered Bacterial Strains Used as a Biological Control Agent in Agriculture Worldwide

Type of Strain	*Bt* Strain/Genetically Engineered Genes	Targeted Pest
Natural Strains	*Bacillus thuringiensis* var. *israelensis*	Lepidoptera
	Bacillus thuringiensis var. *kurstaki*	Lepidoptera
	Bacillus thuringiensis var. *galleriae*	Coleoptera
	Bacillus thuringiensis var. *san diego*	Coleoptera
	Bacillus thuringiensis var. *tenebrionis*	Coleoptera
Improved Strains	*Bacillus thuringiensis* var. *kurstaki* **HD1**	Lepidoptera
	Bacillus thuringiensis var. *kurstaki* **SA-11**	Lepidoptera
	Bacillus thuringiensis var. *kurstaki* **SA-12**	Lepidoptera
Conjugated Strains	*Bacillus thuringiensis* var. *kurstaki* X *Bacillus thuringiensis* var. *tenebrionis* Conjugated—EG2424	Lepidoptera and Coleoptera
	Bacillus thuringiensis var. *kurstaki* X *Bacillus thuringiensis* var. *aizawai* Conjugated—EG2348	Lepidoptera
Recombinant *E. coli* Expressing	*Cry1Ac (x2), Cry3A, Cry3Bb*	Lepidoptera
Recombinant *E. coli* Expressing	*Cry1Ac (x3), Cry2A, Cry1C*	Lepidoptera
Recombinant *E. coli* Expressing	*Cry1Aa, Cry1Ac (x2), Cry2A, Cry IF-1Ac*	Lepidoptera

symbiotic nitrogen fixation. However, the production of chickpea is highly affected by abiotic stresses (drought and salinity, etc.) and biotic (fusarium wilt, Ascochyta blight, gram pod, etc.). The major annual loss to chickpea productivity was done by pod borers that start attacking chickpea from the first Fortnight after sowing avidly while flowering and at the stage of pod growth. Various attempts were made to obtain pest resistance through conventional breeding, but only very fractional accomplishments were obtained because of incompatibility issues with wild types and unavailability of source resistance (Nene and Sheila 1990). Therefore, for the development of pest resistance chickpea variety *Bt*, gene technology becomes a boon. Successful attempts were started with *Bt* genes to develop pest resistance against pod borer (Table 18.4) (Kar et al. 1997; Sanyal et al. 2005; Acharjee et al. 2010; Mehrotra et al. 2011; Ganguly et al. 2014; Chakraborty et al. 2016; Das et al. 2017).

Kar et al. (1997) developed the first *Bt Cry* expressing chickpea transgenic using chickpea cultivars ICCV1 and ICCV6 and introduced *Bt Cry1Ac* gene to develop a defense against pod borer larvae. Sanyal et al. (2005) used *Bt Cry1Ac* gene driven by Camv35S promoter to develop transgenic chickpea via *Agrobacterium*-mediated genetic transformation. A total of 24 T_0 plants were found positive for *Cry1Ac*

TABLE 18.4

Major Studies Carried Out with Main Grain Legumes (Soybean, Chickpea, and Pigeon pea) with Different *B. thuringiensis Cry* Genes

Grain Legumes Studies	*B. thuringiensis Cry* Gene Used	Targeted Pest	References
Chickpea	*Cry1Ac*	Pod Borer	Kar et al. (1997)
	Cry1Ac	Pod Borer	Sanyal et al. (2005)
	Cry1Ac	Pod Borer and *Spodoptera* Larvae	Indurker et al. (2007)
	Cry2Aa	Pod Borer	Acharjee et al. (2010)
	Cry1Aa + Cry1Ac	Pod Borer	Mehrotra et al. (2011)
	Cry1Aa + Cry1Ac	Pod Borer	Ganguly et al. (2014)
	Cry1Ac	Pod Borer	Chakraborty et al. (2016)
	ChimericCry1Aabc	Pod Borer	Das et al. (2017)
Soybean	*Cry1Ab*	Soybean Looper	Parrott et al. (1994)
	Cry1Ab (tic 107)	Soybean Looper	Stewart et al. (1996)
	Cry1Ab (tic 107)	Corn Earworm, Soybean Caterpillar, Soybean Looper, and Cornstalk Borer	Walker et al. (2000)
	Cry1Ac	Hubner	Homrich et al. (2008)
	Cry1Ac	Soybean Caterpillar	Yu et al. (2013)
	Cry1Ac + Cry 1F	Cornstalk Borer, Corn Earworm	Marques et al. (2017)
	Cry1Ac	Soybean Caterpillar, Cornstalk Borer, Corn Earworm	Martins-salles et al. (2017)
Pigeon pea	*Cry1E-C*	Tobacco Cutworm	Surekha et al. (2005)
	ChimericCry1AcF	Pod Borer	Ramu et al. (2012)
	Cry1Ac	Pod Borer	Kaur et al. (2016)
	ChimericCry1Aabc	Pod Borer	Das et al. (2016)
	Cry1Ac + Cry2Aa	Pod Borer	Ghosh et al. (2017)
	Cry2Aa	Pod Borer	Singh et al. (2018)

expression, out of which 11 T_1 plants succeeded to carry out *Cry1Ac* gene to the next progeny. The developed plant showed more than 80% plant protection from pod borer with more than 80% pod borer mortality rate. The amount of*Cry1Ac* protein synthesized also recorded up to 10 ng/mg total soluble proteins (TSP). Indurker et al. (2007) were looking to develop an efficient particle bombardment-mediated gene transfer in chickpea. The developed gene transfer method was later used to generated chickpea expressing *Bt Cry1Ac* gene that was found highly efficient and showed high pod borer and *Spodoptera* larvae mortality. Later on, Acharjee et al. (2010) used *Bt Cry2Aa* gene to develop the transgenic chickpea variety to deliberate the resistance against pod borer. The study was conducted to develop the T_0 and T_1 generation.

A total of 13 transgenic lines successfully expressed the *Cry2Aa* protein in T_0 plants, whereas only 5 were found capable of carrying transgene to T_1 generation following the Mendelian segregation (3:1). The next successful attempt was carried out using *Bt Cry1Aa* and *Bt Cry1Ac* gene, using *Agrobacterium*-mediated genetic transformation. The transgenic T_0 generation expressed *Cry1Aa* and *Cry1Ac* gene either single or pyramided that also followed the Mendelian inheritance (3:1) in T_1 offspring. The leaves of T_0 generation expressed *Cry* toxin vacillating from 5–40 ng/mg TSP. The pyramidal expression of *Cry1Aa* and *Cry1Ac* gene provided higher levels of resistance against pod borer (Mehrotra et al. 2011). After this successful attempt, Ganguly et al. (2014) used the pod-specific MSG promoter from soybean with *Bt Cry1Aa* and *Cry1Ac* genes to produce pest-resistant transgenic chickpea from chickpea var DCP 92-3 and chased the effect of *Bt* toxin produced over pod borer infection. The developed transgenics demonstrated pesticidal activity in both pods and twigs and were found effective more on borer larvae. Chakraborty et al. (2016) made more attempts to develop transgenics from chickpea var. 1CCV89314 by expressing *Bt Cry1Ac* gene under constitutive and tissue-specific manner by targeting cytoplasm and chloroplast. The higher levels of expressed *Cry1Ac* protein provides protection against borer infestation, whereas the chloroplast-targeted system provided much more effectual expression toward hazardous borer (lepidopteran pest).

More recent studies on the development of *Bt* toxin-producing chickpea were performed by Das et al. (2017). The research group used chickpea DCP 92-3 by expressing the chimeric *Bt Cry 1Aabc* gene developed using codon optimization that used three different domains (domains I, II, and III) from *Cry1Aa, Cry1Ab,* and *Cry1Ac* genes, respectively (Das et al. 2017). A total of six putative transgenic chickpeas were established to chase the effect of chimeric *Cry1Aabc* genes on gram pod borer. The study was carried over from T_0 to T_1, T_2, T_3, T_4 generations with prompt positive resistance against pod borer. As discussed earlier, India is the major producer of chickpea and contributes more than 70% of total global demand; therefore, major studies to develop pest-resistant chickpea using *Bt* toxins are done mainly in India. However, all the studies concluded that a major concern of engineering *BtCry* gene is to regulate its expression levels in such a manner that substantial balance should remain in the ecosystem (Das et al. 2017). That can be easily applied using a suitable promoter system to have regulation over transcriptional control.

18.3.1.2 **Bt *Soybean***

Soybean (*Glycine max* (L.). Merrill) is one of the most important and widely used legume crops throughput the globe; it covers almost 6% of the world's total arable land (Hartman et al. 2011; Marques et al. 2017). The United States contributes the majority with 80% of the worldwide production of soybean followed by Brazil (Chang et al. 2015). Like other grain legumes, production of soybean is severely affected by insect pests. Under the umbrella of biological control agent-assisted integrated pest management, several efforts were carried out to develop *Bt* soybean as a capable alternative to control defoliating caterpillars (Table 18.4) (Martins-salles et al. 2017). Parrott et al. (1994) reported the very first attempt to successfully integrate and express the *Bt Cry1Ab* gene in soybean. Later on, transgenic soybean expressing *Bt Cry1Ab (tic 107)* was produced using the plant tissue culture-based transformation, followed by biotic characteristics and molecular characterization (Stewart et al. 1996). Following, Walker et al. (2000) developed the transgenic "Jack" Soybean by recombining *B. thuringiensis* var. *kurstaki* (Jack-*Bt*) *Cry1Ac* gene. This ability of the developed transgenic to provide resistance was evaluated against different lepidopteran pests, viz. corn earworm (*Helicoverpa zea*—boddie), soybean caterpillar (*Agrotis gemmatalis*—Hubner), soybean looper (*C. includens*—walker), and cornstalk borer (*Elasmopalpus lignosellus*—zeller), and demonstrated adequate levels of resistance among them (Walker et al. 2000). Hybrid *Bt* soybean MON 87701 **X** MON 89788 generated has the ability to target a range of insect pest species from soybean looper (*Chrysodeixis includens*), Hubner (Erebidae), and velvetbean caterpillar (*Anticarsia gemmatalis*) and walker (*Nocteuidea*) (Bernardi et al. 2012).

Transgenic soybean (MON 87701) was developed using the synthetic *Bt Cry1Ac (tic 107)*, similarly, Homrich et al. (2008) developed transgenic lineages of "Jack" soybean and obtained resistance against *A. gemmatalis*. The developed transgenics (i) soybean MON 87701 (expressing *BtCry 1Ac* protein) and (i) soybean MON 89788 (expressing pyruvyl shikimate-3-phosphate synthase, EPSPS- for herbicide glyphosate tolerance) were combined to develop a new soybean transgenic conforming resistance against major soybean caterpillars and herbicide glyphosate tolerance (Yu et al. 2013; Martins-salles et al. 2017). Very recently, transgenic soybean expressing *BtCry1Ac* and *BtCry1F* proteins, developed using transgenic event DAS-81419-2 (Conkesta™ technology), was found very effective against primary lepidopteran pests (*E. lignosellus, A. ipsilon,* and *H. armigera*). The developed transgenic significantly reduced the plant death at seedling stage, majorly infected by *E. lignosellus*, and *H. armigera* (Marques et al. 2017). In comparison to *Bt* maize and *Bt* cotton varieties, *Bt* soybean technology provides better control over lepidopteran pests (Bernardi et al. 2012; Martins-salles et al. 2017). Recently, transgenic soybean expressing *BtCry1Ac* gene was approved and cultivated in Brazil (Bernardi et al. 2012; Yano et al. 2015).

18.3.1.3 **Bt *Pigeon pea***

Like soybean and chickpea, pigeon pea (*Cajanus cajan*) is also one of the major grain legumes, specifically in the Asian and Africa regions. Being the source of high protein content and micronutrients, pigeon pea plays a pivotal part in Indian and Asian vegetarian populations. Like chickpea, India also leads in the worldwide production of pigeon pea and contributes 90% of global production (Singh et al. 2018). Regardless of enhanced pigeon pea production of 4.85 million tons over the last decade, the per hectare yield has festered. Lepidopteran pest, gram pod borer (*H. armigera*) is the main problem behind that. The high fertility and robust migration behavior of gram pod borer have made it the utmost problematic pest to control that causes almost 85% loss in yield and $300 million annually (Choudhary et al. 2013; Singh et al. 2018).

The pioneering study to develop the first *Bt* transgenic pigeon pea was carried out by Surekha et al. (2005) by expressing synthetic *Bt Cry 1E-C* gene. The developed variety showed remarkable protection and high mortality of *Spodoptera litura* (tobacco cutworm) larvae. The few developed transgenics were successfully able to carry out integrated transgenes to T_0–T_2 generations (Surekha et al. 2005). One major attempt was made by Ramu et al. (2012) to provide *BtCry* gene-mediated resistance in pigeon pea toward pod bores, where they expressed chimeric *BtCry 1AcF* (encoding *Bt Cry* domains *Cry1Ac* and *Cry1F*). The developed transgenics showed high mortality of pod borer larvae. Only a few developed transgenics were successfully able to carry out integrated transgenes to T_1–T_3 generations (Ramu et al. 2012). Following this, Kaur et al. (2016) used *Bt Cry1Ac* gene to generate pod borer resistance in pigeon pea transgenics, and T_1–T_2 generations were studied for in vitro larvae bioassay on leaves and pods, which were found to be 97.78% efficient. Similar to Ramu et al. (2012), Das et al. (2016) used the chimeric *Bt Cry* genes. The research group used pigeon pea cv. "Asha" by expressing the chimeric *Bt Cry 1Aabc* gene developed using codon optimization that used three different domains (domain I, II, and III) from *Cry1Aa, Cry1Ab*, and *Cry1Ac* genes, respectively. The developed transgenics were found to provide resistance against pod borer with more than 80% larvae. Ghosh et al. (2017) further attempted to generate transgenic pigeon pea from pigeon pea cv. UPAS 120 by expressing *Bt Cry 1Ac* and *Cry 2Aa* genes separately. The transgenics developed with *Cry 2Aa* genes demonstrated 100% pod borer larvae mortality in successive generations, and the integrated genes inherited from T_0 to T_2 generations. The latest study was carried out with *Bt Cry* integrated pigeon pea and was developed by Singh et al. (2018) who reported the expression of *Bt Cry2Aa* gene in pigeon pea transgenics against gram pod borer. The developed *Cry2Aa* gene integrated pigeon pea lineage (T_0–T_3) accumulated *Cry2Aa* protein in the range of 25–80 ug/gm fresh weight, that demonstrated 80%–100% pod borer larvae mortality (Singh et al. 2018). Though being a major grain legume, very few studies were carried out over pigeon pea with the integration of *Bt Cry* genes; there is an urgent need to develop superior *Bt Cry* integrated pigeon pea varieties with commercial potential and agri-field sustainability (Das et al. 2016; Singh et al. 2018).

18.4 Ethical and Environmental Regulations of *Bt* Cry Toxin

Being the natural insecticidal, the filed application (spraying, etc.) of *Bt* as bacterial suspension or in any form of live *Bt* spores and *Bt Cry* toxin is not prohibited, unless the manufacturer ensures the non-inclusion of any nonspecific β-exotoxin. The findings are full with the reports of usage of *Bt* toxin in agriculture on targeted pests and their control as an effective BCA. However, information concerning the effect of *Bt* toxin over nontarget organisms (NTOs) with its other nonspecific effects are scanty (Bøhn et al. 2008; Resende et al. 2016). Initially, government regulations for environmental safety and others were not applied to the practice to test the effect of *Bt* toxin over NTOs until Losey et al. (1999) reported the massive mortality of *Danaus plexippus* (Monarch Butterfly) larvae feeding over pollens of *BtCry1Ab* gene expressing maize. Later on, the role of *Cry1Ab, Cry1Ac, Cry2Ab*, and *Cry3A* was also studied over *Spodoptera littoralis* and *Chrysoperia carnea, Neobelleria bullata*, and Aphids, etc., where dose-dependent-based nonlethal to sublethal effects were observed (Cerstiaens et al. 2001; Hussein et al. 2006; Romeis et al. 2014; Deist et al. 2014; Resende et al. 2016). Several ethical rules, regulations, and safety concerns over GMP have been raised worldwide on appropriate platforms, as in the case of Calgene's Flavr Savr and Ciba-Geigy's *Bt* Corn with *Bt* toxins. Also, the use of antibiotic selection markers is a major concern in genetically modified plants (GMP) with *Bt* toxin that was later undertaken with auxotrophic and food-grade markers (Braun et al. 2001). Therefore, GMP with *Bt* toxins must be meticulously assessed on animal models for toxicity and nonspecific effect before they are released for human consumption and to prevent disturbance in natural insect diversity that will ultimately harm the ecosystem (Maghari and Ardekani 2011). Hence, GMP with *Bt* toxins must be directed under strict government approvals for rules and regulations (Altieri and Rosset 1999).

18.5 Conclusion and Future Perspectives

Insecticidal *Bt* toxins are working as a highly valuable tool as biological pest control in the agriculture sector. Several chemical pesticides have long been used; however, they might have harmful effects on human health and can also harm the native ecosystem, and hence they are restricted in many countries. The *Bt* resistance (resistance toward *Bt* toxin in insects) due to interactions between *Bt* toxin such as *Cry4Aa, Cry4Ba,* and *Cry11Aa*, and *Cyt1Aa* is also a concern for *Bt*-based integrated biological pest management in the agriculture sector. A deep inside study on this important aspect can help researchers to advance the strategies to resolve the *Bt* resistance associated problems. Also, the effect of *Bt* toxin over NTOs and its impact over native ecosystems should be investigated during the development and before release of the *Bt* pesticidal varieties in the agriculture market. The knowledge gathered from the past attributes that the *Bt* toxin insecticidal activity is a complex sum of biological and chemical interactions between *Bt* toxin, viz. *Sip, Vip, Cyt,* and *Cry* with their highly specific biological rectors present inside the insects. Therefore, it is highly required to use emerging modern techniques such as protein engineering to develop new chimeric *Bt* toxin and transgenic crops expressing their different combinations. This will open a new window of *Bt* research.

CONFLICT OF INTEREST

The authors declare that they have no conflict of interest.

REFERENCES

Acharjee S, Sarmah BK, Kumar A et al. (2010) Transgenic chickpeas (*Cicer arietinum* L.) expressing a sequence modified cry2Aa gene. *Plant Sci* 178:333–339.

Altieri MA, Rosset P (1999) Strengthening the case for why biotechnology will not help the developing world: A response to McGloughlin. *AgBioForum* 2:226–236.

Bernardi O, Malvestiti GS, Dourado PM et al. (2012) Assessment of the high-dose concept and level of control provided by MON 87701 X MON 89788 soybean against *Anticarsia gemmatalis* and *Pseudoplusia includens* (Lepidoptera: Noctuidae) in Brazil. *Pest Manag Sci* 68:1083e1091.

Bøhn T, Primicerio R, Hessen DO, Traavik T (2008) Reduced fitness of *Daphnia magna* fed a Bt-transgenic maize variety. *Arch Environ Contam Toxicol* 55:584–592.

Bohra A, Sahrawat KL, Kumar S et al. (2015) Genetics and genomics based interventions for nutritional enhancement of grain legume crops: Status and outlook. *J Appl Genet* 56:151–161

Brasseur K, Auger P, Asselin E et al. (2015) Parasporin-2 from a new *Bacillus thuringiensis* 4R2 strain induces caspases activation and apoptosis in human cancer cells. *PLoS One* 10(8):e0135106.

Braun R, Bennett DJ, Secretariat EFB, Delft O (2001) Antibiotic resistance markers in genetically modified (GM) crops. Task Group on Public Perceptions of Biotechnology, European Federation of Biotechnology.

Cerstiaens A, Verleyen P, van Rie J, van Kerkhove E et al. (2001) Effect of *Bacillus thuringiensis* Cry1 toxins in insect hemolymph and their neurotoxicity in brain cells of *Lymantria dispar. Appl Environ Microbiol* 67:3923–3927.

Chakraborty J, Sen S, Ghosh P et al. (2017). Homologous promoter derived constitutive and chloroplast targeted expression of synthetic cry1Ac in transgenic chickpea confers resistance against *Helicoverpa armigera. Plant Cell Tissue Org Cult* 125:521–535.

Chang WS, Lee HI, Hungria M (2015) Soybean production in the Americas. In: Lugtenberg, B. (Ed.), *Principles of Plant-Microbe Interactions.* Springer International Publishing, Cham, Switzerland, pp. 393e400.

Chattopadhyay P, Banerjee G (2018) Recent advancement on chemical arsenal of *Bt* toxin and its application in pest management system in agricultural field. *3 Biotech* 8:201.

Choudhary AK, Raje RS, Datta S et al. (2013) Conventional and molecular approaches towards genetic improvement in pigeonpea for insects resistance. *Am J Plant Sci* 4:372–385.

Das A, Datta S, Thakur S, Shukla A et al. (2017) Expression of a Chimeric Gene Encoding Insecticidal Crystal Protein Cry1Aabc of *Bacillus thuringiensis* in Chickpea (*Cicer arietinum* L.) Confers Resistance to Gram Pod Borer (*Helicoverpa armigera* Hubner.). *Front Plant Sci* 8:1423.

Das A, Datta S, Sujayanand GK et al. (2016) Expression of chimeric *Bt* gene, Cry1Aabc in transgenic pigeon pea (cv. Asha) confers resistance to gram pod borer (*Helicoverpa armigera* Hubner.). *Plant Cell Tiss Org Cult* 127:705–715.

de Maagd RA, Bravo A, Berry C et al. (2003) Structure, diversity, and evolution of protein toxins from spore forming entomopathogenic bacteria. *Ann Rev Genet* 37:409–433.

Deist BR, Rausch MA, Fernandez-Luna MT et al. (2014) *Bt* toxin modification for enhanced efficacy. *Toxins* 6:3005–3027.

Dhaliwal GS, Jindal V, Dhawan AK (2010) Insect pest problems and crop losses: Changing trends. *Ind J Ecol* 37:1–7.

Ganguly M, Molla KA, Karmakar S et al. (2014) Development of pod borer-resistant transgenic chickpea using a pod-specific and a constitutive promoter-driven fused cry1Ab/Ac gene. *Theor Appl Genet* 127:2555–2565.

Ghosh G, Ganguly S, Purohit A et al. (2017) Transgenic pigeon pea events expressing Cry1Ac and Cry2Aa exhibit resistance to *Helicoverpa armigera*. *Plant Cell Rep* 36:1037–1051.

Global hunger index: The inequalities of hunger (2017) https://www.globalhungerindex.org/pdf/en/2017.pdf

Goyal M, Singh J, Kumar P, Sirohi A (2018) Pulses for human nutritional security. In: Wani, S.H., Jain, M. (Eds.), Pulse *Improvement Physiological, Molecular and Genetic Perspectives*. Springer Nature, Cham, Switzerland, pp. 1–11.

Hartman GL, West ED, Herman TK (2011) Crops that feed the world 2. Soybean worldwide production, use, and constraints caused by pathogens and pests. *Food Secur* 3:5e17.

Homrich MS, Passaglia LMP, Pereira JF et al. (2008) Resistance to *Anticarsia gemmatalis* Hübner (Lepidoptera, Noctuidae) in transgenic soybean (*Glycine max* (L.) Merrill Fabales, Fabaceae) cultivar IAS5 expressing a modified Cry1Ac endotoxin. *Genet Mol Biol* 31:522–531.

Hussein HM, Habuštová O, Turanli F, Sehnal F (2006) Potato expressing beetle–specific *Bacillus thuringiensis* Cry3Aa toxin reduces performance of a moth. *J Chem Ecol* 32:1–13.

Indurker S, Misra HS, Eapen S (2007) Genetic transformation of chickpea (*Cicer arietinum* L.) with insecticidal crystal protein gene using particle gun bombardment. *Plant Cell Rep* 26:755–763.

Kar S, Basu D, Das S, Ramkrishnan NA et al. (1997) Expression of cryIA (c) gene of *Bacillus thuringiensis* in transgenic chickpea plants inhibits development of pod-borer (*Heliothis armigera*) larvae. *Transgenic Res* 6:177–185.

Kaur A, Sharma M, Sharma C et al. (2016) Pod borer resistant transgenic pigeon pea (*Cajanus cajan* L.) expressing cry1Ac transgene generated through simplified *Agrobacterium transformation* of pricked embryo axes. *Plant Cell Tiss Organ Cult* 127:717–727.

Keneni G, Bekele E, Getu E, Imtiaz M et al. (2011) Breeding food legumes for resistance to storage insect pests: Potential and limitations. *Sustainability* 3(9):1399–1415.

Keswani C, Mishra S, Sarma BK et al. (2015) Unraveling the efficient application of secondary metabolites of various Trichoderma. *Appl Microbiol Biotechnol* 98:533–544.

Kumar J, Choudhury AK, Solanki RK, Pratap A (2011) Towards MAS in pulses: A review. *Plant Breed* 130:297–313.

Land M, Miljand M (2014) Biological control of mosquitoes using *Bacillus thuringiensis israelensis:* A pilot study of effects on target organisms, non-target organisms and humans. Mistra EviEM Pilot Study PS4 (www.eviem.se).

Losey JE, Raynor LS, Carter ME (1999) Transgenic pollen harms monarch larvae. *Nature* 399:214.

Maghari BM, Ardekani AM (2011) Genetically modified foods and social concerns. *Avicenna J Med Biotechnol* 3:109.

Margam VM, Coates BS, Ba MN, Sun W et al. (2011) Geographic distribution of phylogenetically-distinct legume pod borer *Maruca vitrata* (Lepidoptera: Pyraloidea: Crambidae). *Mol Biol Rep* 38:893–903.

Marques LH, Santos AC, Castro BA et al. (2017) Field evaluation of soybean transgenic event DAS-81419-2 expressing Cry1F and Cry1Ac proteins for the control of secondary lepidopteran pests in Brazil. *Crop Prot* 96:109–115.

Martins-Salles S, Machado V, Massochin-Pinto L, Fiuza LM (2017) Genetically modified soybean expressing insecticidal protein (Cry1Ac): Management risk and perspectives. *FACETS* 2:496–512.

Mehrotra M, Singh AK, Sanyal I, Altosaar I, Amla DV (2011) Pyramiding of modified cry1Ab and cry1Ac genes of *Bacillus thuringiensis* in transgenic chickpea (*Cicer arietinum* L.) for improved resistance to pod borer insect *Helicoverpa armigera*. *Euphytica* 182:87–102.

Mishra RK, Bohra A, Kamaal N et al. (2018) Utilization of biopesticides as sustainable solutions for management of pests in legume crops: Achievements and prospects. *Egypt J Biol Pest Cont* 28:3.

Naranjo SE (2014) Effects of GM crops on non-target organisms. In: Ricroch, A., Chopra, S., Fleischer, S. (Eds.), *Plant Biotechnology: Experience and Future Prospects*. Springer International Publishing, Cham, Switzerland, pp. 129–142.

Nene YL, Sheila VK (1990) Pigeon pea: Geography and importance, In: Nene, Y.L., Hall, S.D., Sheila, V.K. (Eds.), *The Pigeon Pea*, CAB International, Wallingford, UK, pp. 1–14.

Palma L (2015) Protocol for the fast isolation and identification of insecticidal *Bacillus thuringiensis* strains from soil. *Bt Res* 6:1–3.

Palma L, Muñoz D, Berry C, Murillo J, Caballero P (2014) Draft genome sequences of two *Bacillus thuringiensis* strains and characterization of a putative 41.9-kDa insecticidal toxin. *Toxins* 6:1490–1504.

Parankusam S, Katamreddy S, Bommineni PR, Mathur PB, Sharma KK (2018) Insights into insect resistance in pulse crops: Problems and preventions. In: Wani, S.H., Jain, M. (Eds.), *Pulse Improvement Physiological, Molecular and Genetic Perspectives*. Springer Nature, Cham, Switzerland, pp. 137–173.

Parrott WA, All JN, Adang MJ, Bailey MA, Boerma HR, Stewart CN Jr. (1994) Recovery and evaluation of soybean plants transgenic for a *Bacillus thuringiensis* var. *kurstaki* insecticidal gene. *In Vitro Cell Dev Biol Plant* 30:144–149. doi:10.1007/ BF02632204.

Purkayastha J, Arora R, Singh L (2016) Herbal biomedicines and insecticides: Sustainable and novel eco-friendly approaches towards integrated disease and vector management. In: Veer, V., Gopalakrishnan, R. (Eds.), *Herbal Insecticides, Repellents and Biomedicines: Effectiveness and Commercialization*. Springer, Dordrecht, the Netherlands, pp. 11–23.

Ramu SV, Rohini S, Keshavareddy G et al. (2012) Expression of a synthetic cry1AcF gene in transgenic Pigeon pea confers resistance to *Helicoverpa armigera*. *J App Entomol* 136:675–687.

Reddy AA (2009) Pulses production technology: Status and way forward. *Econ Polit Wkly* 44:73–80.

Resende DC, Mendes SM, Rosangela C, Marucci RC (2016) Does Bt maize cultivation affect the non-target insect community in the agro ecosystem? *Rev Bras Entomol* 60:82–93.

Romeis J, Meissle M, Naranjo SE, Li Y, Bigler F (2014) The end of a myth—Bt (Cry1Ab) maize does not harm green lacewings. *Front Plant Sci* 5:391.

Sanyal I, Singh A, Kaushik M, Amla DV (2005) *Agrobacterium* mediated transformation of chickpea (*Cicer arietinum* L.) with *Bacillus thuringiensis* cry1Ac gene for resistance against Pod borer insect Helicoverpa armigera. *Plant Sci* 168:1135–1146.

Singh RS, Chakravorty S, Chandra M (2013) Diversity of pod associated insect pests and natural enemies in pigeon pea, their relative abundance and crop losses in Bundelkhand region, India. *Flora Fauna* 19(2):294–302.

Singh S, Kumar NR, Maniraj R, Lakshmikanth R et al. (2018) Expression of Cry2Aa, a *Bacillus thuringiensis* insecticidal protein in transgenic pigeon pea confers resistance to gram pod borer, *Helicoverpa armigera*. *Sci Rep* 8:8820.

Stewart CN, Adang MJ, All JN, Boerma HR et al. (1996) Genetic transformation, recovery, and characterization of fertile soybean transgenic for a synthetic *Bacillus thuringiensis* cryIAc gene. *Plant Physiol* 112:121–129.

Surekha C, Beena MR, Arundhati A, Singh PK et al. (2005) *Agrobacterium*-mediated genetic transformation of pigeon pea (*Cajanus cajan* (L.) Millsp.) using embryonal segments and development of transgenic plants for resistance against *Spodoptera*. *Plant Sci* 169:1074–1080.

Walker DR, All JN, McPherson RM et al. (2000) Field evaluation of soybean engineered with a synthetic cry1Ac transgene for resistance to corn earworm, soybean looper, velvet bean caterpillar (Lepidoptera: Noctuidae), and lesser cornstalk borer (Lepidoptera: Pyralidae). *J Econ Entomol* 93:613e622.

Yano SAC, Specht A, Morcardi F, Carvalho RA et al. 2015. High susceptibility and low resistance allele frequency of *Chrysodeixis includens* (Lepidoptera: Noctuidae) field populations to Cry1Ac in Brazil. *Pest Manag Sci* 72:1578e1584.

Yu H, Li Y, Li X et al. (2013) Expression of Cry1Ac in transgenic Bt soybean lines and their efficiency in controlling lepidopteran pests. *Pest Manag Sci* 69:1326e1333.

19

Structural and Functional Diversity of Cry Genes and Their Utilization in Pest Management of Grain Legumes

G. K. Sujayanand and Sonika Pandey

CONTENTS

19.1 Introduction

Insect pests are responsible for more than 20% of crop losses. Chemical insecticides are widely used to control these insect pests effectively. However, these insecticides are very harmful for humans, animals, and environments. Persistent exposure to these insecticides leads to hazardous diseases such as lymphoma, leukemia, breast cancer, asthma, and allergic and immune system diseases. Every year, around 2.5 million tons of chemical insecticides are produced globally and are responsible for about 20,000 human deaths. Concurrently, biological control agents, viz. bacteria, virus, and fungi, can manage the insects very effectively without harming the environment. The most commonly used, reliable, and effective biological control agents are *Bacillus thuringiensis* (*Bt*). The microbial biocontrol agent *Bt* produces numerous insecticidal proteins. These insecticidal proteins are the delta endotoxins named Cry and Cyt, and they also secrete VIP and SIP toxins into the growing medium. These proteins have been used for over 50 years in biopesticide formulations for managing insect pest. Some of these toxins have been expressed in genetically modified crops such as corn, soybeans, and cotton. *Bt* is responsible for 2% of the total insecticidal market; it is effective against larval stages of various insects and kills them by disrupting the midgut tissues and then septicemia. *Bt* toxicity is reliable on the insecticidal toxins and virulence factors. Insecticide compounds include Cry proteins, thuringiensin, vegetative insecticidal proteins (VIP), secreted insecticidal protein (SIP), zwittermicin A, Mtx-like toxin, and Bin-like toxin. In 2013, around 75.9 million hectares were under genetically modified crops, which results in the control of insect pests.

Cry toxins have been mostly studied in lepidopetran pests. Their classification is based on their primary amino acid sequence, and more than 500 different *cry* genes sequences have been studied. These *cry* genes have been categorized into 75 groups (Cryl-Cry75). The *cry* gene sequences have been divided into four phylogenetic families, viz. Mtx, 3 Domain, Cyt, and Bin. Some *Bt* strains produce vegetative insecticidal protein (VIP) or secreted insecticidal protein (SIP), which also has insecticidal action, but these are produced under vegetative growth phase rather than during sporulation. The VIPS have been characterized as VIP1/VIP2, a binary toxin, and VIP3. The VIP1/2 and SIP are found to be toxic to some coleopterans, while VIP 3 is toxic to lepidopterans.

Among the *cry* family, the largest is a three-domain family that includes 40 different groups with more than 200 gene sequences. Cry toxin binding relies on the presence of specific surface proteins of microvillus of larval midgut cells. In the case of lepidopteran insects, Cry 1 binding proteins have been characterized as caderin-like proteins glycosylphosphatidylinositol (GPI)-anchored aminopeptidase-N (APN), and GPI-anchored alkaline phosphatase (ALP). *Manduca sexta*, *Bombyx mori*, *Heliothis virescens*, *Helicoverpa armigera*, *Pectinophora gossypiella*, *Ostrinia nubilalis*, *Lymantria dispar*, and *Plutella xylostella* are known to be affected by Cryl molecules. Cryl toxin binds cadherin proteins with high affinity in the nano-molar range.

19.2 Cry Gene Structural Diversity

During 1981, the cry gene was successfully cloned and sequenced by Schnpef and Whiteley. Initially, the cry genes were classified as *cry* I, *cry* II, *cry* III, *cry* IV, etc., based on their insect toxicity. The primary rank is usually represented in roman numbers. Until now, 75 different primary subgroups (cry 1 to cry 75) were created by *Bacillus thuringiensis* Toxin Nomenclature Committee based on primary ranks in cry nomenclature were established by Crickmore et al. (1998). In this new nomenclature system, a novel toxin is given a four-rank name depending on its degree of pairwise amino acid identity to previously named toxins; additionally, grouping by this criterion does not imply a similar protein structure, host range, or even mode of action. Arabic numbers are used for the first and fourth ranks, and uppercase and lowercase letters are assigned for the second and third ranks, respectively. The length of amino acids of *cry* gene vary from 369 (cry 34) to 1344 (cry 43). Based on protein tertiary structure, the parasporal toxins are classified into three domains—ETX-MTX, Bin, and parasporin. Cry toxins belonging to three domain family display clear differences in their amino acid sequences and overall length of protoxin, but all of them share a conserved three-domain structure (Pardo-Lopez et al. 2013). Five conserved blocks were located in the active core toxin (i.e. protoxin) region and it was identified by multiple sequence alignment of Cry toxin that are released after protein activation by midgut proteases (Höfte and whiteley 1989). Three additional conserved amino acid blocks are also lying outside the active core region (Schnepf et al. 1998) (Figure 19.1).

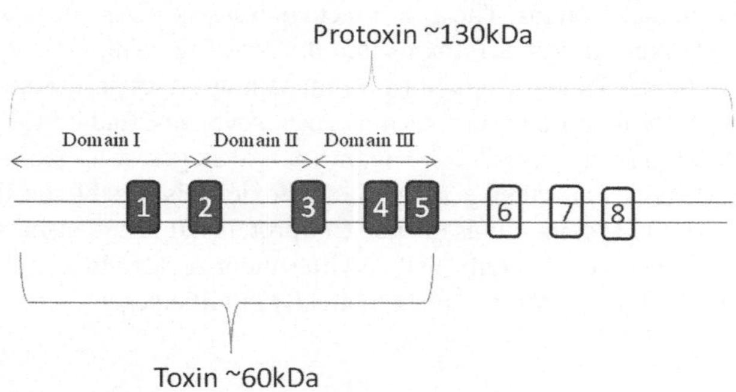

FIGURE 19.1 CrylA protoxin depicting conserved amino acid blocks in active core region.

The Cry toxin has three domains. Domain I is a bundle of seven to eight α-helices, six of which are amphipathic encircling the seventh hydrophobic helix. It is located in N terminus and is subjected to proteolytic cleavage in all three domain cry proteins, which is responsible for membrane insertion and pore formation. Domain I shares structural similarity with pore-forming toxins such as colicin, Ia, N, and diphtheria, supporting the role of this domain in pore formation. Domain II consists of three antiparallel β-sheets with exposed loop regions, and it is responsible for receptor-protein interaction. This domain shares similarity with several carbohydrate-binding proteins such as vitelline, lectin jacalin, and lectin Mpa, which has been reported (De Maagd et al. 2003). Domain III is a galactose-binding domain consisting of two antiparallel β-sandwich and is responsible for receptor binding and pore formation. This domain is similar to carbohydrate-binding proteins such as the cellulose-binding domain of 1,4-β glucanase C, galactose oxidase, sialidase, β-glucoronidase, and the carbohydrate-binding domain of xylanase U and β-galactosidase (De Maagd et al. 2003). Thus, both domains (II and III) confer receptor-binding specificity and help to define the host range (Boonserm et al. 2006). Some cry toxins such as Cry3 and Cry11 lack extended c terminal region and possess 70 kDa.

There are 11 toxins in ETX-MTX group, and its toxin shares the highest homology with *Clostridium perfringens* epsilon toxin. The MTX2 from *Lysinibacillus sphaericus* also belongs to this group (*De* magaad et al. 2003; Berry 2012). There are 13 toxins in Bin-like toxins. The parasporal protein crystals, which don't have any known invertebrate target, are known as parasporins and have both Cry and Ps nomenclature systems. Currently, six classes of parasporins are identified, among them cry 31 (parasporin 1), cry 41 (parasporin 3), and cry 63 (parasporin 6) are three-domain toxins, while cry 45 (parasporin 4) and cry 64 (parasporin 5) are ETX-MTX type toxins. *Cry* 46 (Parasporin-2) doesn't share homology with cry toxin whereas it shares homology with hydralysin, but its structure suggests it may belong to the ETX-MTX group. These parasporins exhibit strong cytocidal activity against human cancer cell lines upon protease activation.

The Cyt (cytolytic) toxins were 28 kDa protein, which was first reported from *B thuringiensis* subsp. *israeliensis*. The Cyt toxin gene family currently consists of 11 holotypes (Crickmore et al. 2019), and they are active against dipterans such as black flies and mosquitoes (Soberon et al. 2013), while some are found to have synergistic effect with other *Bt* proteins (Yu et al. 2012). The toxin consists of two α-helix hairpins flanking a β-sheet core containing seven to eight β-strands (Cohen et al. 2008, 2011). This Cyt protein is similar to volvatoxin A2, a PFT cardiotoxin produced by a straw mushroom *Volvariella volvacea* (Lin et al. 2004).

19.3 Cry Gene Functional/Toxicological Diversity

19.3.1 Three-Domain Cry Toxin

The cry toxin has three domain structures, and its mode of action is widely studied in lepidopteran insects. Three models were proposed to explain their mode of action, viz. classical model, sequential binding model, and signaling pathway model. The "classical model" proposes that after the successful ingestion of crystalline protein inclusion, it was dissolved in alkaline midgut lumen. The next step is activation of protoxin (Cry protein) by proteolysis in midgut leads to formation of a smaller protease-resistant toxic polypeptide. Subsequently, the toxic polypeptide binds with a specific receptor on the surface of midgut epithelial cells that leads to nonselective pore formation permeable to inorganic ions, sugars, and amino acids. This causes the lysis of epithelial cells and hence midgut disarrangements leading to insect death. Eventually, the spores invade hemolymph and cause septicemia (Kirouac et al. 2002; Vachon et al. 2012).

The "sequential binding model" suggests that cry toxin once activated by intestinal protease binds to cadherin-like proteins and leads to proteolytic removal of domain I of α helix and formation of an oligomeric pre-pore structure. The pre-pore structure then binds to a secondary receptor, aminopeptidase, which facilitates insertion of the pre-pore structure into the membrane, leading to insect and cell death (Bravo et al. 2007; Pardo-López et al. 2013).

In contrast, the signaling-pathway model suggests that the toxic activity is mediated by the specific binding to cadherin receptors, leading to undescribed Mg^{2+}-dependent and adenylyl cyclase/protein kinase A signaling pathway produces necrotic cell death (Zhang et al. 2006).

Some of the cry proteins had shown to exhibit novel toxic properties against protozoans, snails, etc. The *Bt* strain B622 and B626 isolated in Japan is found to be toxic to human pathogenic protozoan, *Trichomonas vaginalis*. Similarly, *Bt* strains 977 and NRRL HD-522 displayed moderate but specific molluscicidal activity against Chinese *Oncomelia* snails. These snails are intermediate host of trematode worm, *Schistosoma japonicum*, causing schistosomiasis in China and the Philippines (Ali et al. 2010). Certain cry protein such as cry15A toxin is found to have hemolytic activity, which is undesirable for biopesticide (Naimov et al. 2008). Similarly, cry 13A and cry14A have antibacterial activity that hampers with cloning and expression in *E. coli* (Wei et al. 2003).

19.3.2 Non-three-Domain Cry Toxins

Some of the cry proteins, viz. Cry15, Cry23, Cry33, Cry38, Cry45 (parasporin 4), Cry51, Cry60, and Cry64, show features of the ETX_MTX2 family. The MTX2 toxins are mosquitocidal in nature and predominantly have β sheet-based structure. The mode of action of this toxin is believed to be β barrel pore formation in the target cell membrane (Bokori-Brown et al. 2011; Popoff 2011). The mode of action of parasporin 1 appears to be more complex than simple pore formation. It induces intracellular Ca_{2+} concentration followed by apoptosis by increasing Caspase-3 and PARP cleavage. Parasporin 3 and 6 cause cell swelling and pore formation. Parasporin 4 acts as a cholesterol-independent β-pore-forming toxin that does not induce apoptosis.

The Cry34/Cry35 pair show remarkable structural similarity to another binary toxin, Cry23/Cry37. Another example of binary toxin is cry48/cry49. The activity of these binary toxins against coleopterans still remains to be elucidated.

Cyt 1Aa and Cyt 2Ba 3d structures infer that these proteins are single-domain, three-layer alpha-beta proteins. Cyt toxins are also synthesized as protoxins, and a small portion of N terminus and C terminus are removed to activate the toxin (Li et al. 1996). Cyt proteins are known to synergize the cry and vip3 toxins. Cyt 1Aa inhibits development of resistance against cry 3Aa in *Chrysomela scripta* (Federici and Bauer 1998). Some strains of *Bt* such as *Bacillus thuringiensis* subsp. *morrisoni* harbor *cyt* genes having toxicity against Diptera, Lepidoptera, and Coleoptera (Guerchicoff et al. 2001). Similarly, Cyt 2C is active against Rhabditida nematodes and cancer cells (Van Frankenhuyzen 2009). X-ray crystallography helped in deciphering the three-dimensional structures of cry 3Aa, cry1Aa, Cry2Aa, Cry3Bb, cry4Aa, cry 4Ba, cry8Ea1, cry 5B.

19.4 Cry Gene Receptors

A number of potential receptors other than cadherin have been proposed as cry receptors, e.g., alpha amylases and alpha glycosidases (Zhang et al. 2013), prohibitin (Kuadkitkan et al. 2012), and alkaline phosphatases (Likitvivatanavong et al. 2011). The proteomic analyses of cry protein and its receptors have proved that cry protein binds with alkaline phosphatase (ALPs) present in brush border membrane vesicles of midgut membrane of *Manduca sexta, Heliothis virescens* (Krishnamoorthy et al. 2007), and *Aedes aegypti* (Bayyareddy et al. 2009). RNAi silencing had proved that binding of ALPs is crucial than Amino-peptidase N (APN) for *cry1Ab* toxicity, while *cry 1Ac* relies more on APNs (Flores-Escobar et al. 2013). *cry1A* and *cry3Aa* binds to cadherin receptors in midgut (Fabrick et al. 2009; Gomez et al. 2014). Four models were proposed for their toxicity—sequential binding models, *cry1A –M. sexta model*, penknife model, and umbrella model were proposed.

Apart from the above mentioned cry receptors, other receptors like glycolipids, glycoconjugate, ADAM metalloprotease, amylase, α glucosidase, P252 etc were also reported to bind with Cry toxin. In *Bombyx mori*, P252 binds with cry 1A in GalNAc-independent manner (Pandian et al. 2008).

19.5 Cry Gene Utilization in Pest Management

The *Bt* is well known for possessing the insecticidal crystal protein and its toxicity toward target insect pest. The cry toxin has resulted in controlling insect pests in three different sectors, viz.

1. Control of forest defoliating insect pest,
2. Control of mosquitoes that are human vectors, and
3. Development of transgenic insect-resistant plants.

The lepidopteran defoliators such as gypsy moth, spruce bud worm, nun moth, etc., were effectively controlled by aerial spraying of Btk HD1. Btk was used for the first time in Sweden for managing pine looper, *Bupalus piniara*, in 1997 on ~4000 ha and again in 1998 on 1000 ha against nun moth. Between 1980 and 2014, Btk was applied to a cumulative total of 8.0 million ha of spruce budworm-infested forests, mostly in Canada. The use of ultra-low volume applications of high-potency Btk gained general acceptance in gypsy moth control programs. Currently, doses of 60–95BIU per ha are applied undiluted in volumes of 3.0–4.7 L, usually in a single application. In Canada, the recommended prescription is two applications of 30–50BIU in 2.4–3.9 L per ha (Hajek and van Frankenhuyzen 2017). *Bt* strain *kurstaki HD-1* is widely used for lepidopteran pest management and possesses Cry1Aa, Cy1Ab, Cry1Ac, and Cry2Aa toxins (Van Frankenhuyzen 2000; Bauce et al. 2004). The Btk HD1 is available on the market as Biobit®, Dipel®, and Thuricide®. Other than Btk HD1, Btk SA-11 and Btk SA-12 were also effective against lepidopteran pest. Foil® is an example of the first conjugated *Bt* product, as it contains both lepidopteran active, Cry 1Ac, and coleopteran active, Cry 3A.

Bti is very effective in managing the Dipteran insects such as *Aedes aegypti* (Dengue vector), *Simulium damnosum* (Onchocerciasis vector), and *Anopheles* species (Malaria vector). The reason for preference of Bti as a mosquito control agent is due to lack of resistance, lack of toxicity to nontarget organisms, and appearance of insect-resistant populations to chemical insecticides (Becker 2000). Due to organophosphate resistance in S. *damnosum*, Bti was chosen for an onchocerciasis eradication program in 1 countries of West Africa (Guillet et al. 1990).

Since the identification of the first *cry* gene sequence, i.e., Cry1Aa1 in 1981 (Schnepf and Whitely 1981), many *cry* gene sequences have been submitted in NCBI and other databanks to date. As of now 12 *Bt* genome has been published in public data banks. The advancements in plant transformation such as particle bombardment method and *Agrobacterium*-mediated plant transformation technology help in producing genetically engineered plants. The first genetically modified plant was the tomato, with truncated Cry1Ac, for incorporating resistance against *Helicoverpa zea* (Fischhoff et al. 1987). In tobacco, Vaeck et al. (1987) transformed it with Cry1A for incorporating resistance against *Manduca sexta*. In the United States, the first *Bt* crop to be released was the potato in 1995, followed by corn and cotton in 1996 (Betz et al. 2000).

The *Bt* corn events were approved for cultivation in 1995 in the United States and were commercialized in 1996 for controlling *O. nubilalis* and other stalk-boring Lepidoptera such as southwestern corn borer, *Diatraea grandiosella* Dyer and *D. sacharralis* (Archer et al. 2001; Buntin et al. 2004). Numerous *Bt* insecticidal genes, viz. Cry1Ab (176, MON810 and Bt11), Cry1Ac (DBT418), Cry9C (CBH351), Cry1Fa (TC1507), and Vip3Aa (MIR162), were deployed for generating different *Bt* events (mentioned within the parenthesis behind the corresponding genes above) and were commercialized. Some of the events such as DBT418 (Cry1Ac) and 176 (Cry1Ab) were discontinued based on superior performance of MON810 and Bt11 (Buntin et al. 2004). Similarly, another event named CBH351 (StarLink®) was removed from commercial use due to traces of the product in the food supply (Fox 2001). The event MIR 162 provides excellent control against other stalk-boring Lepidoptera, fall armyworm, black cutworm, western bean cutworm, and *H. zea*; however, it does not provide protection against *O. nubilalis* (Burkness et al. 2010). The first commercial *Bt* corn for managing corn root worm (*Diabrotica* spp.) was released in the United States during 2003 and had Cry 3Bb gene in event MON 863. In high-pest density areas, *Bt* corn—having two different insecticidal genes—were stacked and known as SmartStax®. This product combines events expressing Cry3Bb (MON88017) and Cry34Ab/

Cry35Ab (DAS-59122-7) stacked (Head et al. 2014). Another pyramid for rootworm control in the US market is Duracade™; it is a combination of mCry3A (MIR604) and eCry3.1Ab (5307). Similarly, another event MON89034 is also a stacked *Bt* event containing two lepidopteran-active *Bt* proteins, viz. Cry 1A.105 and Cry2Ab. This combination has a strong performance against many lepidopteran pests, as they possess two different modes of action. This *cry* gene combination also decreased the risk of resistance development in insects.

The *Bt* cotton was first commercially introduced in the United States in 1996, followed by Australia in the same year, for controlling *H. armigera*. At present, more than 18 countries grow *Bt* cotton. In India, the *Bt* cotton-carrying Cry 1Ac (MON531) was commercialized in 2002. The six *Bt* events were approved by the Genetic Engineering Appraisal Committee (GEAC) for commercial cultivation in India (Table 19.1).

In 2009, the pink bollworm was reported to have developed resistance against Bollgard I (*cry 1Ac*) in four districts of Gujarat, India. In 2000, China commercially released *Bt* cotton expressing *cry 1 Ac* along with modified cowpea trypsin inhibitor (*CpTI*). The *Bt* gene stacked events, viz. Bollgard®II and WideStrike®. Recently, Vip 3Aa was combined with cry 1 Ac and cry 1 Fa for managing bollworm complex along with the *Spodoptera* species (Bommireddy et al. 2011) (Table 19.2).

The biolistic method has provided stable transformed soybean plants (McCabe and Christou 1993). However, *Agrobacterium*-mediated transformation of cotyledonary nodes and the microprojectile bombardment of somatic embryos resulted in the growth of less fertile transformed plants (Parott et al. 1994). Years later, through microprojectile bombardment, it was possible to produce a fertile, transformed soybean containing a synthetic *B. thuringiensis* insecticidal crystal protein gene (Cry1Ac). These genetically modified plants showed resistance to *Helicoverpa zea*, the soybean looper *Pseudoplusia includes*, the tobacco burworm (*Heliothis virescens*), and the velvetbean caterpillar (*Anticarsia gemmatalis*) (Stewart et al. 1996). MON 87701 is the only *Bt* event in soybean that has been commercialized in the United States (2011).

TABLE 19.1

Bt Cotton Events Approved for Cultivation in India

S. No.	Event Name	Gene Name	Developing Company	Year Approved
1	Mon 531	*Cry 1 Ac*	Mahyco-Monsanto	2002
2	Mon 531 + Mon 15985	*Cry 1Ac + Cry 2Ab*	Mahyco-Monsanto	2006
3	Event 1	*Cry 1Ac*	IIT, Kharagpur/ J.K.Agri	2006
4	GFM event	*Cry1ab + Cry1Ac*	Nath Seeds	2007
5	BNLA 106	*Cry 1Ac*	ICAR(CICR, UASD, NRCPB)	2008
6	MLS9124	*Cry 1 C*	Metahelix Life Sciences	2008

TABLE 19.2

Bt Cotton Events Approved for Cultivation in the United States

S. No.	Event Name	Gene Name	Developing Company	Year Approved
1	MON531	*Cry1Ac*	Monsanto	1995
2	31807, 31808	*Cry1Ac*	Calgene	1998
3	281-24-236[a]	*Cry 1Fa*	Dow Agro Sciences	2004
4	3006-210-23[a]	*Cry1Ac*	Dow Agro Sciences	2004
5	MON15985[b]	*Cry 1Ac + Cry 2Ab*	Monsanto	2005
6	GHB119[a]	*Cry 2Ae*	Bayer crop sciences	2011
7	T304-40[a]	*Cry 1Ab*	Bayer crop sciences	2011
8	COT 102[a]	*Vip 3Aa*	Syngenta	2011

[a] Commercialized only as a breeding stack.

[b] Created via biolistic transformation of germplasm containing Event MON531 (Cry1Ac) with Cry2Ab2.

The chickpea is the most important grain legume cultivated in Southeast Asia and India. The chickpea pods were damaged by gram pod borer, which inflicts a yield loss of $325 million annually (Sharma 2001). Pigeon pea is the second most important grain legume in India and is also attacked by gram pod borer. More than 50% of insecticides are used to control pod borers in India and China (Gujar et al. 2000). The *cry 1Ac* gene was introgressed into chickpea plant genome to incorporate resistance against the pod borer. A sequence modified gene, *cry 2Aa*, was inserted with a promoter AtSSU in order to delay resistance development in *H. armigera* in chickpea cv "Semsen" and "ICCV 89314" (Acharjee et al. 2010). In order to enhance pod borer resistance in chickpea cv "DCP 92-3," the chimeric *Bt* gene, *cryl Aabc*, was inserted (Das et al. 2017). Similarly, in the case of pigeon pea cv "Asha," the chimeric gene was incorporated to enhance the pod borer resistance (Das et al. 2016). Tabashnik et al. (2011) had reported that hybrid fusion protein such as *cryl Ac/b mod*, *cry3 Bbl* have novel specificities that have increased toxicity over *H. armigera* and may be effective in controlling chickpea pod borer.

A number of *Bt* transgenic initiatives have been explored for management of *M. vitrata* in different legume crops. Transgenic cowpea expressing *Bacillus thuringiensis* (*Bt*) subsp. *kurstaki* Cry1Ab delta endotoxin gene has been experimented by several authors (Higgins et al. 2007; Adesoye et al. 2008; Huesing et al. 2011). The Cry1Ab technology was registered by Monsanto and was given to the African Agricultural Technology Foundation for use in cowpea in Africa. In 2007, the Cry1Ab gene was successfully introgressed into cowpea cultivars in Australia. The first confined field trials (CFTs) of *Bt* cowpea was initiated in August 2008 in Puerto Rico. An advanced breeding line, IT86D-1010, was incorporated with *Cry 1 Ab* gene from TCL-709 and TCL 711 from CSIRO in Canberra, Australia (Mohammed et al. 2014). Recently, cowpea lines expressing Vip3Ba toxin ranging from 155 to 895 ng/mg of TSP were tested against legume pod borer and found to have 100% mortality (Bett et al. 2017). In Brazil, the velvetbean caterpillar has been successfully controlled by *Bt* soybean (Schünemann et al. 2014).

In 1995, *Bt* potato was commercialized by NatureMark (a subsidiary of Monsanto); these potato plants were expressing Cry3A protein against Colorado potato beetles, which caused around a 40% reduction in insecticidal use in 1997. NewLeaf Plus potatoes (possessing Cry3Aa and full-length replicase gene of potato leaf roll) conferred resistance against Colorado potato beetle and potato leaf roll (Lawson et al. 2001). A summary of *cry* genes utilized in various GM crops for insect resistance is listed in Table 19.3.

TABLE 19.3
Widely Used *Cry* Genes in Commercial GM Crop

S. No.	*Cry* Gene Deployed	Crop	Target Insect
1	*Cry 1Ab, Cry 1Ac*	Cotton	Bollworm complex
2	*Cry 1Ac, cry 2Aa*	Rice	Yellow stem borer
3	*Cry 1Ab*	Rice	Lepidopteran pest
4	*Cry 1Ac*	Basmati Rice	Yellow stem borer
5	*Cry 1Ab, cry 9C*	Maize	European corn borer
6	*Cry 3Aa*	Potato	Colorado potato beetle
7	*Cry 1Ab*	Potato	Tuber moth
8	*Cry 1Ac*	Tomato	Fruit borer
9	*Cry 1Ac*	Brinjal	Fruit borer
10	*Cry 1Ac*	Cabbage	Diamondback moth
11	*Cry 1C*	Broccoli	Diamondback moth
12	*Cry 1Ac*	Canola	Diamondback moth
13	*Cry 1C*	Alfalfa	Beet army worm
14	*Cry 1Ac*	coffee	Leaf miner
15	*Cry 1Ac*	Peanut	European corn borer

19.6 Cry Gene Mode of Action

The mode of action of *Bt* toxins is very complex; it kills insects by Cry and cyt endotoxins, which disrupt the cell membranes of midgut epithelium. These endotoxins form transmembrane pores in the target cells. In the case of Cry protein, alpha helix forms the pores while in Cyt toxin, beta barrel is involved in the pore formation. Due to the formation of these pores, ions and water leak into the cells, which leads to swelling, lysis, and eventual death of the insect. Mainly, there are four critical steps:

1. Cry toxin ingestion
2. When cry protein reaches into the digestive tract, it dissolves in the alkaline or the neutral/acidic environment. The bacterium-produced long protoxin is then activated in the midgut lumen by proteolytic removal of a short peptide fragment typically at the N-terminal.
3. This activated 60–70 kDa toxin then binds irreversibly to the insect-specific receptors in the microvilli of the midgut epithelial cells.
4. After binding to the receptors, the activated toxin undergoes conformational changes and oligomerizes into the membranes and forms a pore in the membrane. These pores increase the cell membrane permeability and initiate influx of cations, especially potassium ions. This causes osmotic balance disturbance that prompts the cell to swell and burst, causing the death of larvae within 1 or 2 days.

Bt is a leading biopesticide used in agriculture fields, but it covers only 2% of the insecticidal market; the rest of the market is covered by chemical insecticides. Due to the release of HD-1 strain formulation for the control of lepidopetran insects, *Bt* products are also available nowadays for the control of coleopteran insects. There are many disadvantages related to *Bt* products: the higher costs of commercial *Bt* products than chemical insecticides, lower efficacy of *Bt* compared to chemical insecticides, the limited persistence of *Bt* products and their narrow insect spectrum, and sucking and borer insects are rarely controlled by the spray of *Bt* products.

19.7 Mechanism of Cry Resistance in Insects

The major concern regarding the use of *Bt* is the resistance developed by insect populations against cry toxins. Resistance can be developed by interfering with any of the steps involved in the mode of action of cry toxin. Insects resistant to *Bt* toxin have been mostly developed on the basis of laboratory studies. These studies provide knowledge toward insect resistance. There are many insect species that have developed resistance toward *Bt* toxins, such as the diamondback moth (*P. xylostella*); resistance toward *Bt* spray application *Spodoptera frugiperda*; resistance to Cry1Fa-corn in Puerto Rico and Brazil; *Busseola fusca* resistance to Cry1Ab-corn in South Africa; *Pectinophora gossypiella* resistance to Cry1Ac-cotton in India; *Helicoverpa zea* resistance to Cry1Ac-cotton; and *Diabrotica virgifera virgifera* resistance to Cry3Bb-corn in the United States. Resistance to cry toxins can be developed by blocking any of the steps of *Bt* toxin activity. The following three findings summarize the resistance to *Bt* toxin.

1. Proteolytic activation—Cry protoxins are activated by the midgut proteases. *P. interpunctella* have resistance toward *Bt* toxins; this insect has been found to have lower protease and protoxin activities due to the absence of trypsin such as proteinase in the gut. The result of reduced gut protease results in the lower concentration of activated toxins in the gut.
2. Receptor binding—The defects in cry receptor binding are the major mechanism involved in insect resistance toward cry toxin. Lepidopteran-resistant populations have shown defects in receptor binding in a *P. interpunctella* strain selected for resistance to Cry1Ab; toxin-binding affinity (Kd) was reduced 50-fold. Interestingly, this strain showed an increased susceptibility to Cry1C toxin that correlated with an increased number of binding sites to this toxin. Four *P. xylostella* strains have been reported as developing resistance in the field; they have been

found to show reduced Cry1 A binding correlated with resistance. Binding of Cry 1A toxins to cadherin-like receptors is an important event in the toxin activity. Based on this, it has been found that characterization of cadherin alleles could be worked as a valuable tool in monitoring the insect resistance in the field. Examples of cadherin gene mutations have been found in *Heliothis virescens* (Gahan et al. 2001), *Pectinophora gossypiella* (Morin et al. 2003), and *Helicoverpa armigera* (Xu et al. 2005). In recent years, mutations in an ABCC2 transporter gene have also been linked with resistance to Cry1Ac. ABC transporters mediate the import or export of molecules against their concentration gradient by coupling transport to ATP hydrolysis. This resistance has been found in *H. virescens*, *Plutella xylostella* (Baxter et al. 2011), *T. ni* (Baxter et al. 2011), and *Bombyx mori*.

3. Oligosaccharide synthesis—Domains 2 and 3 of cry toxins show structural similarity to carbohydrate-binding domains and in Cry 1Ac toxin binding to the APN receptor involved in carbohydrate moieties. In the case of the *H. virescens* YHD2Cry1Ac-resistant line, an altered pattern of glycosylation of two microvillar proteins of 63 and 68 kDa was shown to correlate with resistance (Jurat-Fuentes et al. 2002). As mentioned above, YHD2 strain contains several mutations responsible for its Cry1Ac-resistant phenotype. Although the most important resistant allele is the disruption of the cadherin-like gene, altered glycosylation of microvilli membrane proteins increases its resistant phenotype (Jurat-Fuentes et al. 2002).

19.8 Conclusion

Bt produces a wide range of insecticidal proteins with different modes of action. Members of each insecticidal family could have different insect specificities, enabling *Bt* to colonize the insect world. A variety of insecticidal proteins provide an important means to control insect pests. However, only a very few *Bt* strains and Cry, Cyt, and Vip toxins are being commercialized. In the case of GM crops, the main threat is the evolution resistance to these toxins. Understanding their mode of action and the molecular basis of insect specificity led to the development of novel toxins with increased toxicity or toxins that are useful for resistance management. Continuous novel *cry* gene discovery and in vitro genetic evolution of Cry toxins with the aim of enhancing toxicity against specific pests, killing novel targets, or recovering toxicity in the case of resistance to these substances in the field will continue to provide important tools for efficient pest control management with limited use of chemical insecticides.

REFERENCES

Acharjee, Sumita, Bidyut Kumar Sarmah, P. Ananda Kumar, Karen Olsen, Rod Mahon, William J. Moar, Andy Moore, and Higgins, T.J.V. Transgenic chickpeas (*Cicer arietinum* L.) expressing a sequence-modified cry2Aa gene. *Plant Science* 178, no. 3 (2010): 333–339.

Adesoye, A., Machuka, J., and Togun, A. CRY 1AB transgenic cowpea obtained by nodal electroporation. *African Journal of Biotechnology* 7 (2008): 3200–3210.

Ali, B.A., Salem, H.H., Wang, X.M., Huang, T.H., Xie, QD, and Zhang, X.Y. Effect of *Bacillus thuringiensis* var. *israelensis* endotoxin on the intermediate snail host of *Schistosoma japonicum*. *Current Research in Bacteriology* 3, no. 1 (2010): 3–41.

Archer, T.L., Patrick C., Schuster, G., Cronholm, G., Bynum Jr E.D., and Morrison, W.P. Ear and shank damage by corn borers and corn earworms to four events of *Bacillus thuringiensis* transgenic maize. *Crop Protection* 20, no. 2 (2001): 139–144.

Bauce, É., Carisey N., Dupont, A., and Van Frankenhuyzen, K. *Bacillus thuringiensis* subsp. *kurstaki* aerial spray prescriptions for balsam fir stand protection against spruce budworm (Lepidoptera: Tortricidae). *Journal of Economic Entomology* 97, no. 5 (2004): 1624–1634.

Baxter, S.W., Badenes-Pérez, F.R., Morrison, A., Vogel, H., Crickmore, N., Kain, W., Wang, P., Heckel, D.G., and Jiggins, C.D. Parallel evolution of *Bacillus thuringiensis* toxin resistance in Lepidoptera. *Genetics* 189, no. 2 (2011): 675–679.

Bayyareddy, K., Andacht, T.M., Abdullah, M.A., and Adang, M.J. Proteomic identification of *Bacillus thuringiensis* subsp. *israelensis* toxin Cry4Ba binding proteins in midgut membranes from *Aedes (Stegomyia) aegypti* Linnaeus (Diptera, Culicidae) larvae. *Insect Biochemistry and Molecular Biology* 39, no. 4 (2009): 279–286.

Becker, N. Bacterial control of vector mosquitoes and black flies In: *Entomopathogenic Bacteria: From Laboratory to Field Application.* Charles JF, Delecluse A, Neilson-LeRoux C. (Eds.), Springer, Dordrecht, the Netherlands, 2000, pp. 383–398.

Berry, C. The bacterium, *Lysinibacillus sphaericus*, as an insect pathogen. *Journal of Invertebrate Pathology* 109 (2012): 1–10.

Bett, B., Gollasch, S., Moore, A., James, W., Armstrong, J., Walsh, T., Harding, R., and Higgins, T.J.V. Transgenic cowpeas (*Vigna unguiculata* L. Walp) expressing *Bacillus thuringiensis* Vip3Ba protein are protected against the Maruca pod borer (*Maruca vitrata*). *Plant Cell, Tissue and Organ Culture (PCTOC)* 131, no. 2 (2017): 335–345.

Betz, F.S., Hammond, B.G., and Fuchs, R.L. Safety and advantages of *Bacillus thuringiensis* protected plants to control insect pests. *Regulatory Toxicology and Pharmacology* 32, no. 2 (2000): 156–173.

Bokori-Brown, M., Savva, C.G., da Costa, SPF, Naylor, C.E., Basak, A.K., and Titball, R.W. Molecular basis of toxicity of *Clostridium perfringens* epsilon toxin. *The FEBS Journal* 278, no. 23 (2011): 4589–4601.

Bommireddy, P.L., Leonard, B.R., Temple, J., Price, P., Emfinger, K., Cook, D., and Hardke, J.T. Field performance and seasonal efficacy profiles of transgenic cotton lines expressing Vip3A and VipCot against *Helicoverpa zea* (Boddie) and *Heliothis virescens* (F.). *Journal of Cotton Science* 15, no. 3 (2011): 251–259.

Boonserm, P., Moonsom, S., Boonchoy, C., Promdonkoy, B., Parthasarathy, K., and Torres, J. Association of the components of the binary toxin from *Bacillus sphaericus* in solution and with model lipid bilayers. *Biochemical and Biophysical Research Communications* 342(4) (2006): 1273–1278.

Bravo, A., Gill, S.S., and Soberón, M. Mode of action of *Bacillus thuringiensis* Cry and Cyt toxins and their potential for insect control. *Toxicon* 49 (2007): 423–435.

Buntin, G.D., All, J.N., Lee, R.D., and Wilson, D.M. Plant-incorporated *Bacillus thuringiensis* resistance for control of fall armyworm and corn earworm (Lepidoptera: Noctuidae) in corn. *Journal of Economic Entomology* 97, no. 5 (2004): 1603–1611.

Burkness, E.C., Dively, G., Patton, T., Morey, A.C., and Hutchison, W.D. Novel Vip3A *Bacillus thuringiensis* (Bt) maize approaches high-dose efficacy against *Helicoverpa zea* (Lepidoptera: Noctuidae) under field conditions: Implications for resistance management. *GM crops* 1, no. 5 (2010): 337–343.

Cohen, S., Dym, O., Albeck, S., Ben-Dov, E., Cahan, R., Firer, M., and Zaritsky, A. High-resolution crystal structure of activated Cyt2Ba monomer from *Bacillus thuringiensis* subsp. *israelensis. Journal of Molecular Biology* 380, no. 5 (2008): 820–827.

Cohen, S., Albeck, S., Ben-Dov, E., Cahan, R., Firer, M., Zaritsky, A., and Dym, O. Cyt1Aa toxin: Crystal structure reveals implications for its membrane-perforating function. *Journal of Molecular Biology* 413, no. 4 (2011): 804–814.

Crickmore, N., Zeigler, D.R., Feitelson, J., Schnepf, E., van Rie, J., Lereclus, D., Baum, J., and Dean, D.H. Revision of the nomenclature for the *Bacillus thuringiensis* pesticidal crystal proteins. *Microbiology and Molecular Biology Reviews* 62 (1998): 807–813.

Crickmore, N., Zeigler, D.R., Schnepf, E., van Rie, J., Lereclus, D., Baum, J., Bravo, A., and Dean, D.H. *Bacillus thuringiensis* Toxin Nomenclature. Available online: http://www.lifesci.sussex.ac.uk/Home/Neil_Crickmore/Bt/ (Accessed 7 September 2019).

Das, A., Datta, S., Sujayanand. G.K., Kumar, M., Kumar Singh, A., Shukla, A., Ansari, J. et al. Expression of chimeric Bt gene, Cry1Aabc in transgenic pigeonpea (cv. Asha) confers resistance to gram pod borer (*Helicoverpa armigera* Hubner.). *Plant Cell, Tissue and Organ Culture (PCTOC)* 127, no. 3 (2016): 705–715.

Das, A., Datta, S., Thakur, S., Shukla, A., Ansari, J., Sujayanand, G.K., Sushil, K., Chaturvedi, P., Kumar, A., and Singh, N.P. Expression of a chimeric gene encoding insecticidal crystal protein Cry1Aabc of *Bacillus thuringiensis* in chickpea (*Cicer arietinum* L.) confers resistance to gram pod borer (*Helicoverpa armigera* Hubner.). *Frontiers in Plant Science* 8 (2017): 1423.

De Maagd, R.A., Bravo, A., Berry, C., Crickmore, N., and Schnepf, H.E. Structure, diversity, and evolution of protein toxins from spore-forming entomopathogenic bacteria. *Annual Review of Genetics* 37 (2003): 409–433.

Fabrick, J., Oppert, C., Lorenzen, M.D., Morris, K., Oppert, B., and Jurat-Fuentes, J.L. A novel *Tenebrio molitor* cadherin is a functional receptor for *Bacillus thuringiensis* Cry3Aa toxin. *Journal of Biological Chemistry* 284, no. 27 (2009): 18401–18410.

Federici, B.A., and Bauer, L.S. Cyt1Aa protein of *Bacillus thuringiensis* is toxic to the cottonwood leaf beetle, *Chrysomela scripta*, and suppresses high levels of resistance to Cry3Aa. *Applied Environmental Microbiology* 64 (1998): 4368–4371.

Fischhoff, D.A., Bowdish, K.S., Perlak, F.J., Marrone, P.G., McCormick, S.M., Niedermeyer, J.G., Dean, D.A., Kusano-Kretzmer, K., Mayer, E.J., Rochester, D.E., Rogers, S.G., and Fraley, R.T. Insect tolerant transgenic tomato plants. *Nature Biotechnology* 5 (1987): 807–813.

Flores-Escobar, B., Rodríguez-Magadan, H., Bravo, A., Soberón, M., and Gómez, I. Differential role of *Manduca sexta* aminopeptidase-N and alkaline phosphatase in the mode of action of Cry1Aa, Cry1Ab, and Cry1Ac toxins from *Bacillus thuringiensis*. *Applied and Environmental Microbiology* 79, no. 15 (2013): 4543–4550.

Fox, J.L. EPA re-evaluates StarLink license. *Nature Biotechnology* 19, no. 1 (2001): 11.

Gahan, L.J., Gould, F., and Heckel, D.G. Identification of a gene associated with Bt resistance in *Heliothis virescens*. *Science* 293, no. 5531 (2001): 857–860.

Gómez, I., Sánchez, J., Muñoz-Garay, C., Matus, V., Gill, S.S., Soberón, M., and Bravo, A. *Bacillus thuringiensis* Cry1A toxins are versatile proteins with multiple modes of action: Two distinct pre-pores are involved in toxicity. *Biochemical Journal* 459, no. 2 (2014): 383–396.

Guerchicoff, A., Delecluse, A., and Rubinstein, C.P. The *Bacillus thuringiensis cyt* genes for hemolytic endotoxins constitute a gene family. *Applied and Environmental Microbiology* 67 (2001): 1090–1096.

Guillet, P., Kurstack D.C., Philippon, B., Meyer, R. In: De Barjac, H., and Sutherland, D.J. (Eds.), *Bacterial Control of Mosquitoes and Blackflies*. Rutgers University Press, Rutgers, NJ, 1990, 187–190.

Gujar, G.T., Kumari, A., Kalia, V., and Chandrashekar, K. Spatial and temporal variation in susceptibility of the American bollworm, *Helicoverpa armigera* (Hübner) to *Bacillus thuringiensis* var. *kurstaki* in India. *Current Science* 78, no. 8 (2000): 995–1001.

Hajek, A.E., and van Frankenhuyzen, K. Use of entomopathogens against forest pests. In *Microbial Control of Insect and Mite Pests*, pp. 313–330. Ed. L. A. Lacey Academic Press, London, UK, 2017.

Head, G., Carroll, M., Clark, T., Galvan, T., Huckaba, R.M., Price, P., Samuel, L., and Storer, N.P. Efficacy of SmartStax® insect-protected corn hybrids against corn rootworm: The value of pyramiding the Cry3Bb1 and Cry34/35Ab1 proteins. *Crop Protection* 57 (2014): 38–47.

Higgins, T.J., Popelka, C., Ishiyaku, M., Pasquet, R., Mignouna, J., Bokanga, M., Huesing, J. et al. Insect protected cowpeas-transgenics with Bt or alpha-amylase inhibitor genes. *Biotechnology, Breeding and Seed Systems for African Crops* 78 (2007).

Höfte, H., and Whiteley, H.R. Insecticidal Crystal Proteins of *Bacillus thuringiensis*. *Microbiology Reviews* 53 (1989): 242–255.

Huesing, J.J., Romeis, N., Ellstrand, A., Raybould, R., Hellmich, J., Wolf, J., et al., Regulatory considerations surrounding the deployment of Bt-expressing cowpea in Africa: Report of the deliberations of an expert panel. *GM Crops* 2 (2011), 1–14.

Jurat-Fuentes, J.L., Gould, F.L., and Adang, M.L. Altered glycosylation of 63–and 68-kilodalton microvillar proteins in *Heliothis virescens* correlates with reduced Cry1 toxin binding, decreased pore formation, and increased resistance to *Bacillus thuringiensis* Cry1 toxins. *Applied and Environmental Microbiology* 68, no. 11 (2002): 5711–5717.

Kirouac, M., Vachon, V., Noel, J.F., Girard, F., Schwartz, J.L., and Laprade, R. Amino acid and divalent ion permeability of the pores formed by the *Bacillus thuringiensis* toxins Cry1Aa and Cry1Ac in insect midgut brush border membrane vesicles. *Bba-Biomembranes* 1561 (2002): 171–179.

Krishnamoorthy, M., Jurat-Fuentes, J.L., McNall, R.J., Andacht, T., and Adang, M.J. Identification of novel Cry1Ac binding proteins in midgut membranes from *Heliothis virescens* using proteomic analyses. *Insect Biochemistry and Molecular Biology* 37, no. 3 (2007): 189–201.

Kuadkitkan, A., Smith, D.R., and Berry, C. Investigation of the Cry4B-prohibitin interaction in *Aedes aegypti* cells. *Current Microbiology* 65 (2012): 446–454.

Lawson, E.C., Weiss, J.D., Thomas, P.E., and Kaniewski, W.K.. NewLeaf Plus® Russet Burbank potatoes: Replicase-mediated resistance to potato leafroll virus. *Molecular Breeding* 7, no. 1 (2001): 1–12.

Li, Jade, Pandelakis A. Koni, and David J. Ellar. Structure of the mosquitocidal δ-endotoxin CytB from *Bacillus thuringiensis* sp. *kyushuensis* and implications for membrane pore formation. *Journal of Molecular Biology* 257, no. 1 (1996): 129–152.

Likitvivatanavong, S., Chen, J., Evans, A.M., Bravo, A., Soberon, M., and Gill, S.S. Multiple receptors as targets of Cry toxins in mosquitoes. *Journal of Agricultural and Food Chemistry* 59 (2011): 2829–2838.

Lin, S.-C., Lo, Y.-C., Lin, J.-Y., and Liaw, Y.-C. Crystal structures and electron micrographs of fungal volvatoxin A2. *Journal of Molecular Biology* 343, no. 2 (2004): 477–491.

McCabe, D., and Christou, P. Direct DNA transfer using electric discharge particle acceleration (ACCELL™ technology). *Plant Cell, Tissue and Organ Culture* 33, no. 3 (1993): 227–236.

Mohammed, B.S., Ishiyaku, M.F., Abdullahi, U.S., and Katung, M.D., Response of transgenic Bt cowpea lines and their hybrids under field conditions. *Journal of Plant Breeding and Crop Science* 6 (2014): 91–96.

Morin, S., Biggs, R.W., Sisterson, M.S., Shriver, L., Ellers-Kirk, C., Higginson, D., Holley, D. et al. Three cadherin alleles associated with resistance to *Bacillus thuringiensis* in pink bollworm. *Proceedings of the National Academy of Sciences* 100, no. 9 (2003): 5004–5009.

Naimov, S., Boncheva, R., Karlova, R., Dukiandjiev, S., Minkov, I., and de Maagd RA. Solubilization, activation, and insecticidal activity of *Bacillus thuringiensis* serovar *thompsoni* HD542 crystal proteins. *Applied and Environmental Microbiology* 74 (2008): 7145–7151.

Pandian, G.N., Ishikawa, T., Togashi, M., Shitomi, Y., Haginoya, K., Yamamoto, S., Nishiumi, T., and Hori, H. Bombyx mori midgut membrane protein P252, which binds to *Bacillus thuringiensis* Cry1A, is a chlorophyllide-binding protein, and the resulting complex has antimicrobial activity. *Applied and Environmental Microbiology* 74, no. 5 (2008): 1324–1331.

Pardo-López, L., Soberón, M., and Bravo, A. *Bacillus thuringiensis* insecticidal three-domain Cry toxins: Mode of action, insect resistance and consequences for crop protection. *FEMS Microbiology Reviews* 37 (2013): 3–22.

Parott, W.A., J.N. All, M.J. Adang, M.A. Bailey, H.R. Boerma, and C.N. Stewart Jr. Recovery and evaluation of soybean (*Glycine max* (L.) Merr.) plants transgenic for a *Bacillus thuringiensis* var. *kurstaki* insecticidal gene. *In Vitro Cellular & Developmental Biology* 30 (1994): 144–149.

Popoff, M.R. Epsilon toxin: A fascinating pore-forming toxin. *The FEBS Journal* 278 (2011): 4602–4615.

Schnepf, E., Crickmore, N., van Rie, J., Lereclus, D., Baum, J., Feitelson, J., Zeigler, D.R., and Dean, D.H. *Bacillus thuringiensis* and its pesticidal crystal proteins. *Microbiology and Molecular Biology Reviews* 62 (1998): 775–806.

Schnepf, H.E., and Whiteley, H.R. Cloning and expression of the *Bacillus thuringiensis* crystal protein gene in *Escherichia coli. Proceedings of the National Academy of Sciences USA* 78 (1981): 2893–2897.

Schünemann, R., Knaak, N., and Fiuza, L.M. Mode of action and specificity of *Bacillus thuringiensis* toxins in the control of caterpillars and stink bugs in soybean culture. *ISRN Microbiology* 2014 (2014).

Sharma, H.C. Crop protection compendium: *Helicoverpa armigera* (Hub.) in different agro-economical conditions. In *Chickpea in Nineties: Proceedings of the Second International Workshop on Chickpea*, ICRISAT 2001, pp. 181–189. 2001.

Soberon, M., Lopez-Diaz, J.A., and Bravo, A. Cyt toxins produced by *Bacillus thuringiensis*: A protein fold conserved in several pathogenic microorganisms. *Peptides* 41 (2013): 87–93.

Stewart, Jr C.N., Adang, M.J., All, J.N., Boerma, H.R., Cardineau, G., Tucker, D., and Parrott, W.A. Genetic transformation, recovery, and characterization of fertile soybean transgenic for a synthetic *Bacillus thuringiensis* cryIAc gene. *Plant Physiology* 112, no. 1 (1996): 121–129.

Tabashnik, B.E., Huang, F., Ghimire, M.N., Leonard, B.R., Siegfried, B.D., Rangasamy, M., Yang, Y. et al. Efficacy of genetically modified Bt toxins against insects with different genetic mechanisms of resistance. *Nature Biotechnology* 29, no. 12 (2011): 1128.

Vachon, V., Laprade, R., and Schwartz, J.L. Current models of the mode of action of *Bacillus thuringiensis* insecticidal crystal proteins: A critical review. *Journal of Invertebrate Pathology* 111 (2012): 1–12.

Vaeck, M., Reynaerts, A., Hofte, H., Jansens, S., De Beuckeleer, M., Dean, C., Zabeau, M., Montagu, M.V., and Leemans, J., Transgenic plants protected from insect attack. *Nature* 328 (1987): 33–37.

Van Frankenhuyzen, K. Insecticidal activity of *Bacillus thuringiensis* crystal proteins. *Journal of Invertebrate Pathology* 101 (2009): 1–16.

Van Frankenhuyzen, K. Application of *Bacillus thuringiensis* in forestry. In *Entomopathogenic Bacteria: From Laboratory to Field Application*, pp. 371–382. Springer, Dordrecht, the Netherlands, 2000.

Wei, J.Z., Hale, K., Carta, L., Platzer, E., Wong, C., Fang, S.C., and Aroian, R.V. *Bacillus thuringiensis* crystal proteins that target nematodes. *Proceedings of the National Academy of Sciences USA* 100 (2003): 2760–2765.

Xu, X., Yu, L., and Wu, Y. Disruption of a cadherin gene associated with resistance to Cry1Ac δ-endotoxin of *Bacillus thuringiensis* in *Helicoverpa armigera. Applied and Environmental Microbiology* 71, no. 2 (2005): 948–954.

Yu, X., Liu, T., Sun, Z., Guan, P., Zhu, J., Wang, S., Li, S., Deng, Q., Wang, L., Zheng, A. et al. Co-expression and synergism analysis of Vip3Aa29 and Cyt2Aa3 insecticidal proteins from *Bacillus thuringiensis. Current Microbiology* 64 (2012): 326–331.

Zhang, Q., Hua, G., Bayyareddy, K., and Adang, M.J. Analyses of alpha-amylase and alpha-glucosidase in the malaria vector mosquito, *Anopheles gambiae*, as receptors of Cry11Ba toxin of *Bacillus thuringiensis* subsp. *jegathesan. Insect Biochemistry and Molecular Biology* 43 (2013): 907–915.

Zhang, X.B., Candas, M., Griko, N.B., Taussig, R., and Bulla, L.A. A mechanism of cell death involving an adenylyl cyclase/PKA signaling pathway is induced by the Cry1Ab toxin of *Bacillus thuringiensis. Proceedings of the National Academy of Sciences of the United States of America* 103 (2006): 9897–9902.

20

*Microbial Mitigation of Biotic Stresses in Soybean (*Glycine max*)*

E. T. Alori, A. O. Adekiya, A. Joseph, C. O. Aremu, and K. A. Adegbite

CONTENTS

20.1 Introduction

Healthy plants are essential to the survival of humans and animals on earth. Several stress conditions limit crop production worldwide, and among these is biotic stress. Biotic stress agents such as pathogens, parasitic weeds, and harmful insects cause severe damages and losses to agricultural products. Biotic stresses reduce the health, yield, and nutritional value of plants and plant-based products (Fletcher et al. 2010). Pathogens attack virtually all plants, including those cultivated for ornamental purposes and those in natural ecosystems such as forests and rangelands.

Soybean is one of the earliest crops cultivated by man (Stagnari et al. 2017). Its nutritional value includes carbohydrate, which makes 30%; protein (36%); oil (20%); and appreciable amounts of vitamins, minerals, and dietary fiber. Soybean is a relatively cheap protein source of high quality compared to beef, chicken, and eggs. It is an important crop for producing edible oil. However, optimum soybean production is constrained by weeds, insect pests, and diseases.

Conventional agriculture employed to alleviate the menace of biotic agents of crops promotes heavy reliance on the use of agrochemicals such as herbicides, pesticides, and fungicides to control biotic stress agents of plants. Though these had increased food production, it has also been reported that these chemicals have adverse effects on soil, plant, human, and animal health (Aremu et al. 2017, Alori and Babalola 2018). According to Alori and Babalola (2018) and Biswas et al. (2018), hazardous effects associated with the use of agrochemicals to combat biotic stress agents include birth defects, cancer, mutagenicity, neural disorders, and reproductive and developmental anomalies. One of the most devastating consequences of conventional agriculture is soil degradation (Baishya 2015). Inappropriate use of agrochemicals kills helpful soil organisms, increases soil nitrate content, alters soil pH levels, and causes eutrophication of water (Aremu et al. 2017). There is therefore an imperative need for a more reliable and high-yielding agricultural system to meet the increasing demand for food and energy by the ever-increasing global population. More often, plants reduce the burden of biotic stresses with the aid of the inhabitant microorganisms (Turner et al. 2013). Plant-beneficial microorganisms stimulate plant growth and enhance plant resistance to biotic stress (Vimal et al. 2017). Microorganisms enhance seed germination and also form mutual relationships with plants at the root surface or form endophytic relationships within the roots, stems, or leaves (Jalil and Ansari 2018). Microorganisms support plants by securing supplements and fight against or defend infections (Turner et al. 2013). Plant-microbe interaction stimulates plant defense mechanisms against biotic stresses.

Biocontrol technology via microbial inoculants is therefore a promising tool for mitigating the effects of biotic stress agents on crops and thereby minimizing the use of agrochemicals in crop production, while keeping our environment safe (Alori and Babalola 2018). Commercialized microbial inoculant strains used for mitigating biotic stresses in crops include *Azotobacter chroococcum, Pseudomonas syringae, P. chlororaphis, P. fluorescens, P. solanacearum, P. aureofaciens, Agrobacterium radiobacter, Pantoea agglomerans, Azospirillum brasilense, Bacillus fimus, B. licheniformis, B. megaterium, B. pumilus, B. mucilaginous, B. subtilis, B. subtilis* var. *amyloliquefaciens, Delftia acidovorans, Streptomyces griseoviridis, S. lydicus, Burkholderia cepacia, Paenobacillus macerans, Azospirillum lipoferum, Serratia entomophilia*, and several *Rhizobia* spp. (Glick 2012, Alori and Fawole 2017). In the present review, the mitigation of biotic stress due to diseases, insect pests, and weeds in soybean production is discussed. The mechanism of action of microbial inoculants in disease control is also expatiated.

20.2 Biotic Stresses of Soybean

20.2.1 Diseases of Soybean

Soybean is susceptible to many pathogens that cause severe damage to the crop. Wrather et al. (2010) reported a loss of 59.9 million metric tons of soybean to pathogens in the top eight soybean-producing countries in 2006.

Some of these pathogens affect the root, some other ones the stem, while others affect the leaves (Markell and Malvick 2018). One of these pathogens is *Macrophomina phaseolina*, which causes charcoal root rot (Vasebi et al. 2013). Other root diseases of soybean include: *Fusarium* root rot, *Rhizoctonia* root rot, sudden death syndrome, *Phytophthora* root and stem rot, *Phytium* root rot, and soybean cyst nematode (Markell and Malvick 2018). Stem diseases of soybean include anthracnose, stem canker, brown stem rot, white mold, pod and stem blight, while the leaf diseases include bacterial pustule, downy mildew, frogeye leaf spot, powdery mildew, bean pod mottles virus, soybean mosaic virus, *Cercospora* leaf blight, and *Septoria* brown spot (Markell and Malvick 2018).

Another of the most destructive soilborne diseases of soybean is damping off. Damping off diseases of soybean are caused by a number of organisms such as *Rhizoctonia solani, Phytophthora sojae, Aphanomyces euteiches, Sclerotium rolfsii, Pythium* spp., *Fusarium* spp., etc. (Omara et al. 2017).

20.2.2 Pests of Soybean

Insect pest infestation constitutes a major biotic stress to soybean production worldwide. Soybean is attacked by several insect pests that range from beetles, worms, and maggots to aphids and bugs. Some of these pests feed on the leaves, causing defoliation that results in a reduction in yield. Pests also feed on the pods, causing scarring, thereby reducing seed quality, besides exposing the seeds to secondary infection by pathogens that may cause rotting and discoloration. Insect pests such as bean leaf beetles transmit several viruses such as soybean mosaic virus, bean pod mottle virus, cowpea mosaic virus, alfalfa mosaic virus, and southern bean mosaic virus. Other damages by pest infestations include covering of sooty mold, yellow and wrinkled leaves, reduction in gas exchange and photosynthetic rates, stunted plants, reduction in seed size, and aborted pods leading to significant yield loss of up to 40% and more (Wang et al. 2006, Beckendorf et al. 2008, Ragsdale et al. 2011, Tilmon et al. 2011). The larvae of some pests feed on soybean root nodules (Hadi et al. 2012).

The common insect pests of soyben include: beet armyworm (*Spodoptera exigua*), Western striped armyworm (*Spodoptera praefica*), Western striped cucumber beetle (*Acalymma vittata*), Western spotted cucumber beetle (*Diabrotica undecimpunctata*), bean leaf beetle, soybean leaf miner, green clover worm, alfalfa caterpillar, cabbage looper, yellow wooly bear, painted lady, imported longhorned weevil, Dectes stem borer, soybean thrips, whiteflies, soybean aphids, potato leafhopper, two-spotted spider mite, green stink bug, and brown stink bug (Stewart 2016).

20.2.3 Weeds of Soybean

Soybean production is seriously challenged by weed infestation. The prevalence of weed in soybean production is associated with a decline in crop quality and yield. Nave and Wax (1971) reported a reduction in soybean yield of up to 25% and 13% as a result of pigweed (*Amaranthus hybridus* L.) and giant foxtail (*Setaria faberii* Herrm.) infestations, respectively. Weed infestation also causes a decrease in branch number, pod number, node number, and seed number per plant.

Some of the common weeds of soybean crop are: *Conyza bonariensis* (L.) Cronq, *Conyza canadensis* (L.) Cronq, *Eleusine indica* (L.) Gaertn, *Echinochloa crusgalli* (L.) *Eleusine indica* (L.) Beauv, *Sorghum halepense* (L.) Pers., *Bidens pilosa* L. *Ambrosia artemisiifolia* L. *Phyllanthus niruri, Trianthema portulacastrum, Brachiaria reptans, Cleome gynanadra, Amaranthus spinosus Dactyloctenium aegyptium* (L.) *Willd, Leptochloa chinensis* (L.) *Nees*, and *Phyllanthus niruri* (L.) (Keramati et al. 2008, da Silva et al. 2013).

20.3 Microbial Mitigation of Biotic Stresses

Many soil microorganisms exhibit antifungal and antibacterial activities and are therefore used for mitigating stress from plant pathogens (Alori and Babalola 2018). Microbial inoculants in the control of plant pathogens offer an alternative strategy to a chemical control method because they are environmentally friendly and resistance to microbial metabolites has not yet been reported (Alori and Babalola 2018). They do not pose any adverse effect to indigenous microflora or to host plants (Dunne et al. 1996). Table 20.1 shows some microorganisms that have demonstrated antimicrobial properties against phytopathogens of soybean and their mechanisms of action. According to Laditi et al. (2012), no disease incidence was observed in soybean inoculated with *Bacillus* spp. and *Trichoderma* spp.

The application of beneficial soil microorganisms such as fungi, bacteria, and viruses to eradicate weeds has received great attention (Harding and Raizada 2015). Several microbes with herbicidal properties have been identified (Chutia et al. 2007). Examples include several species of *Colletotrichum*, *Xanthomonas*, *Phoma*, *Pseudomonas*, and *Sclerotinia* (Harding and Raizada 2015). Research has shown that soybean inoculated with *Colletotrichum coccodes* significantly reduced velvet leaf seed yield up to 60% (Uremis et al. 2005, Ditommaso et al. 2017).

20.3.1 Mechanism of Action

Biocontrol agents exert their activity in different ways. The various mechanisms of action include:

1. **Competition:** Rhizobacteria, due to their fast chemotactic movement toward root exudates, outcompete pathogen population in the acquisition of nutrients and specific niche and thereby reduce pathogen population. Microbial inoculants such as mycorrhizal fungi and other phosphorus solubilizing microbes increase phosphorus enough to offset symptoms caused by the pathogen (Siddiqui and Akhtar 2007). According to Pal and Gardener (2011), exudates/leachates consumption, siderophore, scavenging, and physical niche occupation are examples

TABLE 20.1

Some Examples of Microbial Mitigation of Biotic Stresses in Soybean

Biotic Stress Agent	Microbial Inoculant	Activity	Author
Sclerotinia sclerotiorum, Rhizoctonia solani, Phomopsis sojae	*Bacillus* sp., *Burkholderia* sp.	Peptides, Bacteriocins, Secondary metabolites	de-Almeida et al. (2018)
Macrophomina phaseolina h-7	*Trichoderma harzianum* (strain 6, 14, 17, 21, 44), *T. asperellum* 26, *T. virens* 32, *T. harzianum* Tj17	Antagonism via production of volatile metabolites	Barari and Foroutan (2016)
Sclerotinia stem rot	*Sporidesmium sclerotivorum*	Parasitism	del Rio et al. (2002)
Sclerotium rolfsii Sacc	*Trichoderma harzianum, T. koningi*	Antagonism	Deb and Dutta (1991)
Sclerotium rolfsii, Pythium, Fusarium oxysporum, Rhizoctonia solani, Fusarium udum, Macrpohomina and *Phytophthora*	*Methylobacterium*	Antagonism	Poorniammal et al. (2009)
Macrophomina phaseolina	*Pantoea agglomerans, Bacillus* sp. BIN, *Trichoderma harzianum* T100	Antagonism	Vasebi et al. (2013)
Macrophomina phaseolina (Tassi)	*Trichoderma harzianum T2 T10 and T12*	Volatile metabolites production (e.g., acetaldehyde, isocyanide) derivatives	Khalili et al. (2016)
Phytophthora sojae Kauf. & Gerd. (*Phytophthora* root rots)	*Streptomyces*	Antagonism	Xiao et al. (2002)
Heterodera glycines	*B. velezensis* strain Bve2, *Bacillus mojavensis* strain Bmo3, *Bacillus safensis* strain Bsa27	Antagonism	Xiang et al. (2017)
Rhizoctonia solani	*Methylobacterium aminovorans* and *Methylobacterium rhodinum, Bradyrhizobium japonicum* (St. 110); *Bacillus megaterium* var. *phosphaticum;* and *Trichoderma viride*	Increased the activities of most soil enzymes	Omara et al. (2017)
Heterodera glycines	*Hirsutella rhossiliensis Hirsutella minnesotensis*	Antagonism	Chen and Liu (2005)
Armyworm	*Bacillus thuringiensis*	Antagonism	CABI (2010)
Fusarium spp.	*Bacillus* spp. and *Trichoderma* spp.	Antagonism	Laditi et al. (2012)
Northern jointvetch (*Aeschynomene virginica*)	*Colletotrichum gloeosporioides* f.sp. *aeschynomene*	Antagonism	Harding and Raizada (2015)
Round leaf mallow (*Malva pusilla*)	*Colletotrichum gloeosporioides* f.sp. *malvae*	Antagonism	Harding and Raizada (2015)
Velvetleaf (*Abutilon theophrasti*)	*Colletotrichum coccodes*	Competition	Ditommaso et al. (2017)

of competition. Siderophores from microbial inoculants inhibit some phytopathogens from acquiring a sufficient amount of iron, thereby limiting their ability to multiply (Glick 2012). Production of a great amount of indole-3-acetic acid (IAA) by *Azospirillum* spp. enhances plant lateral and adventitious root formation, which improves mineral and nutrient uptake (Babalola and Glick 2012).

2. **Antibiosis:** The rhizobacteria having capacity to produce antibacterial and antifungal compounds directly inhibit pathogen growth. Microbial inoculants cause a reduction in galling and

nematode multiplication (Akhtar and Siddiqui 2008). Production of antibiotics such as 2,3-dihydroxybenzoic acid, aminochelin, azotochelin, protochelin, and azotobactin were reported by Kraepiel et al. (2009) and Mali and Bodhankar (2009). These antibiotics have decolonizing efficiency against plant pathogens in rhizoplane soil (Nagaraja et al. 2016). Compound produced by *Bacillus amyloliquefaciens* CNU114001 and identified as iturin, a lipopeptide (LP), exhibit antifungal effect against fungal plant diseases (Ji et al. 2013). *Methylobacterial* spp. produce antibiotics that inhibit the mycelia growth of fungal pathogen such as *Fusarium. oxysporum* and *F. Udum*.

3. **Production of amino acids:** Inoculation with mycorrhizal fungi increase amino acids such as phenylalanine and serine in tomato roots; these amino acids inhibit growth and multiplication of pathogens such as nematodes (Siddiqui and Akhtar 2007).
4. **Production of enzymes:** Enzymes such as β-1,3 glucanases, chitinases, proteases, cellulases, and lipases that can lyse a portion of the cell walls of many plant pathogens are produced by biocontrol bacteria (Glick 2012). Increased production of soil dehydrogenase, urease, and phosphatase activities are also reported (Omara et al. 2017). Some other lytic enzymes produced by microbial inoculants include chitinases, glucanases, and proteases (Pal and Gardener 2011). Dunne et al. (1996) reported production of chitinase and protease enzymes by *Stenotrophomonas maltophilia* strain W81 (P).
5. **Plant immunization:** Colonization of plants by microbial inoculants activate a plant's innate defense system, causing it to respond strongly to the pathogen attack. This mechanism is also referred to as induced resistance (Jain et al. 2013). Microbial inoculants enhance protection against pathogens through augmented elicitation of host defense responses by triggering phenylpropanoid and antioxidant activities and by activating accumulation of total phenol, proline, and pathogenesis-related (PR) proteins (Jain et al. 2012). Siddiqui and Akhtar (2007) also reported systemic resistance induced by *Pseudomonas* as a mechanism for the biocontrol of plant pathogens. Arbuscular mycorrhizal fungi induced systemic resistance in crops attacked by *Meloidogyne incognita* and *Pratylenchus penetrans* (Vos et al. 2012).
6. **Production of volatile secondary metabolites:** Hydrocyanic acid (HCN) is one of the volatile secondary metabolites that inhibit the growth and development of plant pathogens (Ahmad et al. 2008). Other volatile compounds such as albaflavenone and dimethyl disulphide produced by microbial inoculants also inhibit the growth of fungal pathogens (Panpatte et al. 2017). Volatile compounds from microbial inoculants stimulate the antagonistic potential of microorganisms against plant pathogens (Aremu et al. 2017). Hexanedioic and butanoic acids are volatile compounds emitted by *Alcaligenes faecalis* strain JBCS129 (Bhattacharyya et al. 2015). *Pseudomonas fluorescens* F113 was reported to produce 2,4-diacetylphloroglucinol (antifungal secondary metabolite) in its attack against *Pythium ultimum* (Dunne et al. 1996). Table 20.1 shows some biotic stress agents whose actions were mitigated by beneficial microbes and the mechanisms of action exhibited by these microbes.

20.4 Conclusion

The biotic stresses to which crop systems are exposed pose serious challenges to global food security. Plant biotic stresses need to be controlled to maintain abundant and quality food production. This reviewed article provides evidence that beneficial microorganisms have the potential to protect or control soybean diseases, weeds, and insect pests for sustainable food production and hence, food security. Mitigating biotic stresses of soybean via microbial inoculant technology will reduce overall production costs and minimize negative environmental and human exposure effects associated with a chemical control method.

REFERENCES

Ahmad, F., I. Ahmad and M. S. Khan (2008). Screening of free living rhizospheric bacteria for their multiple growth promoting activities. *Microbiological Research* 163, 173–181.

Alori, E. T. and O. B. Fawole (2017). Microbial inoculants-assisted phytoremediation for sustainable soil management. *Phytoremediation: Management of Environmental Contaminants*, A. A. Ansari, S. S. Gill, R. Gill, G. R. Lanza and L. Newman (Eds.), pp. 3–17, Cham, Switzerland: Springer International Publishing.

Alori, E. T. and O. O. Babalola (2018). Microbial inoculant for improving crop quality and human health. *Frontiers in Microbiology* 9(2213), 12.

Aremu, B. R., E. T. Alori, R. F. Kutu and O. O. Babalola (2017). Potentials of microbial inoculants in soil productivity: An outlook on African legumes. *Microorganisms for Green Revolution: Microorganisms for Sustainability*, D. G. Panpatte, Y. K. Jhala, H. N. Shelat, R. V. Vyas (Eds.), pp. 53–75, Singapore: Springer.

Babalola, O. O. and B. R. Glick (2012). The use of microbial inoculants in African agriculture: Current practice and future prospects. *Journal of Food, Agriculture & Environment* 10(3&4), 540–549.

Barari, H. and A. Foroutan (2016). Biocontrol of soybean charcoal root rot disease by using *Trichoderma* spp. *Cercetări Agronomice în Moldova* XLIX, 2(166), 41–51.

Beckendorf, E. A., M. A. Catangui and W. E. Riedell (2008). Soybean aphid feeding injury and soybean yield, yield components, and seed composition. *Agronomy Journal* 100, 237–246.

Bhattacharyya, D., S. M. Yu and Y. H. Lee (2015). Volatile compounds from *Alcaligenes faecalis* JBCS1294 confer salt tolerance in *Arabidopsis thaliana* through the auxin and gibberellin pathways and differential modulation of gene expression in root and shoot tissues. *Plant Growth Regulation* 75, 297–306.

Biswas, S. K., S. Rahman, S. M. A. Kobir, T. Ferdous and N. A. Banu (2018). A review on impact of agrochemicals on human health and environment: Bangladesh perspective. *Plant Environment Development* 3(2), 31–35.

CABI (2010). Crop Protection Compendium *Glycine max* (soyabean) datasheet. http://www.cabi.org/cpc/datasheet/25400. Accessed 10 April 2015.

Chen, S. and X. Liu (2005). Control of the soybean cyst nematode by the fungi *Hirsutella rhossiliensis* and *Hirsutella minnesotensis* in greenhouse studies. *Biological Control* 32(2), 208–219.

da Silva, A. F., L. Galon, I. Aspiazú, E. A. Ferreira, G. Concenço, E. U. J. Ramos, P. R. R. Rocha (2013). Weed management in the soybean crop. *Soybean Pest Resistance*, H. A. El-Shemy (Eds.), Rijeka, Croatia: In Tech.

de-Almeida, L. K. B., V. Carpentieri-Pipolo, D. Fira, P. A. Balatti, S. M. Y. López, T. H. Oro, P. E. Stefani and G. Degrassi (2018). Screening of bacterial endophytes as potential biocontrol agents against soybean diseases. *Journal of Applied Microbiology* 125(5), 1466–1481.

Deb, P. R. and B. K. Dutta (1991). Studies on biological control of foot rot disease of soybean caused by *Sclerotium rolfsii* Sacc. *Journal of Plant Disease and Protection* 98(5), 539–546.

del Rio, L. E., C. A. Martinson and X. B. Yang (2002). Biological control of sclerotinia stem rot of soybean with *Sporidesmium sclerotivorum*. *Plant Disease* 86(9), 999–1004.

Ditommaso, A., A. K. Watson and S. G. Hallett (2017). Infection by the fungal pathogen colletotrichum coccodes affects velvetleaf (*Abutilon theophrasti*)-Soybean competition in the field. *Weed Science* 44(4), 924–933.

Dunne, C., I. Delany, A. Fenton and F. O'Gara (1996). Mechanisms involved in bioconrol by microbial inoculants. *Agronomie* 16, 721–729.

Fletcher, J., D. Luster, R. Bostock, J. Burans, K. Cardwell, T. Gottwald, L. McDaniel, M. Royer and K. Smith (2010). *Emerging Infections*, W. W. Scheld, M. L. Grayson and J. M. Hughes. (Eds.), Washington, DC: ASM Press.

Glick, B. R. (2012). Plant growth-promoting bacteria: Mechanisms and applications. *Scientifica* 2012, 1–15.

Hadi, B. A. R., J. D. Bradshaw, M. E. Rice and J. H. Hill (2012). Bean Leaf Beetle (Coleoptera: Chrysomelidae) and Bean Pod Mottle Virus in Soybean: Biology, ecology, and management. *Journal of Integrated Pest Management* 3(1), B1–B7.

Harding, D. P. and M. N. Raizada (2015). Controlling weeds with fungi, bacteria and viruses: A review. *Frontiers in Plant Science* 6, 659.

Jain, A., A. Singh, S. Singh and H. Singh (2013). Microbial consortium-induced changes in oxidative stress markers in pea plants challenged with *Sclerotinia sclerotiorum. Journal of Plant Growth Regulation* 32(2), 388–398.

Jain, A., S. Singh, B. Kumar Sarma and S. H. Bahadur (2012). Microbial consortium-mediated reprogramming of defence network in pea to enhance tolerance against *Sclerotinia sclerotiorum. Journal of Applied Microbiology* 112(3), 537–550.

Jalil, S. U. and M. I. Ansari (2018). Plant microbiome and its functional mechanism in response to environmental stress. *International Journal of Green Pharmacy* 12(1), S81.

Ji, S. H., N. C. Paul, J. X. Deng, Y. S. Kim, B.-S. Yun and S. H. Yu (2013). Biocontrol activity of *bacillus amyloliquefaciens* CNU114001 against fungal plant diseases. *Mycobiology* 41(4), 234–242.

Keramati, S., H. Pirdashti, M. A. Esmaili, A. Abbasian, M. Habibi (2008). The critical period of weed control in soybean (*Glycine max* (L.) Merr.) in North of Iran conditions. *Pakistan Journal of Biological Sciences* 11(3), 463–467.

Khalili, E., M. A. Javed, F. Huyop, S. Rayatpanah, S. Jamshidi and R. A. Wahab (2016). Evaluation of *Trichoderma* isolates as potential biological control agent against soybean charcoal rot disease caused by Macrophomina phaseolina. *Biotechnology & Biotechnological Equipment* 30(3), 479–488.

Kraepiel, A., J. Bellenger, T. Wichard and F. Morel (2009). Multiple roles of siderophores in free living nitrogen-fixing bacteria. *Biometals* 22, 573–581.

Laditi, M. A., O. C. Nwoke, M. Jemo, R. C. Abaidoo and A. A. Ogunjobi (2012). Evaluation of microbial inoculants as biofertilizers for the improvement of growth and yield of soybean and maize crops in savanna soils. *African Journal of Agricultural Research* 7(3), 405–413.

Mali, G. V. and M. G. Bodhankar (2009). Antifungal and phytohormone potential of *Azotobacter chroococcum* isolates from ground nut (*Arachis hypogeal* L.) rhizosphere. *Asian Journal Experimental Science* 23, 293–297.

Markell, S. and D. Malvick (2018). *Soybean Disease Diagnostic Series*. North Dakota State University Publication, 1867 pages.

Nagaraja, H., G. Chennappa, S. Rakesh, M. K. Naik, Y. S. Amaresh and M. Y. Sreenivasa (2016). Antifungal activity of *Azotobacter nigricans* against trichothecene-producing *Fusarium* species associated with cereals. *Food Science Biotechnology* 25(4), 1197–1204.

Nave, W. R. and L. M. Wax (1971). Effect of weeds on soybean yield and harvesting efficiency. *Weed Science* 19(5), 533–535.

Omara, A., F. Hauka, A. Afify, M. Nour El-Din and M. Kassem (2017). The role of some PGPR strains to biocontrol *Rhizoctonia solani* in soybean and enhancement the growth dynamics and seed yield. *The Environment, Biodiversity & Soil Security* 1, 47–59.

Pal, K. K. and B. M. Gardener (2011). Biological control of plant Pathogens. https://www.apsnet.org/edcenter/disimpactmngmnt/topc/Pages/BiologicalControl.aspx. Accessed December 2018.

Panpatte, D. G., Y. M. Shukla, H. N. Shelat, R. V. Vyas and Y. K. Jhala (2017). Bacterial volatile organic compounds: A new insight for sustainable agriculture. *Microorganisms for Green Revolution: Microorganisms for Sustainable Crop Production*, D. G. Paopatte, Y. K. Jhala, R. V. Vyas and H. N. Shefat (Eds.), Singapore: Springer.

Poorniammal, R., S. P. Sundaram and K. Kumutha (2009). In vitro biocontrol activity of *Methylobacterium* extorquens against *Fungal* pathogens. *International Journal of Plant Protection* 2(1), 59–62.

Ragsdale, D. W., D. A. Landis, J. Brodeur, G. E. Heimpel and N. Desneux (2011). Ecology and management of the soybean aphid in North America. *Annual Review of Entomology* 56, 375–399.

Siddiqui, Z. A. and M. S. Akhtar (2007). Biocontrol of a chickpea root rot disease complex with phosphate solubilizing microorganisms. *Journal of Plant Pathology* 67–77.

Stagnari, F., A. Maggio, A. Galieni and M. Pisante (2017). Multiple benefits of legumes for agriculture sustainability: An overview. *Chemical and Biological Technologies in Agriculture* 4(1), 2.

Stewart, S. (2016). Some soybean insect pests. *UTcrops News Blog*, Institute of Agriculture, The University of Tennessee, 2019.

Tilmon, K. J., E. W. Hodgson, M. E. O'Neal and D. W. Ragsdale (2011). Biology of the soybean *Aphid, Aphis glycines* (*Hemiptera: Aphididae*) in the United States. *Journal of Integrated Pest Management* 2(2), A1–A7.

Turner, T. R., E. K. James and P. S. Poole (2013). The plant microbiome. *Genome Biol* 14, 209.

Uremis, I., M. Arslan and A. Uludag (2005). Effects of weed duration on seed yield and yield components of double-cropped soybean. *Asian Journal of Plant Sciences* 4, 370–373.

Vasebi, Y., N. Safaie and A. Alizadeh (2013). Biological control of soybean charcoal root rot disease using bacterial and fungal antagonists In vitro and greenhouse condition. *Journal of Crop Protection* 2(2), 139–150.

Vimal, S. R., J. S. Singh, N. K. Arora and S. Singh (2017). Soil-plant-microbe interactions in stressed agriculture management: A review. *Pedosphere* 27(2), 177–192.

Vos, C. M., A. F. Tesfahun, B. Panis, D. D. Waele and A. Elsen (2012). Arbuscular mycorrhizal fungi induce systemic resistance in tomato against the sedentary nematode *Meloidogyne incognita* and the migratory nematode *Pratylenchus penetrans*. *Applied Soil Ecology* 61, 1–6.

Wang, R. Y., A. Kritzman, D. E. Hershman, S. A. Ghabrial (2006). *Aphis glycines* as a vector of persistently and nonpersistently transmitted viruses and potential rises for soybean and other crops. *Plant Disease* 90, 920–926.

Wrather, A., G. Shannon, R. Balardin, L. Carregal, R. Escobar, G. K. Gupta, Z. Ma, W. Morel, D. Ploper and A. Tenuta (2010). Effect of diseases on soybean yield in the top eight producing countries in 2006. *Plant Management Network*. https://www.plantmanagementnetwork.org/pub/php/research/2010/loss/.

Xiang, N., K. S. Lawrence, J. W. Kloepper, P. A. Donald and J. A. McInroy (2017). Biological control of *Heterodera glycines* by spore-forming plant growth-promoting rhizobacteria (PGPR) on soybean. *PLoS One* 12(7), e0181201.

Xiao, K., L. L. Kinkel and D. A. Samac (2002). Biological control of phytophthora root rots on Alfalfa and Soybean with Streptomyces. *Biological Control* 23, 285–295. doi:10.1006/bcon.2001.1015.

Index

Note: Page numbers in italic and bold refer to figures and tables, respectively.